ISNM
International Series of
Numerical Mathematics
Vol. 120

Integral Equations

Theory and Numerical Treatment

Wolfgang Hackbusch

Birkhäuser Verlag
Basel · Boston · Berlin

Author:

Wolfgang Hackbusch
Institut für Informatik und Praktische Mathematik
Christian-Albrechts-Universität Kiel
Olshausenstrasse 40
D-24098 Kiel
Germany

Originally published in German under the title "Integralgleichungen. Theorie und Numerik"
© B.G. Teubner, Stuttgart 1989.
This English translation has been authorized by and published in consent with B.G. Teubner Verlag, Stuttgart.

Translation and revisions made by the author.

A CIP catalogue record for this book is available from the Library of Congress, Washington D.C., USA

Deutsche Bibliothek Cataloging-in-Publication Data
Hackbusch, Wolfgang
Integral equations : theory and numerical treatment / Wolfgang
Hackbusch. – Basel ; Boston ; Berlin : Birkhäuser, 1995
(International series of numerical mathematics ; Vol. 120)
Dt. Ausg. u.d.T.: Hackbusch, Wolfgang: Integralgleichungen

ISBN-13: 978-3-0348-9947-5 e-ISBN-13: 978-3-0348-9215-5
DOI: 10.1007/978-3-0348-9215-5

NE: GT

Softcover reprint of the hardcover 1st edition 1995
Camera-ready copy prepared by the author
Printed on acid-free paper produced from chlorine-free pulp
Cover design: Heinz Hiltbrunner, Basel

9 8 7 6 5 4 3 2 1

Preface

The theory of integral equations has been an active research field for many years and is based on analysis, function theory, and functional analysis. On the other hand, integral equations are of practical interest because of the «boundary integral equation method», which transforms partial differential equations on a domain into integral equations over its boundary.

This book grew out of a series of lectures given by the author at the Ruhr-Universität Bochum and the Christian-Albrecht-Universität zu Kiel to students of mathematics. The contents of the first six chapters correspond to an intensive lecture course of four hours per week for a semester. Readers of the book require background from analysis and the foundations of numerical mathematics. Knowledge of functional analysis is helpful, but to begin with some basic facts about Banach and Hilbert spaces are sufficient.

The theoretical part of this book is reduced to a minimum; in Chapters 2, 4, and 5 more importance is attached to the numerical treatment of the integral equations than to their theory. Important parts of functional analysis (e.g., the Riesz-Schauder theory) are presented without proof. We expect the reader either to be already familiar with functional analysis or to become motivated by the practical examples given here to read a book about this topic. We recall that also from a historical point of view, functional analysis was initially stimulated by the investigation of integral equations. Concerning function spaces, we restrict ourselves mostly to the classical ones of continuous or Hölder continuous functions. Sobolev spaces are almost avoided. As a consequence, the integral operators cannot be discussed as pseudo-differential operators in the required generality.

The theory of integral equations is interesting not only in itself, but its results are essential for the analysis of numerical methods. Besides existence and uniqueness statements, the theory concerns, in particular, questions of regularity and stability.

After an introduction which gathers basic facts from analysis, functional analysis, and numerical mathematics, we consider Volterra integral equations first (§2), since these are closely related to ordinary differential equations, which have typically been met in other courses. After these preliminaries, Fredholm integral equations of the second kind are investigated theoretically (§3) and then numerically in the subsequent chapters.

The numerical treatment of Fredholm integral equations of the second kind is discussed both in a chapter on discretisation methods (§4) and in one about the solution of the systems of linear equations which subsequently arise. Concerning the latter topic, §5 discusses in detail the multi-grid method after a brief comment about the method of conjugate gradients.

The fourth and most extensive chapter starts with general statements concerning convergence, consistency, and stability of discretisations (§4.1). The particular discretisation presented first is the kernel approximation (§4.2). Then we study the projection method in general (§4.3) and its most prominent examples: the collocation method (§4.4) and the Galerkin method (§4.5). Further remarks are added in §4.6. Another method outside of the projection method framework is the Nyström discretisation, which is discussed in §4.7. Finally, in §4.8, we give supplementary comments concerning different topics like, e.g., defect correction, extrapolation, and eigenvalue problems.

To introduce weakly and strongly singular integral equations, two examples are presented: Abel's integral equation (§6) and Cauchy integral equations (§7). The boundary integral equation method mentioned above is the subject of Chapter 8. Here, the analytical properties of the equations are discussed first. The numerical treatment leads us to the «boundary element method» (§9).

While the first chapters of the book may be thought of mainly as an introduction to the whole subject of integral equations, some parts of §§4-5 and, in particular, Chapter 9 give concrete hints which might be helpful for practical applications.

The last chapter (§9) is a revision of its counterpart in the original German version of this book. We have added recent results concerning numerical integration and the panel clustering method.

The exercises cited, which may also be understood as remarks without proof, are an integral part of the presentation. Should this book be used as the basis of an academic course, these exercises can be assigned as problems to students. However, the non-student reader should also try to test his comprehension by working on these exercises.

Completeness is not the aim of the bibliography. Furthermore, we tried to restrict ourselves to easily available literature.

The manuscript is written by the text system «Signum». My daughter Jana assisted in the construction of the figures. I am indebted to many students and readers of the first German edition for various helpful hints and corrections. In particular, I would like to thank my collegues I. Graham and A. Spence at the Bath University for polishing the English. I also wish to express my gratitude to Birkhäuser for their friendly cooperation.

Kiel, December 1994 W. Hackbusch

Contents

Notations

Numbers of formulae: Equations in a subchapter x.y are numerated by (x.y.1), (x.y.2), etc. The equation **(3.2.1)** is quoted by **(1)** in the same Section 3.2, while we write **(2.1)** in the other sections of chapter 3.

Numeration of Theorems etc.: All theorems, definitions, and lemmata etc. are enumerated together. The reference to a theorem etc. is analogous to what is said above. **Lemma 3.2.7** is cited as «**Lemma 7**» in Section 3.2, while in the other section of the chapter 3 it is denoted by «**Lemma 2.7**». However, §1 indicates Chapter 1 and never the sections §3.1 or §3.2.1.

Generic Constants: Often a notation like $\|K_n\| \leqslant C$ is used without a definition of C. The meaning is the following: There is a constant C such that $\|K_n\| \leqslant C$. Here, C is independent of the involved parameters (here; n). The value of C may be different at different places. If a fixed value is meant, this is indicated by an index (e.g., $\|K_n\| \leqslant C_K$).

Special Symbols, Abbreviations, and Conventions

a,b	bounds of the interval $I=[a,b]$
$\mathbb{C}$	complex numbers
C	constant (see above)
C, C^k, $C^\varkappa$, $C_L{}^k$, $\hat{C}^\varkappa$	spaces of continuous functions etc. (cf. §1.2)
$C_L{}^k$	space of Lipschitz continuous functions (cf. §1.2.2)
$\hat{C}^\varkappa$	space of Hölder or Lipschitz continuous functions (cf. §1.3.2)
cond	condition number (of a matrix), cf. (1.4.22)
D	domain of definition, domain of integration
det	determinant
dist($\mathbf{x}$,M)	distance of $\mathbf{x}\in\mathbb{R}^d$ from $M\subset\mathbb{R}^d$
f	unknown function of the integral equation (cf. (1.1.3))
f_n	semidiscrete solution of the integral equation
g	known function (inhomogeneity) in the integral equation (cf. (1.1.3))
h	step size of the discretisation
H^k	Sobolev space
I	interval (of integration); identity (identical operator)
$\mathbf{I}$	identity matrix
image(T)	range (image space) of the operator T

k	kernel function in K (cf. (3.1.3))
K	integral operator
$K(X,Y)$	compact linear mapping from X into Y
K_n	semidiscrete integral operator (cf. §4.1.1)
$\mathbf{K}_n$	matrix of the discretisation (cf. §5.1.1)
$K_r(x)$	ball with radius r centred at x
ℓ	level of discretisation (cf. §5.3)
L^2, L^1, L^∞	function spaces (cf. §1.3.3)
L_i, $L_{i,n}$	Lagrange functions
$L(X,Y)$	bounded linear mapping from X into Y
log	natural logarithm
n	discretisation parameter
$\mathbf{n}$, $\mathbf{n}(\mathbf{x})$	normal direction at the point $\mathbf{x}$ of the curve or surface
$\mathbb{N}$	natural numbers $\{1,2,3,...\}$
$\mathbb{N}_0$	$\mathbb{N}\cup\{0\}=\{0,1,2,...\}$
$O(\cdot)$	Landau symbol: $f(\alpha)=O(g(\alpha))$, if $\lvert f(\alpha)\rvert\leqslant C\lvert g(\alpha)\rvert$ with respect to an underlying limit process $\alpha\to 0$ or $\alpha\to\infty$
$o(\cdot)$	Landau symbol: $f(\alpha)=o(g(\alpha))$, if $\lvert f(\alpha)\rvert/\lvert g(\alpha)\rvert\to 0$
P_n, p	prolongation (cf. (5.2.2a), (5.3.3a))
Q, Q_n, $Q_{n,I}$, Q_I, $Q_{[a,b]}$	quadrature formula
$\mathbb{R}$	real numbers
R_n, r	restriction (cf. (5.2.2b), (5.3.3b))
sign(x)	sign of x with sign$(0)=0$
span$\{...\}$	space spanned by $\{...\}$
w_i, $w_{i,n}$	weights in the quadrature formula Q
X, Y	Banach spaces
X',Y'	dual spaces
X_n	subspace (cf. §4.4.1, §4.5.1)
$\mathbb{Z}$	integers
Γ	curve in $\mathbb{R}^2$ (or in $\mathbb{C}$) or surface in $\mathbb{R}^3$
δ_{ij}	Kronecker symbol: $\delta_{ij}=1$ for $i=j$, $\delta_{ij}=0$ otherwise
$\mu(A)$	(Lebesgue) measure of the set A
$\sigma(T)$	spectrum of the operator T (cf. §1.3.9)
Π_n	interpolation, projection
ω_d	surface of the d-dimensional unit sphere
Ω_-, Ω_+	interior and exterior domain (cf. Remark 7.1.11d)
$f(x+0)$, $f(x-0)$	right-sided and left-sided limit
T'	dual operator of T (cf. §1.3.6)
T^*	adjoint operator of T
$\overline{\Omega}$	closure of the set Ω
$\partial\Omega$	boundary of the set Ω
$\frac{\partial}{\partial n}$	normal derivative (cf. (7.4.8b))
$⨏$	Cauchy principal value (cf. §7.1)
∇	gradient
Δ	Laplace operator (cf. §7.4)
$\lvert\cdot\rvert$	absolute value; in §§7-9 also Euclidean norm

1 Introduction

1.1 Integral Equations

We begin by recalling the following example from the analysis of ordinary differential equations. Consider the initial value problem

$$(1.1.1) \qquad y'(x) = f(x,y) \quad \text{for } x \geq x_0, \qquad y(x_0) = y_0.$$

Integration from x_0 to x reduces this to the integral equation

$$(1.1.2) \qquad y(x) = y_0 + \int_{x_0}^{x} f(\xi, y(\xi))\,d\xi \qquad \text{for } x \geq x_0.$$

One reason why the reformulation (2) is of interest is because it is more suitable than (1) for proving existence and uniqueness of a solutions.

Generally speaking, an integral equation is an equation for an unknown function f, where f appears also under the integral sign. The integral equations are classified by various properties, which are listed in the following.

Fredholm integral equation: The integral is taken over a *fixed* interval of $\mathbb{R}^1$ or a more general *fixed* subset of the $\mathbb{R}^d$ including curves and surfaces etc.).

Volterra integral equation: The integral is taken over a domain *varying* with the variable x (cf. (2)).

Another classification is the following:

Integral equation of the first kind: The unknown function appears *only* under the integral sign.

Integral equation of the second kind: The unknown function appears also outside the integral.

We distinguish **linear integral equations** from **nonlinear integral equations** depending on whether the equation is linear with respect to the unknown function or not.

A further characterisation is independent of the previous ones and refers to the kind of the integral:

Regular integral equation: The integral exists as a *proper* integral.

Weakly singular integral equation: The integral exists as an *improper* integral.

Strongly singular integral equation: The integral must be defined by a special regularisation (e.g., Cauchy principal value).

For illustration, we give a number of examples, where g is a given function and f the function to be determined.

linear Fredholm integral equation of the second kind:

$$(1.1.3) \qquad f(x) = g(x) + \int_a^b k(x,y)f(y)dy \qquad \text{for } x\in[a,b],$$

linear Fredholm integral equation of the first kind:

$$(1.1.4) \qquad g(x) = \int_a^b k(x,y)f(y)dy \qquad \text{for } x\in[a,b],$$

linear Volterra integral equation of the second kind:

$$(1.1.5) \qquad f(x) = g(x) + \int_a^x k(x,y)f(y)dy \qquad \text{for } x\geqslant a,$$

linear Volterra integral equation of the first kind:

$$(1.1.6) \qquad g(x) = \int_a^x k(x,y)f(y)dy \qquad \text{for } x\geqslant a,$$

nonlinear Fredholm integral equation of the second kind:

$$(1.1.7a) \qquad f(x) = g(x) + \int_a^b k(x,y,f(x),f(y))dy \quad \text{for } x\in[a,b],$$

or

$$(1.1.7b) \qquad f(x) = F\Big(x, \int_a^b k(x,y,f(x),f(y))dy\Big) \quad \text{for } x\in[a,b].$$

Special subclasses of (7a) are the *Urysohn* equation

$$(1.1.8) \qquad f(x) = g(x) + \int_a^b k(x,y,f(y))dy \qquad \text{for } x\in[a,b]$$

and the *Hammerstein* equation

$$(1.1.9) \qquad f(x) = g(x) + \int_a^b k(x,y)G(y,f(y))dy \quad \text{for } x\in[a,b].$$

A special example of a weakly singular Volterra integral equation of the first kind is *Abel's integral equation*

$$(1.1.10) \qquad g(x) = \int_a^x \frac{f(y)}{\sqrt{x-y}}\,dy \qquad \text{for } x\geqslant a,$$

which will be discussed in §6.

If L is a *curve* in the set of complex numbers $\mathbb{C}$, then

$$(1.1.11) \qquad g(z) = \oint_L \frac{f(\zeta)}{\zeta - z}\,d\zeta \qquad \text{for } z\in L$$

represents an example of a strongly singular Fredholm integral equation of the first kind (defined over of a curve instead of an interval). Equation (11) will be analysed in §7.

Sometimes, integral equations occur with additional derivatives of the unknown function (under the integral or outside). In this case, the problem is called an *integro-differential equation.*

In the case of differential equations, one characterises equations by their order. Since integration is the reverse of differentiation, one might assume that there are also integrations of different order. This is not the case, as the following exercise shows. Nevertheless, later (Remark 8.3.4) we will be able to define an order, which, however, will be unusual since it is not necessarily integer-valued.

Exercise 1.1.1. The mappings $f \mapsto F(\xi) := \int_a^\xi k_1(\xi, y) f(y)\, dy$ and $F \mapsto G(x) := \int_a^x k_2(x, \xi) F(\xi)\, d\xi$ are two successive integrations. Prove that the «twofold» integration $f \mapsto G$ can be represented as «single» integration $G(x) = \int_a^x k(x,y) f(y)\, dy$, where $k(x,y) := \int_y^x k_2(x,\xi) k_1(\xi, y)\, d\xi$ for $x \geqslant y \geqslant a$.

1.2 Basics from Analysis

1.2.1 Continuous Functions

In the following, let D be a finite-dimensional region (e.g., $D \subset \mathbb{R}$ or $D \subset \mathbb{R}^d$ including curves and surfaces). The set of continuous (and real-valued) functions defined on D is denoted by

(1.2.1) $\quad C(D) := \{f: \text{continuous function on } D\}$

or $C^0(D)$ (cf. (2)). The k-fold continuously differentiable functions form the set

(1.2.2) $\quad C^k(D) := \{f: \text{all } \ell\text{-fold derivatives belong to } C(D) \text{ for } 0 \leqslant \ell \leqslant k\},$

where in the higher dimensional case the derivatives are the *partial* ones. In (2), k may take the values $0, 1, \dots$ but also ∞. Accordingly, $C^\infty(D)$ is the set of functions infinitely differentiable on D.

1.2.2 Lipschitz Continuous Functions

A function f is called *globally Lipschitz continuous* on D, if

(1.2.3a) $\quad |f(x') - f(x'')| \leqslant L \|x' - x''\| \qquad \text{for all } x', x'' \in D.$

In this case, L is called the Lipschitz constant of f.

A function f is called *locally Lipschitz continuous* on D, if for any $x \in D$ there is a neighbourhood U_x and a Lipschitz constant L_x such that

(1.2.3b) $\quad |f(x') - f(x'')| \leqslant L_x \|x' - x''\| \qquad \text{for all } x', x'' \in U_x.$

In (3a,b), $\|\cdot\|$ denotes a norm on D (cf. §1.3.1). The choice of the norm is irrelevant because of the equivalence of norms (cf. Remark 3.3), but it influences the size of L or L_x, respectively. The sets of the locally and globally Lipschitz continuous functions are denoted by

(1.2.4a) $\quad C_{L,\text{loc}}(D) := \{f: f \text{ is locally Lipschitz continuous on } D\},$

(1.2.4b) $\quad C_L(D) := \{f: f \text{ is globally Lipschitz continuous on } D\}.$

Exercise 1.2.1. Prove that (a) the function sets introduced above form linear vector spaces with the usual pointwise addition of functions and the multiplication by a scalar factor $\lambda \in \mathbb{R}$.
(b) Global Lipschitz continuity implies local Lipschitz continuity; both imply continuity, i.e.,

$$C_L(D) \subset C_{L,\mathrm{loc}}(D) \subset C(D).$$

(c) $C^1(D) \subset C_{L,\mathrm{loc}}(D)$.
(d) If D is compact (i.e., closed and bounded, cf. §1.3.8), local and global Lipschitz continuity coincide, i.e., $C_{L,\mathrm{loc}}(D) = C_L(D)$.
(e) On $D=\mathbb{R}$, the function $f(x) := |1-e^x|$ is locally Lipschitz continuous but not globally.

Replacing in (2) the spaces $C(D)$ by $C_L(D)$ or $C_{L,\mathrm{loc}}(D)$, respectively, we arrive at the definition of the spaces of the k-fold locally or globally Lipschitz continuously differentiable functions:

(1.2.5a) $$C^k_{L,\mathrm{loc}}(D) := \{f : f^{(\ell)} \in C_{L,\mathrm{loc}}(D) \quad \text{for } 0 \le \ell \le k\},$$

(1.2.5b) $$C^k_L(D) := \{f : f^{(\ell)} \in C_L(D) \quad \text{for } 0 \le \ell \le k\}.$$

Exercise 1.2.2. (a) Rework Exercise 1a-d for the function classes (5a,b).
(b) Show that $f(x) = |x|^\alpha \in C^k_{L,\mathrm{loc}}(\mathbb{R})$ for $\alpha \ge k$.

1.2.3 Hölder Continuous Functions

The estimates (3a,b) can be generalised by

(1.2.6a) $$|f(x')-f(x'')| \le L_x \|x'-x''\|^\lambda \quad \text{for all } x',x'' \in U_x,$$

(1.2.6b) $$|f(x')-f(x'')| \le L \|x'-x''\|^\lambda \quad \text{for all } x',x'' \in D$$

with $0<\lambda<1$. A function f is called *locally Hölder continuous* on D (*with exponent* λ), if for any $x \in D$ there are a neighbourhood U_x and a constant (so-called Hölder constant) L_x satisfying (6a). A function f satisfying (6b) is called *globally Hölder continuous* on D (*with exponent* λ). The sets are abbreviated by

(1.2.7a) $$C^\lambda_{\mathrm{loc}}(D) := \{f : f \text{ is locally Hölder continuously with exp. } \lambda \text{ on } D\},$$

(1.2.7b) $$C^\lambda(D) := \{f : f \text{ is globally Hölder continuously with exp. } \lambda \text{ on } D\}.$$

For $0<\lambda<1$ and $k \in \mathbb{N}_0$, the k-times Hölder continuously differentiable functions form the sets

(1.2.7c) $$C^{k+\lambda}_{\mathrm{loc}}(D) := \{f : f^{(\ell)} \in C^\lambda_{\mathrm{loc}}(D) \quad \text{for } 0 \le \ell \le k\},$$

(1.2.7d) $$C^{k+\lambda}(D) := \{f : f^{(\ell)} \in C^\lambda(D) \quad \text{for } 0 \le \ell \le k\}.$$

$C^\varkappa(D)$ is defined for all real $\varkappa \ge 0$. For $\varkappa \in \mathbb{N}_0$, definition (2) applies. If $\varkappa$ is a non-integer, it admits a unique representation $\varkappa = k+\lambda$ with $k \in \mathbb{N}_0$ and $0<\lambda<1$; hence, $C^\varkappa(D)$ is well-defined by (7d).

Exercise 1.2.3. **(a)** Rework Exercise 1a–d for the function classes (7a–d). In particular, show that $C^{k+\lambda}_{\mathrm{loc}}(D) = C^{k+\lambda}(D)$ for compact D. For all D, $C^{k}_{L,\mathrm{loc}}(D) \subset C^{k+\lambda}_{\mathrm{loc}}(D)$ and $C^{k}_{L}(D) \subset C^{k+\lambda}(D)$ hold.
(b) $f(x) = |x|^{\alpha} \in C^{\alpha}([-1,1])$ for all $0<\alpha \notin \mathbb{N}$.
(c) Let $f(\cdot,y)$ belong to $C^{\lambda}(D)$ for all fixed $y \in D$ and assume that its Hölder constant is independent of y. Similarly, let $f(x,\cdot)$ belong to $C^{\lambda}(D)$ for fixed $x \in D$ with Hölder constant independent of x. Then, $f \in C^{\lambda}(D \times D)$ holds.
(d) Let L be the Hölder constant of $f \in C^{\lambda}(D)$. Then, the second difference $\{[f(x+\varepsilon+\eta)-f(x+\eta)]/\varepsilon^{\alpha}-[f(x+\varepsilon)-f(x)]/\varepsilon^{\alpha}\}/\eta^{\beta}$ can be bounded by $2L\varepsilon^{\lambda-\alpha-\beta}$ for $0 \leqslant \beta \leqslant \lambda \leqslant 1$.

To obtain a normed space, we will modify $C^{\varkappa}(D)$ insignificantly (cf. §1.3.2). In addition, we will introduce a closed related family $\hat{C}^{\varkappa}(D)$ of spaces in (3.4f).

1.3 Basics from Functional Analysis

1.3.1 Banach Spaces

Let X be a linear vector space (of finite or infinite dimension over $\mathbb{R}$ or $\mathbb{C}$). A *norm* $\|\cdot\|_X = \|\cdot\|$ on X is a mapping of X into $[0,\infty)$ satisfying the following norm axioms:

(1.3.1a) $\quad \|x\| = 0 \quad$ only for $x=0$,

(1.3.1b) $\quad \|\lambda x\| = |\lambda|\,\|x\| \quad$ for all $x \in X$ and $\lambda \in \mathbb{R}$ (or $\mathbb{C}$),

(1.3.1c) $\quad \|x+y\| \leqslant \|x\| + \|y\| \quad$ for all $x, y \in X$ (triangle inequality).

The pair $(X, \|\cdot\|)$ denotes a *normed vector space*. If there is more than one vector space or if the choice of the norm not obvious, we use the symbol of the space as subscript for clarity, e.g., $\|\cdot\|_X$, $\|\cdot\|_Y$, etc.

The open ball in X with radius $r>0$ and centre at $x \in X$ is

(1.3.2) $$K_r(x) := \{\xi \in X : \|x-\xi\|_X < r\}.$$

The norm $\|\cdot\|$ defines a topology: $U \subset X$ is a neighbourhood of $x \in X$, if $K_r(x) \subset U$ for some $r>0$. The *convergence* $x_n \to x$ in X is defined as the *convergence in norm*:

$$\|x_n - x\|_X \to 0.$$

A sequence $\{x_n\}$ is called *convergent* in X, if there is some $x \in X$ with $x_n \to x$. The sequence $\{x_n\}$ is called *Cauchy* in X, if for all $\varepsilon>0$ there is some $n \in \mathbb{N}$ such that $\|x_\nu - x_\mu\|_X \leqslant \varepsilon$ for all $\nu, \mu \geqslant n$. A convergent sequence is always Cauchy. The reverse statement leads to

Definition 1.3.1. The normed linear space $(X, \|\cdot\|)$ is called *complete*, if any Cauchy sequence in X is also convergent.

Definition 1.3.2. A complete normed linear space is called a *Banach space*.

Trivial examples for Banach spaces are the finite-dimensional vector spaces $\mathbb{R}^d$ with the *maximum norm* (3a) or the *Euclidean norm* (3b) for the vectors $x=(x_1,\dots,x_d)$:

(1.3.3a) $\|x\|_\infty := \max\{|x_i| : 1\le i\le d\}$,

(1.3.3b) $\|x\|_2 := (\sum_{i=1}^{d} |x_i|^2)^{1/2}$.

The fact that the choice of the norm in the inequalities (2.3a,b) and (2.6a,b) is not important, is a consequence of

Remark 1.3.3. All norms on a finite-dimensional vector space X are *equivalent*, i.e., for each pair $\|\cdot\|$ and $|||\cdot|||$ of norms there is a constant C such that

$$\|x\| \le C\,|||x|||, \qquad |||x||| \le C\,\|x\| \qquad \text{for all } x\in X.$$

Proof. Cf. Hackbusch [3,§2.6.2]. QED

1.3.2 Banach Spaces $C^\varkappa(D)$, $C_L^k(D)$, $\hat{C}^\varkappa(D)$

Let $f\in C(D)$. The *supremum norm* is defined by

(1.3.4a) $\|f\|_\infty := \|f\|_{\infty,D} := \sup\{|f(x)| : x\in D\}$.

Since the supremum in (4a) may take the value ∞, $\|\cdot\|_\infty$ is not yet a norm on $C(D)$. However, it becomes a norm on the subset of the *bounded* continuous functions:

(1.3.5) $C_b(D) := \{f\in C(D) : \|f\|_\infty<\infty\}$.

Exercise 1.3.4. Prove that $C_b(D)=C(D)$ for compact D.

Since for compact D the supremum in (4a) may be replaced by the maximum, $\|\cdot\|_\infty$ is also called the *maximum norm* (on D).

In the following, the normed space $(C_b(D), \|\cdot\|_\infty)$ is again denoted by $(C(D), \|\cdot\|_\infty)$.

Remark 1.3.5. (a) Convergence $f_n\to f$ with respect to the supremum norm $\|\cdot\|_{\infty,D}$ is *uniform convergence* on D.
(b) $(C(D), \|\cdot\|_\infty)$ is a Banach space.

Proof of *(b)*. The limit of a uniformly convergent sequence of continuous functions is again continuous. Hence, $(C(D),\|\cdot\|_\infty)$ is complete. QED

Analogously, $(C^k(D),\|\cdot\|_{C^k(D)})$, $k\in\mathbb{N}_0$, is the Banach space containing those functions from $C^k(D)$ which are bounded with respect to $\|\cdot\|_{C^k(D)}$, where

(1.3.4b) $\|f\|_{C^k(D)} := \max\{\|f^{(\ell)}\|_{\infty,D} : 0\le\ell\le k\}$

describes the uniform convergence of all derivatives $f^{(\ell)}$, $0\le\ell\le k$.

For globally Lipschitz continuous functions, one uses the norm

(1.3.4c) $\|f\|_{C_L(D)} := \max\{\|f\|_{\infty,D},\ L_f\}$,

where L_f is the smallest constant satisfying the Lipschitz estimate (2.3b). Similarly, one defines the norm (4d) for $f \in C_L^k(D)$:

(1.3.4d) $\|f\|_{C_L^k(D)} := \max\{\|f^{(\ell)}\|_{C_L(D)} : 0 \leq \ell \leq k\}$.

The Hölder continuous functions form a Banach space which we denote $(C^{\varkappa}(D), \|\cdot\|_{C^{\varkappa}(D)})$. When $\varkappa \in \mathbb{N}_0$, we use the norm (4b) with $k=\varkappa$. If $\varkappa = k+\lambda$ with $k \in \mathbb{N}_0$ and $0<\lambda<1$, then we proceed as follows. For $k=0$ we define

(1.3.4e$_1$) $\|f\|_{C^{\varkappa}(D)} := \max\{\|f\|_{\infty,D},\ L_{f,\varkappa}\}$ for $0<\varkappa<1$,

where $L = L_{f,\varkappa}$ is the smallest constant fulfilling the estimate (2.6b) for $\lambda=\varkappa$. For $\varkappa = k+\lambda$ with $k \in \mathbb{N}$ and $0<\lambda<1$, one defines

(1.3.4e$_2$) $\|f\|_{C^{\varkappa}(D)} := \max\{\|f\|_{C^k(D)},\ \|f^{(k)}\|_{C^{\lambda}(D)}\}$ for $\varkappa = k+\lambda$.

In our later discussion on error estimates we will see that the bounds do not deteriorate when we use the $C_L^{k-1}(D)$ norm instead of the $C^k(D)$ norm, although $C_L^{k-1}(D)$ is a proper subspace of $C^k(D)$. Therefore, one obtains the same results after replacing $C^{\varkappa}(D)$, $\varkappa \geq 0$, by the spaces

(1.3.4f) $$\hat{C}^{\varkappa}(D) := \begin{cases} C^{\varkappa}(D) & \text{for } \varkappa=0 \text{ and non-integer } \varkappa>0, \\ C_L^{\varkappa-1}(D) & \text{for } \varkappa \in \mathbb{N}. \end{cases}$$

1.3.3 Banach Spaces $L^1(D)$, $L^2(D)$, $L^{\infty}(D)$

Further examples of Banach spaces and associated norms are

(1.3.5a) $L^1(D) := \{f : \text{Lebesgue integrable over } D\}$ with

(1.3.5b) $\|f\|_1 := \int_D |f(x)|\,dx$,

(1.3.6a) $L^2(D) := \{f : f \text{ measurable and } f^2 \in L^1(D)\}$ with

(1.3.6b) $\|f\|_2 := \left[\int_D |f(x)|^2 dx\right]^{1/2}$,

(1.3.7a) $L^{\infty}(D) := \{f : \text{measurable in } D \text{ and } \|f\|_{\infty} < \infty\}$, where

(1.3.7b) $\|f\|_{\infty} := \inf\{\sup\{|f(x)| : x \in S\} : S \subset D \text{ with } \mu(D \setminus S) = 0\}$.

In the spaces mentioned above two functions must be identified if they coincide up to a set of measure zero. Speaking precisely, the elements of the spaces $L^1(D)$, $L^2(D)$ and $L^{\infty}(D)$ are not functions but equivalence classes of functions. The norm $\|\cdot\|_{\infty}$ coincides with the supremum norm (4a) on the subspace $C(D) \subset L^{\infty}(D)$. By «$C(D) \subset L^{\infty}(D)$» we mean: For any $f \in C(D)$ there is an equivalence class $\Phi \in L^{\infty}(D)$ with $f \in \Phi$. The right-hand side in (7b) is also called the «essential supremum» («vrai max» or «sup ess») of f. This takes into consideration that f may be defined arbitrarily on each set $D \setminus S$ of measure $\mu(D \setminus S) = 0$ without changing (the equivalence class of) f.

1.3.4 Dense Subspaces

Definition 1.3.6. Let $A \subset X$ be a subset of a Banach space X. If the closure $\bar{A}$ coincides with X, A is called *dense* in X. If, in addition, A is a subspace, then A is called a *dense subspace*.

Exercise 1.3.7. Prove that A is dense in X, if and only if for all $x \in X$ and $\varepsilon > 0$ there is an $a \in A$ with $\|x - a\| \leqslant \varepsilon$.

Theorem 1.3.8. (Approximation theorem of Weierstraß). Let $D \subset \mathbb{R}^d$ be compact. Then the subspace of all polynomials is dense in $C(D)$.

A further example is contained in

Remark 1.3.9. $C_0(D) := \{f \in C(D) : \operatorname{supp}(f) \text{ compact}\}$ is a dense subspace in $X = L^2(D)$. Here, the *support* of f is defined by

$$\operatorname{supp}(f) := \overline{\{x \in D : f(x) = 0\}}.$$

1.3.5 Banach's Fixed Point Theorem

Theorem 1.3.10. Let X be a Banach space and $\Phi: X \to X$ a *contraction*, i.e., there is a number $0 \leqslant q < 1$ such that

(1.3.8) $$\|\Phi(x) - \Phi(y)\| \leqslant q \|x - y\| \qquad \text{for all } x, y \in X.$$

Then the following holds. (i) The *fixed point equation*

(1.3.9) $$x^* = \Phi(x^*)$$

has exactly one solution $x^* \in X$.
(ii) For each starting value $x^0 \in X$, the sequence $\{x^i\}$ defined by

(1.3.10a) $$x^{i+1} := \Phi(x^i)$$

converges to the solution x^* of (9).
(iii) The following error estimate holds:

(1.3.10b) $$\|x^* - x^i\| \leqslant q^i \|x^1 - x^0\| / (1 - q).$$

Proof. (a) Apply inequality (8) to $x := x^i$ and $y := x^{i+1}$:

(1.3.10c) $$\|x^{i+1} - x^i\| \leqslant q \|x^i - x^{i-1}\| \leqslant q^i \|x^1 - x^0\| \qquad \text{for } i \geqslant 1.$$

The triangle inequality and summation over $j = i, i+1, \ldots, k-1$ yield

(1.3.10d) $$\|x^k - x^i\| \leqslant \|x^k - x^{k-1}\| + \ldots + \|x^{i+1} - x^i\| \leqslant$$
$$\leqslant (q^{k-1} + \ldots + q^i) \|x^1 - x^0\| = (q^i - q^k) \|x^1 - x^0\| / (1 - q).$$

Hence, $\{x^i\}$ is a Cauchy sequence converging to some $x^* \in X$ in the Banach

space X. The Lipschitz condition (8) implies continuity of Φ. This then implies that $x^* = \lim x^{i+1} = \lim \Phi(x^i) = \Phi(x^*)$, i.e., x^* is a solution of (9).
(b) Let x^* and x^{**} be two solutions of (9). From (8) one concludes $\|x^* - x^{**}\| \leqslant q\,\|x^* - x^{**}\|$ with $q<1$; hence, $\|x^* - x^{**}\| = 0$ proving the uniqueness of the solution.
(c) Performing the limiting process $k \to \infty$ in (10d), one obtains (10b). QED

1.3.6 Linear Operators

In the following, we consider linear mappings $A: x \mapsto Ax$ of a Banach space X into a second Banach space Y, where X and Y may coincide. Such an «operator» $A: X \to Y$ is called *bounded*, if the *operator norm*

(1.3.11a) $$\|A\|_{Y \leftarrow X} := \sup\{\|Ax\|_Y / \|x\|_X : 0 \neq x \in X\}$$

is finite. As the terminology indicates, the operator norm $\|\cdot\|_{Y \leftarrow X}$ satisfies the norm axioms (1a-c).

The proofs of the next theorem can be found in the first chapters of any book on functional analysis (e.g., Heuser [1,§10]).

Theorem 1.3.11. (a) A linear operator $A: X \to Y$ is continuous if and only if A is bounded.
(b) The set denoted by

$$L(X,Y) := \{A: X \to Y \text{ linear and continuously}\}$$

forms a linear space. Equipped with the operator norm $\|\cdot\|_{Y \leftarrow X}$, $L(X,Y)$ becomes a Banach space.
(c) It holds that

(1.3.11b) $$\|Ax\|_Y \leqslant \|A\|_{Y \leftarrow X} \|x\|_X \qquad \text{for all } A \in L(X,Y) \text{ and all } x \in X.$$

(d) For $A \in L(X,Y)$ and $B \in L(Y,Z)$, $(BA)x = B(Ax)$ defines the product BA. It satisfies $BA \in L(X,Z)$ and

(1.3.11c) $$\|BA\|_{Z \leftarrow X} \leqslant \|B\|_{Z \leftarrow Y} \|A\|_{Y \leftarrow X}.$$

Exercise 1.3.12. Let $A \in L(X,Y)$.
(a) Prove that $\|A\|_{Y \leftarrow X} = \sup\{\|Ax\|_Y : x \in X,\ \|x\|_X \leqslant 1\}$. (This can be used as an equivalent definition.)
(b) By assuming $\|Ax\|_Y \leqslant C\|x\|_X$ for all $x \in X$, prove that $\|A\|_{Y \leftarrow X} \leqslant C$.
(c) Let C be the smallest constant such that $\|Ax\|_Y \leqslant C\|x\|_X$ for all $x \in X$. Prove that $C = \|A\|_{Y \leftarrow X}$.

The open mapping theorem (cf. Heuser [1,§39] or Yosida [1,§II.5]) implies

Theorem 1.3.13 (Theorem of the continuous inverse). Assume that X and Y are Banach spaces. If $A \in L(X,Y)$ is bijective (i.e., injective and surjective), then the inverse is also continuous: $A^{-1} \in L(Y,X)$.

The following theorem shows that the invertible operators form an open set in $L(X,Y)$.

Lemma 1.3.14. Let X and Y be Banach spaces. Assume $S,T\in L(X,Y)$, $T^{-1}\in L(Y,X)$ and

(1.3.12a) $\|S-T\|_{Y\leftarrow X}\|T^{-1}\|_{X\leftarrow Y} < 1.$

Then S has an inverse satisfying $S^{-1}\in L(Y,X)$ and

(1.3.12b) $\|S^{-1}\|_{X\leftarrow Y} \leqslant \|T^{-1}\|_{X\leftarrow Y}/[1-\|S-T\|_{Y\leftarrow X}\|T^{-1}\|_{X\leftarrow Y}]$

Proof. Let $y\in Y$. The mapping $\Phi(x):=T^{-1}(T-S)x+T^{-1}y$ is a contraction, since $\|\Phi(x)-\Phi(x')\|_X=\|T^{-1}(T-S)(x-x')\|_X\leqslant q\|x-x'\|_X$ holds with the constant $q:=\|T^{-1}\|_{X\leftarrow Y}\|S-T\|_{Y\leftarrow X}<1$. The Banach Fixed Point Theorem (Theorem 10) guarantees the unique solvability of $x=\Phi(x)$. This fixed point equation is equivalent to $Tx=T\Phi(x)=(T-S)x+y$ and $Sx=y$. Therefore, S is bijective and continuous: $S^{-1}\in L(Y,X)$ (cf. Theorem 13). Choose $x^0:=0$ in (10a,b), i.e., $x^1=\Phi(x^0)=T^{-1}y$. Estimate (10b) yields

$$\|S^{-1}y\|_X = \|x\|_X = \|x-x^0\|_X \leqslant q^0\|x^1-x^0\|_X/(1-q) =$$
$$= \|T^{-1}y\|_X/(1-q) \leqslant \|T^{-1}\|_{X\leftarrow Y}\|y\|_Y/(1-q)$$

for an arbitrary $y\in Y$. This proves $\|S^{-1}\|_{X\leftarrow Y} \leqslant \|T^{-1}\|_{X\leftarrow Y}/(1-q)$. QED

Special Banach spaces are the real numbers $\mathbb{R}$ or the complex numbers $\mathbb{C}$ with the absolute value as its norm. For the choice $Y=\mathbb{R}$ or $Y=\mathbb{C}$, the space $X':=L(X,Y)$ is called the (real- or complex-valued) *dual space*. In coincidence with (11a), the *dual norm* is defined by

$$\|\varphi\|_{X'} := \sup\{|\varphi(x)|/\|x\|: 0\neq x\in X\}.$$

Each operator $A\in L(X,Y)$ is associated with a *dual operator* $A'\in L(Y',X')$. The value $A'y'\in X'$ for some $y'\in Y'$ is defined by $(A'y')(x):=y'(Ax)$.

1.3.7 Theorem of Uniform Boundedness

In numerical applications, sequences T_n of operators occur for which one needs a uniform estimate, often provided by the following statement

Theorem 1.3.15 (Theorem of Uniform Boundedness). Assume that X and Y are Banach spaces. Let the sequence $T_n\in L(X,Y)$, $n\in\mathbb{N}$, satisfy

(1.3.13a) $\sup\{\|T_nx\|_Y: n\in\mathbb{N}\}<\infty$ for all $x\in X$.

Then the sequence $\|T_n\|_{Y\leftarrow X}$ is uniformly bounded, i.e.,

(1.3.13b) $\sup\{\|T_n\|_{Y\leftarrow X}: n\in\mathbb{N}\}<\infty$

Proof. Cf. Heuser [1,§40] or Yosida [1,§II.1].

In particular, convergent sequences are bounded. Hence, we obtain

Remark 1.3.16. Let X and Y be Banach spaces. Assume that the sequence $T_n \in L(X,Y)$, $n \in \mathbb{N}$, has a limit $\lim_{n\to\infty} T_n x$ for all $x \in X$. Then (13b) follows.

According to Remark 16, we can define a linear mapping T by

$$Tx := \lim_{n\to\infty} T_n x,$$

for which $T \in L(X,Y)$ can be proved. A variant of the Theorem of Uniform Boundedness, based on the pointwise convergence $T_n x \to Tx$, is the following theorem.

Theorem 1.3.17 (Theorem of Banach-Steinhaus). Let X and Y be Banach spaces. The operators $T_n \in L(X,Y)$ converge pointwise to $T \in L(X,Y)$, i.e.,

(1.3.14a) $\quad T_n x \to Tx$ in Y $\qquad$ for all $x \in X$,

if and only if uniform boundedness (13b) holds and there is a dense set $M \subset X$ such that

(1.3.14b) $\quad \lim_{n\to\infty} T_n x$ exists in Y $\qquad$ for all $x \in M$.

Proof. (i) Due to Remark 16, (14a) implies (13b) as well as (14b).
(ii) Let $\xi \in X$ and $x_k \in M$ be a sequence with $x_k \to \xi$. Split $(T_n - T_m)\xi$ into $T_n(\xi - x_k) + (T_n - T_m)x_k + T_m(x_k - \xi)$. Thanks to (13b), the first and last term can be made sufficiently small by a suitable choice of k independent of n and m. However, for fixed k the middle term tends to zero as $n, m \to \infty$. Hence, $T_n\xi$ is a Cauchy sequence converging in Y to a value $\eta =: T\xi$. QED

Remark 1.3.18. The pointwise convergence (14a) follows from the operator norm convergence $\| T_n - T \|_{Y\leftarrow X} \to 0$, but the converse does not hold. In fact, if T_n converges only pointwise to some T and not with respect to the operator norm, then (as proved below) there are $x \in X$ such that $T_n x$ converges *arbitrarily slowly* to Tx. The latter statement means that there is no positive sequence $\{\alpha_n\}$, $\alpha_n \to 0$, such that $\sup\{\| T_n x - Tx \|_Y / \alpha_n : n \in \mathbb{N}\} < \infty$ (i.e., $\| T_n x - Tx \|_Y \leqslant O(\alpha_n)$).

Proof. If the convergence speed were $O(\alpha_n)$, then the operators $S_n := (T_n - T)/\alpha_n$ satisfy condition (13a). By Theorem 15, $\| S_n \|_{Y\leftarrow X} \leqslant C$ is uniformly bounded, i.e., $\| T_n - T \|_{Y\leftarrow X} \leqslant C\alpha_n \to 0$ converges with respect to the operator norm. QED

1.3.8 Compact Sets and Compact Mappings

Let X be a Banach space. A subset $M \subset X$ is called *precompact*, if all sequences $x_k \in M$ contain a subsequence converging in X. $M \subset X$ is called *compact*, if, in addition, M is closed.

Exercise 1.3.19. (a) In $X=\mathbb{R}^d$, the terms «precompact» and «bounded» are equivalent.
(b) Precompact sets are always bounded.
(c) The image of a [pre]compact set under of a continuous mapping is again [pre]compact.

The important factor in Exercise 19a is that X has finite dimension. In fact, the following holds.

Remark 1.3.20. The unit ball $\{x\in X:\ \|x\|\leq 1\}$ of a Banach space X is compact if and only if $\dim X<\infty$.

Definition 1.3.21. Let X,Y be Banach spaces. An operator $T\in L(X,Y)$ is called *compact*, if the image $\{Tx:\ x\in X,\ \|x\|_X\leq 1\}$ of the X-unit ball is precompact in Y.

Exercise 1.3.22. (a) $T\in L(X,Y)$ is compact, if and only if any bounded sequence x_k from the Banach space X contains a subsequence x_{k_i} such that Tx_{k_i} converges in the Banach space Y.
(b) Let $I\in L(X,X)$ denote the identity: $Ix=x$ for all $x\in X$. I is compact if and only if $\dim X<\infty$.
(c) If $S,T\in L(X,Y)$ are compact, the linear combination $\alpha S+\beta T$ $(\alpha,\beta\in\mathbb{C})$ is compact, too.
(d) The assumption $T\in L(X,Y)$ in Definition 21 is not necessary: compactness implies continuity.
Hints. (a) Discuss $\xi_k:=x_k/C$ for $C:=\sup\{\|x_k\|:\ k\in\mathbb{N}\}$. (b) Cf. Remark 20. (c) Choose a subsequence $\xi_i:=x_{k_i}$ such that $T\xi_i$ converges, and a subsequence ξ_{i_j} of ξ_i such that $S\xi_{i_j}$ converges. (d) Cf. Exercise 19b.

Due to Exercise 22c, the compact linear operators form a linear subspace, which we denote by $K(X,Y)$:

$$K(X,Y) := \{T\in L(X,Y):\ T \text{ compact}\}.$$

Compact operators are also termed *completely continuous.* Compact operators satisfy the following «rules of calculus»:

Theorem 1.3.23. In the following, X,Y,Z are Banach spaces.
(a) The product $ST\in L(X,Z)$ is compact, if one of the factors $T\in L(X,Y)$ or $S\in L(Y,Z)$ is compact.
(b) Let $T_n\in K(X,Y)$ have a limit $T:=\lim T_n$ with respect to the operator norm. Then T is also compact. Hence, $K(X,Y)$ is a closed subspace of $L(X,Y)$.
(c) $T\in L(X,Y)$ is compact, if image(T) has a finite dimension.
(d) $T\in L(X,Y)$ is compact, if and only if the dual operator $T'\in L(Y',X')$ is compact.

The identities $I \in L(Y,Y)$ and $I \in L(X,X)$ are different from the *inclusion* (or *embedding*) $I \in L(Y,X)$ of a subspace $Y \subset X$ in X. $I \in L(Y,X)$ and $I \in L(Y,Y)$ are algebraically equal ($Iy = y$), but the underlying topologies differ.

Definition 1.3.24. Let X and Y be two Banach spaces with the property $Y \subset X$. Then Y is called *continuously embedded* [*compactly embedded*] *in* X, if $I \in L(Y,X)$ [$I \in K(Y,X)$] holds.

Exercise 1.3.25. (a) Y is continuously embedded in X, if $\|\cdot\|_X \leqslant C\|\cdot\|_Y$ holds for some $C \in \mathbb{R}$.
(b) Y is compactly embedded in X, if any sequence $y_k \in Y$ bounded in Y has a convergent subsequence in X.

We return again to the definition of the term «precompact» in order to characterise precompact subsets in the space $C(D)$.

Theorem 1.3.26 (Theorem of Arzelà-Ascoli). Let D be compact and M a subset of $C(D)$. Then M is precompact, if M is *uniformly bounded*, i.e.,

(1.3.15a) $\quad \sup\{\|f\|_\infty : f \in M\} < \infty$

and *equicontinuous*, i.e., for all $\varepsilon > 0$, $x \in D$ there is a δ with

(1.3.15b) $\quad |f(x) - f(y)| \leqslant \varepsilon \qquad$ for all $f \in M$ and $y \in D$ with $|x-y| \leqslant \delta$.

Proof. Cf. Yosida [1,§III]. ∎

Theorem 1.3.27. Assume that D is compact and $0 \leqslant \varkappa < \lambda$. Then the embeddings $C^\lambda(D) \subset C^\varkappa(D)$ and $\hat{C}^\lambda(D) \subset \hat{C}^\varkappa(D)$ are compact.

Proof. (i) First assume $\varkappa = 0$ and without loss of generality $\lambda < 1$. If M is a bounded subset of the Hölder continuous functions $C^\lambda(D)$, then (15b) holds with $\delta := (\varepsilon / L)^{1/\lambda}$, where $L := \sup\{\|f\|_{C^\lambda(D)} : f \in M\}$.
(ii) Now assume $0 < \varkappa < \lambda < 1$ and let $M \subset C^\lambda(D)$ be a bounded set. According to part (i), for an arbitrary sequence we find a subsequence $f_k \in M$ converging in $C(D)$. Set

$$F_k(x,y) := [f_k(x) - f_k(y)] / |x-y|^\varkappa \qquad \text{for } (x,y) \in D \times D \text{ with } x \neq y.$$

For $x = y$ the function F_k is continuously extendable by $F_k(x,x) := 0$. $\{F_k : k \in \mathbb{N}\}$ is a bounded sequence in $C(D \times D)$. One can verify that for fixed y $F_k(x,y)$ belongs to $C^{\lambda-\varkappa}(D)$ with respect to x, where the Hölder constant can be chosen independently of k and y. Analogous statements hold for $F_k(x,\cdot) \in C^{\lambda-\varkappa}(D)$. Exercise 2.3c shows that F_k belongs to $C^{\lambda-\varkappa}(D \times D)$ and that the sequence $\{F_k : k \in \mathbb{N}\}$ is bounded in $C^{\lambda-\varkappa}(D \times D)$. Part (i) ensures the existence of a subsequence F_{k_j} converging in $C(D \times D)$. Since the Hölder constant of $f_i - f_\ell$ coincides with $\|F_i - F_\ell\|_{\infty, D \times D}$ and the latter quantity tends to zero, if f_i and f_ℓ belong to the subsequence f_{k_j}, this subsequence is Cauchy in $C^\varkappa(D)$. ∎

1.3.9 Riesz-Schauder Theory

Let $T \in L(X,X)$ and $\lambda \in \mathbb{C}$. If $\lambda I - T$ is bijective (cf. Theorem 13), λ is called a *regular value* of T; otherwise, a *singular value* of T. We define the *spectrum of T* by

$$\sigma(T) := \{\lambda \in \mathbb{C} : \lambda \text{ is a singular value of } T\}.$$

$\lambda \in \sigma(T)$ is called an *eigenvalue*, if the equation $\lambda e = Te$ has a nontrivial solution (eigenfunction) $e \in X$. Note that, in general, $\sigma(T)$ may contain singular values λ which are not eigenvalues. The following theorem states that for compact operators $T \in K(X,X)$ the same situation holds as for (finite-dimensional) linear systems of equations.

Theorem 1.3.28 (Riesz-Schauder Theory). Let X be a Banach space and $T \in K(X,X)$ compact. Assume that $\lambda \in \mathbb{C}$ is different from zero.
(a) Then either (i) $(\lambda I - T)^{-1} \in L(X,X)$ or (ii) λ is an eigenvalue.
(b) The eigenspace $\mathrm{Ker}(\lambda I - T)$ has the same finite dimension as the quotient space $X/\mathrm{image}(\lambda I - T)$.
(c) The spectrum $\sigma(T)$ consists of at most countably many elements, which have no cluster points except possibly zero.

Part (b) of the theorem expresses the fact that $\lambda I - T$ is injective if and only if it is surjective.

1.3.10 Hilbert Spaces, Orthogonal Complements, Projections

Definition 1.3.29. A Banach space X is called a *Hilbert space* if a *scalar product* $\langle \cdot,\cdot \rangle = \langle \cdot,\cdot \rangle_X : X \times X \mapsto \mathbb{C}$ is defined on $X \times X$, which is connected with the norm by $\|f\|_X = \langle f,f \rangle^{1/2}$.

Remark 1.3.30. The only Hilbert space among the Banach spaces discussed so far is $L^2(D)$. The scalar product on $L^2(D)$ reads

$$\langle f,g \rangle := \int_D f(x)g(x)dx \qquad \text{for all } f,g \in L^2(D). \tag{1.3.16}$$

If complex-valued functions are admitted, one has to replace $g(x)$ in (16) by the complex conjugate value $\overline{g(x)}$.

Two functions $f,g \in X$ of a Hilbert space X are called *orthogonal* (abbreviation: $f \perp g$), if $\langle f,g \rangle = 0$. $f \in X$ is called orthogonal to a set $M \subset X$ (abbreviation: $f \perp M$), if $f \perp g$ holds for all $g \in M$. The *orthogonal complement of* $M \subset X$ is

$$M^{\perp} := \{g \in X : g \perp M\}. \tag{1.3.17}$$

The following theorem proves a decomposition of the Hilbert space.

Theorem 1.3.31. Let V be a closed subspace of a Hilbert space X and $W := V^{\perp}$ the orthogonal complement of V. Then X is the direct sum of V and W, i.e., each $x \in X$ has a unique representation $x = v + w$ with $v \in V$ and $w \in W$.

Definition 1.3.32. (a) A linear mapping $T: X \to X$ is called a *projection*, if $T = T^2$.
(b) Let X be Hilbert space. $T \in L(X, X)$ is called an *orthogonal projection*, if T is a selfadjoint projection, i.e., $T = T^2$ and $\langle Tf, g\rangle = \langle f, Tg\rangle$ for all $f, g \in X$.

The orthogonal projections are uniquely characterised by their images.

Remark 1.3.33. (a) Let $T \in L(X, X)$ be an orthogonal projection with the image space $V := \mathrm{image}(T)$ and $W := V^{\perp}$ the orthogonal complement of V. Then V is closed and the following holds:

(1.3.18a) $\quad Tx = v$ for all $x = v + w$ with $v \in V$, $w \in W$.

(b) Let V be a closed subspace of the Hilbert space X. Then (18a) defines an orthogonal projection.
(c) Another implicit characterisation of T from (18a) reads as follows:

(1.3.18b) $\quad \xi = Tx$ minimises the expression $\{\|x - \xi\|_X : \xi \in V\}$.

1.4 Basics from Numerical Mathematics

1.4.1 Interpolation

In the following, let $D \subset \mathbb{R}^d$ be the domain in which interpolation is performed. On D, n pairwise different *support abscissae* are given:

(1.4.1) $\quad x_i = x_{i,n} \in D \qquad (i = 1, 2, \ldots, n) \qquad (x_i \neq x_j \text{ for } i \neq j).$

Further, a (linear) interpolation is characterised by a linear subspace

(1.4.2) $\quad V = V_n \subset C(D) \qquad$ with $\dim V = n$

of the continuous functions on D. The interpolation problem is the following: Given arbitrary *support ordinates* $y_i \in \mathbb{R}$ $(i = 1, \ldots, n)$, find a function $\varphi \in V$ with

(1.4.3) $\quad \varphi(x_i) = y_i \qquad$ for all $i = 1, \ldots, n$.

In general, problem (3) need not be solvable or uniquely solvable, but solvability and uniqueness are equivalent.

Lemma 1.4.1. The interpolation problem (3) is uniquely solvable, if and only if the zero function is the only solution $\varphi \in V$ of $\varphi(x_i) = 0$ $(1 \le i \le n)$.

In the following, the unique solvability of the interpolation problem is always assumed. Among the possible bases of the linear space V, the following is particularly suited for representation purposes.

Definition 1.4.2. The solution $L_{i,n} = L_i \in V$ of the interpolation problem

(1.4.4a) $\quad L_i(x_j) = \delta_{ij} := \begin{cases} 1 & \text{for } i = j \\ 0 & \text{for } i \neq j \end{cases} \qquad (1 \le i, j \le n)$

is called the *Lagrange function* (belonging to the support abscissa x_i). The basis $\{L_i\}_{i=1,\ldots,n}$ is called the *Lagrange basis*.

Due to linearity (cf. Exercise 7a), one immediately obtains

Lemma 1.4.3. The unique solution of the interpolation problem (3) reads

(1.4.4b) $$\varphi(x) = \sum_{i=1}^{n} y_i L_i(x) \qquad \text{for } x \in D.$$

The classical example of interpolation is

Example 1.4.4 (polynomial interpolation). Let $I \subset \mathbb{R}$. For the polynomial interpolation choose $V := \{P: P \text{ polynomial of degree} \leqslant n-1\}$. Obviously, $\dim V = n$ holds as required in (2). Since the zero polynomial is the only polynomial with n zeros at x_i $(1 \leqslant i \leqslant n)$, Lemma 1 guarantees unique solvability. The Lagrange functions are the well-known *Lagrange polynomials* $L_i(x) = \prod_{j \neq i} [(x - x_j)/(x_i - x_j)]$.

In modern applications, usually *piecewise* polynomial interpolations are applied. For this purpose, in the one-dimensional case, the interval $D = I = [a, b]$ is divided into subintervals $I_k = [x_k, x_{k+1}]$ with the end points

(1.4.5a) $$a = x_{1,n} < x_{2,n} < \ldots < x_{n-1,n} < x_{n,n} = b \qquad \text{(short notation: } x_j = x_{j,n}\text{)}.$$

We associate this interval partition with the *step size* h:

(1.4.5b) $$h := h_n := \max\{|x_{k,n} - x_{k-1,n}|: 2 \leqslant k \leqslant n\}.$$

Concerning the convergence $h_n \to 0$, we assume that there is a number C with

(1.4.5c) $$h_n \leqslant C/n \qquad \text{for } n \in \mathbb{N}.$$

The *equidistant partition* is characterised by $x_{k,n} = a + (k-1)h$ with step size $h = 1/(n-1)$

Example 1.4.5 (piecewise linear interpolation). Partition $I = [a,b]$ according to (5a). Let V consist of all continuous functions being linear on the subintervals $[x_i, x_{i+1}]$ for $i = 1, \ldots, n-1$. The solution $\varphi \in V$ of (3) can be represented explicitly. For $x \in [x_i, x_{i+1}]$ we have $\varphi(x) = \{(x - x_i)y_{i+1} + (x_{i+1} - x)y_i\}/(x_{i+1} - x_i)$.

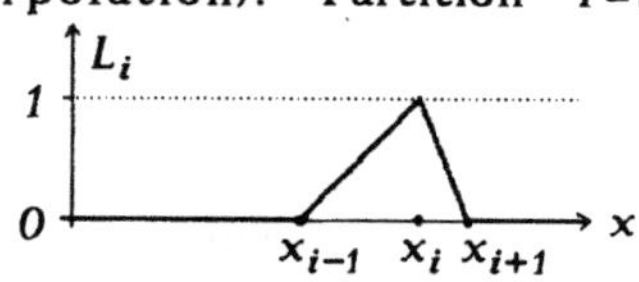

Fig. 1.4.1 hat function L_i

The Lagrange functions L_i («hat functions», cf. Fig. 1) are given by (6):

(1.4.6) $$L_i(x) = \begin{cases} (x_{i+1} - x)/(x_{i+1} - x_i) & \text{for } x \in I_i, \text{ if } i<n, \\ (x - x_i)\ /(x_i - x_{i-1}) & \text{for } x \in I_{i-1}, \text{ if } i>1, \\ 0 & \text{otherwise.} \end{cases}$$

Instead of linear interpolation in each subinterval I_k, one can also use polynomial interpolation of higher degree. Then one needs further support abscissae in each subinterval I_k. Posing additional requirements concerning the smoothness of the interpolating function $(V \subset C^\ell(I))$, one is led to the piecewise Hermite (cf. (2.3.3)) or spline functions.

Up to now, the interpolation problem (3) was understood as a mapping of $\mathbb{R}^n$ onto V: $y=(y_1,\dots,y_n)\mapsto \varphi\in V$. According to the term «interpolation», one should interpret the support ordinates y_i as values $f(x_i)$ of a continuous function $f\in C(D)$. Then, problem (3) becomes

(1.4.3*) $\varphi(x_i) = f(x_i)$ for all $i=1,\dots,n$.

Definition 1.4.6. Assume the interpolation problem (3) to be uniquely solvable. The linear map $\Pi_n=\Pi$: $C(D)\to V\subset C(D)$ mapping $f\in C(D)$ into the solution $\varphi=\Pi f\in V$ of problem (3*) is called the *interpolation operator* (or shortly: *interpolation*).

Parts (b) and (c) of the following exercise show that all information about the interpolation process is contained in Π.

Exercise 1.4.7. (a) The mapping Π defined by $\varphi=\Pi f$ and (3*) is linear.
(b) The subspace V (cf. (2)) is obtained from Π as its image space: $V=\{\varphi=\Pi f:\ f\in C(D)\}$.
(c) The set of points $x\in D$ with $f(x)=(\Pi f)(x)$ for all $f\in C(D)$ equals the set of the support abscissae $\{x_1,\dots,x_n\}$.
(d) Let L_i be the Lagrange functions. Then, Π has the representation

(1.4.7) $$\Pi f := \sum_{i=1}^{n} f(x_i)\,L_i.$$

Theorem 1.4.8. Each interpolation $\Pi=\Pi_n$ is a projection onto $V=V_n$.

Proof. Let $f\in C(I)$ be arbitrary and $\varphi=\Pi f\in V$ the interpolating function. $\psi:=\Pi\varphi\in V$ satisfies $\psi(x_i)=\varphi(x_i)$ $(1\le i\le n)$. Because of the unique solvability of the interpolation problem, $\psi=\varphi$ holds, i.e., $(\Pi\circ\Pi)(f)=\Pi(\Pi f)=\Pi\varphi=\psi=\varphi=\Pi f$. The property $\Pi\circ\Pi=\Pi$ characterises a projection. QED

Lemma 1.4.9. The interpolation Π has the operator norm

(1.4.8) $$\|\Pi\|_{C(D)\leftarrow C(D)} = \Big\|\sum_{i=1}^{n}|L_i(\cdot)|\Big\|_\infty .$$

Proof. (i) Let $L(x):=\sum_{i=1}^{n}|L_i(x)|$. By $|f(x_i)|\le\|f\|_\infty$, one derives from (7) the estimate $|(\Pi f)(x)|\le L(x)\|f\|_\infty\le\|L\|_\infty\|f\|_\infty$. Hence, $\|\Pi\|_{C(D)\leftarrow C(D)}\le\|L\|_\infty$.
(ii) Choose $\xi\in D$ such that $L(\xi)=\|L\|_\infty$. Set $y_i:=\mathrm{sign}(L_i(\xi))$. Let $f\in C(D)$ be the piecewise linear interpolating function from Example 5. It satisfies $\|f\|_\infty=\|y\|_\infty=1$. According to (7), $\Pi f=\sum f(x_i)L_i=\sum y_iL_i$ holds and, in particular, $(\Pi f)(\xi)=\sum y_iL_i(\xi)=\sum|L_i(\xi)|=L(\xi)=\|L\|_\infty=\|L\|_\infty\|f\|_\infty$ implying $\|\Pi\|_{C(D)\leftarrow C(D)}\ge\|L\|_\infty$. QED

The *interpolation error* is the difference $\Pi f-f$. To describe the behaviour of the error, one has to consider a *sequence* of interpolations Π_n defining the *interpolation method* $\{\Pi_n\}=\{\Pi_n\}_{n\in\mathbb{N}}$. The speed by which the error converges to zero as $n\to\infty$ determines the order of the interpolation method.

Definition 1.4.10. Let $\{\Pi_n\}$ be an interpolation method on $I=[a,b]$. It has the *interpolation order* $\lambda>0$, if there is some C_I such that

(1.4.9) $$\|\Pi_n f-f\|_\infty \leqslant C_I[(b-a)/n]^\lambda \|f\|_{\hat{C}^\lambda(I)} \qquad \text{for all } f\in\hat{C}^\lambda(I),\ n\in\mathbb{N}.$$

In the d-dimensional case $D\subset\mathbb{R}^d$, one has to replace (9) by (9'):

(1.4.9') $$\|\Pi_n f-f\|_\infty \leqslant C_I[\mu(D)/n]^{\lambda/d}\|f\|_{\hat{C}^\lambda(D)} \qquad \text{for all } f\in\hat{C}^\lambda(D),\ n\in\mathbb{N}.$$

We remark that the definition does not require that λ is the *maximal* order satisfying (9). Therefore, a method may have more than one order. The following example will have all orders between 0 and 2. Furthermore, note that the constant C_I in (9) must be independent of n. The polynomial interpolation process from Example 4 does not satisfy inequalities (9).

Remark 1.4.11. *Piecewise linear interpolation* with equidistant support abscissae x_i in $I=[a,b]\subset\mathbb{R}$ satisfies (9) with $C_I=1$ for all $\lambda\in(0,2]$, i.e., its maximal order equals 2. For a general partition (5a) with (5b,c), (9) holds with $C_I=C/(b-a)$ and C from (5c).

Proof. An arbitrary $x\in I$ lies in some subinterval $[x_i,x_{i+1}]$. According to the representation from Example 5, the error reads

$$\delta(x)=\{(x-x_i)f(x_{i+1})+(x_{i+1}-x)f(x_i)\}/(x_{i+1}-x_i)-f(x)=$$
$$=\{\alpha(f(x_{i+1})-f(x))+\beta(f(x_i)-f(x))\}/(\alpha+\beta),$$

where $\alpha:=x-x_i$, $\beta:=x_{i+1}-x$. First assume $0\leqslant\lambda\leqslant 1$. Applying the Hölder (respectively, Lipschitz) estimate $|f(x)-f(y)|\leqslant C_f|x-y|^\lambda$, one obtains

$$|\delta(x)| \leqslant C_f[\alpha\beta^\lambda+\alpha^\lambda\beta]/(\alpha+\beta) \leqslant C_f[\alpha(\alpha+\beta)^\lambda+(\alpha+\beta)^\lambda\beta]/(\alpha+\beta) =$$
$$= C_f(\alpha+\beta)^\lambda = C_f(x_{i+1}-x_i)^\lambda = C_f[(b-a)/n]^\lambda;$$

hence, (9) with $C_I=1$ for $0\leqslant\lambda\leqslant 1$. Now consider the case $1<\lambda\leqslant\varkappa=2$. Substitution $\xi=x+\beta t$ yields

$$f(x_{i+1})-f(x)=\int_x^{x_{i+1}} f'(\xi)d\xi = \beta\int_0^1 f'(x+\beta t)dt.$$

The analogous representation for $f(x_i)-f(x)$ yields

$$\delta(x)=\alpha\beta\int_0^1[f'(x+\beta t)-f'(x-\alpha t)]\,dt\,/\,(\alpha+\beta).$$

Using $|f'(\xi)-f'(\eta)|\leqslant C'_f|\xi-\eta|^{\lambda-1}$, one estimates the integral by $|\delta(x)| \leqslant C'_f\alpha\beta(\alpha+\beta)^{\lambda-2}\int t^{\lambda-1}dt \leqslant C'_f\alpha\beta(\alpha+\beta)^{\lambda-2}/\lambda \leqslant C'_f(\alpha+\beta)^\lambda/[4\lambda] \leqslant C'_fC_I[(b-a)/n]^\lambda$ with $C_I:=1/4\lambda$, since $4\alpha\beta\leqslant(\alpha+\beta)^2$. QED

§4.5.9 contains a further discussion of piecewise linear interpolation.

The underlying space $C(D)$ cannot be replaced by $L^\infty(D)$, since the evaluation $f(x_i)$ of $f\in L^\infty(D)$ is undefined (or: all f, which differ only at the point $x=x_i$, belong to the same equivalence class $f\in L^\infty(D)$). Further, it is not sufficient to regard the interpolation Π as a mapping from $C(D)$ into $L^\infty(D)$, since because of the projection property the operator Π^2 must be defined. Occasionally, the choice of the space $C(D)$ is too restrictive. In the following

Example 12, one cannot use $C(D)$, since the interpolating function Πf is not continuous. However, Π is a linear mapping of the Banach space

$$X := \{f \in L^\infty(I) : f \text{ right-sided continuously in } x_i \text{ from (5a)}\}$$

on $V \subset X$. Here, f is called *right-sided continuous* in ξ, if $f(\xi+0) := \lim_{i\to\infty} f(x_i) = f(\xi)$ for all $x_i \to \xi$ with $x_i \geqslant \xi$. The choice of X makes sense as it is shown in

Exercise 1.4.12. Let $I \subset \mathbb{R}$ and $\xi \in I$. Prove that the set $X := \{f \in L^\infty(I) : f$ right-sided continuous in $\xi\}$ is a closed subspace of $L^\infty(I)$.

In the following example, two different variants of piecewise constant interpolation are described. In particular, the second variant will be important for applications in §4.

Example 1.4.13 (piecewise constant interpolation). Let a partition of $I=[a,b]$ into n intervals be given. Assume (5a) with $n+1$ instead of n: $x_{n+1}=b$. *V consists of all functions being constant on the subintervals* $I_i = [x_i, x_{i+1})$ for $i=1,\dots,n$.
(a) The interpolation problem (3*): $\varphi(x_i)=f(x_i)$ for $i=1,\dots,n$ is to be understood in the sense of the right-sided limits: $\varphi(x_i+0)=f(x_i+0)$. The solution of the problem reads $\varphi(x) = f(x_i+0)$ for $x \in [x_i, x_{i+1})$.
(b) Choose the midpoints $\xi_i := (x_i+x_{i+1})/2$ of the subintervals I_i ($1 \leqslant i \leqslant n$) as support abscissae: $\varphi(\xi_i)=f(\xi_i)$ for $i=1,\dots, n$. The solution of the second interpolation problem is given by $\varphi(x)=f(\xi_i)$ for $x \in I_i$.
(c) In *both cases*, the Lagrange function L_i has the value 1 in $[x_i, x_{i+1})$ and 0 otherwise. Under assumption (5c), the piecewise constant interpolation is of *first order*. In the equidistant case, it satisfies (9) with $C_I=1$ and $\varkappa=1$.

1.4.2 Quadrature

Let f be a continuous function on $D \subset \mathbb{R}^d$. Each approximation to $\int_D f(x)dx$ is called a *quadrature formula*. The classical quadrature formulae contain the values $f(x_i)$ of f at the *support abscissae* $x_i \in D$ for $i=1,\dots,n$. We denote the quadrature formula by $Q(f)$ (or by $Q_D(f)$, if the underlying integration domain should be mentioned explicitly):

$$(1.4.10) \qquad Q_D(f) = Q(f) := \sum_{i=1}^{n} w_{i,n} f(x_i).$$

The coefficients $w_i = w_{i,n}$ are called the *weights* of the quadrature rule. The definition of the *quadrature error* $R_D(f) = R(f)$ is given by

$$(1.4.11) \qquad \int_D f(x)dx = Q_D(f) - R_D(f) \qquad \text{for } f \in C(D).$$

Speaking about a *quadrature method*, often, one does not mean a single quadrature formula but a sequence $\{Q_n\}$ or $\{Q_{n,D}\}$ of quadrature formulae. Occasionally, the index n can also be replaced by an equivalent parameter (e.g., by the step size $h=(b-a)/n$ for $D=I=[a,b]$ in the equidistant case).

In the discussion of the rounding error behaviour of a quadrature formula, the quantity $\sum |w_i|$ plays an essential role for the error amplification. This is the subject of

Exercise 1.4.14. Let $Q(\tilde{f})$ be the quadrature result for support ordinates $\tilde{f}(x_i)=f(x_i)+\varepsilon_i$ with an input error bounded by $|\varepsilon_i|\leqslant\varepsilon$. Prove that $|Q(\tilde{f})-Q(f)|\leqslant\varepsilon\sum_{i=1}^{n}|w_i|$ holds, where the equality sign is attained for special values ε_i.

We will see that the error amplification factor is also closely connected with the convergence of a quadrature method.

Definition 1.4.15. A quadrature method $\{Q_n\}$ is called *convergent*, if

(1.4.12) $$\lim_{n\to\infty} Q_n(f) = \int_D f(x)dx \qquad \text{for all } f\in C(D).$$

It is called *consistent*, if there is a dense subset $V\subset C(D)$ of the continuous functions such that the formula converges to the integral: $Q_n(f)\to\int_D f(x)dx$ for all $f\in V$. A quadrature method is called *stable*, if

(1.4.13) $$\sup\{\sum_{i=1}^{n}|w_{i,n}|:\ n\in\mathbb{N}\} < \infty .$$

The quantities in (13) can be interpreted as follows. The maximum norm $\|\cdot\|=\|\cdot\|_\infty$ and the dual norm (operator norm) $\|T\|:=\sup\{|Tf|/\|f\|_\infty:\ 0\neq f\in C(D)\}$ are defined on $C(D)$ and $C(D)'=L(C(D),\mathbb{R})$, respectively. Since Q_n is a continuous, linear mapping of $C(D)$ into $\mathbb{R}$, $\|Q_n\|$ is defined.

Lemma 1.4.16. Q_n from (10) has the norm $\|Q_n\| = \sum_{i=1}^{n}|w_{i,n}|$. The stability condition (13) takes the following form:

(1.4.14) $$\sup\{\|Q_n\|:\ n\in\mathbb{N}\} < \infty .$$

Proof. Analogous considerations as in Exercise 14. QED

Theorem 1.4.17 (convergence theorem). A quadrature method is convergent, if and only if it is stable and consistent.

Proof. Direct application of Theorem 3.17 of Banach-Steinhaus. QED

The proof of consistency uses the approximation theorem of Weierstraß (Theorem 3.8). All classical quadrature methods (Newton-Cotes, Gauß etc.) satisfy the following

Criterion 1.4.18. Let there be degrees γ_n with $\gamma_n\to\infty$ $(n\to\infty)$ such that the quadrature formula Q_n is exact for all polynomials of degree $\leqslant\gamma_n$, i.e., $Q_n(P)=\int_D P(x)dx$ for polynomials with $\mathrm{grad}(P)\leqslant\gamma_n$. Let D be bounded. Then, is $\{Q_n\}$ consistent.

Proof. Choose $V:=\{\text{polynomials}\}$ for V from Definition 15 and apply Theorem 3.8. QED

The concept of consistency can be quantified as follows.

Definition 1.4.19. A quadrature method $\{Q_n\}$ in the interval $I=[a,b]$ is called *consistent of the order* $\lambda>0$, if the quadrature error $R_n(f)=R(f)$ (cf. (11)) satisfies the estimate (15) with a $C_Q>0$:

(1.4.15) $$|R_n(f)| \leqslant (b-a)^{1+\lambda} C_Q n^{-\lambda} \|f\|_{\hat{C}^\lambda(I)} \qquad \text{for all } f\in \hat{C}^\lambda(I),\ n\in\mathbb{N}.$$

Since $V=\hat{C}^\lambda(D)$ is dense in $C(D)$, consistency of positive order implies consistency in the sense of Definition 15. In the higher dimensional case $D\subset\mathbb{R}^d$, one has to replace (15) by (15'):

(1.4.15') $$|R_n(f)| \leqslant \mu(D)^{1+\lambda/d} C_Q n^{-\lambda/d} \|f\|_{\hat{C}^\lambda(D)} \quad \text{for all } f\in \hat{C}^\lambda(D).$$

The classical *Newton-Cotes formulae* (cf. Stoer [1]) are based on polynomial interpolation with equidistant support abscissae in I. Generally speaking, each interpolation method Π_n: $C(D)\to C(D)$ induces a so-called «interpolatory» quadrature method defined by means of the formula

(1.4.16a) $$Q_n(f) := \int_D (\Pi_n f)(x)dx.$$

Exercise 1.4.20. Prove that **(a)** the interpolatory quadrature method (16a) has the representation (10) with weights

(1.4.16b) $$w_{i,n} := \int_D L_{i,n}(x)dx \qquad (L_{i,n}\text{: Lagrange functions}).$$

(b) If $\varkappa$ is the order of the interpolation Π_n, the induced quadrature method (16a) has at least the consistency order $\varkappa$.

An alternative to (16a) consists in approximating the integrand only partially. For this purpose, we write the integrand f as product

(1.4.16c) $$f(x) = \omega(x)\varphi(x).$$

The integral $\int_D f(x)dx$ becomes $\int_D \omega(x)\varphi(x)dx$. Here, $\omega(x)$ is called the «weight function». In Gauß quadrature it is assumed that $\omega(x)>0$ for almost all $x\in D$. However, in the following development, this assumption is *not* needed. For *interpolatory quadrature methods* Q_n^ω *with a weight function*, one has to approximate only the factor φ by $\Pi_n\varphi$:

(1.4.16d) $$Q_n^\omega(\varphi) := \int_D \omega(x)(\Pi_n\varphi)(x)dx.$$

As in Exercise 20a, one proves representation (16e) for Q_n^ω from (16d):

(1.4.16e) $$Q_n^\omega(\varphi) = \sum_{i=1}^n w_{i,n}\,\varphi(x_i) \qquad \text{with } w_{i,n} := \int_D \omega(x) L_{i,n}(x)dx.$$

The description of a quadrature formula can be made independent of the interval by a transformation onto a standard interval (e.g., $[0,1]$ or $[-1,1]$).

Remark 1.4.21. Let $I=[a,b]$ and $I'=[a',b']$ be two intervals. The affine mapping $\Phi(x) := a'+(x-a)(b'-a')/(b-a)$ maps I onto I'. Since $\int_{I'} f(x')dx' = c\int_I f(\Phi(x))dx$ holds with $c:=(b'-a')/(b-a)$, the quadrature formula Q_I from (10) defined over I induces the quadrature formula $Q_{I'}$ over I' with the weights $w'_{i,n} := c\,w_{i,n}$ and the support abscissae $x'_i := \Phi(x_i)$.

By definition, the norm $\|f\|_{\hat{C}^\lambda(I)}$ is a maximum of some quantities, e.g., the Hölder or Lipschitz constant $L_{f^{(\ell)}}$ of the ℓ-fold derivative $f^{(\ell)}$, where $\ell := \ell(\lambda)$ is defined by $\ell \in \mathbb{N}_0$ and $\ell < \lambda \leqslant \ell + 1$ (cf. (3.4e-f)). This proves $L_{f^{(\ell)}} \leqslant \|f\|_{\hat{C}^\lambda(I)}$. Many quadrature methods (e.g., the Newton-Cotes formulae) satisfy the error estimate (15) with $L_{f^{(\ell)}}$ instead of $\|f\|_{\hat{C}^\lambda(I)}$:

(1.4.17) $$|R_n(f)| \leqslant (b-a)^{1+\lambda} C_Q n^{-\lambda} L_{f^{(\ell)}} \qquad \text{with } \ell := \ell(\lambda) \text{ for all } f \in \hat{C}^\lambda(I).$$

The constant C_Q in (17) is invariant under the transformation of a quadrature formula Q_I on I into the induced formula $Q_{I'}$ on I' by means of the affine mapping Φ from Remark 21. For the proof check how $L_{f^{(\ell)}}$ changes under the substitution $f(x) \mapsto F(x) := f(\Phi(x))$. Since the constant C_Q is independent of the interval I, the following remark holds.

Remark 1.4.22. For the proof of estimate (17), and therefore also (15), it suffices to prove (17) for a standard interval I'.

Let Q_I be a quadrature formula over an interval I. Define $I' = I_k := [a+kh, a+(k+1)h]$ for $h = (b-a)/n$, $0 \leqslant k \leqslant n-1$ and choose $Q^{(k)}$ as the quadrature formula on the interval I_k corresponding to Q_I. The sum $Q_n(f) = \sum_k Q^{(k)}(f)$ is called the *composite quadrature formula*. For the sake of convenience, the index n may be different from the number of support abscissae in Q_n. The latter is $n(m-1)+1$ or nm, depending on whether both end points of I are also support abscissae of Q_I or not.

Lemma 1.4.23. Let I be a fixed interval of the length $|I|$ in $\mathbb{R}$ (e.g., $I = [0,1]$). Let the quadrature formula Q_I satisfy the error estimate

$$|R_I(\varphi)| \leqslant C_I L_{\varphi^{(\ell)}} \quad \text{with } \ell := \ell(\lambda) \text{ (see above)} \quad \text{for all } \varphi \in \hat{C}^\lambda(I).$$

The corresponding composite quadrature method Q_n is of order λ and satisfies the estimate (17) with $C_Q := C_I/|I|^{\lambda+1}$. Instead of $\varphi \in \hat{C}^\lambda(I)$, the *piecewise* smoothness $\varphi \in \hat{C}^\lambda(I_k)$ for all subintervals is sufficient.

Proof. Let I_k be one of the subintervals of length h and Φ: $I \to I_k$ the affine mapping from I onto I_k. $R_k(f)$ denotes the quadrature error of $Q^{(k)}$. The function $f \in \hat{C}^\lambda(I_k)$ is transformed into $F(\xi) := f(\Phi(\xi)) \in \hat{C}^\lambda(I)$. R_k and R_I are connected by $R_k(f) = R_I(F)h/|I|$. The above-mentioned substitution rule for the Hölder/Lipschitz constant yields $L_{f^{(\ell)}} = (h/|I|)^\lambda L_{F^{(\ell)}}$. Therefore, $|R_I(\varphi)| \leqslant C_I L_{\varphi^{(\ell)}}$ implies the estimate $|R_k(f)| \leqslant (h/|I|)^{\lambda+1} C_I L_{f^{(\ell)}} \leqslant [(b-a)/|I|]^{\lambda+1} n^{-\lambda-1} C_I L_{F^{(\ell)}}$. Summation over k yields the desired result. ∎

Exercise 1.4.24. Interpret the composite Newton-Cotes formulae as interpolatory quadratures induced by a piecewise polynomial interpolation on I.

The piecewise constant interpolation from Example 13 induces the *left-sided rectangular rule*

(1.4.18a) $$Q_n^{\mathrm{lR}}(f) := h \sum_{i=0}^{n-1} f(x_i) \qquad \text{with } x_i := a + ih,\ h := (b-a)/n,\ n \in \mathbb{N}.$$

Analogously, one obtains the *right-sided rectangular rule*

(1.4.18b) $Q_n^{\mathrm{rR}}(f) := h \sum_{i=1}^{n} f(x_i)$ with x_i and h as in (18a).

Their mean value is the *composite (secant) trapezoidal rule*, which is also induced by the piecewise linear interpolation from Example 3:

(1.4.18c) $w_{0,n} := w_{n,n} := \frac{1}{2}h,\ w_{i,n} := h$ for $1 \leqslant i \leqslant n-1$ in (10).

In the following, the arising formula is written

(1.4.18c') $Q_n^{\mathrm{ST}}(f) = h \sum{}'' f(x_i)$ with $x_i := a + ih,\ h := (b-a)/n$,

where the notation $\sum''$ denotes the sum from $i=0$ to n with the first and last terms halved.

The *composite Simpson rule* Q_n^{Simpson} is defined only for even n and has the weights

(1.4.18d) $w_{0,n} := w_{n,n} := h/3,\quad w_{i,n} := 4h/3 \quad (i = 1, 3, \ldots, n-1),$
$w_{i,n} := 2h/3\ (i = 2, 4, \ldots, n-2),$ where $h = (b-a)/n$.

Exercise 1.4.25. Prove that the consistency orders are 1 for the rectangular rules (18a,b), 2 for the composite secant-trapezoidal rule (18c) and 4 for the composite Simpson rule (18d). The associated constants from inequality (15) are $C_Q = \frac{1}{2}$, $1/12$, and $32/90$, respectively. What is the order of the *composite tangential trapezoidal rule* (18e)?

(1.4.18e) $Q_n^{\mathrm{TT}}(f) := h \sum_{i=0}^{n-1} f(x_i)$ with $x_i := a + (i + \frac{1}{2})\,h,\ h := (b-a)/n$.

1.4.3 Condition Number of a System of Equations

In the following, A is an $n \times n$ matrix and a and b are n-vectors:

$$A = (a_{jk})_{j,k=1,\ldots,n}, \qquad a = (a_k)_{k=1,\ldots,n}, \qquad b = (b_j)_{j=1,\ldots,n}.$$

For given A and b, let a be the solution of the system of equations

(1.4.19) $A\,a = b.$

The following characterisation is well-known from Linear Algebra.

Remark 1.4.26. The following statements are equivalent: **(a)** A is non-singular; **(b)** Equation (19) is solvable for all $b \in \mathbb{R}^n$; **(c)** Equation (19) with a special right-hand side b is uniquely solvable; **(d)** $\mathrm{Ker}(A) = \{0\}$; **(e)** $a = 0$ is the only solution for $b = 0$; **(f)** $\mathrm{image}(A) = \mathbb{R}^n$.

In (3.3a,b), the maximum and the Euclidean norm were mentioned as *norms* for vectors from $\mathbb{R}^n$. Since a matrix A can be regarded as an element of $L(X,X)$ with $X = \mathbb{R}^n$, the operator norm from (11a) can be defined for A. In this connection, the norm is called the *matrix norm* (more precisely, the matrix norm associated with the vector norm $\|\cdot\|$):

(1.4.20) $\|A\| := \max\{\|Ax\| / \|x\| : 0 \neq x \in \mathbb{R}^n\}.$

Notation 1.4.27. The matrix norm associated with the vector norm $\|\cdot\|_2$ or $\|\cdot\|_\infty$ (cf. (3.3a,b)) is again denoted by $\|\cdot\|_2$ or $\|\cdot\|_\infty$, respectively. Here, $\|\cdot\|_2$ is called the *spectral norm* and $\|\cdot\|_\infty$ the *row-sum norm.*

Remark 1.4.28. **(a)** $\|A\|_2$ equals $\lambda^{1/2}$, where the non-negative number λ is the largest eigenvalue of the matrices AA^T or A^TA.
(b) The name of the row-sum norm $\|A\|_\infty$ stems from the characterisation

(1.4.21) $$\|A\|_\infty = \max\{\sum_{k=1}^{n} |a_{jk}| : 0 \leqslant j \leqslant n\}.$$

(c) For arbitrary matrices we have $\|A\|_2^2 \leqslant \|A\|_\infty \|A^T\|_\infty$; in particular, symmetric matrices A satisfy $\|A\|_2 \leqslant \|A\|_\infty$.

Definition 1.4.29. Let $\|\cdot\|$ be a matrix norm (associated with some vector norm). The associated *condition number* of a non-singular matrix is

(1.4.22) $$\mathrm{cond}(A) := \|A\|\,\|A^{-1}\|.$$

The notation of the norm carries over to the condition number: $\mathrm{cond}_2(A)$ is the *spectral condition*, $\|\cdot\|_\infty$ corresponds to the condition number $\mathrm{cond}_\infty(A)$. The importance of the condition number is shown by the following theorem.

Theorem 1.4.30. Let A be a non-singular matrix and a the solution of the system (19). $\|\cdot\|$ denotes the vector and matrix norm.
(a) The relative error δa of the solution $a' = a + \delta a$ of the equation

(1.4.23a) $$Aa' = b' := b + \delta b$$

can be estimated by

(1.4.23b) $$\frac{\|\delta a\|}{\|a\|} \leqslant \mathrm{cond}(A) \frac{\|\delta b\|}{\|b\|}.$$

(b) If $A' := A + \delta A$ is a perturbed matrix with the relative error

(1.4.23c) $$\frac{\|\delta A\|}{\|A\|} \mathrm{cond}(A) =: q < 1,$$

then A' is also non-singular, and the error of the solution $a' = a + \delta a$ of

(1.4.23d) $$A'a' = b \qquad (A' := A + \delta A)$$

is bounded by

(1.4.23e) $$\frac{\|\delta a\|}{\|a\|} \leqslant \frac{\mathrm{cond}(A)}{1-q} \frac{\|\delta A\|}{\|A\|}.$$

Proof. Cf. Stoer [1,§4.4]. By Lemma 3.14 A' is non-singular. ∎

Exercise 1.4.31. Prove that for particular b and δb the equality sign can hold in (23b).

The introduction of the condition is by no means restricted to the finite-dimensional case. For a Banach space X, one defines the associated condition number («X-condition») of a bijective operator $A \in L(X,X)$ analogously to (22) by $\mathrm{cond}_X(A) := \|A\|_{X \leftarrow X} \|A^{-1}\|_{X \leftarrow X}$ (cf. Exercise 4.1.6).

2 Volterra Integral Equations

2.1 Theory of Volterra Integral Equations of the Second Kind

As shown by equations (1.1.1-2), there is a close relationship between ordinary differential equations and Volterra integral equations. First, we discuss the unique solvability. Afterwards, in §2.1.2, we discuss the regularity of the solution.

2.1.1 Existence and Uniqueness of the Solution

Depending on whether the solution of the Volterra integral equation is to be defined in a half-infinite interval $[a,\infty)$ or in a finite interval $[a,b]$, we set

(2.1.1) $$I := [a,b] \qquad \text{or} \qquad I := [a,\infty).$$

I is the domain of definition of the given function g and the desired function f. f is to be determined from the following equation:

Volterra integral equation of the second kind:

(2.1.2) $$f(x) = g(x) + \int_a^x k(x,y,f(y))dy \qquad \text{for } x\in I.$$

The function $k(x,y,z)$ is called the *kernel* (or *kernel function*). It is defined on the domain $\{(x,y,z)\colon x\in I,\ y\in[a,x],\ z\in\mathbb{R}\}\subset I\times I\times\mathbb{R}$. We introduce the abbreviation

(2.1.3) $$G := \{(x,y)\colon x\in I,\ y\in[a,x]\} \subset I\times I.$$

Then, $G\times\mathbb{R}$ is the domain of definition of the kernel.

Theorem 2.1.1. We assume $g\in C(I)$, $k\in C(G\times\mathbb{R})$. Furthermore, let the kernel k satisfy the following Lipschitz condition: There is a function $L\in C(G)$ such that

(2.1.4) $$|k(x,y,z)-k(x,y,z')| \leqslant L(x,y)|z-z'| \quad \text{for all } (x,y)\in G,\ z,z'\in\mathbb{R}.$$

Then, there is exactly one solution $f\in C(I)$ of the integral equation (2).

The assumption of Lipschitz continuity of k with respect to z corresponds to the analogous assumption of the uniqueness theorem for ordinary differential equations. Before presenting the proof, we mention a difference between integral equations and ordinary initial-value problems (1.1.1).

Remark 2.1.2. (a) In contrast to ordinary differential equations, one needs no initial value at $x=a$. The integral equation (2) implies the initial value

(2.1.5a) $$f(a) = g(a).$$

(b) The formulation of the Lipschitz condition (4) with the function L is tailored for the case of an *infinite* interval $I=[a,\infty)$. For a *bounded* I, one can replace $L(x,y)$ in (4) by the constant $\max\{L(x,y): (x,y)\in G\}$.

Exercise 2.1.3. Let g be differentiable at $x=a$ and suppose the kernel k is continuous at $(a,a,g(a))$. Show that the derivative $f'(a)$ exists and satisfies

(2.1.5b) $$f'(a) = g'(a)+k(a,a,g(a)).$$

Proof of Theorem 1. (i) The theorem holds for $I=[a,\infty)$, if it holds for finite intervals $I=[a,b]$ with arbitrary $b>a$. Therefore, it is sufficient to consider $I=[a,b]$. Note that boundedness of I also implies boundedness of G.
(ii) Since, by (i), G is compact, the continuous function $L(x,y)$ defined on G has a finite maximum L_0: $0\le L(x,y)\le L_0$ $(x\in I)$. We apply the Banach Fixed Point Theorem (Theorem 1.3.10), where the space $X=C(I)$ of continuous functions is equipped with the special norm

$$\|\varphi\| := \max\{|e^{-\beta L_0 x}\varphi(x)|: x\in I\} \qquad (\beta\ge 1).$$

which is equivalent to the maximum norm $\|\cdot\|_\infty$. The involved number $\beta\ge 1$ can be chosen arbitrarily. We write the integral equation (2) as fixed point equation $f=T(f)$, where

$$(T(\varphi))(x) := g(x)+\int_a^x k(x,y,\varphi(y))\,dy.$$

(iii) Since g and k are continuous, $T(\varphi)$ belongs to $C(I)$ for all $\varphi\in C(I)$, i.e., $T: X\to X$. To prove the contraction we write the difference

$$\big(T(\varphi)-T(\psi)\big)(x) = \int_a^x [k(x,y,\varphi(y))-k(x,y,\psi(y))]\,dy.$$

Hence,

$$\big|\big(T(\varphi)-T(\psi)\big)(x)\big| \le \int_a^x L_0|\varphi(y)-\psi(y)|\,dy,$$

using (4) and the choice of L_0. The particular definition of the norm implies

$$|\varphi(y)-\psi(y)| = e^{\beta L_0 y}|e^{-\beta L_0 y}(\varphi-\psi)(y)| \le$$
$$\le e^{\beta L_0 y}\max\{|e^{-\beta L_0 \xi}(\varphi-\psi)(\xi)|: \xi\in I\} = e^{\beta L_0 y}\|\varphi-\psi\|$$

for all $y\in I$. Inserting this estimate into the previous inequality, we obtain

$$\big|\big(T(\varphi)-T(\psi)\big)(x)\big| \le \int_a^x L_0 e^{\beta L_0 y}\|\varphi-\psi\|\,dy \le$$
$$\le \|\varphi-\psi\|\, e^{\beta L_0 x}[1-e^{\beta L_0(a-x)}]/\beta \le$$
$$\le \|\varphi-\psi\|\, e^{\beta L_0 x}[1-e^{\beta L_0(a-b)}]/\beta,$$

i.e., $e^{-\beta L_0 x}\left|\left(T(\varphi)-T(\psi)\right)(x)\right| \leqslant q\|\varphi-\psi\|$ with $q := [1-e^{\beta L_0(a-b)}]/\beta$ for all $x\in I=[a,b]$. Hence, $\|T(\varphi)-T(\psi)\| \leqslant q\|\varphi-\psi\|$. Since $\beta \geqslant 1$ implies $q<1$, T is a contraction. The Banach Fixed Point Theorem then guarantees the existence and uniqueness of a solution $f\in X=C(I)$ of the fixed point equation $f=T(f)$, which is equivalent to (2). QED

Corollary 2.1.4. Let the assumptions of Theorem 1 hold. Furthermore, assume that there are constants C, L_0, and $\beta \geqslant 1$ such that

$$|g(x)+\textstyle\int_a^x k(x,y,0)\,dy| \leqslant Ce^{\beta L_0 x} \qquad \text{and} \qquad L(x)\leqslant L_0 \quad \text{for } x\in I,$$

where $\beta>1$ is to be chosen in the case of an infinite interval. Then the solution f of the integral equation (2) satisfies the estimate

$$|f(x)| \leqslant \begin{cases} Ce^{L_0(b-a+x)}, & \text{if } \beta=1 \text{ and } x\in I=[a,b], \\ C\frac{\beta}{\beta-1}e^{\beta L_0 x}, & \text{if } \beta>1 \text{ and } x\in I. \end{cases} \tag{2.1.6}$$

Proof. Start the fixed point iteration $f_{\nu+1}=T(f_\nu)$ with $f_0:=0$ and apply estimate (1.3.10b) for $i=0$. QED

Remark 2.1.5. **(a)** The continuity of g is not needed for the existence and uniqueness theorem. The assumption $g\in L^\infty(I)$ is sufficient. Then, however, the solution also belongs to this space, i.e., $f\in L^\infty(I)$.
(b) The continuity of $k(\cdot,\cdot,\cdot)$ can be replaced by integrability of $k(x,\cdot,\varphi(\cdot))$ for $\varphi\in C(I)$ and continuity of $\int_a^x k(x,y,f(y))\,dy$.
(c) The Lipschitz condition (4) is required for all $z,z'\in\mathbb{R}$. If k is only locally Lipschitz continuous, one can prove existence of a solution only in a neighbourhood of $x=a$.

2.1.2 Regularity of the Solution

By *regularity* one means smoothness properties of a function. Theorem 1 ensures nothing more than continuity of the solution f. Next, we look for conditions under which $f\in C^m(I)$ holds for some $m\geqslant 1$. Such a property plays an essential role in estimating the discretisation error (see §2.2).

Theorem 2.1.6. Let $m\in\mathbb{N}$ and $g\in C^m(I)$. Concerning k, we assume

$$\frac{\partial}{\partial x}k\in C^{m-1}(G\times\mathbb{R}), \qquad k(x,x,z)\in C^{m-1}(\{(x,x,z):\ (x,z)\in I\times\mathbb{R}\}).$$

Furthermore, let the integral equation (2) have a continuous solution $f\in C(I)$ (a sufficient condition would be (4) from Theorem 1). Then, $f\in C^m(I)$ holds.

Proof by induction on m. For $m=0$, the assertion coincides with the assumption $f\in C(I)$. For $m\geqslant 1$, the right-hand side of the integral equation (2) [and therefore also f] is once differentiable, and f' has the representation

$$f'(x) = g'(x)+k(x,x,f(x))+\int_a^x \frac{\partial}{\partial x}k(x,y,f(y))\,dy \qquad \text{for } x\in I.$$

By inductive hypothesis, $f\in C^{m-1}(I)$. Hence, g' and $k(x,x,f(x))$ also belong to $C^{m-1}(I)$. Since the integrand $\partial k/\partial x$ is an element of $C^{m-1}(G\times\mathbb{R})$, the integral is $(m-1)$-times continuously differentiable, so that $f'\in C^{m-1}(I)$ follows. This statement is identical to $f\in C^m(I)$. QED

We remark that the condition $\frac{\partial}{\partial x}k \in C^{m-1}(G\times\mathbb{R})$ is sufficient but not necessary for $\int_a^x \frac{\partial}{\partial x}k(x,y,f(y))\,dy \in C^{m-1}(I)$. Therefore, weaker assumptions are possible; e.g., $\partial^\ell k/\partial x^\ell \in C(G\times\mathbb{R})$ for $1\leqslant\ell\leqslant m-1$ is sufficient.

When estimating the discretisation error, one often can replace the assumption $f\in C^m(I)$ by Lipschitz continuity of the $(m-1)$st derivative: $f\in C_L^{m-1}(I)$. First, we discuss the local Lipschitz continuity of f.

Theorem 2.1.7. Let the integral equation (2) have a solution $f\in C(I)$. Assume that g is locally Lipschitz continuous: $g\in C_{L,\mathrm{loc}}(I)$, and that for each bounded subset $Z\subset\mathbb{R}$, $k\in C(G\times\mathbb{R})$ satisfies a Lipschitz condition with respect to the first argument:

$$|k(x',y,z)-k(x'',y,z)| \leqslant C(x,Z)\,|x'-x''|$$

for all $x,y\in I$, $y\leqslant x'\leqslant x''\leqslant x$, $z\in Z$. Then, $f\in C_{L,\mathrm{loc}}(I)$.

Proof. The integrand in the first integral of the representation

$$f(x')-f(x'') = g(x')-g(x'') + \\ +\int_a^{x'} [k(x',y,f(y))-k(x'',y,f(y))]\,dy - \int_{x'}^{x''} k(x'',y,f(y))\,dy$$

is bounded by $C(x,Z)|x'-x''|$, where $Z:=\{f(y):\ a\leqslant y\leqslant x\}$. Further, the second integral is bounded by $C_0|x'-x''|$, where C_0 is the maximum of the continuous function $|k(x'',y,f(y))|$ on the compact set $a\leqslant x'\leqslant y\leqslant x''\leqslant x$. Using $g\in C_{L,\mathrm{loc}}(I)$, one obtains the local Lipschitz property of f in $[a,x]$. QED

The proof of Theorem 7 directly shows what can also be concluded from Exercise 1.2.1d:

Corollary 2.1.8. In addition to the assumptions of Theorem 7, assume that $I=[a,b]$ is finite. Then, f is *globally* Lipschitz continuous: $f\in C_L(I)$.

A generalisation of the statement from $f\in C_L(I)$ to $f\in C_L^{m-1}(I)$ for arbitrary $m\geqslant 2$ is left to the reader:

Exercise 2.1.9. Assume $I=[a,b]$ and $m\geqslant 2$. Replace the spaces C^m and C^{m-1} involved in the assumption of Theorem 6 by C_L^{m-1} and C_L^{m-2}, respectively, and prove that one obtains a solution $f\in C_L^{m-1}(I)$.

Sometimes, *piecewise* smooth functions are of interest. For this purpose, let a partitioning of the interval I be given:

$$a=a_0<a_1<\ldots<a_\ell=b,\ \text{if } I=[a,b], \quad \text{or} \quad \lim a_\ell=\infty,\ \text{if } I=[a,\infty).$$

Theorem 2.1.10. Let the integral equation (2) have a solution $f\in C(I)$. Assume $g\in C(I)$ and $k\in C(G\times\mathbb{R})$. Suppose the assumptions of Theorem 6, 7, or Exercise 9 are satisfied on each of the intervals $[a_{k-1},a_k]$. Then, $f\in C^m([a_{k-1},a_k])$, $f\in C_L([a_{k-1},a_k])$, or $f\in C_L^{m-1}([a_{k-1},a_k])$ holds, respectively, for each k.

Proof. The assertion is valid for the interval $[a_0,a_1]$. In $I_2:=[a_1,a_2]$, the function f satisfies the new integral equation

$$f(x)=g_2(x)+\int_{a_1}^{x} k(x,y,\varphi(y))dy \quad \text{with} \quad g_2(x):=g(x)+\int_{a}^{a_1} k(x,y,\varphi(y))dy;$$

hence, it also follows for $[a_1,a_2]$. Etc. ∎

2.2 Numerical Solution by Quadrature Methods

2.2.1 Derivation of the Discretisation

An obvious idea is to approximate the integral $\int_a^x k(x,y,f(y))dy$ in the Volterra equation (1.2) by a quadrature formula. The function to be integrated is $\varphi(y):=k(x,y,f(y))$ or in short notation $\varphi=k(x,\cdot,f(\cdot))$. The integration interval $[a,x]$ depends on x. The quadrature formula for this interval is denoted by $Q_{[a,x]}$. Replacing the integral by quadrature, we obtain the equation

$$\tilde{f}(x) = g(x) + Q_{[a,x]}(k(x,\cdot,\tilde{f}(\cdot))) \qquad (x\in I) \tag{2.2.1}$$

for a function $\tilde{f}$, which we expect to approximate f. The quadrature formula $Q_{[a,x]}$ contains n support abscissae x_i, whose number and values depend on x; therefore, we use the notation $n=n(x)$ and $x_i=x_i(x)$. Concerning the choice of these support abscissae there are severe restrictions:

Remark 2.2.1. For each support abscissa $\xi=x_i(x)$ of the quadrature formula $Q_{[a,x]}$, the function $k(x,\cdot,\tilde{f}(\cdot))$, which is to be integrated, must be known at $y=\xi$. In particular, the value $\tilde{f}(\xi)$ must be available. The computation of $\tilde{f}(\xi)$ requires the evaluation of the right-hand side of equation (1) at $x=\xi$. This shows that a quadrature over $[a,x]$ involving a support abscissa ξ requires a preceding quadrature over $[a,\xi]$ with support abscissae $x_i(\xi)$.

In order to get a practicable method, one has to choose the support abscissae x_0, x_1, x_2, ... such that all arising quadrature formulae involve only intervals of the form $[a,x_i]$ and use no more support abscissae than x_0, $x_1,\ldots, x_i$. Then it suffices to evaluate equation (1) at the discrete values $x=x_0$, $x_1,\ldots$. The problem remains, how to approximate $\tilde{f}(x_0)$ at the initial support abscissa x_0. However, this question is solved by equation (1.5a), when we choose $x_0:=a$: $f(x_0)=f(a)=g(a)$. The arising algorithm reads as follows.

	discretisation by a quadrature method
	support abscissae: $a = x_0<x_1<x_2<\ldots<x_{\ell-1}<x_\ell<\ldots$
	quadrature formulae: $Q_i:=Q_{[a,x_i]}$ with abscissae x_ℓ, $0\le\ell\le i$.
	start:
(2.2.2a)	$f_0 := g(a)$
	recursion for $i:=1,2,\ldots$:
(2.2.2b)	$f_i := g(x_i) + Q_{[a,x_i]}(k(x_i,\cdot,\tilde{f}(\cdot)))$, where $\tilde{f}(x_\ell):=f_\ell$.

According to (1.4.10), we write the quadrature formulae $Q_i := Q_{[a,x_i]}$ as

(2.2.2c) $$Q_{[a,x_i]}(f) = \sum_{\ell=0}^{i} w_{\ell,i} f(x_\ell).$$

Therefore, (2b) reads

(2.2.2b') $$f_i := g(x_i) + \sum_{\ell=0}^{i} w_{\ell,i} k(x_i, x_\ell, f_\ell).$$

Remark 2.2.2. (2a,b) is a staggered system of equations. In general, each single equation (2b') is nonlinear with respect to f_i. However, they become an explicit equation for f_i, if $w_{i,i}=0$. In the latter case, the discretisation is called *explicit*, otherwise *implicit*.

Remark 2.2.3. Algorithm (2a,b) elucidates the twofold character of the values x_ℓ $(\ell \geq 0)$. First, they serve as support abscissae in the discretisation of the integrals. Second, they are - independent of the treatment of the integrals - the discrete x-values at which the unknown function f is to be approximated.

The described requirements prevent us from using the support abscissae of Gauß quadrature, which is usually preferred because of its stability and accuracy, since the Gauß points needed in $[a,x_i]$ are completely different from those belonging to $[a,x_{i-1}]$ (but cf. §2.3). The obvious choice of the support abscissae are the equidistant points

(2.2.3) $$x_i := a + ih \qquad \text{with some step size } h>0.$$

Remark 2 implies that the computation becomes particularly easy when the weights are determined such that $w_{i,i}=0$. This excludes the standard Newton-Cotes formulae. An example satisfying $w_{i,i}=0$ is the left-sided rectangular rule (1.4.17a):

(2.2.4) $$Q_n^{\mathrm{lR}}(f) := h \sum_{i=0}^{n-1} f(x_i).$$

Inserted into (2b) or (2b'), it yields the following equation:

(2.2.4') $$f_i := g(x_i) + h \sum_{\ell=0}^{i-1} k(x_i, x_\ell, f_\ell).$$

Remark 2.2.4. In contrast to the situation for ordinary differential equations, the *amount of computational work* is not proportional to the number of function evaluations. The computation of $f_0, f_1, \ldots, f_i$ requires $i(i+1)/2$ additions and, in particular, $i(i+1)/2$ evaluations of the kernel k, together with $i+1$ g-evaluations and i multiplications. Since all values f_ℓ for $0 \leq \ell \leq i$ are involved in the ith step, one has to store these quantities during the computation.

2.2.2 Error Estimate

The values f_i determined by (2b') should approximate the unknown solution f at the points x_i. Therefore, we have to discuss the error

(2.2.5) $$d_i := f_i - f(x_i).$$

Does the error tend to zero when we refine the discretisation? If so, what is the speed of convergence? The discretisation parameter from Example (4') is the step size h. In the general case (2a,b), the support abscissae are not necessarily equidistant, but we can define

(2.2.6) $$h := \sup\{(x_i-a)/i : i \geq 1\}$$

as discretisation parameter. For the equidistant case, h from (6) coincides with h from (3). Due to (2a), the initial error vanishes:

(2.2.7a) $$d_0 = 0,$$

while for $i>0$ one concludes from (2b') and the integral equation (1.2) that

(2.2.7b) $$\begin{aligned} d_i &= g(x_i)+\sum_{\ell=0}^{i} w_{\ell,i}\, k(x_i,x_\ell,f_\ell)-[g(x_i)+\int_a^{x_i} k(x_i,y,f(y))dy] \\ &= \sum_{\ell=0}^{i} w_{\ell,i}\,[k(x_i,x_\ell,f_\ell)-k(x_i,x_\ell,f(x_\ell))] + R(q_i), \end{aligned}$$

where $R(q_i)$ is the quadrature error for the integrand

(2.2.7c) $$q_i(y) := k(x_i,y,f(y))$$

with respect to the interval $[a,x_i]$ (cf. (1.4.11)). Under the assumptions $g\in\hat{C}^m(I)$ and $k\in\hat{C}^m(G\times\mathbb{R})$ (or weaker conditions), Theorem 1.6 or Exercise 1.9 ensure $f\in\hat{C}^m(I)$, provided a continuous solution f exists. In addition, $k\in\hat{C}^m(G\times\mathbb{R})$ and $f\in\hat{C}^m(I)$ imply $q_i\in\hat{C}^m([a,x_i])$; furthermore,

(2.2.7d) $$\|q_i\|_{\hat{C}^m([a,x_i])} \leq C_I \quad \text{with} \quad C_I := \max\{\|k(x,\cdot,f(\cdot))\|_{\hat{C}^m([a,x])} : x\in I\}$$

holds for

(2.2.7e) $$x_i \in I=[a,b].$$

Remark 2.2.5. If the quadrature method $\{Q_i=Q_{[a,x_i]}\}$ (cf. (2a,b)) has consistency order m and (7d) holds, then the quadrature error can be estimated by

(2.2.7f) $$|R(q_i)| \leq (x_i-a)^{1+m} C_Q C_I\, i^{-m} \leq (x_i-a) C_Q C_I h^m.$$

Proof. Replace b in (1.4.15) by x_i. QED

In (7b), $k(x_i,x_\ell,f_\ell)-k(x_i,x_\ell,f(x_\ell))$ is still to be estimated. For this purpose, one needs a Lipschitz condition similar to (1.4), which we required for existence and uniqueness in Theorem 1.1:

(2.2.8) $$|k(x,y,z)-k(x,y,z')| \leq L|z-z'| \qquad \text{for all } (x,y)\in G,\ z,z'\in\mathbb{R}$$

(G given by (1.3)). In (1.4), the Lipschitz constant L could depend on x. In the present case of a finite interval I, (8) and (1.4) are equivalent (cf. Remark 1.2b). Then using (8), we have the estimate

(2.2.7g) $$|k(x_i,x_\ell,f_\ell)-k(x_i,x_\ell,f(x_\ell))| \leq L|f_\ell-f(x_\ell)| = L|d_\ell|.$$

If the quadrature formula Q_i is exact for constants, it follows that $\sum_\ell w_{\ell,i} = x_i - a \leqslant ih$. If, in addition, the weights are equally distributed in the following sense:

$$|w_{\ell,i}| \leqslant C_W h \qquad \text{for all } 0 \leqslant \ell \leqslant i, \tag{2.2.7h}$$

then, from (7b,f,g,h), one finally obtains the inequality

$$|d_i| \leqslant C_W L h \sum_{\ell=0}^{i-1} |d_\ell| + L|w_{i,i}||d_i| + (b-a) C_Q C_I h^m. \tag{2.2.7i}$$

Example 2.2.6. For the left-sided rectangular formula (4), the constants are $C_W = 1$, $C_Q = \frac{1}{2}$, and $w_{i,i} = 0$.

For the discussion of inequality (7i), we simplify the notation by introducing new constants. Let $M_1 = C_W L$, $M_0 = L|w_{i,i}|$, $M_2 = (b-a) C_Q C_I h^m$, $E_i = |d_i|$. Then (7i) may be written

$$E_0 = 0, \qquad 0 \leqslant E_i \leqslant M_0 E_i + h M_1 \sum_{\ell=0}^{i-1} E_\ell + M_2 \qquad (i \geqslant 1). \tag{2.2.9a}$$

Lemma 2.2.7 (Lemma of Gronwall). **(a)** Assume that there are numbers $\alpha, \beta_\ell \geqslant 0$ and $0 \leqslant M_0 < 1$ such that

$$0 \leqslant E_i \leqslant \alpha + \sum_{\ell=0}^{i-1} \beta_\ell E_\ell + M_0 E_i \qquad \text{for } i \geqslant 0. \tag{2.2.9b}$$

Then the quantities E_i fulfil the estimate (10b) below for all $i \geqslant 0$:

$$E_i \leqslant \frac{\alpha}{1-M_0} \prod_{\ell=0}^{i-1} (1 + \beta_\ell/(1-M_0)) \leqslant \frac{\alpha}{1-M_0} \exp\{\sum_{\ell=0}^{i-1} \beta_\ell/(1-M_0)\}. \tag{2.2.10b}$$

(b) Any the sequence $\{E_i\}$ satisfying inequality (9a) with $M_0, M_1, M_2, h \geqslant 0$ and $M_0 < 1$ is bounded by

$$\begin{aligned} &E_i \leqslant M_2 C_2 e^{C_1 ih} && \text{for all } i \geqslant 0 \text{ with} \\ &C_1 := M_1/(1-M_0), && C_2 := 1/(1-M_0) = C_1/M_1. \end{aligned} \tag{2.2.10a}$$

Proof. (i) (9a) is the particular case $\alpha := M_2$, $\beta_\ell := hM_1$ of (9b). Inserting these values into (10b), we obtain (10a). Part (a) remains to be shown.
(ii) Since $0 \leqslant M_0 < 1$, the term $M_0 E_i$ in the right-hand side of (9b) can be moved onto the other side. After division by $1 - M_0 > 0$, one obtains the modified inequality (9b) with $\alpha' := \alpha/(1-M_0)$ instead of α, with $\beta'_\ell := \beta_\ell/(1-M_0)$ instead of β_ℓ, and without the term $M_0 E_i$. Estimate (10b) becomes $E_i \leqslant \alpha' \prod_{\ell=0}^{i-1} (1 + \beta'_\ell)$. Its proof is by induction.
(iii) Start of induction: For $i = 0$, $E_0 \leqslant \alpha'$ is a consequence of the empty sum in (9b). This inequality coincides with $E_i \leqslant \alpha' \prod_{\ell=0}^{i-1} (1 + \beta'_\ell)$ for $i = 0$ because of the convention for empty products.
(iv) Let the estimate $E_j \leqslant \alpha' \prod_{\ell=0}^{j-1} (1 + \beta'_\ell)$ hold for $0 \leqslant j \leqslant i-1$. Insert $E_0 \leqslant \alpha'$ from (iii) into (9b) and obtain

$$E_i \leqslant \alpha' + \beta'_0 E_0 + \sum_{\ell=1}^{i-1} \beta'_\ell E_\ell \leqslant \alpha'(1 + \beta'_0) + \sum_{\ell=1}^{i-1} \beta'_\ell E_\ell \qquad \text{for } i \geqslant 1.$$

This estimate is of the form (9b) and involves only $i-1$ terms (with shifted index), so that the inductive hypothesis applies yielding the assertion $E_i \leqslant$

$\leqslant \alpha'(1+\beta'_0)\prod_{\ell=1}^{i-1}(1+\beta'_\ell) = \alpha'\prod_{\ell=0}^{i-1}(1+\beta'_\ell)$ and therefore the first part of (10b).

(v) An elementary analysis of the function $e^x - 1 - x$ yields the inequality

$$(2.2.11) \qquad e^x \geqslant 1 + x \qquad \text{for all } x \in \mathbb{R}.$$

Applying (11) for $1+\beta'_\ell \leqslant \exp\{\beta'_\ell\}$, one obtains the second inequality in (10b). QED

In (4'), we defined the discretisation by means of the left-sided rectangular formula. The corresponding error estimate is contained in

Theorem 2.2.8 (quadrature by the *rectangular rule*). Assume $g \in C_L(I)$ and $k \in C_L(G \times \mathbb{R})$ for an interval $I = [a,b]$ and let the integral equation (1.2) have a continuous solution f. Then, the error estimate

$$(2.2.12) \qquad |f_i - f(x_i)| \leqslant \frac{b-a}{2} C_I e^{L(x_i-a)} h \qquad (x_i = a + ih,\ C_I,\ L \text{ from (7d), (8)})$$

holds for the approximations $\{f_i\}$ defined by (4').

Proof. Theorem 1.7 and Corollary 1.8 imply that $f \in C_L(I)$. The assumption on k ensures $k(x,y,f(y)) \in C_L(G) = \hat{C}^1(G)$, so that (7d) holds for $m=1$. Remark 5 yields (7f) with the consistency order $m=1$ of the rectangular formula (cf. Exercise 1.4.25). Via (9a) and Lemma 7, estimate (7i) leads us to the inequality

$$|d_i| = |f_i - f(x_i)| \leqslant \tfrac{h}{2}(b-a)C_I e^{Lih} = \tfrac{h}{2}(b-a)C_I e^{L(x_i-a)},$$

since $C_W = 1$, $w_{ii} = 0$, and $C_Q = \frac{1}{2}$ (cf. Example 6). QED

Estimate (12) describes *convergence of first order*. More cannot be expected from a quadrature method of first order. To achieve faster convergence, one can apply the *composite trapezoidal rule*, which is of second order (cf. Exercise 1.4.25). Using the notation $\sum''$ from (1.4.18c'), we rewrite the corresponding discretisation as

$$(2.2.13) \qquad f_0 := g(a), \qquad f_i := g(x_i) + h\sum_{j=0}^{i}{}'' k(x_i, x_j, f_j) \qquad (i = 1, 2, \ldots).$$

Since the last weight is $w_{i,i} = \frac{1}{2}h \neq 0$, the discretisation is implicit; (13) represents a nonlinear equation for all $i \geqslant 1$ (cf. Remark 2). This poses the question about existence and uniqueness of the solution f_i and about the numerical solution method.

Lemma 2.2.9. Let the kernel function k satisfy the Lipschitz condition

$$(2.2.14\text{a}) \qquad |k(x,x,z) - k(x,x,z')| \leqslant L|z - z'| \qquad \text{for all } x \in I,\ z, z' \in \mathbb{R},$$

which, e.g., follows from (8). The step size h must be sufficiently small:

$$(2.2.14\text{b}) \qquad h < 2/L \qquad (L \text{ from (14a)}).$$

Then the nonlinear equations (13) have unique solutions f_i.

Proof. To simplify the notation, set

$$(2.2.14\text{c}) \qquad c_i := g(x_i) + \tfrac{h}{2}k(x_i, x_0, f_0) + h\sum_{j=1}^{i-1} k(x_i, x_j, f_j).$$

Then write equation (13) for f_i in the form of the fixed point equation

(2.2.14d) $\quad f_i = T(f_i) \qquad$ with $T(\zeta) := c_i + \frac{h}{2}k(x_i, x_i, \zeta)$.

Since

(2.2.14e) $\quad |T(\xi)-T(\zeta)| = \frac{h}{2}|k(x_i,x_i,\xi)-k(x_i,x_i,\zeta)| \leq \frac{hL}{2}|\xi-\zeta|$,

it follows that T is a contraction in the Banach space $X=(\mathbb{R},|\cdot|)$ with $q := hL/2 < 1$. The Banach Fixed Point Theorem (Theorem 1.3.10) completes the proof.

QED

Furthermore, the Banach Fixed Point Theorem offers a *computational method* for the approximative solution of the equations (13):

(2.2.15a$_1$) $\quad f_i^{(0)} := f_{i-1} \qquad$ (starting value)

or

(2.2.15a$_2$) $\quad f_i^{(0)} := c_i + \frac{h}{2}k(x_i, x_{i-1}, f_{i-1}) \qquad$ (starting value);

(2.2.15b) $\quad f_i^{(\nu+1)} := c_i + \frac{h}{2}k(x_i, x_i, f_i^{(\nu)}) \qquad$ (iteration).

Exercise 2.2.10. Assume (14b), $g \in C_L(I)$, and $k \in C_L(G\times\mathbb{R})$. Prove the following: **(a)** The starting values from (15a$_{1,2}$) have the error $|f_i^{(0)} - f_i| = O(h)$ or $O(h^2)$, respectively.
(b) The iterates from (15b) satisfy the estimate

$$|f_i^{(\nu)} - f_i| \leq (Lh/2)^\nu |f_i^{(0)} - f_i| \qquad (\nu \geq 0),$$

so that $|f_i^{(\nu)} - f_i| = O(h^{\nu+1})$ for (15a$_1$) and $|f_i^{(\nu)} - f_i| = O(h^{\nu+2})$ for (15a$_2$).

The iteration (15b) may be recommended if $Lh/2$ is small. But also other methods are possible, e.g., Newton's method (cf. Stoer [1,§5]).

Exercise 2.2.11. Assume that k has a derivative $\partial k(x,x,z)/\partial z$ and satisfies (14a,b). Show that the derivative involved in Newton's method is globally $\neq 0$.

In general, the *exact* solution of the nonlinear equation (13) is not available, but also *not necessary*. Since the approximations f_i are expected to have an error $O(h^2)$ in any case, one can terminate the iterative process (15a,b) as soon as the iteration error $O(h^2)$ is reached.

Theorem 2.2.12 (quadrature by the *trapezoidal rule*). For $I=[a,b]$, assume $g \in C_L^1(I) = \hat{C}^2(I)$ and $k \in C_L^1(G\times\mathbb{R})$. Let L be the Lipschitz constant from (8). The step size h is assumed to satisfy (14b), i.e., $h < 2/L$. As approximation to the nonlinear equation (13) use the first iterate $\tilde{f}_i := f_i^{(1)}$ from (15b) with starting value (15a$_1$). Then the following error estimate holds for $\tilde{f}_i$:

(2.2.16) $\quad |\tilde{f}_i - f(x_i)| \leq Ch^2 e^{L(x_i-a)}$.

Proof. (i) The assumptions guarantee the regularity $f \in C_L^1(I)$.
(ii) The approximations to f_i are

$$\tilde{f}_0 := f_0 = g(a), \quad f_i^{(0)} := \tilde{f}_{i-1}, \quad \tilde{f}_i := f_i^{(1)} \quad \text{according to (15b).}$$

We set $d_i := \tilde{f}_i - f(x_i)$ and $d_i^{(\nu)} := f_i^{(\nu)} - f(x_i)$ for $\nu = 0, 1$:

$$|d_i^{(0)}| = |f_i^{(0)} - f(x_i)| \leqslant |\tilde{f}_{i-1} - f(x_{i-1})| + |f(x_{i-1}) - f(x_i)| \leqslant$$
$$\leqslant d_{i-1} + C_f h,$$

where C_f is the Lipschitz constant of f. As in (7b), one obtains

$$d_i^{(1)} = R(q_i) + \Big\{ h\sum_{j=0}^{i-1}{}' k(x_i, x_j, \tilde{f}_j) + \tfrac{h}{2} k(x_i, x_i, f_i^{(0)}) - h\sum_{j=0}^{i}{}'' k(x_i, x_j, f(x_j)) \Big\}.$$

Here, Σ' denotes the sum with the weights 1 for $1 \leqslant j \leqslant i-1$ and $\frac{1}{2}$ for $j = 0$. The quadrature error of the composite trapezoidal rule is bounded by $|R(q_i)| \leqslant C_T h^2$. Because of (8), the expression in braces can be estimated by $hL[\Sigma' |d_j| + \frac{1}{2}|d_i^{(0)}|]$. Together with the bound for $|d_i^{(0)}|$, we obtain

$$|d_i| = |d_i^{(1)}| \leqslant hL\,[\sum_{j=0}^{i-1}{}' |d_j| + \tfrac{1}{2}|d_{i-1}|] + (C_T + \tfrac{1}{2} L C_f) h^2.$$

Then Lemma 7 is applicable with $M_0 = 0$, $\sum_{j=0}^{i-1} \beta_j \leqslant ihL$, $\alpha = (C_T + \frac{1}{2} L C_f) h^2$ in (9b) or (10b), respectively. This yields the desired result (16) with $C = C_T + \frac{1}{2} L C_f$. QED

Second order convergence can also be achieved by an *explicit* method. For this purpose, one modifies the quadrature method as follows.

Remark 2.2.13. Let Q_I be a composite quadrature method $Q_I = \sum_{k=1}^{i} Q^{(k)}$, where $Q^{(k)}$ is a method of order $\varkappa$ over the subinterval I_k of the length $\leqslant$const h. According to Lemma 1.4.23, Q_I is again of order $\varkappa$. Changing in Q_I one (or, more generally, a *fixed* number of) terms $Q^{(k)}$ into $\tilde{Q}^{(k)}$, where $\tilde{Q}^{(k)}$ is of order $\varkappa - 1$, we gain a composite method $\tilde{Q}_I$ of the same order $\varkappa$.

Proof. The quadrature error of $\tilde{Q}^{(k)}$ on the interval I_k is $O(h^{\varkappa})$. This error together with the errors $O(h^{\varkappa+1})$ of $Q^{(k)}$ over the remaining intervals yields the global error $O(h^{\varkappa})$. QED

We can apply Remark 13 to the trapezoidal rule ($\varkappa = 2$) by replacing the term corresponding to the last interval $[x_{i-1}, x_i]$ by the left-sided rectangular formula, whose order is $\varkappa - 1 = 1$. The arising discretisation reads

(2.2.17a) $\quad f_0 := g(a),$

(2.2.17b) $\quad f_i := g(x_i) + \frac{1}{2} h k(x_i, x_0, f_0) + h \sum_{j=1}^{i-2} k(x_i, x_j, f_j) + \frac{3}{2} h k(x_i, x_{i-1}, f_{i-1}) \quad$ for $i = 1, 2, \ldots$

Remark 2.2.14. Under the assumptions of Theorem 12, the discretisation (17a,b) is convergent of *second order*:

(2.2.18) $\quad |f_i - f(x_i)| \leqslant C h^2 e^{L(x_i - a)}.$

Under suitable smoothness assumptions on $k(x, y, f(y))$, we obtain a *method of third order* by the following quadrature formulae $Q_{[a, x_i]}$:

(2.2.19a) $\quad i$ even: $\quad Q_{[a,x_i]} := Q^{\text{Simpson}}_{[a,x_i]},$

(2.2.19b) $\quad i$ odd: $\quad Q_{[a,x_i]} := Q^{\text{Simpson}}_{[a,x_{i-1}]} + Q^{\text{Trapez}}_{[x_{i-1},x_i]},$

where Q_I^{Simpson} denotes the composite Simpson rule over I and $Q_{[x_{i-1},x_i]}^{\text{Trapez}}$ is the (non-composite) trapezoidal rule over $[x_{i-1},x_i]$. Simpson's rule is of fourth order, but here we only use the fact that is it also of third order. Although the trapezoidal rule is of second order, due to Remark 13 the total formula is of third order, since the trapezoidal rule is applied to only one subinterval. Analogously to Theorem 12, one can prove that the implicit method (19a,b) has third order. Again, the values f_i can be approximated by a Banach fixed point iteration similar to (15a,b). It suffices to choose $\tilde{f}_i = f_i^{(2)}$ (cf. Exercise 10b).

The construction of a *method of fourth order* leads to a completely new problem: Due to Remark 13, we may choose a method of third (instead of fourth) order on one subinterval. However, there is no quadrature formula of third order for the interval $[a, a+h]$ using only the support ordinates at $x_0 = a$ and $x_1 = a+h$. The best formula with these input data is the trapezoidal rule, which we apply for $i=1$:

(2.2.20a) $\quad i=1: \qquad Q_{[a,x_i]} := Q_{[a,a+h]}^{\text{Trapez}}.$

For $i \geq 2$, formulae of fourth order are available, e.g.,

(2.2.20b) $\quad i$ even: $\qquad Q_{[a,x_i]} := Q_{[a,x_i]}^{\text{Simpson}},$

(2.2.20c) $\quad i \geq 3$ odd: $\qquad Q_{[a,x_i]} := Q_{[a,x_{i-3}]}^{\text{Simpson}} + Q_{[x_{i-3},x_i]}^{3/8 \text{ rule}}.$

The weights of the (non-composite) *3/8 rule* are $w_1 = w_4 = 3/8$, $w_2 = w_3 = 9/8$.

Remark 2.2.15. Assume g and k to be sufficiently smooth. The quadrature in (2b) is defined by (20a-c). The Banach fixed point iteration (15b) for solving (2b') has to be performed at least until $\nu = 3$. Let h be sufficiently small. Then, the approximations $\tilde{f}_i$ satisfy the error estimates

(2.2.20d) $\quad |\tilde{f}_1 - f(x_1)| \leq Ch^3,$

(2.2.20e) $\quad |\tilde{f}_i - f(x_i)| \leq Ch^4 \qquad$ for $i \geq 2$.

Proof. (20d) holds in accordance to Remark 13. It remains to show the surprising fourth order estimate that for $i \geq 2$. The reason for this is that the sum $hM_1 \sum_\ell E_\ell$ in (9a) is of order $O(h^m)$ even if $E_1 = O(h^{m-1})$ and $E_\ell = O(h^m)$ for $\ell \geq 2$. ∎

The difficulty for $i=1$ can be circumvented by a technique which is similar to the Runge-Kutta methods for ordinary differential equations. Compute an *intermediate value* $f_{1/2} \approx f(a + \frac{1}{2}h)$ by means of the trapezoidal rule. By analogy with (20d), one has $|f_{1/2} - f(a + \frac{1}{2}h)| = O(h^3)$. Finally, compute f_1 by Simpson's rule (support abscissae: a, $a + \frac{1}{2}h$, $a+h$). The accuracy of the results is $f_1 = f(x_1) + O(h^4)$, so that (20e) holds for all $i \geq 1$.

This method generalises to the construction of methods of arbitrary order by introducing suitable intermediate support abscissae $\xi_{i,k}$ and auxiliary values $f_{i,k}$.

2.3 Further Numerical Methods

The method presented above is analogous to the *one-step methods* for ordinary differential equations. Analogues of *multi-step methods* can be obtained by replacing the term f_i on the left-hand side of equation (2.2b) by $k+1$ terms $\alpha_0 f_i + \alpha_1 f_{i-1} + \ldots + \alpha_k f_{i-k}$. In addition, approximations of the so-called «delay terms» $g(x) + \int_a^{x_\ell} k(x,y,f(y))dy$ are added for $x_\ell \leqslant x$. The analogue of the Runge-Kutta techniques developed for ordinary differential equations yields the *Volterra-Runge-Kutta formulae*. These methods are extensively described in the monograph of Brunner - van der Houwen [1].

As an alternative approach, we present a special *collocation method*. The direct discretisation by a quadrature method forced us to choose equidistant support abscissae $x_j = a + jh$ (cf. (2.3)). A remedy is the following: Interpolate the values f_j at the support abscissae $x_j = a + jh$ for $0 \leqslant j \leqslant i$ by a function $\Phi(x)$:

(2.3.1) $$\Phi(x_j) = f_j \qquad \text{for } 0 \leqslant j \leqslant i.$$

Hence, the integral $\int_a^{x_i} k(x,y,f(y))dy$, which is needed for the computation of the coefficients of the equations for the unknowns f_i, can be approximated by $\int_a^{x_i} k(x,y,\Phi(y))dy$. Since $k(x,y,\Phi(y))$ is known for all $y \in [a, x_i]$, arbitrary quadrature formulae with support abscissae in $[a, x_i]$ can be used for the approximation of the integral. The support abscissae of the quadrature formula $Q_{[a,x_i]}$ may be different from the support abscissae x_j in (1). For example, the *Gauß quadrature* with its non-equidistant support abscissae is a possible choice.

The simplest example of Φ is the *piecewise linear interpolation*:

(2.3.2) $$\Phi(x) = \tfrac{1}{h}\,[(x - x_j) f_{j+1} + (x_{j+1} - x) f_j] \qquad \text{for } x_j \leqslant x \leqslant x_{j+1}$$

(cf. Example 1.4.3). Since the interpolation error of Φ is of order $O(h^2)$, it makes no sense to apply quadrature methods of higher order. Choosing again the composite trapezoidal rule, one regains method (2.13). The interpolation order increases to 4, if one applies, e.g., the *piecewise cubic Hermite ansatz* :

(2.3.3) on each subinterval $[x_j, x_{j+1}]$
$\Phi \in C^1([a, x_i])$ coincides with a cubic polynomial.

The condition $\Phi \in C^1([a, x_i])$ implies, that the left- and right-sided derivatives $\Phi'(x_j \pm 0)$ coincide for all $1 \leqslant j \leqslant i-1$.

In the following, the cubic polynomials Φ (instead of the support ordinates f_i) are to be determined successively for the subintervals $[a, a+h] = [x_0, x_1]$, $[x_1, x_2], \ldots, [x_{i-1}, x_i], \ldots$. At the left end-point x_{i-1} of the interval, the zeroth and first derivative of Φ are known. Here, one has to treat the starting point $i=1$ in (4a,b) different from the subsequent computations in (4c) $(i>1)$:

(2.3.4a) $$\Phi(a) = g(a),$$

(2.3.4b) $$\Phi'(a) = f'(a) = g'(a) + k(a,a,g(a)) \qquad \text{(cf. (1.5b))}$$

(2.3.4c) $$\Phi(x_{i-1}+0) = \Phi(x_{i-1}-0), \qquad \Phi'(x_{i-1}+0) = \Phi'(x_{i-1}-0).$$

The equality of the left- and right-sided limits and derivatives given in

(4c) is necessary to ensure $\Phi \in C^1$. The common value $\Phi(x_{i-1}+0)=\Phi(x_{i-1}-0)$ replaces the previous support ordinate f_{i-1}. Since the cubic polynomial Φ (restricted to $[x_{i-1},x_i]$) contains four parameters, one also needs four equations to determine Φ on $[x_{i-1},x_i]$, uniquely. Besides (4a,b) or (4c), respectively, two additional equations are required, which can be derived from the integral as follows.

Suppose Φ is known on $[a,x_{i-1}]$. Together with the extension of Φ onto $[x_{i-1},x_i]$, which is still to be determined, one approximates f on $[a,x_i]$. The *collocation method* consists in satisfying the integral equation (1.2) at two points

(2.3.5) $$x_{i,1}=x_{i-1}+\Theta_1 h, \quad x_{i,2}=x_{i-1}+\Theta_2 h, \qquad \text{where } \Theta_1,\Theta_2 \in (0,1],$$

with the ansatz function Φ instead of f:

(2.3.6) $$\Phi(x_{i,j}) = g(x_{i,j}) + \int_a^{x_{i,j}} k(x_{i,j},y,\Phi(y))dy \qquad \text{for } j=1,2.$$

Remark 2.3.1. In each interval $[x_{i-1},x_i]$, we have to satisfy four equations: (4a,b) and (6) for $i=1$, and (4c/6) for $i>1$. The two equations (6) are nonlinear in the four coefficients of the cubic polynomial Φ on $[x_{i-1},x_i]$. For $I=[a,b]$, we define G by (1.3). If $k\in C(G\times\mathbb{R})$, there is a bound h_0 such that the nonlinear equations are uniquely solvable for all $h\leqslant h_0$ and all $x_i\leqslant b$. Hence, the piecewise cubic Hermite function $\Phi=\Phi_h$ of the step size h is uniquely determined.

The described discretisation is not yet practicable, since the equations (6) require an exact integration. Instead, the integrals over $[a,x_{i,j}]$ have to be replaced by suitable quadratures:

(2.3.6') $$\Phi(x_{i,j})=g(x_{i,j})+Q_{[a,x_{i,j}]}(k(x_{i,j},\cdot,\Phi(\cdot))) \qquad \text{for } j=1,2.$$

For example, we can recommend a composite Gauß quadrature over the intervals $[x_{j-1},x_j]$ $(j=1,\dots,i-1)$ and $[x_{i-1},x_{i,j}]$ with two support abscissae. Remark 1 remains true, when equations (6) are replaced by (6').

Since the interpolation by the piecewise cubic Hermite functions is of fourth order and the quadrature proposed above has order 4, one expects an error estimate of the form

(2.3.7) $$\|\Phi-f\|_{\infty,[a,b]} \leqslant C h^4 \|g\|_{\hat{C}^4([a,b])}$$

for $h\leqslant h_0$ under suitable conditions on g, k, and h_0. However, it turns out that the collocation points $x_{i,j}$ from (5) have to fulfil a further *stability condition*. If the stability condition is violated, the error increases with $h\to 0$. A reasonable choice for the second collocation point is

(2.3.8a) $$\Theta_2 = 1 \qquad \text{(i.e., } x_{i,2}=x_i\text{)}.$$

Then, the collocation method defined by (4a-c), (5), (6'), (8a) is stable if

(2.3.8b) $$\tfrac{1}{2} < \Theta_1 < 1,$$

while it diverges for $0\leqslant\Theta_1<\frac{1}{2}$. A good choice for Θ_1 is $\Theta_1=0.6$. For the proof of this stability statement, the convergence (7), and an extensive discussion of the presented collocation method we refer to Esser [1]. Furthermore, we mention chapter 5 in Brunner - van der Houwen [1], which is devoted to collocation methods.

The order $2m=4$ of the Hermite collocation methods described above can be increased. For this purpose, replace definition (3) by

(2.3.9) on each subinterval $[x_j, x_{j+1}]$, $\Phi \in C^{m-1}([a, x_i])$ coincides with a polynomial of degree $\leq 2m-1$.

The collocation (6) or (6'), respectively, has to be performed at $m-1$ points $x_{i,j}$ $(1 \leq j \leq m-1)$. The error estimate (7) improves. The right-hand side becomes $Ch^{2m} \| g \|_{C^{2m}([a,b])}$ (cf. Esser [1]).

2.4 Linear Volterra Integral Equations of Convolution Type

The *linear* Volterra integral equation of the second kind reads

$$f(x) = g(x) + \int_a^x k(x,y) f(y)\, dy \qquad \text{for } x \in I. \tag{2.4.1}$$

Existence and uniqueness of the solution in shown in §2.1.1 (at least for $k \in C(G)$). The Lipschitz condition (1.4) is satisfied by the function $L(x,y) := |k(x,y)|$.

In general, (1) may not be solved in a closed form. However, if the kernel is of *convolution type*, i.e.,

$$k(x,y) = \varkappa(x-y), \tag{2.4.2}$$

then the closed form solution can be obtained as follows. Writing again $k(x-y)$ for $\varkappa(x-y)$, one obtains the equation

$$f(x) = g(x) + \int_a^x k(x-y) f(y)\, dy \qquad \text{for } x \in I. \tag{2.4.3}$$

For the special choice $g=1$, one defines the function R by

$$R(x) = 1 + \int_a^x k(x-y) R(y)\, dy \qquad \text{for } x \in I. \tag{2.4.4}$$

Definition 2.4.1. The solution R of equation (4) is called the *resolvent* (of the convolution kernel k).

Exercise 2.4.2. In the case of a constant convolution kernel $k=$ const, one obtains $R(x) = e^{k(x-a)}$.

By means of the resolvent one can explicitly represent the solution of the general equation (3).

Theorem 2.4.3. Let R be the resolvent associated with $k \in C([0, b-a])$ and assume $g \in C^1(I)$. Then, the solution of the integral equation (3) is given by

$$f(x) = R(x) g(a) + \int_a^x g'(a+x-y) R(y)\, dy \qquad \text{for } x \in I. \tag{2.4.5}$$

Proof. Inserting f from (5) into the right-hand side of integral equation (3), one obtains

$$g(x)+\int_a^x k(x-y)R(y)g(a)dy+I_1 \underset{(4)}{=} g(x)+R(x)g(a)-g(a)+I_1$$

with $$I_1 := \int_a^x k(x-y)\int_a^y g'(a+y-t)R(t)dt\,dy.$$

The left-hand side $f(x)$ from (3) can be written by means of (5) and (4) as

$$R(x)g(a)+\int_a^x g'(a+x-y)[1+\int_a^y k(y-t)R(t)dt]dy =$$
$$= R(x)g(a)+g(x)-g(a)+I_2$$

with $$I_2 := \int_a^x g'(a+x-y)\int_a^y k(y-t)R(t)dt\,dy.$$

Obviously, both sides coincide if $I_1=I_2$. For the proof of $I_1=I_2$, we formally define $k(t):=0$ for $t<0$ and $g'(t):=0$ for $t<a$. Let $a\leqslant t\leqslant x$. Since $g'(a+x-s-t)=0$ for $x-t<s\leqslant x-a$ and $k=0$ for $a-t\leqslant s<0$, the integral $J(t):=\int_0^{x-t}k(s)g'(a+x-s-t)ds$ can also be written as $J(t)=\int_0^{x-a}k(s)g'(a+x-s-t)ds$ or $J(t)=\int_{a-t}^{x-t}k(s)g'(a+x-s-t)ds$, . The substitutions $s=x-y$ in the first integral and $s=y-t$ in the second one yield

$$J(t) = \int_a^x k(x-y)g'(a+y-t)dy = \int_a^x k(y-t)g'(a+x-y)dy.$$

The first representation of $J(t)$ shows

$$\int_a^x J(t)R(t)dt = \int_a^x\int_a^x k(x-y)g'(a+y-t)R(t)dt\,dy =$$
$$=\int_a^x k(x-y)\int_a^x g'(a+y-t)R(t)dt\,dy = I_1,$$

since $g'=0$ for $y<t\leqslant x$. Analogously, one concludes from the second representation that

$$\int_a^x J(t)R(t)dt = \int_a^x\int_a^x g'(a+x-y)k(y-t)R(t)dt\,dy =$$
$$=\int_a^x g'(a+x-y)\int_a^x k(y-t)R(t)dt\,dy = I_2.$$

Since $I_1=I_2$, it follows that f from (5) satisfies (3). ∎

The assumption $g\in C^1(I)$ of Theorem 3 can be weakened.

Lemma 2.4.4. A continuous convolution kernel leads to a continuously differentiable resolvent R. The derivative $r:=R'$ solves the integral equation

$$r(x) = k(x-a)+\int_a^x k(x-y)\,r(y)\,dy. \tag{2.4.6}$$

Proof. Since k is continuous, (6) has a continuous solution r. Put $R(x):=1+\int_a^x r(y)dy$. Similar manipulations as in the previous proof show that R satisfies equation (4). Hence, R is differentiable with r being its derivative. ∎

Integration by parts in the integral of (5) leads to

$$f(x) = g(x)+\int_a^x g(a+x-y)\,r(y)dy. \tag{2.4.7}$$

Corollary 2.4.5. Let the convolution kernel k and the function g be continuous. Then, the solution f of (3) can be represented by (7).

2.5 Volterra Integral Equations of the First Kind

These are of the general form

(2.5.1) $$g(x) = \int_a^x k(x,y,f(y))\,dy \qquad \text{for } x \in I = [a,b],$$

where g is a given function and f is the solution to find. Assuming k, k_x, and f to be continuous, we know that the right-hand side in (1) is differentiable. Therefore, g necessarily satisfies the following condition:

(2.5.2) $$g \in C^1(I), \quad g(a)=0.$$

Differentiating both sides of the integral equation (1), one obtains

(2.5.3) $$g'(x) = k(x,x,f(x)) + \int_a^x k_x(x,y,f(y))\,dy.$$

Assume the mapping $z \mapsto k(x,x,z)$ to be invertible for all $x \in I$ (at least locally). The inverse function is denoted by $\Phi(x,\cdot)$:

$$z = \Phi(x,k(x,x,z)) \qquad \text{for } x \in I,\ z \in \mathbb{R}.$$

In the linear case, i.e., $k(x,x,z) = k(x,x)z$, the invertibility reduces to the requirement $k(x,x) \neq 0$ and $\Phi(x,\eta)$ is given by $\eta/k(x,x)$. By means of Φ, one can solve equation (3) for f:

(2.5.4) $$f(x) = \Phi\Big(x, g'(x) - \int_a^x k_x(x,y,f(y))\,dy\Big).$$

Equation (4) can be regarded as a nonlinear *Volterra integral equation of the second kind*. To see the connection with a standard equation of the second kind, we introduce $\varphi(x) := k(x,x,f(x))$ as the new unknown function. One regains f from φ by $f(x) = \Phi(x,\varphi(x))$. Equation (3) becomes

(2.5.5) $$g'(x) = \varphi(x) + \int_a^x k_x(x,y,\Phi(y,\varphi(y)))\,dy.$$

This is the usual Volterra integral equation of the second kind with g' instead of g and $\hat{k}(x,y,z) := -k_x(x,y,\Phi(y,z))$ instead of k. For another transformation involving k_y, compare Pipkin [1, §5.3].

Exercise 2.5.1. Show that the solution f of equation (1) has the initial value $f(a) = \Phi(a,g'(a))$.

It is obvious how to formulate the assumptions for existence and uniqueness of the solution of (3): Besides (2) and the continuity of k, k_x, one needs a Lipschitz condition on k_x with respect to the last argument as well as the existence of the inverse function Φ. If Φ is globally unique, uniqueness of the solution f can be ensured. If, however, multiple local inverses exist, one also obtains multiple solutions of equation (3).

In §6 we will study a special first kind Volterra integral equation: Abel's integral equation, involving a kernel k for which neither k_x nor k itself is continuous. In this case, condition (2), i.e., $g \in C^1(I)$, will prove to be not necessary.

A obvious *discretisation* of the integral equation (1) of the first kind consists in solving the new equation (5) of the second kind by the methods mentioned above. In principle, a direct discretisation of equation (1) by a quadrature method (in the form $g_i = \sum_{j=0}^{i} w_{i,j}\, k(x_i, x_j, f_j)$ for $i = 1,2,\dots$) is applicable, but it might give rise to instability and lead to divergence.

3 Theory of Fredholm Integral Equations of the Second Kind

In the very end of the last century, Erik Ivar Fredholm (Stockholm) investigated those equations, which are now named in honour of him. Together with results of Hilbert, his theory led to the development of functional analysis, which took shape in the beginning of this century.

3.1 Fredholm Integral Equations of the Second Kind

In (1.1.3), we have already introduced the Fredholm integral equation of the second kind:

(3.1.1) $$f(x) = g(x) + \int_a^b k(x,y)f(y)dy \qquad \text{for } x \in [a,b].$$

As before, we abbreviate the interval by

(3.1.2a) $$I = [a,b]$$

and write

(3.1.1') $$f(x) = g(x) + \int_I k(x,y)f(y)dy \qquad \text{for } x \in I.$$

Occasionally, we admit infinite intervals, e.g.,

(3.1.2b) $$I = [a,\infty) \quad \text{or} \quad I = (-\infty,\infty).$$

The domain of integration may also be a curve in $\mathbb{R}^2$, a subset of $\mathbb{R}^d$, or a surface of a domain. The general symbol of the domain of integration, including the case $D = I$, is

(3.1.2c) $$D \subset \mathbb{R}^d \ (1 \leqslant d < \infty) \qquad \text{or} \qquad D = \partial\Omega \text{ for a domain } \Omega \subset \mathbb{R}^d.$$

Hence, the formulation of the problem reads as follows:

> Fredholm integral equation of the second kind
>
> (3.1.1") $$f(x) = g(x) + \int_D k(x,y)f(y)dy \qquad \text{for } x \in D.$$

In case D is a surface, $\int_D \ldots dy$ denotes the surface integral, which usually is written in the form $\int_D \ldots d\Gamma_y$ (cf. §8.1.2.2).

The function $k(x,y)$ is called the *kernel function* or the *kernel*. The integral in (1), (1'), or (1"), respectively, defines a linear mapping, which we call the *integral operator* K:

(3.1.3) $K: f \mapsto \int_I k(\cdot,y)f(y)dy$ or $\int_D k(\cdot,y)f(y)dy$.

The integral equations (1), (1'), and (1") become

(3.1.4) $f = g + Kf$.

An apparent generalisation of (4) is obtained by inserting a further factor $\lambda \in \mathbb{C}$:

(3.1.5) $\lambda f = g + Kf$.

Remark 3.1.1. Equation (4) is the particular case $\lambda = 1$ of (5). On the hand, for each $\lambda \neq 0$ one can divide equation (5) by λ and obtains (4) with $\frac{1}{\lambda}g$ and $\frac{1}{\lambda}K$ instead of g and K. In this chapter, the case $\lambda = 0$ is generally excluded, since then (5) represents a Fredholm integral equation *of the first kind*.

Given the integral operator K, one has to look for a suitable Banach space X such that

(3.1.6) $K \in L(X,X)$.

In the following, we will discuss the simplest spaces $X = C(D)$ and $X = L^2(D)$.

3.2 Compactness of the Integral Operator *K*

3.2.1 General Considerations

Under suitable conditions, the integral operators K turn out to be compact. Then, (1.6) can be replaced by the stronger statement

(3.2.1) $K \in K(X,X)$

(cf. §1.3.8). Compactness enables the application of the Riesz-Schauder theory, which leads to the following existence and uniqueness (cf. Theorem 1.3.28).

Theorem 3.2.1. Let a Fredholm integral equation $\lambda f = g + Kf$ of the second kind be given, i.e., $\lambda \neq 0$, and assume K to be compact in some Banach space X (assumption (1)). Then either (i) or (ii) below holds:
(i) The operator $\lambda I - K$ has a bounded inverse

(3.2.2) $(\lambda I - K)^{-1} \in L(X,X)$,

and the equation $\lambda f = g + Kf$ has a unique solution for each $g \in X$ given by

(3.2.3) $f = (\lambda I - K)^{-1} g \in X$.

(ii) λ is one of the at most countably many eigenvalues, which may only cluster at 0. Then, the eigenvalue problem

(3.2.4) $\lambda e = Ke$

has n linear independent eigenfunctions $e_1, \dots, e_n$ as solutions, where

(3.2.5) $1 \leqslant n := \dim \text{kernel}\,(\lambda I - K) = \dim\,(X/\text{image}(\lambda I - K)) < \infty.$

In the following subsections we discuss the compactness statement (1) for the spaces $X = C(D)$ and $X = L^2(D)$.

This subsection concludes with characterisations of the boundedness (1.6). In the case $X = C(D)$, one can explicitly describe the operator norm of K by means of the kernel k.

Lemma 3.2.2. Assume that $f \in C(D)$ implies $Kf \in C(D)$. If the right-hand side in (6) below is finite, then K belongs to $L(C(D), C(D))$ and the operator norm satisfies

(3.2.6) $\| K \|_{C(D) \leftarrow C(D)} = \sup_{x \in D} \int_D |k(x,y)|\,dy.$

Proof. (i) Set $C := \sup\{ \int_D |k(x,y)|\,dy : x \in D \} < \infty$. For each $x \in D$, $|(Kf)(x)| \leqslant \int_D |k(x,y)|\,|f(y)|\,dy \leqslant C \|f\|_\infty$ and consequently $\|Kf\|_\infty \leqslant C\|f\|_\infty$. This implies the inequality $\| K \|_{C(D) \leftarrow C(D)} \leqslant C$.
(ii) Let $\varepsilon > 0$ be arbitrary. By definition of the supremum, there is a $\xi \in D$ with $C \leqslant \int_D |k(\xi,y)|\,dy + \varepsilon$. The function $\Phi(y) := \text{sign}\, k(\xi,y)$ is bounded, i.e., $\Phi \in L^\infty$, but not necessarily continuous. If $C < \infty$, $K\Phi$ is defined and, because of the identity $|z| = z\,\text{sign} z$, it satisfies the estimate

$$\| K\Phi \|_\infty \geqslant |(K\Phi)(\xi)| = \int_D |k(\xi,y)|\,dy \geqslant C - \varepsilon.$$

In the following part, we will look for a *continuous* function φ leading to almost the same inequality.
(iii) For $\varepsilon > 0$ there is an $\eta \in (0,1]$ such that with $D_0 := \{ y \in D : |k(\xi,y)| \leqslant \eta \}$ we have $\int_{D_0} |k(\xi,y)|\,dy \leqslant \varepsilon$. Let $U \subset D$ be a neighbourhood of D_0 such that $\int_U |k(\xi,y)|\,dy \leqslant 2\varepsilon$. We set $D_1 := U \setminus D_0$ and $D_2 := D \setminus U$. Therefore, D_0, D_1, D_2 are disjoint sets whose union is D. Define the function φ by $\varphi = 0$ on D_0 and $\varphi = \Phi = \text{sign}\, k(\xi,y)$ on D_2. In the gap D_1 between D_0 and D_2, one can define φ in such a way that it becomes a continuous function with values only in $[-1, 1]$. Again, we have

$$\begin{aligned} \| K\varphi \|_\infty &\geqslant |(K\varphi)(\xi)| = \left| \int_U k(\xi,y)\varphi(y)\,dy + \int_{D_2} k(\xi,y)\varphi(y)\,dy \right| \geqslant \\ &\geqslant \int_{D_2} |k(\xi,y)|\,dy - \int_U |k(\xi,y)\varphi(y)|\,dy \geqslant \int_{D_2} |k(\xi,y)|\,dy - 2\varepsilon = \\ &= \int_D |k(\xi,y)|\,dy - \int_U |k(\xi,y)|\,dy - 2\varepsilon \geqslant C - \varepsilon - 2\varepsilon - 2\varepsilon \geqslant C - 5\varepsilon = \\ &= (C - 5\varepsilon)\,\| \varphi \|_\infty. \end{aligned}$$

Since ε is arbitrary, C is the smallest constant with $\| K\varphi \|_\infty \leqslant C \| \varphi \|_\infty$ for all $\varphi \in C(D)$. Hence, (6) follows from Exercise 1.3.12b. QED

Lemma 3.2.3. When $X = L^2(D)$, a sufficient condition for $K \in L(X,X)$ is $k \in L^2(D \times D)$. Then the following estimate holds:

(3.2.7) $\| K \|_{L^2(D) \leftarrow L^2(D)} \leqslant \left[\int_D \int_D |k(x,y)|^2\,dx\,dy \right]^{1/2}.$

Proof. The Schwarz inequality for integrals yields

$$\|Kf\|^2_{L^2(D)} = \int_D |(Kf)(x)|^2 dx = \int_D \Big|\int_D k(x,y)f(y)dy\Big|^2 dx \leqslant$$

$$\leqslant \int_D \Big[\int_D |k(x,y)|^2 dy\Big]\Big[\int_D |f(y)|^2 dy\Big] dx = \Big[\int_D\int_D |k(x,y)|^2 dx dy\Big] \|f\|^2_{L^2(D)}.$$ ∎

Definition 3.2.4. A kernel $k \in L^2(D \times D)$ is called a *Hilbert-Schmidt kernel.*

3.2.2 The Case $X = C(D)$

For the introduction, we start with a simple assumption upon k: The kernel function is assumed to be continuous.

Theorem 3.2.5. Let D be a compact domain of definition and assume $k \in C(D \times D)$. Then, K is compact in $X = C(D)$.

Proof. The assumption $k \in C(D \times D)$ with a compact set D immediately implies the weaker conditions (8a,b) of the following Theorem 3.2.6. Another proof will be mentioned in §3.3.2. ∎

The assumption that D is compact, excludes, e.g., unbounded intervals like $I = \mathbb{R}$. In §3.2.4 we will return to this case and see examples, where K cannot be compact. Continuity of k implies integrability of $k(x,y)f(y)$ with respect to y for $f \in C(D)$. Obviously, this can also be achieved under weaker assumptions. It suffices that $k(x,y)$ can be integrated in the sense of an improper integral (more about improper integrals in §6.1.3). The following theorem describes sufficient conditions of this kind implying compactness of K. Condition (8a) can also be written as $k(x,\cdot) \in L^1(D)$ for all $x \in D$. Here, we do not discuss the interpretation of $k(x,y)dy$ as a (signed) measure $\mu_x(dy)$. We only mention that conditions (8a,b) could be understood as statements about the total variation of the measures $\mu_x(dy)$ or $\mu_\xi(dy) - \mu_x(dy)$, respectively.

Theorem 3.2.6. Let D be the compact domain of definition. Suppose the kernel function k of the integral operator K satisfies

(3.2.8a) $\int_D |k(x,y)| dy < \infty$ for all $x \in D$,

(3.2.8b) $\lim_{\xi \to x} \int_D |k(\xi,y) - k(x,y)| dy = 0$ for all $x \in D$.

Then, K is compact in $X = C(D)$, i.e., $K \in K(X,X)$. If, vice versa, $K \in K(X,X)$ is a compact integral operator with a kernel $k(x,\cdot) \in L^1(D)$, then the conditions (8a,b) hold.

Proof. (i) Let $B := \{Kf : f \in C(D),\ \|f\|_\infty \leqslant 1\}$ be the image of the unit ball $K_1(0) \subset X = C(D)$. It is be shown that B is a precompact set in X. Due to the theorem of Arzelà-Ascoli (cf. Theorem 1.3.26), the uniform boundedness and the equicontinuity of B is to be checked.

(ii) Define

$$\varphi(x) := \int_D |k(x,y)|\,dy \quad\text{and}\quad \Phi(\xi,x) := \int_D |k(\xi,y)-k(x,y)|\,dy.$$

By the reverse triangle inequality $||\alpha|-|\beta|| \leqslant |\alpha-\beta|$, we obtain the estimate

$$|\varphi(\xi)-\varphi(x)| =$$
$$= \left|\int_D |k(\xi,y)|-|k(x,y)|\,dy\right| \leqslant \int_D |k(\xi,y)-k(x,y)|\,dy = \Phi(\xi,x).$$

Assumption (8b) guarantees continuity of φ, i.e., $\varphi\in C(D)$. The same argument shows $|\Phi(\xi,x)-\Phi(\xi,x')| \leqslant \Phi(x,x')$, $|\Phi(\xi,x)-\Phi(\xi',x)| \leqslant \Phi(\xi,\xi')$ and altogether $|\Phi(\xi,x)-\Phi(\xi',x')| \leqslant \Phi(x,x')+\Phi(\xi,\xi')$. Because of (8b), Φ is continuous on $D\times D$ and satisfies $\Phi(x,x)=0$. Since D is compact, so is $D\times D$. Hence, φ is bounded, i.e., $\|\varphi\|_\infty<\infty$ and $\Phi\in C(D\times D)$ uniformly continuous.
(iii) Each $g=Kf\in B$ is bounded by $\|g\|_\infty \leqslant \|\varphi\|_\infty$, as we conclude from

$$|g(x)| = |(Kf)(x)| = \left|\int_D k(x,y)f(y)\,dy\right| \leqslant$$
$$\leqslant \int_D |k(x,y)|\,|f(y)|\,dy \leqslant \int_D |k(x,y)|\,\|f\|_\infty\,dy = \varphi(x)\underbrace{\|f\|_\infty}_{\leqslant 1} \leqslant \|\varphi\|_\infty$$

for all $x\in D$. Therefore, B is uniformly bounded.
(iv) Let $\varepsilon>0$ be given. Since Φ is uniformly continuous on $D\times D$ (cf. (ii)), there is some δ such that $\Phi(\xi,x)=|\Phi(\xi,x)-\Phi(x,x)|\leqslant\varepsilon$ for all $\xi,x\in D$ with $|\xi-x|\leqslant\delta$. Let $g=Kf\in B$ be arbitrary and $|\xi-x|\leqslant\delta$. The estimate

$$|g(\xi)-g(x)| = \left|\int_D k(\xi,y)f(y)\,dy - \int_D k(x,y)f(y)\,dy\right| \leqslant$$
$$\leqslant \left|\int_D \{k(\xi,y)-k(x,y)\}f(y)\,dy\right| \leqslant \int_D |k(\xi,y)-k(x,y)|\,\underbrace{|f(y)|}_{\leqslant 1}\,dy \leqslant$$
$$\leqslant \int_D |k(\xi,y)-k(x,y)|\,dy = \Phi(\xi,x) \leqslant \varepsilon$$

proves the equicontinuity of functions $g\in B$.
(v) Compactness implies boundedness of K: $K\in L(X,X)$. Therefore, because of (6), condition (8a) is necessary.
(vi) It remains to show the necessity of condition (8b). Let $x_k\in D$ be a sequence with the limit $\xi\in D$. The functional $T_kf := (Kf)(x_k)-(Kf)(\xi)$ has the representation $T_kf = \int_D t_k(y)f(y)\,dy$ with $t_k(y) := k(x_k,y)-k(\xi,y)$. As in the proof of Lemma 2, one shows $\sup\{|T_kf|: f\in C(D),\ \|f\|_\infty\leqslant 1\} = \int_D |t_k(y)|\,dy$. Hence, for all $\varepsilon>0$ there are functions $f_k\in C(D)$ with $\|f_k\|_\infty\leqslant 1$ and

$$\left|T_kf_k - \int_D |t_k(y)|\,dy\right| \leqslant \varepsilon.$$

Since K is compact, there is a subsequence of $g_k := Kf_k$ converging uniformly to $g\in C(D)$. For a sufficiently large k, the estimate $\|g_k-g\|_\infty\leqslant\varepsilon$ and, since g is continuous in ξ, also $|g(x_k)-g(\xi)|\leqslant\varepsilon$ holds. Therefore,

$$|T_kf_k| = |g_k(x_k)-g_k(\xi)| \leqslant$$
$$\leqslant |g_k(x_k)-g(x_k)| + |g(x_k)-g(\xi)| + |g(\xi)-g_k(\xi)| \leqslant 3\varepsilon.$$

Together with the previous inequality, $\int_D |k(x_k,y)-k(\xi,y)|\,dy = \int_D |t_k(y)|\,dy \leqslant 4\varepsilon$ holds for the subsequence of $x_k\to\xi$. This proves (8b). ∎

In §3.4, we shall obtain compactness from regularity results.

3.2.3 The Case $X = L^2(D)$

The condition $k \in L^2(D \times D)$, which by Lemma 3 is sufficient for boundedness, is also sufficient for compactness in $X = L^2(D)$.

Theorem 3.2.7. An operator K with a Hilbert-Schmidt kernel is compact in $X = L^2(D)$; i.e., $k \in L^2(D \times D)$ implies $K \in K(L^2(D), L^2(D))$.

Proof. The proof will be given latter in §3.3.2. Compare also Heuser [1,§87] or Yosida [1, §X2]. QED

3.2.4 The Case of an Unbounded Interval I

Theorem 7 might give the impression that all bounded operators are already compact. Therefore, we discuss a counterexample. The Fourier integral transformation $f \mapsto \hat{f}$ is defined by

$$(3.2.9) \qquad \hat{f}(\xi) := (2\pi)^{-1/2} \int_{\mathbb{R}} e^{-i\xi x} f(x)\,dx,$$

i.e., by an integral operator K with the continuous kernel $k(\xi, x) := (2\pi)^{-1/2} e^{-i\xi x}$ over the infinite interval $I = \mathbb{R}$.

Remark 3.2.8. The integral operator K defined by the Fourier transformation (9) belongs to $L(X,X)$ for $X = L^2(\mathbb{R})$, but it is not compact on any Banach space.

Proof. As known from Fourier theory (cf. Meister [1]), the Fourier transformation is norm preserving in $L^2(\mathbb{R})$, so that $\|K\|_{X \leftarrow X} = 1$ proves $K \in L(X, X)$ for $X = L^2(\mathbb{R})$. A second application of K to $\hat{f}$ yields $g := K\hat{f} = K^2 f$ with $g(x) = f(-x)$ for all $x \in \mathbb{R}$. g is the reflected version of the function f. By a further reflection, we regain f: $K^4 f = K^2 g = f$, i.e., $K^4 = I$ (I: identity). If K were compact with respect to some Banach space X, also $I = K^4$ were compact. Since these function spaces are infinite-dimensional, Exercise 1.3.22b shows the contradiction. QED

Theorem 6 has been proved by means of the theorem of Arzelà-Ascoli, which cannot be generalised to an unbounded domain of definition D. In the following, we will consider the one-sided unbounded interval

$$(3.2.10) \qquad I = [0, \infty).$$

The Banach space $X = C(I) = C([0,\infty))$ of the continuous and bounded functions is too large. Instead, we consider the proper subspace

$$(3.2.11) \qquad X = C_{\lim}([0,\infty)) := \{f: \text{ continuous on } [0,\infty), \lim_{x\to\infty} f(x) \text{ exists}\}$$

equipped with the supremum norm $\|\cdot\|_\infty$.

Exercise 3.2.9. Show that **(a)** all $f \in X$ are bounded.
(b) $(C_{\lim}([0,\infty)), \|\cdot\|_\infty)$ is a Banach space.

The limit condition in the definition (11) ensures continuity at $x=\infty$. One may regard X as the space $C([0,\infty])$ over the compactified interval $[0,\infty]$. The direct translation of the conditions (8a,b) reads

(3.2.12a) $\int_0^\infty |k(x,y)|\,dy < \infty$ for all $x \in I = [0,\infty)$,

(3.2.12b) $\lim_{\xi\to x} \int_0^\infty |k(\xi,y)-k(x,y)|\,dy = 0$ for all $x \in I$.

The analogue of condition (12b) at $x=\infty$ is

(3.2.12c) $\lim_{x\to\infty} \sup_{\xi > x} \int_0^\infty |k(\xi,y)-k(x,y)|\,dy = 0.$

The proof of the following theorem can be found in Sloan [1].

Theorem 3.2.10. The integral operator $(Kf)(x) = \int_0^\infty k(x,y)f(y)\,dy$ is compact in X (cf. (11)), if and only if (12a-c) hold.

Exercise 3.2.11. Let (12a-c) be valid and show that the left-hand side in (12a) is *uniformly* bounded with respect to $x \in [0,\infty)$.

The previous statements can directly be generalised to the two-sided unbounded interval $I=\mathbb{R}$. Note that the conditions (12a-c) are restrictive. The kernel $k(x,y)$ of the Fourier transformation only depends on the product xy. Such operators cannot be compact in X because of the following

Remark 3.2.12. A kernel of the form $k(x,y)=\varkappa(xy)$ with $\varkappa$ not vanishing identically cannot satisfy (12a-c).

Proof. The substitution $t=xy$ yields

$$\int_0^\infty |k(x,y)|\,dy = \int_0^\infty |\varkappa(xy)|\,dy = \frac{1}{x}\int_0^\infty |\varkappa(t)|\,dt \to \infty \qquad \text{for } x \to 0$$

in contradiction to the statement of Exercise 11. QED

K is called an operator of convolution type if $k(x,y)=\varkappa(x-y)$ depends only on the difference of the arguments. Also this convolution kernel, which, e.g., appears in the Wiener-Hopf integral equation, is excluded. For details of the Wiener-Hopf equation compare, e.g., Dautray - Lions [1, chapter XI,§1] or Pipkin [1, §8].

Exercise 3.2.13. A kernel of the form $k(x,y)=\varkappa(x-y)$ not vanishing identically cannot satisfy (12c).

3.3 Finite Approximability of the Integral Operator K

3.3.1 Convergence with Respect to the Operator Norm

In the following chapter 4, we shall discretise the integral equation, i.e., we replace the equation by a *finite-dimensional* problem. The integral operator K will be replaced by an «approximation» K_n, where n, e.g., may be the dimension of the image of K_n. Theorem 1.3.23c proves

Remark 3.3.1. Let $K_n \in L(X,X)$ and $\dim \mathrm{image}(K_n) < \infty$. Then K_n is compact, i.e., $K_n \in K(X,X)$.

Hence, all discrete analogues of K are compact. An obvious (but not the only) possibility of an approximation of K by K_n is

$$(3.3.1) \qquad \|K - K_n\|_{X \leftarrow X} \to 0 \qquad \text{for } n \to \infty.$$

In this case, K_n is called *convergent* (to K) *with respect to the operator norm.* Theorem 1.3.23b and Remark 1 imply

Remark 3.3.2. If the operators $K_n \in L(X,X)$ with $\dim \mathrm{image}(K_n) < \infty$ converge to K with respect to the operator norm, K is compact, i.e., $K \in K(X,X)$.

One learns from Remark 2 that *only* compact operators can be approximated with respect to the operator norm. Therefore, the assumption of compactness is not only a tool for proving existence and uniqueness (cf. Theorem 2.1). If K is not compact, one has to approximate it by K_n in another (weaker) sense than in (1).

3.3.2 Degenerate Kernels

Let X be the underlying Banach space. A particularly simple case is given if the (x,y)-dependence can be factorised as follows.

Definition 3.3.3. A kernel function k_n $(n \in \mathbb{N})$ is called *degenerate* (or separable), if

$$(3.3.2) \qquad k_n(x,y) = \sum_{j=1}^{n} a_j(x)\, b_j(y).$$

Here, $a_j \in X$ and $b_j \in X'$ (X': dual space of X).

The functions $a_j(x)$ and $b_j(y)$ must be such that the integrals $\int a_j(x) b_j(y) f(y) dy = a_j(x) \int b_j(y) f(y) dy$ exist for all $f \in X$ and the images belong again to X. The latter requirement is guaranteed by $a_j \in X$. The first condition asks for the existence of the integrals $\int b_j(y) f(y) dy$ for all $f \in X$. Identifying b_j with the functional B_j: $f \mapsto \int b_j(y) f(y) dy$, one is led to the requirement $b_j \in X'$. The dual space of $X = L^2(D)$ is again $X' = L^2(D)$, so that $a_j, b_j \in L^2(D)$ must hold. In the case of $X = C(D)$, the dual space X' contains generalised functions. In order to use only classical functions, we will restrict ourselves to the subspace $L^1(D) \subset X'$ and require $b_j \in L^1(D)$. If D has a finite measure (e.g., if it is bounded), $X = C(D) \subset L^1(D) \subset X'$ is valid and shows that $b_j \in C(D)$ is sufficient.

Exercise 3.3.4. Suppose K_n is an integral operator with degenerate kernel k_n as in (2). Show that **(a)** $K_n \in L(X,X)$. **(b)** $\dim(\mathrm{image}(K_n)) \le n$ holds. **(c)** Let $X = C(D)$. Given another integral operator $K \in L(X,X)$ with a kernel k, (3) holds (cf. (2.6)):

$$(3.3.3) \qquad \|K - K_n\|_{C(D) \leftarrow C(D)} = \sup_{x \in D} \int_D |k(x,y) - k_n(x,y)|\, dy.$$

The approximation of K by a sequence of operators K_n with degenerate kernels can be used for the compactness proof as follows.

Second *Proof of Theorem 2.5.* If $D \subset \mathbb{R}^d$ is compact, so is $D \times D$ in $\mathbb{R}^{2d}$. Let $\varepsilon > 0$. By the theorem of Weierstraß (Theorem 1.3.8), the continuous kernel k can be approximated by a polynomial k_ε, so that the right-hand side in (3) becomes $\leqslant \varepsilon$. Therefore, the operators K_ε defined by means of the kernel k_ε converge to K with respect to the operator norm. However, a polynomial $k_\varepsilon(x,y)$ is a degenerate kernel, since each monomial $\alpha_{\nu\mu} x^\nu y^\mu$ has the representation $a(x)b(y)$ with $a(x) = \alpha_{\nu\mu} x^\nu \in C(D)$ and $b(y) = y^\mu \in C(D) = X \subset X'$. The embedding $X \subset X'$ is based on the convention that each $\varphi \in X = C(D)$ is identified with the functional $\Phi(\psi) := \int_D \varphi \psi \, dx$. Since D is compact, φ also belongs to $L^1(D)$ and, hence, satisfies the dual norm $\|\Phi\|_{X'} = \|\varphi\|_{L^1(D)} \leqslant \mu(D) \|\varphi\|_\infty$. Using Exercise 4b and Remark 2, one obtains the compactness of K. QED

Proof of Theorem 2.7. Let $D \subset \mathbb{R}^d$ and $\varepsilon > 0$. First, we approximate the kernel $k \in L^2(D \times D)$ by a continuous function $\varkappa \in C(D \times D)$ with compact support so that $\|k - \varkappa\|_{L^2(D \times D)} \leqslant \frac{\varepsilon}{2}$ (cf. Remark 1.3.9a). Choose $R := [-L, L]^{2d} \cap (D \times D)$ such that R contains the support of $\varkappa$. Outside of R, set $\varkappa := 0$. Following Weierstraß (Theorem 1.3.8), $\varkappa \in C(R)$ can be approximated by a polynomial $P(x,y) := \sum \alpha_{ij} x^i y^j$ (i, j are multi-indices, if $d > 1$), so that $\|\varkappa - P\|_\infty \leqslant \varepsilon / (2^{d+1} L^d)$. Let $\chi(\xi)$ be the characteristic function of the interval $[-L, L]$, i.e., $\chi = 1$ for $|\xi| \leqslant L$ and $\chi = 0$ otherwise. For $d > 1$, $\chi(x)$ denotes the product $\chi(x_1) \cdot \ldots \cdot \chi(x_d)$. Obviously, the kernel $k_\varepsilon(x,y) := \sum \alpha_{ij} \chi(x) x^i \chi(y) y^j$ is degenerate and satisfies $|\varkappa(x,y) - k_\varepsilon(x,y)| \leqslant \varepsilon / (2^{d+1} L^d)$ in R, while $\varkappa = k_\varepsilon = 0$ holds outside of R. Because of $\|\varphi\|_{L^2(R)} \leqslant C \|\varphi\|_\infty$ with $C^2 = \mu(R) \leqslant (2L)^{2d}$ for all $\varphi \in C(R)$ (Schwarz inequality), one has $\|\varkappa - k_\varepsilon\|_{L^2(D \times D)} \leqslant \|\varkappa - k_\varepsilon\|_{L^2(R)} \leqslant \varepsilon/2$, implying $\|k - k_\varepsilon\|_{L^2(D \times D)} \leqslant \varepsilon$. Estimate (2.7) yields the convergence of the degenerate integral operators to K with respect to the operator norm. As above, compactness follows. QED

3.4 The Image Space of K

In the following, we are going to characterise the properties of the image Kf. Often, Kf for $f \in X$ belongs to a subspace $Y \subset X$ which has a ***stronger*** norm $\|\cdot\|_Y$ than X. Here, $\|\cdot\|_Y$ is called stronger than $\|\cdot\|_X$, if $\|f\|_X \leqslant C \|f\|_Y$ for all $f \in Y$ and $\{\|f\|_Y / \|f\|_X : f \in Y \subset X\}$ is unbounded.

3.4.1 Smooth Kernels $k(x,y)$

Consider $X = C(D)$ with compact D. In Theorem 2.5, continuity $k \in C(D \times D) = C^0(D \times D)$ has been assumed. Stronger smoothness conditions would be $k \in C^\lambda(D \times D)$ for some $\lambda > 0$, $k \in C_L(D \times D)$. or $k \in \hat{C}^\lambda(D \times D)$. In these cases, for each $f \in C(D)$ the image Kf belongs to the spaces $C^\lambda(D)$, $C_L(D)$, or $\hat{C}^\lambda(D)$, respectively.

Theorem 3.4.1. Let D be compact and assume that the kernel k of the operators K satisfies either (i) $k\in C^{\lambda}(D\times D)$ or (ii) $k\in \hat{C}^{\lambda}(D\times D)$ for some $\lambda\geqslant 0$. Then, (i) $K\in L(C(D),C^{\lambda}(D))$ or (ii) $K\in L(C(D),\hat{C}^{\lambda}(D))$ holds, respectively. Even without compactness of D, $K\in L(L^1(D),C^{\lambda}(D))$ or $K\in L(L^1(D),\hat{C}^{\lambda}(D))$ are valid.

Proof. (a) For $\lambda=0$, apply Theorem 2.5. Now, consider $0<\lambda<1$. Let $L\leqslant \leqslant\|k\|_{C^{\lambda}(D\times D)}$ be the constant in $|k(x,y)-k(x',y)|\leqslant L|x-x'|^{\lambda}$. The inequality

$$\begin{aligned}|(Kf)(x)-(Kf)(x')| &\leqslant \int_D |k(x,y)-k(x',y)|\,|f(y)|\,dy \leqslant \\ &\leqslant \mu(D)L|x-x'|^{\lambda}\,\|f\|_{\infty} \qquad (\mu(D)\text{: measure of } D)\end{aligned} \tag{3.4.1}$$

proves the Hölder continuity of Kf and the estimate $\|Kf\|_{C^{\lambda}(D)}\leqslant C\|f\|_{\infty}$ with $C:=\max\{\mu(D)L,\ \sup_x \int_D |k(x,y)|\,dy\}$.
(b) Let $\lambda=1$. Since D is compact, the derivative d/dx may be interchanged with the integral to obtain $d(Kf)(x)/dx = \int_D \partial k(x,y)/\partial x\, f(y)\,dy$. The estimate $\|(Kf)'\|_{\infty}\leqslant \max\{\int_D|\partial k(x,y)/\partial x|dy:\ x\in D\}\,\|f\|_{\infty}$ implies $\|Kf\|_{C^1(D)}\leqslant C\|f\|_{\infty}$ with $C:=\max\{\max_x \int_D|k(x,y)|dy,\ \max_x \int_D|k_x(x,y)|dy\}$.
(c) For general $\lambda>1$, the considerations are analogous: Derivatives of the integral become derivatives of the kernel function. As in (a), Hölder estimates of higher derivatives of Kf lead to Hölder estimates of higher x-derivatives of $k(x,y)$. In the case of $\hat{C}^{\lambda}(D)$ with λ integer, the Hölder estimate is replaced by the Lipschitz estimate. QED

The proof shows that smoothness of the kernel is only needed with respect to x. We do not formulate the corresponding weaker assumptions, since function spaces with different properties with respect to the different arguments would be necessary. The following theorem formulates the conditions corresponding to those in Theorem 2.6 and yields $Kf\in C^{\lambda}(D)$.

Theorem 3.4.2. Let D be a compact domain of definition and $0<\lambda<1$. Assume that the kernel function k of the integral operator K satisfies

$$\int_D |k(x,y)|\,dy < \infty \qquad \text{for all } x\in D, \tag{3.4.2a}$$

$$\int_D |k(\xi,y)-k(x,y)|\,dy \leqslant C|\xi-x|^{\lambda} \qquad \text{for all } \xi,x\in D \tag{3.4.2b}$$

for some constant C. Then, K is a bounded mapping from $X=C(D)$ into $C^{\lambda}(D)$, i.e., $K\in L(C(D),C^{\lambda}(D))$.

Proof. Inequality (1) shows the global Hölder continuity of $\varphi:=Kf$: $|\varphi(x)-\varphi(\xi)|\leqslant C\|f\|_{\infty}$. Since (2a,b) also implies (2.8a,b), $K\in L(X,X)$ holds, i.e., $\|Kf\|_{\infty}\leqslant C'\|f\|_{\infty}$. Therefore, the Hölder norm of Kf is bounded by $\|Kf\|_{C^{\lambda}(D)}\leqslant\max\{C,C'\}\,\|f\|_{\infty}$. QED

Exercise 3.4.3. Show that in Theorem 2 the compactness of D can be replaced by the *uniform* estimate $\int_D|k(x,y)|\,dy\leqslant C_0$ for all $x\in D$ instead of (2a).

Exercise 3.4.4. Assume $k(x,y)=k_1(x,y)\,k_2(x,y)$ and $k_1\in C^{\lambda}(D\times D)$ for a compact D. Let k_2 satisfy the conditions (2a,b). Prove that $K\in L(C(D),C^{\lambda}(D))$ holds for the operator with kernel k.

3.4.2 The Image Kf for $f \in C^{\lambda}(I)$

In §3.4.1, we used the smoothness of k to infer $Kf \in C^{\lambda}(I)$ for $f \in C(I)$. Can the smoothness of k be replaced by that of f? In general, the answer is negative. For a proof, consider the integral operator with the kernel $k(x,y)=a(x)\,b(y)$ on the interval $I=[a,b]$, where $a \in C(I)$ is continuous but not Hölder continuous. Let f be arbitrarily smooth, e.g., $f \in C^{\infty}(I)$. One verifies that $Kf=\beta a$ with $\beta=\int_I b(y)f(y)dy$. Unless $\beta=0$, Kf has the same smoothness properties as the function a and is, in particular, not Hölder continuous.

However, smoothness of f leads to a smooth image Kf, if the kernel is of convolution type, i.e., of the form

$$k(x,y) = \tilde{k}(x-y).$$

In the following, we write again k instead of $\tilde{k}$, so that

(3.4.3) $$(Kf)(x) = \int_I k(x-y)f(y)dy$$

is the integral operator.

Theorem 3.4.5. K is given by (3) with $I=[a,b]$, $a<b$. **(a)** Let the kernel be integrable: $k \in L^1([a-b,b-a])$, i.e.,

(3.4.4a) $$\int_{a-b}^{b-a} |k(\xi)|d\xi =: C_0 < \infty;$$

furthermore, assume that there are $\lambda \in (0,1]$ and a constant C_1 such that

(3.4.4b) $$\int_x^{\xi} |k(t)|dt \le C_1(\xi-x)^{\lambda} \qquad \text{for all } a-b \le x \le \xi \le b-a.$$

Then, K belongs to $L(\hat{C}^{\mu}(I), \hat{C}^{\mu}(I))$ for all exponents $0 \le \mu \le \lambda$.
(b) If $k \in C([a-b,b-a])$, even $K \in L(C^1(I), C^1(I))$ holds. For $f \in C^1(I)$, the image $\varphi=Kf$ has the derivative

(3.4.5) $$\varphi'(x) = \int_I k(x-y)f'(y)dy + k(x-a)f(a) - k(x-b)f(b).$$

Proof. (i) If (4b) holds for λ, then it also holds for any $\mu \in (0,\lambda)$, with the constant C_1 replaced by $C_1(2(b-a))^{\lambda-\mu}$. Therefore, it suffices to consider only the cases $\mu=\lambda$ and $\mu=0$. The latter is left to the reader.
(ii) So let us consider the case $\mu=\lambda$. Set $\varphi:=Kf$ for some $f \in \hat{C}^{\lambda}(I)$, i.e., $f \in C^{\lambda}(I)$ for $0<\lambda<1$ or $f \in C_L(I)$ for $\lambda=1$. Let $x', x \in I$ be two arguments with $x'=x+\delta>x$. Substitution of $\eta:=y-\delta$ yields

$$\varphi(x') = \int_I k(x'-y)f(y)dy = \int_a^b k(x+\delta-y)f(y)dy =$$
$$= \int_{a-\delta}^{a} k(x-\eta)f(\eta+\delta)d\eta + \int_a^{b-\delta} k(x-\eta)f(\eta+\delta)d\eta$$

and, analogously,

$$\varphi(x) = \int_a^{b-\delta} k(x-\eta)f(\eta)d\eta + \int_{b-\delta}^{b} k(x-\eta)f(\eta)d\eta.$$

The difference can be split into

$$\varphi(x')-\varphi(x) = D_0 + D_1 + D_2$$

with $$D_0 := \int_a^{b-\delta} k(x-\eta)[f(\eta+\delta)-f(\eta)]\,d\eta,$$

$$D_1 := \int_{a-\delta}^{a} k(x-\eta) f(\eta+\delta) d\eta, \qquad D_2 := -\int_{b-\delta}^{b} k(x-\eta) f(\eta) d\eta.$$

By $|f(\eta+\delta)-f(\eta)| \leqslant C_f \delta^\lambda$, D_0 can be estimated by $C_f \delta^\lambda \int_a^{b-\delta} |k(x-\eta)| d\eta \leqslant C_0 C_f \delta^\lambda$. Due to assumption (4b), D_1 and D_2 are bounded by $C_1 \delta^\lambda \|f\|_\infty$. Together, one obtains the Hölder or Lipschitz continuity of $\varphi = Kf$.
(iii) One can use the representation of $\varphi(x') - \varphi(x)$ from (ii) to show that (5) is the limit of $[\varphi(x')-\varphi(x)]/(x'-x)$ for $x' \to x$. ∎

The representation (5) allows further statements such as, e.g., $K \in L(C^1(I), C^{1+\lambda}(I))$ for $k \in C^\lambda([a-b, b-a])$. The assumptions (4a,b) signify that $|k(t)|$ has a Hölder continuous (or, in the case of $\lambda = 1$, a Lipschitz continuous) primitive. Modifications of Theorem 5 are given in

Exercise 3.4.6. Prove: **(a)** If the kernel k and f are periodic with the period $b-a$, condition (4b) is not needed.
(b) Theorem 5a remains valid for unbounded intervals I, if the assertion is restricted to $\mu = \lambda$. For $I = \mathbb{R}$, (4b) can be omitted.

Often, $k(\cdot)$ is singular only at one (or finitely many) points, but smooth elsewhere. In the next theorem, we consider one singularity of $k(t)$ at $t=0$. A more general singularity expansion is developed by Graham [1].

Theorem 3.4.7. Let $k \in L^1([a-b, b-a])$ and $k \in C^\infty((0, b-a]) \cap C^\infty([a-b, 0))$. For $\lambda \geqslant 0$ and f in either of the spaces $C^\lambda(I)$ or $\hat{C}^\lambda(I)$, the image Kf belongs to the same space with I replaced by the *open* interval (a, b).

Proof. For $x \in (a, b)$, choose $\varepsilon > 0$ such that $a \leqslant x - \varepsilon \leqslant x + \varepsilon \leqslant b$. Split the integral $(Kf)(x) = \int_a^b k(x-y) f(y) dy$ into $\int_a^{x-\varepsilon} \ldots + \int_{x-\varepsilon}^{x+\varepsilon} \ldots + \int_{x-\varepsilon}^{b} \ldots$. The first and third integrals are infinitely differentiable with respect to x. The middle integral becomes $\int_{-\varepsilon}^{\varepsilon} k(t) f(x-t) dt$ after the substitution $t = x - y$ and satisfies all Hölder/Lipschitz estimates or differentiability conditions which are satisfied by f. ∎

The previous considerations can be extended onto a larger class of kernels than $k(x, y) = k(x-y)$ from (3).

Remark 3.4.8. Suppose the kernel defined on $I = [a, b]$ is of the form

$$k(x, y) = \ell(x, y) \varkappa(x-y), \tag{3.4.6}$$

where $\ell \in C^\lambda(I \times I)$ and $\varkappa$ satisfies the conditions (4a,b) for some $0 < \lambda \leqslant 1$. Then the assertion of Theorem 5 holds for the integral operator K associated with k.

Proof. $\varphi(x') = (Kf)(x')$ at $x' = x + \delta$ is the sum of

$$\int_I [\ell(x', y) - \ell(x. y)] \varkappa(x'-y) f(y) dy \quad \text{and} \quad \int_I \ell(x, y) \varkappa(x'-y) f(y) dy.$$

The first integral can be estimated by const$\cdot \delta^\lambda$. The difference between the second integral and $\varphi(x)$ reads $\int_I [\varkappa(x'-y) - \varkappa(x-y)][\ell(x, y) f(y)] dy$ and can be treated as in Theorem 5. ∎

3.4.3 Kernels with Integrable Singularity

The considerations of §3.4.2 apply to the following kernel:

(3.4.7a) $k(x,y) = \ell(x,y)/|x-y|^{1-\lambda}$ for a $\lambda\in(0,1)$,

where the numerator ℓ satisfies, e.g., the Hölder condition

(3.4.7b) $\ell\in C^{\lambda}(I\times I)$.

Even the uniform Hölder continuity only with respect to x is sufficient:

(3.4.7c) $\|\ell(\cdot,y)\|_{C^{\lambda}(I)} \leqslant C$ for all $y\in I$.

A further weakening reads as follows: There are constants C, C_1, C_2 with

(3.4.7d$_1$) $\|\ell\|_{\infty,I\times I} \leqslant C_1$,

(3.4.7d$_2$) $|\ell(x,y)-\ell(\xi,y)| \leqslant C_2|x-\xi|^{\lambda}$ for all $|y-x|>C|x-\xi|$.

Also the following non-uniform Lipschitz estimate suffices:

(3.4.7e) (7d$_1$) and $|\ell(x,y)-\ell(\xi,y)|\leqslant C_2\frac{|x-\xi|}{|y-x|}$ for $|y-x|>C|x-\xi|$.

By the mean value theorem, the latter condition follows from

(3.4.7f) (7d$_1$) and $|\frac{d}{dx}\ell(x,y)| \leqslant \frac{C_2}{|y-x|}$ for all $x\neq y$.

The kernel k has the form (6) with $\varkappa(t):=|t|^{\lambda-1}$. Since $\varkappa$ is improperly integrable for $\lambda>0$ (cf. §6.1.3) and the primitive $|t|^{\lambda}/\lambda$ is Hölder continuous with exponent λ, the assumptions (4a,b) are satisfied. Theorem 5 or its respective generalisation in Remark 8 yield $K\in L(C^{\mu}(I),C^{\mu}(I))$ for all $0\leqslant\mu\leqslant\lambda$. This statement can be strengthened to $K\in L(C(I),C^{\lambda}(I))$, as formulated in

Theorem 3.4.9. Let I be compact. Assume that the weakly singular kernel k of the integral operators K satisfies (7a) and one of the properties (7b) to (7f). Then, K belongs to $L(C(I),C^{\lambda}(I))$.

Proof. (i) Because of the implications (7b) ⟹ (7c) ⟹ (7d) and (7f) ⟹ (7e), it suffices to discuss (7d$_{1,2}$) and (7e).
(ii) Condition (2a) is satisfied, since $\int_I|k(x,y)|dy\leqslant\|\ell\|_{\infty}\int_I|x-y|^{1-\lambda}dy<\infty$.
(iii) Without loss of generality, the constant C in (7d,e) can be assumed to satisfy $C\geqslant 2$. For fixed $x,\xi\in I$, set $\varepsilon:=|x-\xi|$. To prove (2b), we split the integral over $I=D$ into integrals over

$$I_0 := \{y\in I:\ |y-x|\leqslant C\varepsilon\} \quad\text{and}\quad I_1:=I\backslash I_0=\{y\in I:\ |y-x|>C\varepsilon\}.$$

Because $C\geqslant 2$, ξ lies in I_0. On I_0, one estimates the integral as follows:

$$\int_{I_0}|k(\xi,y)-k(x,y)|dy \leqslant \int_{I_0}[|k(\xi,y)|+|k(x,y)|]\,dy \leqslant$$
$$\leqslant C_1\int_{I_0}[|\xi-y|^{\lambda-1}+|x-y|^{\lambda-1}]\,dy \leqslant$$

$$\leqslant \tfrac{1}{\lambda} C_1 [(C\varepsilon-\varepsilon)^\lambda + (C\varepsilon+\varepsilon)^\lambda + 2(C\varepsilon)^\lambda] \leqslant \tfrac{4}{\lambda} C_1 C^\lambda \varepsilon^\lambda = O(|x-\xi|^\lambda).$$

(iv) In the case of ($7d_{1,2}$), we estimate the integral over I_1 by

$$\int_{I_1} |k(\xi,y)-k(x,y)|\,dy \leqslant$$

$$\leqslant \int_{I_1} \Big\{ \frac{|\ell(\xi,y)-\ell(x,y)|}{|y-\xi|^{1-\lambda}} + |\ell(x,y)|\,\big||y-\xi|^{\lambda-1} - |y-x|^{\lambda-1}\big| \Big\} dy \leqslant$$

$$\text{(3.4.7g)}\qquad \leqslant C_2 \int_{I_1} \frac{|x-\xi|^\lambda}{|y-\xi|^{1-\lambda}}\,dy + C_1 \int_{I_1} \big||y-\xi|^{\lambda-1} - |y-x|^{\lambda-1}\big|\,dy.$$

The first integral in (7g) is of the order

$$C_2 |x-\xi|^\lambda \int_{I_1} |y-\xi|^{\lambda-1} dy = O(|x-\xi|^\lambda).$$

In (v) we show that also the second integral equals $O(|x-\xi|^\lambda)$. Hence, condition (2b) is satisfied and Theorem 2 proves the assertion.

(v) Let $I=[a,b]$. Then, I_1 splits into the intervals $[a, x-C\varepsilon]$ and $[x+C\varepsilon, b]$, which we may enlarge to $(-\infty, x-C\varepsilon]$ and $[x+C\varepsilon, \infty)$. For $y \geqslant x+C\varepsilon$, the mean value theorem yields the representation

$$(y-\xi)^{\lambda-1} - (y-x)^{\lambda-1} = (\xi-x)(\lambda-1)(y-\zeta)^{\lambda-2}$$

with an intermediate value ζ between ξ and x. Since $|y-\zeta| = y-\zeta \geqslant y-(x+\varepsilon)$, we have

$$\int_{x+C\varepsilon}^{\infty} \big||y-\xi|^{\lambda-1} - |y-x|^{\lambda-1}\big|\,dy \leqslant \varepsilon|\lambda-1| \int_{x+C\varepsilon}^{\infty} (y-x-\varepsilon)^{\lambda-2} dy =$$

$$= \varepsilon(1-\lambda) \int_{(C-1)\varepsilon}^{\infty} t^{\lambda-2} dy = -\varepsilon\,[t^{\lambda-1}]_{(C-1)\varepsilon}^{\infty} = \varepsilon^\lambda (C-1)^{\lambda-1}.$$

The same bound is valid for the integral over $(-\infty, x-C\varepsilon]$, so that

$$\text{(3.4.7h)}\qquad \int_{I_1} \big||y-\xi|^{\lambda-1} - |y-x|^{\lambda-1}\big|\,dy \leqslant 2(C-1)^{\lambda-1}\varepsilon^\lambda.$$

(vi) In the case of (7e), one estimates the integrand of the first integral in (7g) by

$$\text{(3.4.7i)}\qquad \frac{|\ell(\xi,y)-\ell(x,y)|}{|y-\xi|^{1-\lambda}} \leqslant C_2 \frac{|x-\xi|}{|x-y|\,|y-\xi|^{1-\lambda}} \leqslant C_2 \frac{\varepsilon}{|x-y|\,|y-\xi|^{1-\lambda}}.$$

If $I=[a,b]$, I_1 has the representation $I_1 = [a, x-C\varepsilon) \cap (x+C\varepsilon, b]$. Assume $b \geqslant x-C\varepsilon$, since otherwise the second interval is empty. In $(x+C\varepsilon, b]$, the following inequality holds because $C \geqslant 2$:

$$|y-\xi| \geqslant |y-y_0|,\ |y-x| \geqslant |y-y_0| \qquad \text{for } y_0 := x+(C-1)\varepsilon.$$

Hence,
$$\int_{x+C\varepsilon}^{b} \frac{dy}{|x-y|\,|y-\xi|^{1-\lambda}} \leqslant \int_{x+C\varepsilon}^{b} \frac{dy}{|y-y_0|^{2-\lambda}} = \int_{\varepsilon}^{b-x-C\varepsilon-\varepsilon} t^{\lambda-2}\,dt = O(\varepsilon^{\lambda-1}).$$

The integral over $[a, x-C\varepsilon)$ can be bounded, similarly. This proves

$$\int_{I_1} \frac{|\ell(\xi,y)-\ell(x,y)|}{|y-\xi|^{1-\lambda}}\,dy \leqslant C_2\varepsilon\, O(\varepsilon^{\lambda-1}) = O(\varepsilon^\lambda) = O(|x-\xi|^\lambda).$$

Estimating the second integral in (7g) by Part (v), we obtain condition (2b), so that Theorem 2 is applicable. QED

Another weakly singular kernel of practical interest is

$$(3.4.8) \qquad k(x,y) = \ell(x,y)\log|x-y| \qquad \text{with } \ell\in C_L(I\times I).$$

Theorem 3.4.10. Assume that I is compact and the kernel k satisfies (8). Then, $K\in L(C(I),C^\lambda(I))$ holds and even $K\in K(C(I),C^\lambda(I))$ for all $0<\lambda<1$.

Proof. The proof of Theorem 9 can be adapted to this case and yields the estimate $|\varphi(\xi)-\varphi(x)|\leqslant C|\xi-x|\log|\xi-x|\,\|f\|_\infty$ for $\varphi=Kf$, i.e., φ is Hölder continuous for any exponent $0<\lambda<1$: $K\in L(C(I),C^\lambda(I))$. The compactness property $K\in K(C(I),C^\lambda(I))$ will be proved in §3.4.4. QED

A strengthening of Theorem 10 to $K\in L(C(I),C_L(I))$ is *not* possible.

Example 3.4.11 (*parabolic boundary control problem*, cf. Hackbusch [4]). Let the function $y(u)=y(x,t;u)$ describe the solution of the parabolic initial-boundary value problem (heat equation)

$$(3.4.9a) \qquad y_t(x,t)-y_{xx}(x,t) = f(x,t) \quad \text{for } x>0,\ 0<t<T, \qquad \text{(pde)}$$

$$(3.4.9b) \qquad -y_x(0,t) = u(t) \quad \text{for } 0<t<T \qquad \text{(boundary values)},$$

$$(3.4.9c) \qquad y(x,0) = y_0(x) \quad \text{for } x>0 \qquad \text{(initial values)},$$

where the right-hand sides f and y_0 are fixed quantities, while $u\in L^2([0,T])$ is a function used for controlling the «state» $y=y(u)$. By $z\in L^2([0,\infty))$ we denote the desired value which $y(\cdot,T)$ should approximate at the terminal time $t=T$ as well as possible. More precisely, the «cost function»

$$(3.4.9d) \qquad J(u) := \int_0^\infty |y(x,T;u)-z(x)|^2\,dx + \pi^{-1/2}\int_0^T |u(t)|^2\,dt$$

is to be minimised over all $u\in L^2([0,T])$. The «optimal control» u^* minimising $J(u)$ turns out to be the solution of the integral equation

$$(3.4.9e) \qquad u(t) = g(t) - \int_0^T u(\tau)/\sqrt{2T-t-\tau}\,d\tau \qquad (0\leqslant t\leqslant T),$$

where g can be determined as follows. Let $\hat{y}(x,t)$ be the solution of the equations (9a-c) with $u=0$ and $\hat{p}(x,t)$ the solution of the differential equation $\hat{p}_t+\hat{p}_{xx}=0$ with the starting value $\hat{p}(x,T)=\hat{y}(x,T)-z(x)$ and boundary values $\hat{p}_x(0,t)=0$. Then, $g(t):=\hat{p}(0,t)$.

Exercise 3.4.12. The integral operator K defined by (9e) belongs to $L(C(I),C^{1/2}(I))$ with $I=[0,T]$, where the exponent $\frac{1}{2}$ cannot be improved. Nevertheless, $Kf\in C^\infty([0,T))$ holds for the half-open interval $[0,T)$.

3.4.4 Compactness

The results concerning image(K) yield a new proof of compactness.

Remark 3.4.13. An operator $K\in L(C(D),C^{\lambda}(D))$ with $\lambda>0$ is compact in the space $X=C^{\mu}(D)$ for any $0\leqslant\mu\leqslant\lambda$: $K\in K(C^{\mu}(D),C^{\mu}(D))$. Furthermore, $K\in K(C^{\varkappa}(D),C^{\mu}(D))$ holds for all $\varkappa,\mu\in[0,\lambda]$ with $\varkappa>0$ or $\mu<\lambda$.

Proof. We express K: $C^{\varkappa}(D)\to C^{\mu}(D)$ as the product I_2KI_1 with I_1: $C^{\varkappa}(D)\to C(D)$ and I_2: $C^{\lambda}(D)\to C^{\mu}(D)$ being the embeddings. I_1 is continuous, i.e., $I_1\in L(C^{\varkappa}(D),C(D))$ and for $\varkappa>0$ even compact (cf. Theorem 1.3.27). Analogously, I_2: $C^{\lambda}(D)\to C^{\mu}(D)$ is continuous and for $\mu<\lambda$ compact. Therefore, the product $K=I_2KI_1$: $C^{\varkappa}(D)\to C(D)\to C^{\lambda}(D)\to C^{\mu}(D)$ contains at least one compact factor. By Theorem 1.3.23a, K: $C^{\varkappa}(D)\to C^{\mu}(D)$ is compact. QED

3.4.5 Volterra Integral Equation

The Volterra integral operator $\int_a^x k(x,y)f(y)dy$ $(a\leqslant x\leqslant b)$ can formally be regarded as Fredholm operator on $I=[a,b]$, where

$$k(x,y) = 0 \qquad \text{for } a\leqslant x<y\leqslant b. \tag{3.4.10}$$

In general, this Fredholm kernel k is discontinuous at the diagonal $x=y$. Continuity would require $k(x,x)=0$. However, such a condition is undesirable for Volterra integral equations of the first kind. On the contrary, we shall see in §6 that k may even be singular at $x=y$.

Remark 3.4.14. Let $\ell\in\mathbb{N}$. $G:=\{(x,y): a\leqslant y\leqslant x\leqslant b\}$ is the domain of definition of the Volterra kernel k. Suppose $\partial^{\nu}k/\partial x^{\nu}\in C(G)$ for all $0\leqslant\nu\leqslant\ell$. Then, for $p=0,1,\dots,\ell-1$, the integral operator belongs to $L(C^{p}(I),C^{p+1}(I))$.

Proof by induction on ℓ. Let $\varphi:=Kf$. The derivative equals

$$\varphi'(x) = \int_a^x k_x(x,y)f(y)dy + k(x,x)f(x).$$

Since $k_x(x,y)f(y)\in C(G)$ and $k(x,x)f(x)\in C(I)$, the assertion for $\ell:=1$ follows. Let the assertion hold for $\ell-1$. Since k_x satisfies the assumptions for $\ell-1$, $\int_a^x k_x(x,y)f(y)dy$ belongs to $C^{p+1}(I)$, $0\leqslant p\leqslant\ell-2$, provided that $f\in C^{p+1}(I)$. Since also $k(x,x)f(x)\in C^{p+1}(I)$, $\varphi'\in C^{p+1}(I)$ holds implying $\varphi\in C^{p+2}(I)$ for $0\leqslant p\leqslant\ell-2$ and $f\in C^{p+1}(I)$. An index shift $p\to p-1$ yields $\varphi\in C^{p+1}(I)$ for all $0\leqslant p\leqslant\ell-1$ and all $f\in C^{p}(I)$. QED

3.4.6 K as Mapping Defined on $L^{\infty}(D)$

For most of the proofs the continuity of $f\in C(D)$ has not been used. For proving boundedness, one only needs $f\in L^{\infty}(D)$.

Remark 3.4.15. In the Theorems 2.5, 2.6, 1, 2, 9, 10, and Remark 13, one can replace the respective spaces $L(C(D),Y)$ and $K(C(D),Y)$ by $L(L^{\infty}(D),Y)$ and $K(L^{\infty}(D),Y)$. In the case of Theorem 2.5, K regarded, e.g., as a mapping from $L^{\infty}(D)$ into $Y=C(D)$ is compact.

3.5 Solution of the Fredholm Integral Equation of the Second Kind

3.5.1 Existence and Uniqueness

Because of the compactness of K, for almost all λ the Fredholm integral equation $\lambda f = g + Kf$ of the second kind has a unique solution, since by Theorem 2.1 the exceptional set consists of only countably many eigenvalues λ, which in $\mathbb{C}$ has measure zero. However, for numerical applications, the existence of a solution is not sufficient. It must also be sufficiently regular (i.e., smooth) in order to lead to reasonable approximations.

3.5.2 Regularity

Since $Y = C^{\lambda}(D)$ is even compactly embedded in $X = C^{\varkappa}(D)$ with $\varkappa < \lambda$, $C^{\lambda}(D)$ is also continuously embedded in $C^{\varkappa}(D)$. The following theorem shows that the solution f belongs to a space Y with higher differentiability properties, if the inhomogeneity g of the integral equation $\lambda f = g + Kf$ belongs to Y and the image of K is in Y.

Theorem 3.5.1. Let Y be continuously embedded in X and assume $K \in L(X,Y)$, $\lambda \neq 0$, and $(\lambda I - K)^{-1} \in L(X,X)$. Then,

(3.5.1) $$(\lambda I - K)^{-1} \in L(Y,Y).$$

Hence, for $g \in Y$, the Fredholm integral equation $\lambda f = g + Kf$ has a unique solution $f = (\lambda I - K)^{-1} g \in Y$.

Proof. Since $(\lambda I - K)^{-1} \in L(X,X)$, there is a unique solution $f \in X$. $K \in L(X,Y)$ implies $Kf \in Y$. Together with $g \in Y$, $\lambda f = g + Kf \in Y$ follows. Thanks to $\lambda \neq 0$, $f \in Y$ is an immediate consequence. To estimate f in Y, we use

$$\|f\|_Y \leq \frac{1}{|\lambda|} \left(\|K\|_{Y \leftarrow X} \|f\|_X + \|g\|_Y \right).$$

By $\|f\|_X = \|(\lambda I - K)^{-1} g\|_X \leq \|(\lambda I - K)^{-1}\|_{X \leftarrow X} \|I\|_{X \leftarrow Y} \|g\|_Y$, we obtain

$$\|(\lambda I - K)^{-1}\|_{Y \leftarrow Y} \leq \frac{1}{|\lambda|} \left(\|K\|_{Y \leftarrow X} \|(\lambda I - K)^{-1}\|_{X \leftarrow X} \|I\|_{X \leftarrow Y} + 1 \right).$$ QED

The solution f can be in Y without the assumption $K \in L(X,Y)$.

Corollary 3.5.2. Let $X = X_0 \supset X_1 \supset X_2 \supset \ldots \supset X_n = Y$ be a sequence of continuously embedded spaces. Furthermore, $(\lambda I - K)^{-1} \in L(X,X)$ and $K \in L(X_{i-1}, X_i)$ are assumed for $1 \leq i \leq n$. Then, $(\lambda I - K)^{-1} \in L(X_i, X_i)$ for all $1 \leq i \leq n$ and, in particular, (1) hold: $(\lambda I - K)^{-1} \in L(Y,Y)$.

Proof. Apply Theorem 1 n-times. QED

The generalisation in Corollary 2 is applied, e.g., in Remark 4.14 (compare also the example in Hackbusch [1,Example 16.1.3]).

4 Numerical Treatment of Fredholm Integral Equations of the Second Kind

4.1 General Considerations

In §§4.4.2-5 and §4.4.7, several possibilities for constructing the discrete analogue K_n of an operator K will be described. The general properties of such approximations will be investigated in the following section.

4.1.1 Notation of the Semidiscrete Problem

Let X be the underlying Banach space and K an operator from $L(X,X)$. The following statements remain true, even if K is not an *integral* operator. For ease of notation the norms are written without subscript, i.e., $\|f\| = \|f\|_X$ for $f \in X$. Furthermore, the symbol $\|\cdot\|$ is also used for the *operator* norm of $L(X,X)$:

$$\|A\| = \|A\|_{X \leftarrow X} \qquad \text{for } A \in L(X,X).$$

For the approximation of K, a sequence $(K_n)_{n \in \mathbb{N}}$ of operators $K_n \in L(X,X)$, $n \in \mathbb{N}$, is used. Besides K, also the function g in

$$(4.1.1) \qquad \lambda f = g + Kf \qquad (\lambda \neq 0)$$

can be approximated by $(g_n)_{n \in \mathbb{N}}$. The approximate equation reads (4.1.2a,b):

semidiscrete equation			
(4.1.2a)	$\lambda f_n = g + K_n f_n,$	$n \in \mathbb{N},$	or
(4.1.2b)	$\lambda f_n = g_n + K_n f_n,$	$n \in \mathbb{N},$	

where existence and uniqueness of $f_n \in X$ is still to be discussed. Since, in the applications, (2a/b) represents a finite-dimensional problem, (2a/b) is called the *semidiscrete* (integral-)equation, while (1) represents the *continuous* problem. The *discretisations* (2a/b) are uniquely characterised by the sequences $(K_n)_{n \in \mathbb{N}}$ or $(K_n, g_n)_{n \in \mathbb{N}}$, respectively. The term «*semi*discrete» is chosen, since the problem is posed in the same infinite-dimensional space X which was used to describe the original problem.

4.1.2 Consistency and Stability

By *consistency* we mean a suitable condition guaranteeing the proximity of the discretisation (K_n, g_n) to the original data (K, g). For most of the later applications, the following definition is appropriate.

Definition 4.1.1 (consistency). The discretisation $(K_n)_{n\in\mathbb{N}}$ is called *consistent* (in X), if

(4.1.3a) $$\lim_{n\to\infty} \| K\varphi - K_n\varphi \| = 0 \qquad \text{for all } \varphi\in X.$$

The consistency condition (3a), which we can also write in the form

(4.1.3a') $$K_n\varphi \to K\varphi \qquad \text{for all } \varphi\in X,$$

describes the *pointwise convergence* of K_n to K.

Remark 4.1.2. (a) In the case of discretisation (2b), one has also to assume:

(4.1.3b) $$\lim_{n\to\infty} \| g - g_n \| = 0.$$

The conditions (3a,b) are equivalent to

(4.1.3c') $$g_n + K_n\varphi \to g + K\varphi \text{ in } X \qquad \text{for all } \varphi\in X.$$

(b) The consistency condition (3a) is equivalent to the pointwise convergence $K_n\varphi \to K\varphi$ for all $\varphi\in M$ from a dense subset M of X together with the uniform boundedness

(4.1.3c) $$\sup\{ \| K_n \| : n\in\mathbb{N} \} < \infty.$$

(c) A sufficient condition for (3a) is convergence $\lim_{n\to\infty} K_n = K$ in the sense of the operator norm:

(4.1.4) $$\| K - K_n \| \to 0 \qquad \text{for } n\to\infty.$$

Proof. (b) Apply the Theorem of Banach-Steinhaus (Theorem 1.3.17).
(c) (4) implies $\| K\varphi - K_n\varphi \| \le \| K - K_n \| \, \| \varphi \| \to 0$. QED

The consistency condition (3a) does not guarantee solvability of the semi-discrete problem (2a). Even if $\lambda I - K_n$ has an inverse, the solution f_n of (3a) need not converge to f. The inverse $(\lambda I - K_n)^{-1}$ must remain uniformly bounded for all $n\in\mathbb{N}$. This property is called «stability» (here, λ is *fixed*).

Definition 4.1.3 (stability). The discretisation $(K_n)_{n\in\mathbb{N}}$ is called *stable* (in X), if there are $n_0\in\mathbb{N}$ and a constant C such that, for all $n\ge n_0$, $\lambda I - K_n$ has an inverse $(\lambda I - K_n)^{-1}\in L(X,X)$ satisfying

(4.1.5) $$\| (\lambda I - K_n)^{-1} \| \le C \qquad \text{for all } n\ge n_0.$$

The next theorem shows that condition (4), which is already sufficient for the consistency property (3a), also guarantees (5), provided that $\lambda I - K$ has an inverse (d.h. $\lambda \notin \sigma(K)$).

Theorem 4.1.4. Let $(\lambda I-K)^{-1} \in L(X,X)$. Further, assume (4): $\|K-K_n\| \to 0$. Then the discretisation is stable. More precisely, the estimate

(4.1.6a) $$\|(\lambda I-K_n)^{-1}\| \leqslant \|(\lambda I-K)^{-1}\| / [1-\|(\lambda I-K)^{-1}\| \, \|K-K_n\|]$$

holds for all n with the property

(4.1.6b) $$\|K-K_n\| < 1/\|(\lambda I-K)^{-1}\|.$$

Proof. Because of convergence (4), inequality (6b) holds for sufficient large $n \geqslant n_0$. (6a) follows from Lemma 1.3.14 with $S := \lambda I - K_n$ and $T := \lambda I - K$. QED

Theorem 4 can be reversed as follows.

Corollary 4.1.5. Let $K, K_n \in L(X,X)$, $n \in \mathbb{N}$. For sufficient large $n \geqslant n_0$ assume

(4.1.7a) $$\|K-K_n\| < 1/\|(\lambda I-K_n)^{-1}\|.$$

Then $\lambda I - K$ is also invertible and satisfies the estimate

(4.1.7b) $$\|(\lambda I-K)^{-1}\| \leqslant \|(\lambda I-K_n)^{-1}\| / [1-\|(\lambda I-K_n)^{-1}\| \, \|K-K_n\|]$$

for $n \geqslant n_0$. The operator norm convergence (4) together with the stability (5) is sufficient for (7a).

Proof. Apply Lemma 1.3.14 with $S := \lambda I - K$ and $T := \lambda I - K_n$. QED

Exercise 4.1.6. For bijective operators $A \in L(X,X)$, the condition number (with respect to X) is defined by

(4.1.8) $$\mathrm{cond}_X(A) := \|A\| \, \|A^{-1}\|$$

(cf. Definition 1.4.29). For $K, K_n \in L(X,X)$ with invertible $\lambda I - K$ and $\alpha := \|(\lambda I-K)^{-1}\| \, \|K-K_n\| < 1$ prove that

(4.1.9) $$\mathrm{cond}_X(\lambda I-K_n) \leqslant [\mathrm{cond}_X(\lambda I-K) + \alpha] / (1-\alpha).$$

Further remarks concerning the condition number $\mathrm{cond}_X(\lambda I-K_n)$ can be found in §4.1.6.

4.1.3 Convergence

The obvious definition of convergence of a discretisation reads as follows.

Definition 4.1.7 (convergence). The discretisation $\{K_n\}$ is called *convergent* (in X), if $n_0 \in \mathbb{N}$ exists so that for all $g \in X$ the semidiscrete equation (2a) for $f_n \in X$ is uniquely solvable for all $n \geqslant n_0$ and the solution f_n has the property that $\lim_{n\to\infty} f_n$ exists (in X).

The convergence from Definition 7 does not prescribe the limit of the f_n. The latter is determined by the consistency condition connecting K_n and K.

Lemma 4.1.8. Let the discretisation $\{K_n\}$ be consistent and convergent. Then the limit $f := \lim_{n\to\infty} f_n$ (which depends on $g \in X$) satisfies equation (1), i.e., $\lambda f = Kf + g$. In particular, $\lambda I - K$ is *surjective*.

Proof. Consistency (3a) implies uniform boundedness of the norms $\|K_n\|$ by a constant C (cf. Remark 2b). Split $K_n f_n - Kf$ into $K_n(f_n - f) + (K_n - K)f$. The first term tends to zero, since convergence implies $\|K_n(f_n - f)\| \leq C\|f_n - f\| \to 0$. The second term vanishes for $n \to \infty$ because of the consistency (3a). Hence, we can perform the limit in (2a), $\lambda f_n = g + K_n f_n$, and obtain the result (1): $\lambda f = g + Kf$. Since for each $g \in X$ there exists a solution f, the operator $\lambda I - K$ is surjective. ∎

Exercise 4.1.9. **(a)** Let the discretisation $\{K_n\}$ be consistent and convergent. Consider equation (2b) with g_n satisfying (3b). Then there is an $n_0 \in \mathbb{N}$ so that the semidiscrete equation (2b) has a unique solution $f_n \in X$ for all $n \geq n_0$ and $\lim_{n\to\infty} f_n$ exists.
(b) If K is compact, then both, consistency and convergence yield

(4.1.10) $$(\lambda I - K)^{-1} \in L(X, X).$$

The next lemma proves the *injectivity* of the operators $\lambda I - K$.

Lemma 4.1.10. Let the discretisation $(K_n)_{n\in\mathbb{N}}$ be stable and consistent. Then $\lambda I - K$ is injective and $(\lambda I - K)^{-1} \in L(\text{image}(\lambda I - K), X)$.

Proof. The assertion holds if $\|(\lambda I - K)\varphi\| \geq \eta\|\varphi\|$ with $\eta > 0$ for all $\varphi \in X$. For the indirect proof assume that there is a sequence $\varphi_n \in X$ with $\|\varphi_n\| = 1$ and $\psi_n := (\lambda I - K)\varphi_n$ satisfies $\|\psi_n\| \leq 1/n$. Because of consistency, $K_m\varphi_n$ converges to $K\varphi_n$ for $m \to \infty$ and fixed n. Hence, an index $m = m_n$ exists so that $\|K_m\varphi_n - K\varphi_n\| \leq 1/n$. $\zeta_n := (\lambda I - K_m)\varphi_n = \psi_n - (K_m\varphi_n - K\varphi_n)$ can be estimated by $\|\zeta_n\| \leq 2/n$. From stability (5) and the representation $\varphi_n = (\lambda I - K_m)^{-1}\zeta_n$ we conclude that $1 = \|\varphi_n\| \leq C\|\zeta_n\| \leq 2C/n$. The contradiction follows for $n > 2C$. ∎

4.1.4 Stability and Convergence Theorem

While convergence (as in Definition 7) seems to be an obvious requirement of the approximation scheme (2a), it is not so obvious why stability (5) should be required. The following theorem shows that stability is necessary.

Theorem 4.1.11 (stability theorem). **(a)** *Convergence implies stability.* **(b)** Convergence *and* consistency yield *existence of the inverse operator* $(\lambda I - K)^{-1} \in L(X, X)$ as well as stability.

Proof. (a) Because of convergence, the operators $T_n := (\lambda I - K_n)^{-1}$ satisfy Remark 1.3.16 for $n \geq n_0$ and, therefore, they are uniformly bounded for all $n \geq n_0$.

(b) According to Lemma 8, $\lambda I - K$ is surjective. By Part (a) of this proof stability holds and Lemma 10 guarantees injectivity of $\lambda I - K$. A bijective operator has an inverse in $L(X,X)$ (cf. Theorem 1.3.13). QED

More important than the statement of Theorem 11 would be the reverse statement ensuring convergence of the approximations f_n to the desired solution of equation (1). For this purpose, we have to add minor assumptions about the solvability of equation (1) besides stability and consistency.

Theorem 4.1.12 (convergence theorem). **(a)** Assume stability (5) and consistency (3a). Further, assume that either (i) $\lambda I - K$ is surjective or (ii) $\lambda \neq 0$ and K compact. Then the discretisation (2a) is *convergent* to the solution of Equation (1).
(b) If, in addition, (3b) holds, the convergence statement $f_n \to f$ is also valid for discretisation (2b).

Proof. (i) Lemma 10 guarantees injectivity of $\lambda I - K$. Assuming surjectivity, one obtains $(\lambda I - K)^{-1} \in L(X,X)$ (cf. step (b) in the proof of Theorem 11). For a compact operator K, injectivity and surjectivity coincide, provided that $\lambda \neq 0$ (cf. Theorem 1.3.28b). Hence, in all cases, $\lambda I - K$ is surjective and $(\lambda I - K)^{-1} \in L(X,X)$ is guaranteed.
(ii) Let $g \in X$ be arbitrary and set $f := (\lambda I - K)^{-1} g$. Define

$$(4.1.11a) \qquad d_n := \lambda f - K_n f - g_n,$$

where $g_n := g$ holds in the case of discretisation (2a). Consistency condition (3a) and (where necessary) (3b) imply $d_n \to 0$. (11a) can be rewritten as

$$(4.1.11b) \qquad (\lambda I - K_n) f = g_n + d_n.$$

Subtracting equation (2b): $(\lambda I - K_n) f_n = g_n$, one obtains

$$(4.1.11c) \qquad (\lambda I - K_n)(f - f_n) = d_n$$

and $f - f_n = (\lambda I - K_n)^{-1} d_n$ for sufficiently large n. Stability leads to the estimate

$$(4.1.11d) \qquad \|f - f_n\| \leqslant C \|d_n\| \qquad \text{with } C \text{ from (5).}$$

Inequality (11d) together with $d_n \to 0$ ensures the convergence $f_n \to f$. QED

Assuming the stronger operator norm convergence $K_n \to K$ from (4), all desired properties follow, as shown in the next theorem.

Theorem 4.1.13. Assume (4): $\|K_n - K\| \to 0$ and let λ be a regular value of K: $(\lambda I - K)^{-1} \in L(X,X)$. Then the discretisation (2a) is *consistent, stable*, and *convergent*.

Proof. Remark 2b and Theorem 4 guarantee consistency and stability. Assumption (i) of Theorem 12 is satisfied and yields convergence. QED

4.1.5 Error Estimates

The practical usefulness of f_n as an approximation to f will depend on the *speed of convergence*. If the convergence can be arbitrarily slow (cf. Remark 1.3.18), the approximation of f by f_n may not be practicable. Since, in practice, the index n coincides with the dimension of the finite-dimensional problem (2a/b), n has an upper bound determined by the available computing time and storage. This situation leads to another question: Which error bound of $\|f-f_n\|$ can be guaranteed for a *fixed* n? The assertions of the following lemma hold for a fixed n. Note that by §4.1.4 its hypotheses can be satisfied for a sufficiently large $n \geq n_0$.

Lemma 4.1.14. For a fixed n assume $(\lambda I - K_n)^{-1} \in L(X,X)$ and $(\lambda I - K)^{-1} \in L(X,X)$. Let f and f_n be the solutions of equations (1) or (2b), respectively. Then the error $f-f_n$ has the representations

(4.1.12a) $$f - f_n = (\lambda I - K_n)^{-1}[(K-K_n)f + g - g_n],$$

(4.1.12b) $$f - f_n = (\lambda I - K)^{-1}[(K-K_n)f_n + g - g_n]$$

and fulfils the *error estimates* (12c-e):

(4.1.12c) $$\|f-f_n\| \leq \|(\lambda I - K_n)^{-1}\| \, \|(K-K_n)f + g - g_n\|,$$

(4.1.12d) $$\|f-f_n\| \leq \|(\lambda I - K_n)^{-1}\| \, [\|(K-K_n)f\| + \|g-g_n\|],$$

(4.1.12e) $$\|f-f_n\| \leq \|(\lambda I - K_n)^{-1}\| \, [\|K-K_n\| \, \|f\| + \|g-g_n\|].$$

Proof. (11c) yields $f - f_n = (\lambda I - K_n)^{-1} d_n$ with $d_n := \lambda f - K_n f - g_n$ from (11a). Replacing λf by $g + Kf$, we obtain $d_n = (K-K_n)f + g - g_n$ proving (12a). The proof of (12b) is similar. (12c-e) are an immediate consequence of (12a). ∎

4.1.6 Condition Numbers

The sensitivity of the semidiscrete equations $\lambda f_n = g_n + K_n f_n$ with respect to perturbations of g_n or K_n is described by the *condition numbers*

(4.1.13) $$\mathrm{cond}_X(\lambda I - K_n) = \|\lambda I - K_n\| \, \|(\lambda I - K_n)^{-1}\|$$

(cf. Theorem 1.4.30, which also holds for operators from $L(X,X)$). For the numerical purposes it is essential that these numbers are not too large. In particular, is must be guaranteed that the condition number does not become arbitrarily large as $n \to \infty$. Fortunately, consistency and stability are sufficient for the uniform boundedness of $\mathrm{cond}_X(\lambda I - K_n)$.

Theorem 4.1.15. Let the discretisation $\{K_n\}$ be consistent and stable. Then there are an index n_0 and a constant C such that

$$\mathrm{cond}_X(\lambda I - K_n) \leq C \qquad \text{for all } n \geq n_0.$$

Proof. By means of the Theorem of Banach-Steinhaus (Theorem 1.3.17), the consistency (3a) implies the uniform boundedness of $\|\lambda I - K_n\|$. Because of stability, we have $\|(\lambda I - K_n)^{-1}\|$ defined for all $n \geq n_0$ with suitable n_0 and remaining uniformly bounded (with respect to n). ∎

We recall that thanks to the stability theorem 11, the assumption of stability can be replaced by convergence.

The condition number (13) is not the final quantity describing the numerical behaviour. To solve the problem in practice, the semidiscrete equation $\lambda f_n = g_n + K_n f_n$ must be converted into a linear system of n equations of n unknowns. An improper choice of the equations could lead to a system with a worse condition number than $\mathrm{cond}_X(\lambda I - K_n)$. This question will be taken up in Sections 4.2.7, 4.4.4, 4.5.5.

4.2 Discretisation by Kernel Approximation

4.2.1 Degenerate Kernels

We have already encountered the use of degenerate kernels in the compactness proof (§3.3.2). Now we ask whether degenerate kernels are useful for practical computation. We recall Definition 3.3.3: A kernel $k_n(x,y)$ is called *degenerate* if it is of the form

$$k_n(x,y) = \sum_{j=1}^{n} a_j(x) b_j(y) \qquad \text{with } a_j = a_{j,n} \text{ and } b_j = b_{j,n}, \tag{4.2.1}$$

where, in the most general case, $a_j \in X$ and $b_j \in X'$ must hold, in order to ensure that the associated integral operator K_n belongs to $L(X,X)$. The notation $a_{j,n}$ and $b_{j,n}$ should indicate that these functions may vary with the «discretisation parameter» n. As stated in Exercise 3.3.4, the following holds.

Remark 4.2.1. (a) Let $K_n \in L(X,X)$ have the kernel (1). Then, obviously,

$$\mathrm{range}(K_n) = \mathrm{span}\{a_1, a_2, \ldots, a_n\} \tag{4.2.2}$$

is finite-dimensional and therefore K_n is compact: $K_n \in K(X,X)$.
(b) If $k_n(x,\cdot)$ converges uniformly in $L^1(D)$ for all $x \in D$ to $k(x,\cdot)$, then K_n converges with respect to the operator norm of $L(C(D), C(D))$ to the integral operator K with the kernel k:

$$\|K - K_n\|_{C(D)\leftarrow C(D)} = \sup\{\|k_n(x,\cdot) - k(x,\cdot)\|_{L^1(D)} : x \in D\} = \sup_{x \in D} \int_D |k_n(x,y) - k(x,y)|\,dy. \tag{4.2.3}$$

Remark 4.2.2. Without loss of generality, we may assume that the functions $\{a_1, a_2, \ldots, a_n\}$ as well as $\{b_1, b_2, \ldots, b_n\}$ are linearly independent. Hence, n is the dimension of the image space of K_n.

Proof. If, e.g., $a_n(x) = \sum_{j=1}^{n-1} \alpha_j a_j(x)$, one can insert this representation into (1) to obtain $k_n(x,y) = \sum_{j=1}^{n-1} a_j(x)[b_j(y) + \alpha_j b_n(y)]$. The number of terms in this sum can be reduced further if the functions are still linearly dependent. ∎

The *discretisation by kernel approximation* leads us to the equation

(4.2.4) $\lambda f_n = g + K_n f_n,$ where K_n has the kernel (1).

Hence, we obtain a special case of equation (1.2a). Theorem 1.13 and equation (3) yield

Corollary 4.2.3. Under the assumption $\sup_{x\in D}\int_D |k_n(x,y)-k(x,y)|\,dy \to 0$, the discretisation is *consistent* (with respect to $X=C(D)$). Further, it is *stable* and *convergent*, if λ is a regular value of K, i.e., $(\lambda I-K)^{-1}\in L(X,X)$.

4.2.2 Setting Up the System of Equations

Thanks to Remark 2, in the following the functions $\{a_1, a_2, \dots, a_n\}$ may be assumed to be linearly independent and, therefore, they can be taken as a basis of the image space of K_n. If equation (4) has a solution f_n, then $K_n f_n \in \mathrm{image}(K_n)$ is a linear combination of the form $K_n f_n = \sum \gamma_j a_j$ with $\gamma_j := \int_D b_j(y) f_n(y)\,dy$. Inserting this representation into (4), one obtains

$$\lambda f_n = g + \sum_{j=1}^{n} \gamma_j a_j.$$

Dividing by λ ($\lambda \neq 0$, since the equation is *of the second kind*) and setting $\alpha_j := \gamma_j/\lambda$, one obtains following statement.

Lemma 4.2.4. If (4) has a solution f_n, it admits the representation

(4.2.5) $$f_n = \frac{1}{\lambda} g + \sum_{j=1}^{n} \alpha_j a_j .$$

The *numerical task* consists in determining the coefficients α_j in (5). We insert the ansatz (5) into (4) and obtain

$$g+\lambda\sum_{j=1}^{n}\alpha_j a_j \underset{(5)}{=} \lambda f_n \underset{(4)}{=} g+K_n f_n \underset{(5)}{=} g+K_n\Big(\frac{1}{\lambda}g+\sum_{k=1}^{n}\alpha_k a_k\Big) =$$
$$= g+\frac{1}{\lambda}K_n g + \sum_{k=1}^{n}\alpha_k K_n a_k.$$

The function g cancels on both sides. $\frac{1}{\lambda}K_n g$ is represented by

(4.2.6a) $$\frac{1}{\lambda}(K_n g)(x) = \frac{1}{\lambda}\int_D \sum_{j=1}^{n} a_j(x) b_j(y) g(y)\,dy = \sum_{j=1}^{n} \beta_j a_j(x) \qquad \text{with}$$

(4.2.6b) $$\beta_j := \frac{1}{\lambda}\int_D b_j(y) g(y)\,dy \qquad \text{for } 1\le j\le n.$$

Analogously, we obtain the following representation for $K_n a_k$:

(4.2.7a) $$(K_n a_k)(x) = \int_D \sum_{j=1}^{n} a_j(x) b_j(y) a_k(y)\,dy = \sum_{j=1}^{n} \beta_{jk} a_j(x) \qquad \text{with}$$

(4.2.7b) $$\beta_{jk} := \int_D b_j(y) a_k(y)\,dy \qquad (1\le j,k\le n).$$

Together, we have

$$\lambda\sum_{j=1}^{n}\alpha_j a_j = \frac{1}{\lambda}K_n g + \sum_{k=1}^{n}\alpha_k K_n a_k = \sum_{j=1}^{n}\beta_j a_j + \sum_{k=1}^{n}\alpha_k \sum_{j=1}^{n}\beta_{jk} a_j =$$
$$= \sum_{j=1}^{n}\Big\{\beta_j + \sum_{k=1}^{n}\beta_{jk}\alpha_k\Big\} a_j .$$

Because of the linear independence of the a_j, we can equate coefficients to obtain the following equation:

(4.2.8) $$\lambda \alpha_j = \beta_j + \sum_{k=1}^{n} \beta_{jk} \alpha_k \qquad \text{for } 1 \leq j \leq n.$$

In matrix notation, the *system of equations* to be solved reads as follows:

(4.2.9a) $$(\lambda I - B) a = b \quad \text{or} \quad (\lambda I - B_n) a_n = b_n$$

with

(4.2.9b) $$a := a_n := \begin{bmatrix} \alpha_1 \\ \vdots \\ \alpha_n \end{bmatrix}, \quad b := b_n := \begin{bmatrix} \beta_1 \\ \vdots \\ \beta_n \end{bmatrix} \qquad (\alpha_j \text{ from (5)}, \beta_j \text{ from (6b)}),$$

(4.2.9c) $$B := B_n := \begin{bmatrix} \beta_{11} & \beta_{12} & \cdots & \beta_{1n} \\ \vdots & \vdots & & \vdots \\ \beta_{n1} & \beta_{n2} & \cdots & \beta_{nn} \end{bmatrix} \qquad \text{with } \beta_{jk} \text{ from (7b)}.$$

System (9a) and discretisation (4) are equivalent in the following sense.

Theorem 4.2.5. Let $\lambda \neq 0$. Suppose the functions $\{a_1, \ldots, a_n\}$ appearing in (1) are linearly independent.
(a) The matrix $\lambda I - B$ is non-singular if and only if λ is a regular value of K_n, i.e., if $(\lambda I - K_n)^{-1} \in L(X, X)$.
(b) A solution a of the system (9a-c) yields a solution f_n of (4): $\lambda f_n = g + K_n f_n$ via the defining equation (5).
(c) Suppose in addition that f_n is the solution of (4). Then the coefficient vector a from the representation (5) of f_n solves the system (9a-c).

Proof. Part (c) was already shown, when we set up the system of equations (9a-c). The conclusions can be reversed and prove Part (b). Since the functions a_j are linearly independent, the correspondence of a and f_n by (5) is one-to-one. Following (b) and (c), the unique solvability of the equations (9a-c) and (4) are equivalent. On the other hand, the unique solvability of (9a-c) or (4), respectively, is equivalent to the non-singularity of $\lambda I - B$ or, respectively, to the existence of $(\lambda I - K_n)^{-1} \in L(X, X)$. QED

Exercise 4.2.6. If the functions a_j are linearly dependent, the matrix $\lambda I - B$ must be singular.

4.2.3 Kernel Approximation by Interpolation

As seen in Exercise 1.4.7, the interpolant of $\varphi \in C(D)$ can be described by its Lagrange representation (1.4.7), i.e., $\sum \varphi(x_j) L_j(x)$. Since the Lagrange functions are linearly independent, they are candidates for the system of functions $\{a_1, \ldots, a_n\}$. Interpolation of $k(\cdot, y)$ with respect to x yields the approximation

(4.2.10a) $$k_n(x, y) = \sum_{j=1}^{n} a_j(x) k(x_j, y) \qquad (a_j = L_j\text{: Lagrange function}).$$

The functions b_j are given by $k(x_j, \cdot)$. The coefficients of the system, which are to be computed, are

(4.2.10b) $$\beta_j = \frac{1}{\lambda} \int_D k(x_j, y) g(y) dy, \quad \beta_{jk} = \int_D k(x_j, y) a_k(y) dy \qquad (1 \leq j, k \leq n).$$

4.2.4 Tensor Approximation of k

A further, practically convenient approximation of the kernel k is described by

(4.2.11a) $$k_n(x,y) = \sum_{k=1}^{n} \sum_{j=1}^{n} c_{jk} a_j(x) b_k(y)$$

with coefficients $c_{jk}=c_{jk,n}$ depending on n and a linearly independent system $\{a_1, a_2, \ldots, a_n\}$, where $a_j=a_{j,n}$, $b_k=b_{k,n}$. The coefficients c_{jk} from (11a) form the $n \times n$ matrix

(4.2.11b) $$C := C_n := (c_{jk,n})_{j,k=1,\ldots,n}.$$

In particular, a representation of k_n of the form (11a) arises, when k is approximated by a twofold interpolation with respect to x and y (cf. Lemma 10).

Since the double sum in (11a) has n^2 terms, while only n linearly independent a_j exist, the representation (11a) is not a special case of (1). However, the range of the operator K_n with the kernel k_n again satisfies (2), so that the ansatz (5) for the solution f_n is still valid. The reader may repeat the arguments of section 4.2.2 and thus prove the following analogous result.

Theorem 4.2.7. The *system of equations* for determining the coefficient vectors $a=a_n$ for the representation (5) of f_n reads

(4.2.11c) $$(\lambda I - B')a = b' \quad \text{with} \quad B' = B_n' := CB \quad \text{and} \quad b' = b_n' := Cb,$$

where B and b are the quantities defined in (9b,c). Under the same assumptions as in Theorem 4, the solvability of problem (4) for all $g \in X$ coincides with the non-singularity of the matrix $\lambda I - B'$. The corresponding solutions are connected via (5).

Note that in the case of the approximation of k by k_n according to (11a), the functions a_j and b_j can be chosen independently of k. To obtain a good approximation, one has to choose suitable coefficients c_{jk} (i.e., a suitable matrix C). In the case of (1), the quality of the approximation hinges upon the definition of a_j and b_j.

The advantages of the tensor approximation (11a) compared with the approximation (10a) described in §4.2.3 are summarised in

Remark 4.2.8. The matrix C is explicitly given by (11b). For computing the matrix B, one only need to perform the quadratures in $\beta_{jk} = \int b_j(y) a_k(y) dy$, which depend on the choice of the systems of functions a_j, b_j but not on k as in the case (10a). Because of the usually smaller supports of b_j, the computation of $\beta_j := \frac{1}{\lambda} \int_D b_j(y) g(y) dy$ (cf. (6b)) is much more convenient than the evaluation of $\beta_j = \frac{1}{\lambda} \int_D k(x_j, y) g(y) dy$ which are required in (10b).

4.2.5 Examples of Kernel Approximations

For theoretical purposes, *polynomials* $k_n(x,y)=\sum c_{ij}x^i y^j$ play an important role. However, for numerical purposes they are often *unsuitable*, particularly, when they are of a high degree. Often, the polynomials used in theory are those from the approximation theorem of Weierstraß. They are not available for numerical purposes. Special polynomials as, e.g., Taylor's polynomials of the function $k(x,y)$ or interpolatory polynomials, often lead to numerical instabilities. Furthermore, the monomials $a_i(x)=x^i$ and $b_j(y)=y^j$ are functions with a global support: $\mathrm{supp}(a_i) = \mathrm{supp}(b_j) = D$.

Remark 4.2.9. If the integrals in (6b) and (7b) determining the coefficients β_j and β_{jk} are not explicitly known, one has to approximate them by *numerical quadrature*. The quadrature has only to be performed over the *support* of the integrand. Usually, the amount of computational work of the quadrature formula can be reduced if the size of the integration domain is mad smaller. Furthermore, small supports of a_k and b_j appearing in formula (7b) for β_{jk} might imply that the supports are disjoint and, therefore, the product $a_k b_j$ and as well as β_{jk} vanish.

The indicated difficulties can be circumvented by choosing a system $a_j = b_j$ of *orthogonal functions*. In the case of $I=[0,2\pi]$, one can use the trigonometric functions $\{\frac{1}{2}$, sin x, cos x, sin 2x, cos 2x,...$\}$. Then, the integrals (7b) for β_{jk} take the value π for $j=k$ and zero, otherwise. The computation of the coefficients c_{jk} and β_j requires a Fourier analysis of $k(x,y)$ and g. One may argue that the convergence $k_n \to k$ for non-doubly periodic kernels $k(x,y)$ may be slow. Furthermore, no orthogonal systems are available for general domains $D \subset \mathbb{R}^d$.

In consideration of Remark 9, *piecewise linear interpolation* is an obvious choice. For $D=I=[0,1]$ and $n\in\mathbb{N}$ choose equidistant support abscissae $\xi_j := j/n$ $(0 \leqslant j \leqslant n)$. The desired functions $a_j = b_j$ are the Lagrange functions (cf. Def. 1.4.4), which for the piecewise linear interpolation (cf. Example 1.4.5) take the shape

$$(4.2.12a) \qquad a_j(x) = b_j(x) = \begin{cases} 0 & \text{for } |x-j/n| > 1/n, \\ 1-n|x-j/n| & \text{for } |x-j/n| \leqslant 1/n. \end{cases}$$

The function interpolating $k(x,\cdot)$ with respect to y for fixed x is $\sum_m k(x,m/n)\,b_m(y)$ (cf. Lemma 1.4.3). Interpolating again with respect to x, one obtains the *tensor product interpolation*

$$k_n(x,y) = \sum_{j=1}^{n}\sum_{m=1}^{n} k(j/n,m/n)\,a_j(x)\,b_m(y).$$

This proves the first part of the following Lemma. The reader may check (12c). (12d) is analogous to Remark 1.4.11 (although more complicated).

Lemma 4.2.10. (a) In the case of piecewise linear interpolation, the coefficients c_{jm} from (11a) are

$$(4.2.12b) \qquad c_{jm} := k(j/n,m/n) \qquad (0 \leqslant j,m \leqslant n).$$

In contrast to (5), the index range is $0,1,\dots,n$ (instead of $1,\dots,n$).

(b) The support of the $a_j(x) = b_j(x)$ are the intervals $[(j-1)/n, (j+1)/n] \cap I$. The matrix B is tridiagonal. Its coefficients $\beta_{jk} := \int_D b_j(y)\, a_k(y)\, dy$ are

(4.2.12c) $$\beta_{00} = \beta_{nn} = \tfrac{1}{3n}, \quad \beta_{jj} = \tfrac{2}{3n} \ (1 \leqslant j < n), \quad \beta_{j,j\pm 1} = \tfrac{1}{6n}, \quad \beta_{jk} = 0 \ \text{ for } |j-k| \geqslant 2.$$

(c) If $k \in C^2(I \times I)$ or $k \in C_L^1(I \times I)$, then

(4.2.12d) $$\| K - K_n \|_{C(I) \leftarrow C(I)} \leqslant \| k - k_n \|_\infty \leqslant 2n^{-2} \| k \|_{C_L^1(I \times I)}.$$

Estimate (12d) shows convergence of the order 2, which is confirmed by the numerical results reported in §4.2.8. To attain a higher order, e.g., order 4, one may use piecewise cubic Hermite interpolation (cf. §2.3, (2.3.3) and (2.3.9)) or cubic spline interpolation. The latter ansatz is, e.g., described by Hämmerlin-Schumaker [1]. Higher convergence rates are obtained by the «blending splines» (cf. Bamberger-Hämmerlin [1]).

4.2.6 A Variant of the Kernel Approximation

Instead of approximating only the kernel k by (11a), we can also replace the inhomogeneity g by

(4.2.13a) $$g_n := \sum_{j=1}^{n} \gamma_j a_j,$$

resulting in the equation

(4.2.13b) $$f_n = g_n + K_n f_n \quad \text{with } K_n \text{ as in (4)} \qquad \text{(cf. (1.2b))}.$$

Remark 4.2.11. If $g_n \to g$ in X, then all convergence, consistency, and stability properties of the discretisation (4) carry over to (13b).

The ansatz (5) for f_n now becomes simpler:

(4.2.13c) $$f_n := \sum_{j=1}^{n} \hat{\alpha}_j a_j \qquad \text{with } \hat{\alpha}_j := \hat{\alpha}_{j,n} := \alpha_j + \tfrac{1}{\lambda} \gamma_j.$$

The coefficients $\hat{\alpha}_j$ and γ_j form the vectors

(4.2.13d) $$\hat{\alpha} := \hat{\alpha}_n := (\hat{\alpha}_j)_{j=1,\dots,n}, \qquad c := c_n := (\gamma_j)_{j=1,\dots,n}.$$

Exercise 4.2.12. For determining the coefficients $\hat{\alpha}_j$ from (13c), one has to solve the following system of equations:

(4.2.14) $$(\lambda I - B')\, \hat{\alpha} = c \qquad \text{with } B' = C_n B_n \text{ from (11b)}.$$

4.2.7 Analysis of the System of Equations

In constructing the data for the system (11c) or (14), one has to expect perturbations of the right-hand side and the matrix. One reason may be rounding errors of the input data or the floating point arithmetic. However, the (usually unavoidable) quadrature errors committed during the computation of the values β_j in (6b) can also be regarded as perturbation of the right-hand side of the linear system. From Section 1.4.3, we know that the *condition number* of the matrix of the corresponding system of equations (9a), (11c), or (14), respectively, is responsible for the amplification of the input error.

The following positive result can be obtained, when we introduce the following vector norm (15) in $\mathbb{R}^n$ specially tailored for the problem:

$$(4.2.15) \qquad |||a||| := \|\sum_{j=1}^{n} \alpha_j a_j\|_X \qquad \text{for } a=(\alpha_j)\in\mathbb{R}^n \text{ with } a_j \text{ from (11a).}$$

Exercise 4.2.13. Let $\{a_1,\dots,a_n\}$ be linearly independent. Prove that (15) is a norm in $\mathbb{R}^n$.

The definition of $|||\cdot|||$ reflects the fact that the coefficients α_j are less interesting than $\Sigma\alpha_j a_j \in X$. The vector norm $|||\cdot|||$ induces a matrix norm $|||\cdot|||$ and, thereby, a condition number (1.4.22), which we denote by $\mathrm{cond}_{|||}(A)$.

Theorem 4.2.14. The $|||\cdot|||$-condition number of the matrices $\lambda I - C_n B_n$ from (11c) and (14) satisfy equation (16) below where K_n is given by (4):

$$(4.2.16) \qquad \mathrm{cond}_{|||}(\lambda I - C_n B_n) = \mathrm{cond}_X(\lambda I - K_n).$$

Proof. (i) We use equation (14), i.e., $(\lambda I - B')\hat{a} = c$, since then the identities $|||\hat{a}||| = \|f_n\|_X$ and $|||c||| = \|g_n\|_X$ hold (f_n, g_n from (13b)).
(ii) Let $c=(\gamma_j)\in\mathbb{R}^n$ be arbitrary and f_n solution of (13b) with $g_n := \sum_j \gamma_j a_j$. Combining $|||\hat{a}||| = \|f_n\|$ and $|||c||| = \|g_n\|$ with the inequalities $\|f_n\| \leqslant \|(\lambda I - K_n)^{-1}\| \|g_n\|$ and $|||\hat{a}||| \leqslant |||(\lambda I - C_n B_n)^{-1}||| \, |||c|||$ (cf. Exercise 12), one obtains $\|(\lambda I - K_n)^{-1}\| = |||(\lambda I - C_n B_n)^{-1}|||$.
(iii) Analogously, we can also show $|||\lambda I - C_n B_n||| = \|\lambda I - K_n\|$. Together, we have the claimed estimate (16). QED

Following Theorem 1.15 and Corollary 3, the right-hand side in (16) is uniformly bounded. Hence, the systems (11c) and (14) are well-conditioned for all $n \geqslant n_0$, provided that we use the norm $|||\cdot|||$.

The following exercises show that the norm $|||\cdot|||$ may coincide with well-known vector norms.

Exercise 4.2.15. **(a)** Define the discretisation and the system of equations as in Lemma 10. The underlying Banach space is $X=C(I)$. Prove that $|||\cdot|||$ is the maximum norm:

$$(4.2.17a) \qquad |||\cdot||| = \|\cdot\|_\infty .$$

(b) Assume that the functions $a_j = b_j$ involved in the kernel approximation (11a) are *orthonormal* and choose $X=L^2(D)$ as underlying Banach space. Prove that $|||\cdot|||$ is the Euclidean norm $\|\cdot\|_2$:

$$(4.2.17b) \qquad |||\cdot||| = \|\cdot\|_2 .$$

The result in Theorem 14 may be be considered somewhat surprising, since it says that in the case of an «almost linearly dependent» system $\{a_1,\dots,a_n\}$, the condition number of the associated matrix has the same bound as for a «reasonably» chosen basis $\{a_1,\dots,a_n\}$. In fact this is not true, as we now show.

Rounding errors, which may originate from

(i) the rounding of the exact coefficients $c_{jk}, \alpha_{jk}, \beta_j$,

(ii) floating point errors during the computation of the matrix $\lambda I - C_n B_n$ or of the vector $b' = Cb$,

(iii) floating point errors during the solution of the system (11c) or (14),

can be estimated by means of vector norms (e.g., $\|\cdot\|_\infty$, $\|\cdot\|_2$, $\|\cdot\|_F$, cf. Wilkinson [1]), suited to the origin of the errors, but not (primarily) by $|||\cdot|||$. For the comparison of the norms one needs estimates of the form

(4.2.18a) $\|\cdot\| \leqslant C_1 |||\cdot|||, \qquad |||\cdot||| \leqslant C_2 \|\cdot\|,$

where $\|\cdot\|$ is one of the mentioned norms $\|\cdot\|_\infty$, $\|\cdot\|_2$, $\|\cdot\|_F$. For the rounding error analysis, the product

(4.2.18b) $C_0 := C_{0,n} := C_1 C_2$ (C_1, C_2 minimal constants from (18a))

can be considered as a condition number.

Exercise 4.2.16. Let $\|\cdot\|$ and $|||\cdot|||$ denote the *vector norms* as well as the *matrix norms* and assume (18a,b). Prove that

(4.2.18c) $\|A\| \leqslant C_0 |||A||| \quad$ and $\quad |||A||| \leqslant C_0 \|A\| \qquad$ for matrices A.

If, e.g., the numerical computation of $A_n := \lambda I - C_n B_n$ yields $\tilde{A}_n = A_n + \Delta$, where the perturbation Δ can be estimated with respect to the $\|\cdot\|$-matrix norm by $\|\Delta\| \leqslant \eta \|A_n\|$, Exercise 16 admits the estimate (in the $|||\cdot|||$-matrix norm) by $|||\Delta||| \leqslant C_0^2 \eta |||A_n|||$. According to Theorem 1.4.30b, the relative error of the solution $\hat{a}$ of system (14) induced by Δ is bounded by $C_0^2 \operatorname{cond}_{|||}(\lambda I - C_n B_n)\eta + O(\eta^2)$. This elucidates the error amplification by the factor C_0.

Exercise 4.2.17. Assume $a_j = b_j$ in (11a). Let $\|\cdot\| = \|\cdot\|_2$ be the Euclidean norm, leading to the condition number $\operatorname{cond}_2(A) = \|A\|_2 \|A^{-1}\|_2$. $X = L^2(D)$ is equipped with the usual norm (1.3.6b). **(a)** Prove

(4.2.18d) $C_{0,n} = \sqrt{\operatorname{cond}_2(B_n)}$ (B_n from (9c)).

(b) For the cases mentioned in Exercise 15, the *optimal* result $C_0 = 1$ holds.

A quadrature error associated with the integrals $\beta_j := \frac{1}{\lambda}\int_D b_j(y) g(y)dy$ in (6b) can be treated as an error of $b = (\beta_j)$. In this case, the analysis leads again to an error amplification by the factor $C_0 \operatorname{cond}_{|||}(\lambda I - C_n B_n)$. This can be avoided, if one succeeds in interpreting the numerical quadratures by means of a backward analysis (cf. Wilkinson [1]).

Remark 4.2.18. Let $\tilde{\beta}_j$ $(1 \leqslant j \leqslant n)$ be the results of the *numerical quadrature* of the integrals $\beta_j := \frac{1}{\lambda}\int_D b_j(y) g(y)dy$ in (6b). Assume that $\tilde{g} \in X$ is a function such that $\tilde{\beta}_j$ is the exact integral of $\tilde{g}$: $\tilde{\beta}_j := \frac{1}{\lambda}\int_D b_j(y)\tilde{g}(y)dy$ for all $1 \leqslant j \leqslant n$, where $\|g - \tilde{g}\| \leqslant \eta$. This leads to a perturbed solution $\tilde{f}_n$ with $\|\tilde{f}_n - f_n\|_X \leqslant \eta \|(\lambda I - K_n)^{-1}\|$.

4.2.8 Numerical Examples

Let $D=I=[0,1]$. As first example, we choose the smooth kernel

(4.2.19a) $k(x,y) = \cos(\pi x y)$.

The solution is prescribed by

(4.2.19b) $f(x) = 1$.

For $\lambda=1$, one obtains the following inhomogeneity $g=\lambda f-Kf$:

(4.2.19c) $g(x) = 1 - \frac{\sin \pi x}{\pi x}$.

We approximate the kernel K_n by *piecewise linear interpolation* as described in (11a), (12a), and Lemma 10. The matrices B and C are already defined in (12b,c). The integrals $\beta_j := \frac{1}{\lambda}\int_D b_j(y)g(y)dy$ with b_j from (12a), needed for determining the vector b, cannot be expressed explicitly (without the sine-integral function). Therefore, the values β_j are approximated by means of the quadrature formula (4.6.26a,b), for which the estimates $|\tilde\beta_j-\beta_j|=O(h^5)$ $(1\le j\le n-1)$ and $|\tilde\beta_j-\beta_j|=O(h^3)$ $(j=0,\ j=n)$ hold. Here, $h=1/n$ is the step size of the equidistant support abscissae. The vector a is determined by solving the system (11c). Then, (5) is used to obtain f_n. The following Table 1 shows the maximal error

(4.2.20) $e_h := \max\{|f_n(\nu h)-f(\nu h)|:\ \nu=0,1,\dots,n\}$.

h	e_h	
1/2	0.086 717 6	
		2.86
1/4	0.030 298 8	
		3.47
1/8	0.008 731 6	
		3.78
1/16	0.002 307 6	
		3.90
1/32	0.000 591 0	
		3.96
1/64	0.000 149 4	
		3.98
1/128	0.000 037 6	

Table 4.2.1 Example (19a-c) with tensor approximation (11a)

h	e_h	
1/2	1.225 361 8	
		5.08
1/4	0.241 311 6	
		7.38
1/8	0.032 678 9	
		8.10
1/16	0.004 032 4	
		8.20
1/32	0.000 492 0	
		8.15
1/64	0.000 060 4	
		5.87
1/128	0.000 010 3	

Table 4.2.2 Example (19a), (21a,b) with tensor approximation (11a)

Remark 4.2.19. Since a_j are the Lagrange functions, the representation (20') for e_h follows from (5):

(4.2.20') $e_h := \max\{|\frac{1}{\lambda}g(\nu h)+\alpha_\nu-f(\nu h)|:\ \nu=0,1,\dots,n\}$.

The last column of Table 1 contains the ratios e_{2h}/e_h of two succeeding errors. The values approximate 4, which suggests second order convergence, in agreement with estimate (12d),.

The second example uses the same smooth kernel (19a). However, the solution is (as proposed by Atkinson [1], [2], [3, p.59])

(4.2.21a) $f(x) = e^x\cos(7x)$.

$g=\lambda f-Kf$ becomes

(4.2.21b) $$g(x) = \lambda f(x) - \frac{1}{2}\frac{e[\cos(\pi x+7)+(\pi x+7)\sin(\pi x+7)]-1}{1+(\pi x+7)^2} - \frac{1}{2}\frac{e[\cos(\pi x-7)+(\pi x-7)\sin(\pi x-7)]-1}{1+(\pi x-7)^2}$$

(here: $\lambda=1$). Again, the exact computation of the integrals β_j is not possible and so they are approximated as above. The numerical results for different step sizes of the piecewise linear interpolation are reported in Table 2. The ratios in the last column seem to indicate third order convergence. A comparison with the later Table 4 and the quotient 5.87 between $h=1/64$ and $1/128$ suggest that until $h=1/64$ the quadrature error $O(h^3)$ dominates the discretisation error $O(h^2)$.

An alternative to the approximation of k by the tensor product (11a) is the kernel approximation (10a) with the Lagrange functions a_j from (12a) and $b_j(y)=k(x_j,y)$. The approximation by (10a) is more accurate, since no interpolation error arises with respect to y. The difficulties with this choice are mentioned in Remark 8. In the following, the integrals β_{jk} are approximated with the same numerical quadrature as used above for determining β_j. Now, the coefficients β_j are integrals of the function $k(x_j,y)g(y)$. For these we apply the composite Simpson rule with $2n+1$ support abscissae (cf. (1.4.18d)). For both of the examples, ratios of the errors show second order convergence. Note that the errors from Table 4 are smaller in spite of a worse convergence order than for the tensor approximation from Table 2. The comparison of the values of Tables 1 and 3 shows no essential differences.

h	e_h	
1/2	0.114 595 1	
		4.20
1/4	0.027 285 5	
		4.28
1/8	0.006 375 5	
		4.18
1/16	0.001 525 4	
		4.10
1/32	0.000 375 1	
		4.05
1/64	0.000 091 8	

Table 4.2.3 Example (19a-c) with kernel approximation

h	e_h	
1/2	0.049 578 2	
		3.93
1/4	0.012 605 3	
		4.57
1/8	0.002 759 0	
		4.25
1/16	0.000 648 7	
		4.07
1/32	0.000 159 4	
		4.04
1/64	0.000 039 5	

Table 4.2.4 Example (19a), (21a,b) with kernel approximation

Unfortunately, a weakly singular kernel like, e.g., $k(x,y)=|x-y|^{-1/2}$ cannot be handled by the kernel approximation without difficulty. The approximation (11a), (12a), based on a tensor product of interpolations, cannot be applied because of the required coefficients $c_{ij}=k(ih,jh)$. Also, a reasonable approximation (different from an interpolation) is not obvious. One possibility is the approximation of k by k_n given in (10a). Then, one has to evaluate the integrals (10b), i.e., $\beta_j=\frac{1}{\lambda}\int_D b_j(y)\,g(y)dy$ for $b_j(y)=|x_j-y|^{-1/2}$. In principle, there are available numerical methods for these integral such as, e.g., the product integration explained in §4.7.8 and §4.6.7. However, in our example ((4.22c) below), the function g is not smooth and so large quadrature errors seem to be unavoidable during the computation of the vector b. However, the next chapters will present successful discretisations of equations with weakly singular kernel.

4.3 Projection Methods in General

4.3.1 Subspaces

In §4.2, we have simplified the kernel in order to obtain a finite-dimensional problem. Now, we shall replace the (infinite-dimensional) space X ($X=C(D)$ or $X=L^2(D)$) by a finite-dimensional subspace X_n. These subspaces will be the ranges of projections, from which this method derives its name. In the case of the *collocation method* (§4.4), the projections play a primary role, while the *Galerkin method* (§4.5) is completely characterised by the subspaces. For the general (abstract) representation, we start by assuming that

(4.3.1a) $\quad X_n \subset X$ are subspaces $\qquad (n \in \mathbb{N})$.

The *distance* of a «point» $x \in X$ from X_n is given by

$$\operatorname{dist}(x, X_n) := \inf\{\|x-\xi\| : \xi \in X_n\}.$$

We shall see that the convergence of the method requires

(4.3.1b) $\quad \lim_{n\to\infty} \operatorname{dist}(x, X_n) = 0 \qquad$ for all $x \in X$

(cf. Remark 5). Condition (1b) says that the spaces X_n comes close (in some sense) to X.

Exercise 4.3.1. Prove the following: **(a)** If $\bigcup_{j\in\mathbb{N}} X_j$ is dense in X, a sufficient condition for (1b) is that $X_1 \subset X_2 \subset \dots$.
(b) A necessary condition for (1b) is $\dim X_n \to \infty$.

Because of practicability, we only consider finite-dimensional X_n:

(4.3.1c) $\quad \dim X_n < \infty$.

4.3.2 Projections

We recall that a linear mapping $\Pi_n: X \to X$ is a *projection*, if $\Pi_n^2 = \Pi_n$. Each projection Π_n defines a subspace as image space:

(4.3.2a) $\quad X_n := \operatorname{range}(\Pi_n)$.

In this case, we say that Π_n is a projection *onto* X_n.

Exercise 4.3.2. (a) Let Π_n be a projection onto X_n. Prove that X_n can be characterised as a set of *fixed points*:

(4.3.2a') $\quad X_n = \{x \in X : \Pi_n x = x\}$.

(b) Let D be bounded and $a, b \in X = C(D)$ functions with $\int_D a(x)b(x)dx = 1$. Define Π_1 by $(\Pi_1 f)(x) = a(x)\int_D b(y)f(y)dy$ and prove that Π_1 is a projection onto $X_1 = \operatorname{span}\{a\}$.

Given a projection Π_n, there is only one subspace X_n associated with Π_n (cf. (2a)). However, the converse is not true. In general, there are infinitely many projections onto the same subspace X_n. For example, different $b \in X$ in

Exercise 2b yield different projections, although the range is fixed.

For a *sequence* $\{\Pi_n\}_{n\in\mathbb{N}}$ *of projections*, we introduce the following terms:

Definition 4.3.3. Let X be a Banach space and $\{\Pi_n\}_{n\in\mathbb{N}}$ be a sequence of projections $\Pi_n \in L(X,X)$. **(a)** $\{\Pi_n\}_{n\in\mathbb{N}}$ is called *convergent*, if

(4.3.3a) $\quad \Pi_n x \to x \quad$ in X for all $x\in X$.

(b) $\{\Pi_n\}_{n\in\mathbb{N}}$ is called *consistent*, if there is a dense subset $M\subset X$ with

(4.3.3b) $\quad \Pi_n x \to x \quad$ in X for all $x\in M$.

(c) $\{\Pi_n\}_{n\in\mathbb{N}}$ is called *stable*, if $\|\Pi_n\|$ is uniformly bounded:

(4.3.3c) $\quad \sup\{\|\Pi_n\| : n\in\mathbb{N}\} < \infty$.

The theorem of Banach-Steinhaus (Theorem 1.3.17) proves the following.

Remark 4.3.4. $\{\Pi_n\}_{n\in\mathbb{N}}$ is convergent, if and only if it is consistent and stable.

The requirement (1b) is a result of the following remark.

Remark 4.3.5. **(a)** Suppose $\{\Pi_n\}_{n\in\mathbb{N}}$ is consistent (and even convergent). Then (1b) holds, i.e., $\lim_{n\to\infty} \operatorname{dist}(x, X_n) = 0$ for all $x\in X$.
(b) If $\{\Pi_n\}_{n\in\mathbb{N}}$ is stable and if (1b) holds, then $\{\Pi_n\}_{n\in\mathbb{N}}$ is convergent.

Proof. (a) Let $\varepsilon > 0$ and $x\in X$ arbitrary. Since $M\subset X$ is dense, there is an $x'\in M$ with $\|x - x'\| \le \varepsilon/2$. According to (3b), there is an n such that $\|\xi - x'\| \le \varepsilon/2$ for $\xi := \Pi_n x' \in X_n$. Hence, $\operatorname{dist}(x, X_n) \le \|x - \xi\| \le \|x - x'\| + \|x' - \xi\| \le \varepsilon$.
(b) Let $x\in X$ be arbitrary. Because of (1b), there is a sequence $\xi_n \in X_n$ with $\xi_n \to x$. The splitting $x - \Pi_n x = x - \xi_n - \Pi_n(x - \xi_n)$ together with (3c) proves (3a). QED

4.3.3 Lemmata

The property of Π_n given in (3a) is known as *pointwise* convergence to the identity I. However, pointwise convergence does not imply operator norm convergence (cf. Exercise 6 below). The study of convergence in §4.3.4 will require statements concerning the convergence behaviour of products $\Pi_n K$, which is provided by Lemmata 7 and 8.

Exercise 4.3.6. Prove that the convergence $\|\Pi_n - I\| \to 0$ for projections Π_n implies $\Pi_n = I$ for sufficiently large $n \ge n_0$.

Lemma 4.3.7. Let $M\subset X$ be a precompact subset of a Banach space X. The operators $A_n \in L(X,X)$ are assumed to be pointwise convergent to A, i.e., $A_n x \to Ax$ for all $x\in X$. Then the sequences $\{A_n x\}$ converge uniformly for all $x\in M$, i.e.,

(4.3.4) $$\sup_{x\in M} \|A_n x - Ax\| \to 0 \qquad \text{for } n\to\infty.$$

Proof. (i) $C := \sup\{\|A_n\| : n \in \mathbb{N}\}$ is finite («stability» of $\{A_n\}$, cf. Remark 1.3.16). Moreover, $\|A\| \leq C$ holds.
(ii) If (4) were not valid, an $\varepsilon > 0$ and a subsequence $\mathbb{N}' \subset \mathbb{N}$ would exist such that $\sup_{x \in M} \|A_n x - Ax\| \geq \varepsilon$ for all $n \in \mathbb{N}'$. Hence, there exist $x_n \in M$ with $\|A_n x_n - Ax_n\| \geq \varepsilon/2$ for all $n \in \mathbb{N}'$. Since M is precompact, there is a further subsequence $\mathbb{N}'' \subset \mathbb{N}'$ such that $\lim_{n \in \mathbb{N}''} x_n = x$ exists. Choose $n \in \mathbb{N}''$ with $\|x_n - x\| < \varepsilon/(8C)$ and $\|A_n x - Ax\| < \varepsilon/4$. Then we have $\|A_n x_n - Ax_n\| \leq \leq \|(A_n - A)(x_n - x)\| + \|(A_n - A)x\| < (\|A_n\| + \|A\|)\|x_n - x\| + \varepsilon/4 < \varepsilon/2$ which contradicts the inequality above. QED

Lemma 4.3.8. Let X be a Banach space. Assume that $A_n \in L(X,X)$ converges *pointwise* to A. For any compact operator $K \in K(X,X)$, the product $A_n K$ converges with respect to the opera*tor norm* to AK: $\|A_n K - AK\| \to 0$.

Proof. Set $B := \{Kx : \|x\| \leq 1\}$. Since K is compact, B is precompact. (4) shows that $\|A_n K - AK\| := \sup_{\|x\| \leq 1} \|A_n Kx - AKx\| = \sup_{y \in B} \|A_n y - Ay\| \to 0$. QED

4.3.4 Discretisation by means of a Projection

Let projections Π_n and their ranges X_n be given. We want to determine a discrete solution f_n in the subspace X_n. In general, the equation $\lambda f = g + Kf$ admits *no* solution in the subspace X_n, but each $f_n \in X_n$ generates an error

$$d_n := (\lambda I - K) f_n - g.$$

We cannot expect that d_n vanishes, but requiring that its projection must vanish (i.e., $\Pi_n d_n = 0$), we obtain

(4.3.5a) $\qquad \Pi_n[(\lambda I - K) f_n - g] = 0$

or

(4.3.5b) $\qquad \lambda \Pi_n f_n = \Pi_n g + \Pi_n K f_n.$

Since the solution f_n must belong to X_n, we have $\Pi_n f_n = f_n$ (cf. (2a')) and so equation (5b) can be written in the form

(4.3.5c) $\qquad \lambda f_n = \Pi_n g + \Pi_n K f_n.$

This is an example of the general form (1.2b):

semidiscrete projection equation

(4.3.6) $\qquad \lambda f_n = g_n + K_n f_n \quad$ with $K_n := \Pi_n K$ and $g_n := \Pi_n g$.

The next lemma reduces the solvability of equation (6) in X_n to that in X.

Lemma 4.3.9. Let $\lambda \neq 0$. Any solution $f_n \in X$ of equation (6) belongs to X_n.

Proof. Let $f_n \in X$ be a solution of (6). Since $g_n = \Pi_n g \in X_n$ and $K_n f_n = \Pi_n K f_n \in X_n$, also the sum λf_n must belong to X_n. $\lambda \neq 0$ yields $f_n \in X_n$. QED

Note that the statement of Lemma 9 is no longer valid if in (6) we use approximations $g_n \notin X_n$ instead of $g_n = \Pi_n g$. Furthermore, Lemma 9 does not hold for equations of the first kind. i.e., for $\lambda = 0$.

Further, almost equivalent *representations* of the problem are

(4.3.7) $$\lambda f_n = g_n + \hat{K}_n f_n \quad \text{with } \hat{K}_n := \Pi_n K \Pi_n \text{ and } g_n := \Pi_n g$$

(4.3.8) $$\lambda f'_n = g + K'_n f'_n \quad \text{with } K'_n := K\Pi_n .$$

Remark 4.3.10. Let $\lambda \neq 0$. **(a)** The problems (6) and (7) are equivalent, i.e., a solution of the first problem is a solution of the second one and vice versa.
(b) The problems (6) and (8) are equivalent in the following sense: If $f'_n \in X$ is a solution of (8), then $f_n := \Pi_n f'_n \in X_n$ is a solution of (6). If, on the other hand, $f_n \in X_n$ solves problem (6), then

(4.3.9) $$f'_n := \tfrac{1}{\lambda}(g + K f_n)$$

satisfies equation (8).
(c) If one of the operators $\lambda I - K_n$, $\lambda I - \hat{K}_n$ or $\lambda I - K'_n$ has an inverse, then all inverses exist and are connected by

(4.3.10a) $$(\lambda I - K'_n)^{-1} = \tfrac{1}{\lambda}[I + K(\lambda I - K_n)^{-1}\Pi_n] = \tfrac{1}{\lambda}[I + K(\lambda I - \hat{K}_n)^{-1} + \tfrac{1}{\lambda}K(\Pi_n - I)],$$

(4.3.10b) $$(\lambda I - \hat{K}_n)^{-1} = \tfrac{1}{\lambda}(I - \Pi_n) + (\lambda I - K'_n)^{-1} = \tfrac{1}{\lambda}(I - \Pi_n) + \Pi_n(\lambda I - K'_n)^{-1},$$

(4.3.10c) $$(\lambda I - K_n)^{-1} = \tfrac{1}{\lambda}[I + \Pi_n(\lambda I - K'_n)^{-1}] = \tfrac{1}{\lambda}[I + (\lambda I - \hat{K}_n)^{-1}K + \tfrac{1}{\lambda}(\Pi_n - I)K].$$

Proof. (a) The solution f_n of (6) belongs to X_n (cf. Lemma 9). Hence, $\Pi_n f_n = f_n$ and $K_n f_n = \Pi_n K f_n = \Pi_n K \Pi_n f_n = \hat{K}_n f_n$. Therefore, f_n solves (7) and vice versa.
(b) Let f'_n be a solution of equation (8). Application of Π_n yields $\lambda \Pi_n f'_n = \Pi_n K'_n f'_n + \Pi_n g$. For $f_n := \Pi_n f'_n$ one verifies that $\lambda f_n = \Pi_n(\lambda f'_n) = \Pi_n(g + K'_n f'_n) = \Pi_n K \Pi_n f'_n + \Pi_n g = \Pi_n K f_n + g_n = K_n f_n + g_n$. For the reverse conclusion, let f_n be a solution of (6), and define f'_n by (9). From $\Pi_n f'_n = \frac{1}{\lambda}(\Pi_n g + \Pi_n K f_n) = \frac{1}{\lambda}(g_n + K_n f_n) = f_n$ one concludes that $\lambda f'_n - g = \lambda \frac{1}{\lambda}(g + K f_n) - g = K f_n = K \Pi_n f'_n = K'_n f'_n$, i.e., equation (8) is fulfilled.
(c) We abbreviate the terms in (10a) by $T'_n := \lambda I - K'_n$ and $A_n := \frac{1}{\lambda}[I + K(\lambda I - K_n)^{-1}\Pi_n]$, assuming existence of the inverse $(\lambda I - K_n)^{-1}$. The identity $\Pi_n K'_n = K_n \Pi_n$ shows

$$A_n T'_n = \tfrac{1}{\lambda}[I + K(\lambda I - K_n)^{-1}\Pi_n](\lambda I - K'_n) =$$

$$= I - \tfrac{1}{\lambda}K'_n + \tfrac{1}{\lambda}K(\lambda I - K_n)^{-1}(\lambda \Pi_n - \Pi_n K'_n) =$$

$$= I - \tfrac{1}{\lambda}K'_n + \tfrac{1}{\lambda}K(\lambda I - K_n)^{-1}(\lambda \Pi_n - K_n \Pi_n) =$$

$$= I - \tfrac{1}{\lambda}K'_n + \tfrac{1}{\lambda}K(\lambda I - K_n)^{-1}(\lambda I - K_n)\Pi_n = I - \tfrac{1}{\lambda}K'_n + \tfrac{1}{\lambda}K\Pi_n = I.$$

Analogously, one proves $T'_n A_n = I$, which implies that A_n is the inverse of T'_n. All other cases are analogous. ∎

4.3.5 Convergence Analysis

We return to the main variant (6) of the projection method and prove as a central result the following theorem.

Theorem 4.3.11. Let the sequence of projections $\{\Pi_n\}$ be convergent (cf. (3a)) and K be compact, i.e., $K \in \mathcal{K}(X,X)$. Then the operators $K_n = \Pi_n K$ from (6) *converge* with respect to the operator norm to K:

(4.3.11a) $\quad \|K_n - K\| \to 0 \qquad$ for $n \to \infty$.

If λ is a regular value, i.e., $(\lambda I - K)^{-1} \in L(X,X)$, the projection discretisation (6) is *convergent, stable*, and *consistent*. The solutions f_n of the projection discretisation satisfy

(4.3.11b) $\quad f_n \to f = (\lambda I - K)^{-1} g \qquad$ in X.

Proof. (i) (11a) is the consequence of Lemma 8 with $A_n := \Pi_n$ and $A := I$.
(ii) Theorem 1.13 yields consistency, stability, and convergence.
(iii) The convergence of $\{\Pi_n\}$ guarantees the convergence $g_n \to g$ according to (1.3b), so that $f_n \to f$ can be derived from Theorem 1.12b. ∎

For the variants (7) and (8) involving the respective approximations $\{\hat{K}_n\}$ and $\{K'_n\}$, (11a) cannot be shown under the same assumptions. Nevertheless, the following similar statement holds.

Theorem 4.3.12. Under the assumptions of Theorem 11 and, in addition, $\lambda \neq 0$, the discretisations $\{\hat{K}_n\}$ and $\{K'_n\}$ are convergent, consistent and stable. The associated solutions satisfy (11b).

Proof. (i) $K'_n = K\Pi_n$ is *consistent* because of the convergence of $\{\Pi_n\}$: $K'_n x = K\Pi_n x \to Kx$ for all $x \in X$. Following Remark 4, $\{\Pi_n\}$ is stable, while Theorem 11 guarantees the stability of K_n. The uniform boundedness of $\{\Pi_n\}$ and $\{\|(\lambda I - K_n)^{-1}\|\}$ together with $\lambda \neq 0$ and (10a) proves that $\{\|(\lambda I - K'_n)^{-1}\|\}$ is uniformly bounded, too. Hence, $\{K'_n\}$ is *stable*. Using Theorem 1.12, we obtain *convergence*. Finally, $g_n = \Pi_n g \to g$ implies condition (1.3b) ensuring (11b).
(ii) For the proof of the convergence $\hat{K}_n x \to Kx$, split $(\hat{K}_n - K)x$ into $[\Pi_n Kx - Kx] + \Pi_n K[\Pi_n x - x]$ and note that $\Pi_n x \to x$, $\Pi_n Kx \to Kx$ and that $\|\Pi_n\|$ is uniformly bounded. Hence, $\{\hat{K}_n\}$ is consistent. As in (i), we obtain stability and convergence. ∎

4.3.6 Error Estimate

As already emphasized in §4.1.5, *quantitative* estimates of the error $f_n - f$ for a *fixed* discretisation parameter n are much more important than qualitative statements about the limit process $n \to \infty$.

Theorem 4.3.13. For a fixed index n assume $(\lambda I-K_n)^{-1}\in L(X,X)$ and $(\lambda I-K)^{-1}\in L(X,X)$. Let f and f_n be the solutions of the integral equation (1.1), i.e., $\lambda f=g+Kf$ and of its discretisation (6), respectively. Then the error $f-f_n$ admits the representation

(4.3.12a) $$f-f_n = \lambda(\lambda I-K_n)^{-1}(I-\Pi_n)f$$

and satisfies the *error estimate*

(4.3.12b) $$\|f-f_n\| \leqslant |\lambda|\,\|(\lambda I-K_n)^{-1}\|\,\|f-\Pi_n f\|.$$

The error in f_n compared to the projection $\Pi_n f$ of the exact solution is given by

(4.3.12c) $$\Pi_n f-f_n = (\lambda I-K_n)^{-1}K_n(I-\Pi_n)f$$

and satisfies the estimate

(4.3.12d) $$\|\Pi_n f-f_n\| \leqslant \|(\lambda I-K_n)^{-1}\|\,\|\Pi_n\|\,\|K(f-\Pi_n f)\|.$$

Proof. (i) After inserting the definitions $K_n=\Pi_n K$ and $g_n=\Pi_n g$ into the general estimate (1.12a) from Lemma 1.14, the square bracket in (1.12a) becomes $(I-\Pi_n)(g+Kf)$. From $\lambda f=g+Kf$ one concludes the claimed representation (12a). Estimate (12b) is a trivial consequence of (12a).
(ii) $\lambda I=(\lambda I-K_n)+K_n$ implies $\lambda(\lambda I-K_n)^{-1}=I+(\lambda I-K_n)^{-1}K_n$. Hence, (12a) can also be written as

(4.3.12e) $$f-f_n = (I-\Pi_n)f + (\lambda I-K_n)^{-1}K_n(I-\Pi_n)f,$$

whence (12c) and, because of $K_n=\Pi_n K$, also (12d) follow. QED

In Theorem 13, neither the assumption of convergence nor of stability appears. However, one needs these conditions implicitly to ensure the existence of the inverse $(\lambda I-K_n)^{-1}$ (for sufficiently large n). Analogously, the importance of *stability* - i.e., of the uniform boundedness of $\|(\lambda I-K_n)^{-1}\|$ - becomes obvious. If this quantity tended to ∞, the right-hand side of estimate (12b) would contain an unbounded factor which may destroy the overall convergence of the right-hand side. Therefore, assume $\{K_n\}$ to be stable. The *stability constant* is denoted by C_S:

(4.3.13) $$\|(\lambda I-K_n)^{-1}\| \leqslant C_S \qquad (n\in\mathbb{N}).$$

The error estimate (12b) takes the form

(4.3.14a) $$\|f-f_n\| \leqslant |\lambda|\,C_S\,\|f-\Pi_n f\|.$$

If the projection Π_n is well-chosen, $\tilde f_n := \Pi_n f$ should be an acceptable approximation of f. Inequality (14a) implies that the true error $\|f-f_n\|$ is comparable with this acceptable error $\|f-\tilde f_n\|$ up to the fixed factor $|\lambda|\,C_S$.

Remark 4.3.14. If X is a Hilbert space and Π_n the *orthogonal projection onto* X_n, (14a) becomes a *quasi-optimal* estimate

(4.3.14b) $$\|f-f_n\| \leqslant |\lambda|\,C_S\,\mathrm{dist}(f,X_n)$$

(cf. (1b)). Here, one calls the estimate (14b) quasi-optimal, since it equals the smallest possible error up to a fixed factor.

By estimate (14a), the error analysis of the discrete equation (6) reduces to the error analysis of the projection Π_n. The following theorem shows one of the possible applications.

Theorem 4.3.15. Let $X=C(D)$. Assume that the integral equation $\lambda f=g+Kf$ has a solution $f\in C^{\tau}(D)$ with $\tau>0$ and the discretisation (6) is stable. If the projection Π_n satisfies

(4.3.15) $$\|f-\Pi_n f\|_{\infty} \leqslant C_{\Pi}\, n^{-\tau}\, \|f\|_{C^{\tau}(D)},$$

the projection method is *convergent of the order* τ:

(4.3.16) $$\|f-f_n\|_{\infty} = O(n^{-\tau}).$$

Proof. Insert (15) into (14a). QED

In §4.6.4, we shall see that often the difference $\Pi_n f-f_n$ is small in comparison to $\Pi_n f-f$, so that, by (12e), the error $f-f_n$ almost equals the projection error $\Pi_n f-f$.

4.4 Collocation Method

4.4.1 Definition of the Projection by Interpolation

In the following,

(4.4.1) $$X = C(D)$$

is the underlying Banach space. As explained in §1.4.1, an interpolation in $C(D)$ can be characterised by an projection $\Pi_n\in L(X,X)$. *The projection method (3.6) with this choice of the projection* Π_n *defines the collocation method.*

For the interpolation we choose the following disjoint support abscissae

$$\xi_{1,n},\ \xi_{2,n},\ \ldots,\ \xi_{n,n} \in D.$$

They form the set

(4.4.2) $$\Xi := \{\xi_{1,n},\ \xi_{2,n},\ \ldots,\ \xi_{n,n}\} \subset D.$$

Let $X_n\subset X$ be the subspace which contains the interpolating functions (*ansatz functions, trial function*). For the numerical application, it is essential to be able to represent functions of X_n by a means of a *basis*

(4.4.3a) $$\{\Phi_{1,n},\ \Phi_{2,n},\ldots,\ \Phi_{n,n}\} \qquad \text{with}$$

(4.4.3b) $$X_n = \operatorname{span}\{\Phi_{1,n},\ \Phi_{2,n},\ \ldots,\ \Phi_{n,n}\}.$$

The interpolation of a function $\varphi\in X$ by $\varphi_n\in X_n$ with

(4.4.4) $\varphi_n(\xi_{j,n}) = \varphi(\xi_{j,n})$ for all $1 \leq j \leq n$

exists and is unique, if

(4.4.5) $\left(\Phi_{j,n}(\xi_{i,n})\right)_{i,j=1,\dots,n}$ is non-singular.

Exercise 4.4.1. Prove: **(a)** Condition (5) implies that the «ansatz functions» (3a) are linearly independent in X.
(b) Two functions $\varphi_n, \psi_n \in X_n$ are identical if and only if they coincide at all support abscissae: $\varphi_n(\xi) = \psi_n(\xi)$ for all $\xi \in \Xi$.

Under assumption (5), the interpolation defines the projection

(4.4.6) $\Pi_n: \varphi \in X \mapsto \varphi_n \in X_n$ (φ_n defined by (4)).

4.4.2 Setting up the System of Linear Equations

By Lemma 3.9, both sides of the semidiscrete equation (3.6),

(4.4.7a) $\lambda f_n = g_n + K_n f_n$ with $K_n := \Pi_n K$ and $g_n := \Pi_n g$,

belong to the space X_n of ansatz functions. Exercise 1b shows that the semidiscrete projection equation (7a) is equivalent to the n equations

(4.4.7b) $\lambda f_n(\xi_{j,n}) = g_n(\xi_{j,n}) + (K_n f_n)(\xi_{j,n})$ for $1 \leq j \leq n$.

From the interpolation property $g_n(\xi_{j,n}) = g(\xi_{j,n})$ and $(K_n f_n)(\xi_{j,n}) = (K f_n)(\xi_{j,n})$ (cf. (4)) one concludes that (7a) and (7b) are equivalent to

collocation equations:

(4.4.7c) $\lambda f_n(\xi_{j,n}) = g(\xi_{j,n}) + (K f_n)(\xi_{j,n})$ for $1 \leq j \leq n$.

It it clear from equation (7c) that the integral equation $\lambda f = g + Kf$ is not required to be satisfied in all points $x \in D$ but only at the support abscissae $\xi \in \Xi$. Because of this property, these $\xi \in \Xi$ are called the *collocation points*.

After the choice of a basis (3a,b) for X_n, the numerical problem consists in determining the coefficients α_k of the representation

(4.4.8) $f_n = \sum_{k=1}^{n} \alpha_k \Phi_{k,n}$.

Insert (8) into (7c):

(4.4.7d)
$$\lambda \sum_{k=1}^{n} \alpha_k \Phi_{k,n}(\xi_{j,n}) = g(\xi_{j,n}) + \Big(K \sum_{k=1}^{n} \alpha_k \Phi_{k,n}\Big)(\xi_{j,n}) =$$
$$= g(\xi_{j,n}) + \sum_{k=1}^{n} \alpha_k (K\Phi_{k,n})(\xi_{j,n}) \qquad \text{for } 1 \leq j \leq n.$$

(7d) is a system of n equations for the n unknowns $\{\alpha_k\}$. We define the vectors

(4.4.9a) $$a := a_n := \begin{bmatrix} \alpha_1 \\ \vdots \\ \alpha_n \end{bmatrix}, \qquad b := b_n := \begin{bmatrix} g(\xi_{1,n}) \\ \vdots \\ g(\xi_{n,n}) \end{bmatrix}$$

and matrices

(4.4.9b) $$A := A_n := \begin{bmatrix} \alpha_{11} & \alpha_{12} & \cdots & \alpha_{1n} \\ \vdots & \vdots & & \vdots \\ \alpha_{n1} & \alpha_{n2} & \cdots & \alpha_{nn} \end{bmatrix} \qquad \text{with } \alpha_{jk} := \Phi_{k,n}(\xi_{j,n})$$

and

(4.4.9c) $$B := B_n := \begin{bmatrix} \beta_{11} & \beta_{12} & \cdots & \beta_{1n} \\ \vdots & \vdots & & \vdots \\ \beta_{n1} & \beta_{n2} & \cdots & \beta_{nn} \end{bmatrix} \qquad \text{with } \beta_{jk} := (K\Phi_{k,n})(\xi_{j,n}).$$

Using the vectors (9a) and the matrices A and B to express (7d), one obtains the representation (10) below, which is the final *discrete* problem.

> The collocation method with the collocation points Ξ and the basis $\{\Phi_{1,n}, \Phi_{2,n}, \ldots, \Phi_{n,n}\}$ from (3a) leads to the system
>
> (4.4.10) $$(\lambda A - B)a = b \qquad (a, b, A, B \text{ from (9a-c)})$$
>
> for the coefficients α_k of the representation (8).

To express the dependence of the quantities A, B, a, b upon the discretisation parameter n explicitly, we write instead of (10):

(4.4.10') $$(\lambda A_n - B_n)a_n = b_n.$$

Analogously to Theorem 2.5, the following holds (with a similar proof).

Remark 4.4.2. The matrix $\lambda A_n - B_n$ is non-singular if and only if Equation (7a) is solvable for all $g \in X$ (hence, also for all $g_n \in X_n$). The solutions of (7a) and (10) are connected via (8).

In the following, we discuss the *amount of computational work* needed for setting up the system (10). In the case of the bases recommended in §4.4.3, the matrix A is very easy to compute and, in particular, sparse. The simplest and even frequent case is the following.

Remark 4.4.3. If the basis $\{\Phi_{1,n}, \ldots, \Phi_{n,n}\}$ (cf. (3a)) consists of Lagrange functions (i.e., $\Phi_j(\xi_k) = \delta_{jk}$, cf. (1.4.5)), then

(4.4.11) $$A = I.$$

The main work is needed for the matrix B, which, in general, is not sparse.

Remark 4.4.4. For computing $B = B_n$, we have to evaluate n^2 integrals

(4.4.12) $$\beta_{jk} := \int_D k(\xi_j, y)\Phi_k(y)\,dy \qquad (1 \leq j, k \leq n).$$

If the integrals are not known explicitly, the integral must be approximated by numerical quadrature.

4.4.3 Examples for Interpolations

The (global) *interpolation by polynomials* should not be used because of its instability. In general, $\|\Pi_n\| \geq O(\log n)$ holds for this choice of interpolation, cf. Faber [1]. In fact, the optimal case $\|\Pi_n\| = O(\log n)$ is reached for the choice of the Chebyshev knots $\xi_{i,n} := a + \frac{1}{2}(b-a)(1+\cos((2i-1)/(2n)))$ in $I=[a,b]$. Despite this, a convergence theory for this case can be given (see Theorem 6.10 below).

Another choice of global interpolation uses the *trigonometric* functions (cf. Stoer [1]). In particular, this is an obvious choice when the integration domain D is a closed curve in $\mathbb{R}^2$ since then the solutions and kernels are periodic. The integration in (12) reduces to the Fourier analysis problem. However, again $\|\Pi_n\| = O(\log n)$ holds with respect to the maximum norm.

Exercise 4.4.5. Let D be a closed curve in $\mathbb{R}^2$, parametrised by $t \in [0,2\pi] \mapsto \varphi(t) = (\varphi_1(t), \varphi_2(t)) \in D$ with $d\varphi(t)/dt \neq 0$. Show that the integral equation $f(x) = g(x) + \int_D k(x,y) f(y) dy$ is equivalent to

$$\hat{f}(s) = \hat{g}(s) + \int_0^{2\pi} \hat{k}(s,t) \hat{f}(t) dt, \tag{4.4.13}$$

where

$$\hat{f}(s) := f(\varphi(s)),\ \hat{g}(s) := g(\varphi(s)),$$

$$\hat{k}(s,t) := k(\varphi(s),\varphi(t))\,|\varphi'(t)|$$

with $\quad |\varphi'(t)| := [|d\varphi_1(t)/dt|^2 + |d\varphi_2(t)/dt|^2]^{1/2}.$

Show also that the new kernel $\hat{k}(s,t)$ is 2π-periodic in both variables s and t.

The most flexible approach to interpolation uses *piecewise polynomial functions.* Besides their stability properties, they also offer computational advantages. In particular, this approach allows to use basis functions $\{\Phi_{1,n}, \ldots, \Phi_{n,n}\}$ in (3a) with *local support* (cf. Remark 2.9).

In §1.4 and §4.2.4, we have become acquainted with *piecewise linear interpolation* on intervals $D=I=[a,b]$. As basis functions Φ_k we choose the Lagrange functions (1.4.6), i.e., the piecewise linear functions with $\Phi_k(\xi_j) = \delta_{jk}$. By Remark 3, $A=I$ holds. Concerning the computation of the matrix B, we refer to

Remark 4.4.6. Let $D=I=[a,b]$. Let the collocation points (support abscissae) be $a=\xi_1 < \xi_2 < \ldots < \xi_n = b$. The integrals β_{jk} from (12) reduce to integrals over the two subintervals $[\xi_{k-1}, \xi_k]$, $[\xi_k, \xi_{k+1}]$:

$$\beta_{jk} = \frac{1}{\xi_k - \xi_{k-1}} \int_{\xi_{k-1}}^{\xi_k} k(\xi_j, y)(y - \xi_{k-1}) dy + \tag{4.4.14}$$
$$+ \frac{1}{\xi_{k+1} - \xi_k} \int_{\xi_k}^{\xi_{k+1}} k(\xi_j, y)(\xi_{k+1} - y) dy.$$

Here, the first integral has to be ignored for $k=1$ and the last for $k=n$.

The step size of an equidistant partition is $h = h_n = \xi_k - \xi_{k-1} = (b-a)/n$.

Theorem 4.4.7 (convergence order). Let f_n be the collocation solution with *piecewise linear interpolation* in $D=I=[a,b]$ and equidistant collocation points $\xi_k := a+(b-a)(k-1)/(n-1)$, $1 \leq k \leq n$. Assume that λ is a regular value of the compact operator K and the solution f of $\lambda f = g+Kf$ belongs to $C_L^1(D)$. Then, for sufficiently large n_0, solutions f_n $(n \geq n_0)$ exist and are *second order convergent*, i.e.,

(4.4.15) $\quad \|f-f_n\|_\infty \leq \text{const}\, h_n^2.$

Proof. Theorem 3.15 can be applied with $\tau=2$. Stability holds because of $K \in K(X,X)$ (cf. Theorem 3.11). QED

The result of Theorem 7 remains valid even for non-equidistant collocation points $\xi_k \in \Xi$, if one requires (1.4.5c): There is some constant C with $\xi_{k+1}-\xi_k \leq C/n$ $(1 \leq k < n)$.

Higher piecewise approaches may be of Hermite (cf. §2.3) or spline type. In the latter case, one has to choose the B-spline basis $\{\Phi_{1,n},\dots\}$ (cf. Stoer [1]) in order to obtain local supports. The piecewise ansatz is also suited for higher dimensional domains $D \subset \mathbb{R}^d$. In the two-dimensional case, the subintervals have to be replaced, e.g., by triangles (cf. §9.2.3).

However, also lower orders than the piecewise linear ones can be used.

Remark 4.4.8 (piecewise constant interpolation). Partition the interval I into n subintervals

$$I_1 = [x_0, x_1), \quad I_2 = [x_2, x_3), \quad \dots, \quad I_n = [x_{n-1}, x_n]$$

with $a = x_0 < x_1 < \dots < x_n = b$. Define the collocation points by

$$\xi_k := (x_k + x_{k-1})/2 \in I_k \qquad \text{(midpoint of } I_k).$$

$X_n \subset L^\infty(I)$ is the subspace of the functions being constant on I_k. The Lagrange functions are the characteristic functions of I_k:

(4.4.16a) $\quad \Phi_{k,n}(x) = 1$ for $x \in I_k$, $\qquad \Phi_{k,n}(x) = 0$, otherwise,

so that $\text{supp}(\Phi_{k,n}) = I_k$. Remark 3 shows $A=I$. The entries of B are

(4.4.16b) $\quad \beta_{jk} := \int_{I_k} k(\xi_j, y)\,dy \qquad (1 \leq j,k \leq n).$

Under assumption (1.4.5c), i.e., $x_k - x_{k-1} \leq h_n \leq C/n$, piecewise constant interpolation is of first order and so $\{\Pi_n\}$ converges in $C(I)$. Hence, for compact K and sufficiently large n_0, solutions f_n $(n \geq n_0)$ exist and are *first order convergent*:

(4.4.16c) $\quad \|f-f_n\|_\infty \leq C h_n.$

The numerical results (Table 2) will even show second order convergence: $\|f-f_n\|_\infty = O(h_n^2)$. This «superconvergence» will be explained in Theorem 4.6.17a. Concerning estimates of the interpolation error with respect to the $L^2(I)$ norm, we refer to (6.8a).

Although, all examples discussed in this chapter involve a bounded domain of definition D, unbounded domains are not excluded (cf. §3.2.4). For the numerical treatment of such domains compare Sloan - Spence [1] and Amini - Sloan [1].

4.4.4 Condition Number of the System of Equations

For a basis $\Phi := \{\Phi_{1,n}, \Phi_{2,n}, \ldots, \Phi_{n,n}\}$ of $X_n \subset X = C(D)$ (or $X_n \subset X = L^\infty(D)$) we define the following constant depending on n.

(4.4.17) $$C_n(\Phi) := \sup\{\sum_{j=1}^{n} |\Phi_{j,n}(x)| : x \in D\}.$$

If $\Phi_{j,n}$ are the Lagrange functions (cf. Definition 1.4.4), the following result shows that $C_n(\Phi)$ coincides with the operator norm $\|\Pi_n\| = \|\Pi_n\|_{C(D)\leftarrow C(D)}$:

Lemma 4.4.9. A basis $\{\Phi_{1,n},\ldots,\Phi_{n,n}\}$ of Lagrange functions satisfies

(4.4.18) $$\|\Pi_n\|_{C(D)\leftarrow C(D)} = C_n(\Phi).$$

Proof. (i) The interpolating function of $f \in C(D)$ is $\Pi_n f = \sum f(\xi_j)\Phi_{j,n}$. From $|(\Pi_n f)(x)| = |\sum f(\xi_j)\Phi_{j,n}(x)| \leq \sup\{|f(\xi_j)| : 1 \leq j \leq n\} \sum |\Phi_{j,n}(x)| \leq \|f\|_\infty C_n(\Phi)$, one concludes $\|\Pi_n\|_{C(D)\leftarrow C(D)} \leq C_n(\Phi)$.
(ii) Let $x_0 \in D$ be an argument, for which the supremum is approximated in the right-hand side of (17): $C_n(\Phi) \leq \sum |\Phi_{j,n}(x_0)| + \varepsilon$. Choose the piecewise linear function $f \in C(D)$ with $f(\xi_j) = \text{sign}(\Phi_j(x_0))$. The equations $\|f\|_\infty = 1$ and $\Pi_n f(x_0) = \sum |\Phi_{j,n}(x_0)| \geq C_n(\Phi) - \varepsilon$ imply that $\|\Pi_n\| \geq C_n(\Phi) - \varepsilon$. QED

The norm $\|\cdot\|_\infty$ in $C(D)$ corresponds to the maximum norm $\|\cdot\|_\infty$ of $\mathbb{R}^n$ (cf. (1.3.3a)). Therefore, we want to determine the $\|\cdot\|_\infty$ condition number cond_∞ of the system (10): $(\lambda A_n - B_n)\, a_n = b_n$. For this purpose, we estimate $\|\lambda A_n - B_n\|_\infty$ in Lemma 11 and $\|(\lambda A_n - B_n)^{-1}\|_\infty$ in Lemma 12 and present the result in Theorem 13.

Exercise 4.4.10. Prove the following: **(a)** $\|A_n\|_\infty \leq C_n(\Phi)$ with A_n from (9b).
(b) Let $c = (\gamma_j) \in \mathbb{R}^n$ and $g = \sum_j \gamma_j \Phi_{j,n}$. Then $\|g\|_\infty \leq C_n(\Phi)\|c\|_\infty$.
(c) Let c and g be as in (b). Further, define $q := (q_i) \in \mathbb{R}^n$ by $q_i := g(\xi_i)$. Prove that $q = A_n c$, $\|q\|_\infty \leq \|g\|_\infty$, and $\|c\|_\infty \leq \|A_n^{-1}\|_\infty \|g\|_\infty$.

Lemma 4.4.11. With $C_n(\Phi)$ from (17) we have

(4.4.19) $$\|\lambda A_n - B_n\|_\infty \leq C_n(\Phi)\,\|(\lambda I - K_n)\| \qquad \text{with } K_n = \Pi_n K.$$

Proof. Let $a = (\alpha_j) \in \mathbb{R}^n$ be arbitrary. Put $f_n := \sum_j \alpha_j \Phi_{j,n} \in X_n$ and $g_n := (\lambda I - K_n) f_n$. According to Remark 2, $b = (\beta_j) \in \mathbb{R}^n$ with $\beta_j = g_n(\xi_j)$ is the right-hand side in system (10): $(\lambda A_n - B_n) a = b$. Exercise 10b,c allows us to estimate b by $\|b\|_\infty \leq \|g_n\|_\infty \leq \|\lambda I - K_n\| \|f_n\|_\infty \leq \|\lambda I - K_n\|\, C_n(\Phi) \|a\|_\infty$. This proves (19). QED

Lemma 4.4.12. Assume that the matrix $\lambda A_n - B_n$ from (10) is non-singular. Then

(4.4.20) $\quad \|(\lambda A_n - B_n)^{-1}\|_\infty \le \|A_n^{-1}\|_\infty \|\Pi_n\| \, \|(\lambda I - K_n)^{-1}\| \qquad$ with $K_n = \Pi_n K$.

Proof. Let $b = (\beta_i) \in \mathbb{R}^n$ be an arbitrary vector. As in the proof of Lemma 9, we choose a function $g \in C(D)$ with the properties

$$g(\xi_i) = \beta_i, \qquad \|g\|_\infty \le \|b\|_\infty$$

and consider the system $(\lambda A_n - B_n)a = b$ for $a = (\alpha_j)$. Again, $f_n := \sum \alpha_j \Phi_j$ is the solution of $(\lambda I - K_n) f_n = \Pi_n g$. Hence,

$$\|f_n\|_\infty \le \|(\lambda I - K_n)^{-1}\| \, \|\Pi_n g\|_\infty \le \|(\lambda I - K_n)^{-1}\| \, \|\Pi_n\| \, \|g\|_\infty \le$$
$$\le \|(\lambda I - K_n)^{-1}\| \, \|\Pi_n\| \, \|b\|_\infty .$$

Exercise 10c shows $\|a\|_\infty = \|A_n^{-1} A_n a\|_\infty \le \|A_n^{-1}\|_\infty \|f_n\|_\infty$. Together, both estimates prove (20). QED

Theorem 4.4.13. The condition number of system (10) equals

(4.4.21a) $\quad \mathrm{cond}_\infty(\lambda A_n - B_n) \le C_n(\Phi) \|A_n^{-1}\|_\infty \|\Pi_n\| \, \mathrm{cond}_{C(D)}(\lambda I - K_n)$.

If the basis $\{\Phi_{1,n}, \dots, \Phi_{n,n}\}$ consists of Lagrange functions, the estimate (21a) simplifies to

(4.4.21b) $\quad \mathrm{cond}_\infty(\lambda I - B_n) \le \|\Pi_n\|^2 \, \mathrm{cond}_{C(D)}(\lambda I - K_n)$.

Proof. (21a) is obtained as product of the inequalities (19), (20). Inequality (21b) is a consequence of (11) (i.e., $A_n = I$) and (18) (i.e., $\|\Pi_n\| = C_n(\Phi)$). QED

Exercise 4.4.14. Prove for piecewise linear interpolation with Lagrange basis that $\sum_{j=1}^n |\Phi_{j,n}(x)| = 1$ for all $x \in D$ and hence that $C_n(\Phi) = 1$.

Inserting the result of Exercise 14 into (21b), one obtains

Remark 4.4.15. Piecewise linear interpolation yields the optimal condition number estimate $\mathrm{cond}_\infty(\lambda I - B) \le \mathrm{cond}_{C(D)}(\lambda I - K_n)$.

In the general case (21a), consistency and stability of $\{K_n\}$ as well as stability of $\{\Pi_n\}$ ensure uniform boundedness of $\|\Pi_n\| \, \mathrm{cond}_{C(D)}(\lambda I - K_n)$ (cf. Theorem 1.15). (21a) reduces to

(4.4.21c) $\quad \mathrm{cond}_\infty(\lambda A_n - B_n) \le \mathrm{const}\, C_n(\Phi) \|A_n^{-1}\|_\infty$,

where on the right-hand side only $C_n(\Phi)$ and $\|A_n^{-1}\|_\infty$ depend on the choice of the basis $\{\Phi_{1,n}, \dots, \Phi_{n,n}\}$. Since $C_n(\Phi)$ is closely connected with the norm $\|A_n\|_\infty$ (cf. Exercise 10a), the right-hand side in (21c) is very similar to $\mathrm{const} \cdot \mathrm{cond}_\infty(A_n)$.

Occasionally, one needs statements about the *spectral condition number* of A_n, i.e. the condition number with respect to the Euclidean vector norm $\|\cdot\|_2$. An example is the analysis of the cg-method in §5.1.4. The following theorem requires an anticipation of the Galerkin method.

Theorem 4.4.16. Let $\lambda A_n - B_n$ be non-singular and $K \in L(L^2(D), C(D))$. A_n^G is the Galerkin matrix defined later in (5.7b) (there called A_n). With respect to the spectral (matrix) norm and the spectral condition number corresponding to the Euclidean vector norm, the following inequalities hold

$$\|\lambda A_n - B_n\|_2 \leq \sqrt{\text{cond}_2(A_n^G)}\, \|A_n\|_2\, \|\lambda I - K_n\|_2,$$
$$\|(\lambda A_n - B_n)^{-1}\|_2 \leq \sqrt{\text{cond}_2(A_n^G)}\, \|A_n^{-1}\|_2\, \|(\lambda I - K_n)^{-1}\|_2,$$
$$\text{cond}_2(\lambda A_n - B_n) \leq \text{cond}_2(A_n^G)\,\text{cond}_2(A_n)\,\text{cond}_{L^2(D)}(\lambda I - K_n).$$

Proof. Analogous to Theorem 13 and Theorem 5.10. QED

The conclusions which we can draw from these condition number estimates, are discussed in the Sections 4.6.6 and 5.1.4.

4.4.5 Numerical Examples

We return to the examples from Section 4.2.8. The integral equation defined by (2.19a–c) has the exact solution $f = 1$. As can be seen from the following remark, this problem is not a reasonable test for the collocation method; nevertheless, it may serve as a first check of the computer program.

Remark 4.4.17. If the exact solution f of the integral equation lies in the subspace X_n, it is also the collocation solution, i.e., $f_n = f$.

The second example from §4.2.8 is described by (2.19a) and (2.21a,b):

$$k(x,y) = \cos(\pi x y), \qquad f(x) = e^x \cos(7x), \qquad I = [0,1], \qquad \lambda = 1.$$

First, we use collocation by *piecewise linear interpolation* at the equidistant support abscissae $\xi_k := kh$, $h := 1/n$, $k = 0, 1, \ldots, n$. The vector b results from pointwise evaluation of the function g from (2.21b) at the collocation points ξ_k. Since we choose the Lagrange functions (2.12) as basis functions, it follows that $A = I$. The matrix B has the coefficients β_{jk} from (14). Evaluation of these integrals yields

$$\beta_{jk} = h\cos(\pi x \xi_k)\left[\frac{\sin x\pi h/2}{x\pi h/2}\right]^2 \qquad (1 \leq k \leq n-1),$$
$$\beta_{j0} = \tfrac{1}{2} h \left[\frac{\sin x\pi h/2}{x\pi h/2}\right]^2,$$
$$\beta_{jn} = \frac{\sin x\pi}{x\pi}\left(1 - \frac{\sin x\pi h}{x\pi h}\right) + \tfrac{1}{2} h \cos(\pi x)\left[\frac{\sin x\pi h/2}{x\pi h/2}\right]^2,$$

h	e_h	
1/2	0.558 853 9	
		7.18
1/4	0.077 801 0	
		4.54
1/8	0.017 152 1	
		4.13
1/16	0.004 155 0	
		4.03
1/32	0.001 030 6	
		4.01
1/64	0.000 257 1	

Table 4.4.1 Example (2.19a), (2.21a,b). Collocation solution with piecewise *linear* interpolation

h	e_h	
1/2	0.583 336 6	
		7.13
1/4	0.081 817 6	
		4.37
1/8	0.018 739 5	
		4.08
1/16	0.004 594 2	
		4.02
1/32	0.001 143 1	
		4.00
1/64	0.000 285 5	

Table 4.4.2 Same example. Collocation solution with piecewise *constant* interpolation

where $x := \xi_j$ $(j=0,\dots,n)$. The errors $f_n - f$ evaluated in the support and collocation points ξ_k define the maximum e_h from (2.20). The comparison of the error e_h for h between $1/2$ and $1/64$ (cf. Table 1) clearly confirms the *second order convergence* of the method.

Next, we consider the use of *piecewise constant* interpolation with equidistant subintervals $I_k = [(k-1)h, kh]$, $k=1,\dots,n$. Following Remark 8, we choose the midpoints $\xi_k := (k-\frac{1}{2})h$ to be the collocation points. The computation of the matrix B is more convenient, since, according to (16b), the coefficients are given by the simpler integrals $\beta_{jk} = \int_{I_k} \cos \pi x y \, dy$ with $x = \xi_j$:

$$\beta_{jk} = h \frac{\sin x\pi h/2}{x\pi h/2} \cos \pi \xi_j \xi_k \qquad \text{for example (2.19a).}$$

Due to estimate (16c), one should expect linear convergence. However, the numerical results from Table 2 clearly show an error improvement by the factor $e_{2h}/e_h \approx 4$ indicating *second order convergence*. Furthermore, even with respect to their absolute magnitude, the errors almost coincide with those obtained for the linear interpolation. Since piecewise constant interpolation is easier to implement, so far this method is preferable to linear interpolation. The discrepancy between the linear convergence statement in (16c) and the observed second order convergence will be discussed in §4.6.4. There we shall see that the better convergence order («superconvergence») is caused by the fact that e_h involves the error $f_n - f$ only at the collocation points ξ_k.

The previous examples have a very smooth kernel k. In the following, we test the collocation method for the weakly singular kernel

$$(4.4.22a) \qquad k(x,y) = 1/\sqrt{|x-y|}, \qquad x,y \in I := [0,1].$$

The regularity considerations from Theorem 3.4.9 together with Theorem 3.5.1 show that, in general, sufficiently smooth functions g lead to a solution f belonging to the Hölder space $C^{1/2}(I)$. Therefore, we choose f by

$$(4.4.22b) \qquad f(x) = \sqrt{x} + \sqrt{1-x} \qquad (x \in I).$$

Then, f belongs to $C^{1/2}(I)$. The integration needed for computing $g := f - Kf$ is left to the reader. The result is

$$(4.4.22c) \qquad g(x) = -\tfrac{\pi}{2} - x\log(1+\sqrt{1-x})/\sqrt{x}) - (1-x)\log(1+\sqrt{x})/\sqrt{1-x}).$$

g belongs to $C^{\lambda}(I)$ for all $\lambda < 1$.

First, we apply collocation by *piecewise linear interpolation* with the same equidistant support abscissae as before. The coefficients of B can be computed according to (14) as

$$(4.4.23) \qquad \beta_{jk} = \beta_{jk,1} + \beta_{jk,2} \quad \text{with } \beta_{j0,1} = \beta_{jn,2} = 0 \text{ and}$$

$$\beta_{jk,1} = \pm\tfrac{4}{3}\sqrt{h}\,[(\tfrac{3}{2}+j-k)\sqrt{|k-j|} - (j-k+1)\sqrt{|k-j-1|}] \qquad \text{for} \begin{cases} 1 \le j \le k-1, \\ k \le j \le n, \end{cases}$$

$$\beta_{jk,2} = \mp\tfrac{4}{3}\sqrt{h}\,[(\tfrac{3}{2}+k-j)\sqrt{|k-j|} - (k-j+1)\sqrt{|k-j-1|}] \qquad \text{for} \begin{cases} j \le k \le n-1, \\ 0 \le j \le k. \end{cases}$$

The error e_h is now denoted by $e_{h,\infty}$:

(4.4.24a) $\quad e_{h,\infty} := \max\{|f_n(\nu h)-f(\nu h)|:\ \nu=0,1,\dots,n\}.$

In addition, we use the Euclidean error norm

(4.4.24b) $\quad e_{h,2} := \left[\sum_{\nu=0}^{n}|f_n(\nu h)-f(\nu h)|^2/(n+1)\right]^{1/2},$

which is a discrete analogue of the L^2 norm $\|f_n-f\|_2$. Both errors shown in Table 3 are clearly larger than for the previous examples with smooth kernel. Since the ratios in Table 3 range between 2 and $2\sqrt{2}$, a convergence order between *1* and *3/2* appears to have been achieved. Because of Theorem 3.15, one only expects the order $\frac{1}{2}$. The reason for the observed higher order is that the errors $e_{h,\infty}$, $e_{h,2}$ only involve the values of f_n-f at the support abscissae νh. One can then apply the later inequality (6.15) from §4.6, which uses estimates of $K(f_n-f)$, to obtain higher order convergence at the support abscissae.

h	$e_{h,\infty}$		$e_{h,2}$	
1/2	0.326 07	1.80	0.190 27	2.02
1/4	0.180 95	1.52	0.094 34	1.28
1/8	0.118 96	1.88	0.073 91	1.74
1/16	0.063 36	2.35	0.042 57	2.28
1/32	0.026 97	2.38	0.018 68	2.60
1/64	0.011 33	2.41	0.007 17	2.75
1/128	0.004 71		0.002 61	

Table 4.4.3 Example (22a-c). Collocation with piecewise *linear* interpolation

h	$e_{h,\infty}$		$e_{h,2}$	
1/2	0.031 18	0.26	0.031 18	0.26
1/4	0.117 83	2.26	0.113 06	2.86
1/8	0.052 15	2.43	0.039 53	2.75
1/16	0.021 48	2.51	0.014 36	2.81
1/32	0.008 57	2.44	0.005 11	2.81
1/64	0.003 52	2.35	0.001 82	2.79
1/128	0.001 50		0.000 65	

Table 4.4.4 Example (22a, c). Collocation with piecewise *constant* interpolation

In the case of collocation with *piecewise constant interpolation* (as already used for Table 2), the coefficients (16b) become

(4.4.25) $\quad \beta_{kk} = 2\sqrt{2h}, \qquad \beta_{jk} = 2\sqrt{h}\,/\left(\sqrt{|k-j+\tfrac{1}{2}|}+\sqrt{|k-j-\tfrac{1}{2}|}\right) \qquad \text{for } j\neq k.$

In spite of the lower interpolation order, the associated errors $e_{h,\infty}$ and $e_{h,2}$ from Table 4 show not only a similar convergence behaviour as for linear interpolation but are even smaller. With respect to the Euclidean norm, the convergence is of the order $\frac{3}{2}$: $e_{h,2}=O(h^{3/2})$, since $2^{3/2}=2.81\dots$.

For smooth g, according to Theorem 3.4.7, the solution f should be smooth except the interval ends, where it behaves as the antiderivative of $k(t)=1/\sqrt{|t|}$. This enables us, even without knowing the solution (22b), to predict a singularity of the form $\sqrt{x}$ at $x=0$ and $\sqrt{1-x}$ at $x=1$. Since the discretisation error can be derived from the projection error, we want to apply a *non-equidistant* interval partition leading to an optimal projection error. For this purpose, we choose the subintervals $I_j := [x_{j-1}, x_j]$ with $x_j = 1-x_{n-j} = \frac{1}{2}(2j/n)^{7/4}$ for $0\leq j\leq \frac{n}{2}$. The interval lengths vary between $O(n^{-7/4})$ and $O(1/n)$. We use *piecewise constant interpolation onto* I_j with collocation points $\xi_j := \frac{1}{2}(x_{j-1}+x_j)$. The results reported in Table 5 show that piecewise constant collocation with this problem-fitted interval partition (so-called «graded mesh») is better than equidistant piecewise linear collocation. The accuracy obtained in the latter case for $n=128$ is now attainable by $n=32$ collocation points. The ratios of the error norms vary strongly, but asymptotically they almost indicate second order convergence.

n	$e_{n,\infty}$		$e_{n,2}$	
2	0.031 18	0.59	0.031 18	0.70
4	0.052 47	1.44	0.044 44	1.50
8	0.036 31	2.62	0.029 55	2.77
16	0.013 86	3.14	0.010 67	3.27
32	0.004 41	3.54	0.003 26	3.57
64	0.001 24	3.59	0.000 91	3.73
128	0 000 83		0 000 25	

Table 4.4.5 Example (22a-c). Collocation with piecewise *constant* interpolation on a graded mesh

4.5 Galerkin Method

4.5.1 Subspace, Orthogonal Projection

The following considerations are based on the assumption that the space X is a *Hilbert space* whose scalar product is denoted by $\langle\cdot,\cdot\rangle$. The natural choice is $X=L^2(D)$. Let X_n be a subspace (*ansatz space*) or, more precisely, an element of a sequence of subspaces:

(4.5.1) $\qquad X_n \subset X.$

A particular projection onto X_n is the orthogonal projection Π_n onto X_n (cf. Definition 1.3.32). *The Galerkin method is the projection method from §4.3 based on the orthogonal projection Π_n onto X_n.* The solution of the Galerkin method with subspace X_n is called the ***Galerkin solution*** in X_n.

Since the orthogonal projection is uniquely defined by X_n (cf. Remark 1.3.33), the convergence of the sequence $\{\Pi_n\}$ of projections is also determined by the sequence of subspaces $\{X_n\}$. Then (1.3.18') proves

Remark 4.5.1. (a) Let $\{\Pi_n\}$ be a sequence of orthogonal projections onto the subspaces X_n. $\{\Pi_n\}$ is *convergent* (i.e., $\Pi_n x\to x$ for all $x\in X$) if and only if (3.1b) holds: $\mathrm{dist}(x,X_n)\to 0$ for all $x\in X$.
(b) Stability of the sequence $\{\Pi_n\}$ is trivial because of $\|\Pi_n\|\leq 1$.

Lemma 4.5.2. Two elements $f,g\in X_n$ coincide if and only if

$$\langle f,\varphi\rangle = \langle g,\varphi\rangle \qquad \text{for all } \varphi\in X_n.$$

Proof. The choice $\varphi:=f-g$ yields $0=\langle f-g,\varphi\rangle=\|f-g\|^2$; hence, $f=g$. QED

For the numerical implementation, it is again necessary to select a *basis* of the subspace X_n, i.e., let us suppose

(4.5.2) $\qquad X_n = \mathrm{span}\{\Phi_{1,n}, \Phi_{2,n}, \ldots, \Phi_{n,n}\}.$

Lemma 4.5.3. Two elements $f,g\in X_n$ coincide if and only if

$$\langle f,\Phi_{j,n}\rangle = \langle g,\Phi_{j,n}\rangle \qquad \text{for all } 1\leq j\leq n.$$

Proof. Let $\varphi \in X_n$ be arbitrary. It has a representation $\varphi = \sum \alpha_j \Phi_{j,n}$. The identity $\langle f, \Phi_{j,n}\rangle = \langle g, \Phi_{j,n}\rangle$ yields $\langle f, \varphi\rangle = \sum \alpha_j \langle f, \Phi_{j,n}\rangle = \sum \alpha_j \langle g, \Phi_{j,n}\rangle = \langle g, \varphi\rangle$ and via Lemma 2 the coincidence $f = g$. QED

Exercise 4.5.4. The functions $\Phi_1, \Phi_2, \ldots, \Phi_n \in X$ are linearly independent if and only if the *Gram matrix*

$$A := \left(\langle \Phi_j, \Phi_k\rangle\right)_{j,k=1,\ldots,n}$$

is non-singular. If A is non-singular, it is also *positive definite*, i.e., $A = A^T$ and $(Ac, c) > 0$ for all $0 \neq c \in \mathbb{R}^n$, where $(\cdot,\cdot)$ is the scalar product in $\mathbb{R}^n$.

4.5.2 Derivation of the System of Equations

From Lemma 3.9 we know that both sides of the projection equation

$$\lambda f_n = g_n + K_n f_n \quad \text{with } K_n := \Pi_n K \quad \text{and } g_n := \Pi_n g \tag{4.5.3}$$

belong to X_n. Lemma 3 enables the reduction of the semidiscrete equation (3) to n scalar equations

$$\lambda \langle f_n, \Phi_{j,n}\rangle = \langle \lambda f_n, \Phi_{j,n}\rangle = \langle g_n + K_n f_n, \Phi_{j,n}\rangle =$$
$$= \langle g_n, \Phi_{j,n}\rangle + \langle K_n f_n, \Phi_{j,n}\rangle.$$

By definition of g_n, we have $\langle g_n, \Phi_{j,n}\rangle = \langle \Pi_n g, \Phi_{j,n}\rangle$. An orthogonal projection is selfadjoint, i.e., $\langle \Pi_n g, \Phi_{j,n}\rangle = \langle g, \Pi_n \Phi_{j,n}\rangle$. Since $\Phi_{j,n} \in X_n$, we obtain $\Pi_n \Phi_{j,n} = \Phi_{j,n}$. Together, we have $\langle g_n, \Phi_{j,n}\rangle = \langle g, \Phi_{j,n}\rangle$. Analogously, the second term becomes

$$\langle K_n f_n, \Phi_{j,n}\rangle = \langle \Pi_n K f_n, \Phi_{j,n}\rangle = \langle K f_n, \Pi_n \Phi_{j,n}\rangle = \langle K f_n, \Phi_{j,n}\rangle.$$

This allows us to simplify the n equations:

discrete Galerkin equation

$$\lambda \langle f_n, \Phi_{j,n}\rangle = \langle g, \Phi_{j,n}\rangle + \langle K f_n, \Phi_{j,n}\rangle \quad \text{for all } 1 \leq j \leq n. \tag{4.5.4}$$

Other authors (e.g., Atkinson - Bogomolny [1]) use the term «discrete Galerkin» to describe the method obtained by approximating the integrals arising from (4) by quadrature.

As in (4.8), we use the ansatz

$$f_n = \sum_{k=1}^{n} \alpha_k \Phi_{k,n}, \tag{4.5.5}$$

which we can insert into (4). The left-hand side becomes

$$\lambda \langle f_n, \Phi_{j,n}\rangle = \lambda \langle \sum_{k=1}^{n} \alpha_k \Phi_{k,n}, \Phi_{j,n}\rangle = \lambda \sum_{k=1}^{n} \alpha_k \langle \Phi_{k,n}, \Phi_{j,n}\rangle,$$

while the right-hand side becomes

$$\langle g,\Phi_{j,n}\rangle + \langle Kf_n,\Phi_{j,n}\rangle = \langle g,\Phi_{j,n}\rangle + \sum_{k=1}^{n}\alpha_k\langle K\Phi_{k,n},\Phi_{j,n}\rangle .$$

The coefficients α_k from (5) are the solution of the system

$$\lambda\sum_{k=1}^{n}\alpha_k\langle\Phi_{k,n},\Phi_{j,n}\rangle = \langle g,\Phi_{j,n}\rangle + \sum_{k=1}^{n}\alpha_k\langle K\Phi_{k,n},\Phi_{j,n}\rangle \qquad (1\le j\le n). \tag{4.5.6}$$

As above (cf. (4.9a-c)), we introduce vectors

$$a := a_n := \begin{bmatrix}\alpha_1\\ \vdots\\ \alpha_n\end{bmatrix}, \qquad b := b_n := \begin{bmatrix}\langle g,\Phi_{1,n}\rangle\\ \vdots\\ \langle g,\Phi_{n,n}\rangle\end{bmatrix} \tag{4.5.7a}$$

and matrices (7b,c):

$$A := A_n := \begin{bmatrix}\alpha_{11} & \alpha_{12} & \cdots & \alpha_{1n}\\ \vdots & \vdots & & \vdots\\ \alpha_{n1} & \alpha_{n2} & \cdots & \alpha_{nn}\end{bmatrix} \qquad \text{with } \alpha_{jk} := \langle\Phi_{k,n},\Phi_{j,n}\rangle \tag{4.5.7b}$$

and

$$B := B_n := \begin{bmatrix}\beta_{11} & \beta_{12} & \cdots & \beta_{1n}\\ \vdots & \vdots & & \vdots\\ \beta_{n1} & \beta_{n2} & \cdots & \beta_{nn}\end{bmatrix} \qquad \text{with } \beta_{jk} := \langle K\Phi_{k,n},\Phi_{j,n}\rangle . \tag{4.5.7c}$$

Remark 4.5.5. (a) A is the positive definite Gram matrix (cf. Exercise 4).
(b) For an orthonormal basis $\{\Phi_{1,n},\dots,\Phi_{n,n}\}$, the matrix becomes $A=I$.
(c) For a *symmetric kernel*, i.e., $k(x,y)=k(y,x)$, different from the collocation method, B is a symmetric matrix as one reads from the representation

$$\beta_{jk} = \int_D\int_D \Phi_{j,n}(x)k(x,y)\Phi_{k,n}(y)\,dx\,dy . \tag{4.5.7c'}$$

Expressing equation (6) be means of the vectors a, b and the matrices A and B, we obtain the representation (8).

> The Galerkin method based on the subspace X_n with the basis $\{\Phi_{1,n},\Phi_{2,n},\dots,\Phi_{n,n}\}$ leads to the linear system
>
> (4.5.8) $\quad (\lambda A_n - B_n)a_n = b_n \quad$ or $\quad (\lambda A - B)a = b$, respectively.

Note that the computation of the data $\{A_n, B_n, b_n\}$ is more expensive than in the case of the collocation method. The coefficients $\alpha_{jk} := \Phi_{k,n}(\xi_{j,n})$ and $g_j = g(\xi_{j,n})$ of the collocation method require only a *function evaluation*, while the coefficients $\alpha_{jk} = \langle\Phi_{k,n},\Phi_{j,n}\rangle$ and $\beta_j = \langle g,\Phi_{j,n}\rangle$ of the Galerkin method involve an *integration*. The largest amount of work arises from the computation of the coefficients β_{jk} from (7c') which require a *double integration*. In contrast, collocation involves only one integration.

4.5.3 Convergence in $L^2(D)$ and $L^\infty(D)$

According to Theorem 3.11, the convergence of the solution f_n of the projection method is convergent in X to the solution $f=(\lambda I-K)^{-1}g$, if the sequence of projections $\{\Pi_n\}$ is convergent and λ is a regular value of the compact operator K. Using Remark 1a and Theorem 3.2.7, we obtain the following sufficient conditions for the case $X=L^2(D)$.

Theorem 4.5.6. Let $X=L^2(D)$ and suppose K is compact, i.e., $K\in K(L^2(D),L^2(D))$ (sufficient for this is $k\in L^2(D\times D)$). Assume that λ is a regular value of K (i.e., $(\lambda I-K)^{-1}\in L(X,X)$). Let the sequence of subspaces satisfy (3.1b): $\operatorname{dist}(x,X_n)\to 0$ for all $x\in X$. *Then the Galerkin method is convergent*: $f_n\to f=(\lambda I-K)^{-1}g$ in X. Furthermore, $K_n\to K$ converges with respect to the operator norm.

Although the choice $X=L^2(D)$ seems to be natural, it is not the only possibility. In particular, the uniform convergence of f_n to f is interesting, which corresponds to the choice $X=C(D)$ or $X=L^\infty(D)$. First, we give a negative result.

Remark 4.5.7. A sequence of projections $\{\Pi_n\}$ converging in $L^2(D)$ need not be convergent in $C(D)$ or $L^\infty(D)$.

Proof. Choose, e.g., $D=[0,2\pi]$ and define $\Pi_n f$ as nth partial sum of the Fourier series of f. Fourier series converge in $L^2([0,2\pi])$, but it is known that $\Pi_n f$ does not converge *uniformly* for all continuous f. ∎

However, uniform convergence of the Galerkin method can be obtained if we assume that the basis $\Phi:=\{\Phi_{1,n},\dots,\Phi_{n,n}\}$ from (2) spanning the subspace X_n leads to a uniformly bounded constant

(4.5.9) $$C_n(\Phi):=\|A^{-1}\|_\infty\Big\|\sum_{j=1}^{n}|\Phi_{j,n}(\cdot)|\Big\|_\infty \max\{\textstyle\int_D|\Phi_{j,n}(x)|dx:\ 1\le j\le n\},$$

where A is the matrix from (7b). This is proved in

Lemma 4.5.8. Let $X_n:=\operatorname{span}\{\Phi_{1,n},\dots,\Phi_{n,n}\}\subset L^1(D)\cap L^\infty(D)$. **(a)** Then

(4.5.10a) $$\|\Pi_n\|_{L^\infty(D)\leftarrow L^\infty(D)}\le C_n(\Phi)$$

holds for the orthogonal projection Π_n onto X_n with $C_n(\Phi)$ from (9). If

(4.5.10b) $$\sup\{C_n(\Phi):\ n\in\mathbb{N}\}<\infty,$$

the sequence of the orthogonal projections $\{\Pi_n\}$ is *stable* in $L^\infty(D)$.
(b) If $\Phi_{j,n}\in C(D)$, statement (a) holds with (10c) instead of (10a):

(4.5.10c) $$\|\Pi_n\|_{C(D)\leftarrow C(D)}\le C_n(\Phi).$$

(c) If there exists a convergent interpolation $\Pi_n':\ C(D)\to(X_n,\|\cdot\|_\infty)$, then $\operatorname{dist}(\varphi,X_n)\to 0$ with respect to the supremum norm of $L^\infty(D)$ for all $\varphi\in C(D)$.
(d) Suppose $\operatorname{dist}(\varphi,X_n)\to 0$ with respect to $\|\cdot\|_\infty$ for all $\varphi\in C(D)$ and suppose (10b) holds, Then the *convergence* $\|\Pi_n\varphi-\varphi\|_\infty\to 0$ follows for all $\varphi\in C(D)$.
(e) Suppose $\|\Pi_n\varphi-\varphi\|_\infty\to 0$ and $K\in K(C(D),C(D))$. Then $\|K-K_n\|_{L^\infty(D)\leftarrow C(D)}\to 0$ in the case (10a) and $\|K-K_n\|_{C(D)\leftarrow C(D)}\to 0$ in the case (10c). If, furthermore, λ is a regular value, then the discretisation $\{K_n\}$ is convergent, stable and consistent with respect to the supremum norm. In particular, if f_n denotes the Galerkin solution, then $\|f_n-f\|_\infty\to 0$.

Proof. (i) Let $\varphi \in L^\infty(D)$ be arbitrary. Let the orthogonal projection $\Pi_n\varphi$ have the representation $\Pi_n\varphi = \sum \alpha_j \Phi_j$. The coefficients α_j form the vector $a \in \mathbb{R}^n$. For each $x \in D$, $|(\Pi_n\varphi)(x)| \leqslant \|a\|_\infty \sum |\Phi_j(x)|$ holds. Hence, we have proved

$$\|\Pi_n f\|_\infty \leqslant \|a\|_\infty \|\sum_{j=1}^{n} |\Phi_{j,n}(\cdot)|\|_\infty .$$

The coefficients α_j are determined from the identity $\langle \Pi_n \varphi, \Phi_j\rangle = \langle \varphi, \Phi_j\rangle$ for all $1 \leqslant j \leqslant n$ (cf. §4.5.2). This yields the equation $Aa = b$ with A from (7b) and $b = (\beta_j)$, $\beta_j := \langle \varphi, \Phi_j\rangle$. Hence,

$$\|a\|_\infty \leqslant \|A^{-1}\|_\infty \|b\|_\infty .$$

From $|\beta_j| = |\langle \varphi, \Phi_j\rangle| \leqslant \|\varphi\|_\infty \|\Phi_j\|_{L^1(D)}$ one concludes that

$$\|b\|_\infty \leqslant \|\varphi\|_\infty \max\{\|\Phi_j\|_{L^1(D)} : 1 \leqslant j \leqslant n\}.$$

Together, the inequalities lead to (10a).
(ii) Let $\Pi_n' : C(D) \to X_n$ be a convergent interpolation. Then $\mathrm{dist}(f, X_n) \leqslant \leqslant \|f - \Pi_n' f\|_\infty \to 0$ follows and proves part (c).
(iii) Part (d) follows from Remark 3.5b, and part (e) from Theorem 3.11. QED

It remains to discuss whether uniform boundedness of the quantities $C_n(\Phi)$ can be satisfied, as required in (10b). We will see that subspaces X_n defined by basis functions $\Phi_{j,n}$ with local support not only lead to numerical advantages but also to good analytical properties.

Criterion 4.5.9. **(a)** Let $\{\Phi_{1,n}, \ldots, \Phi_{n,n}\} \subset L^1(D) \cap L^\infty(D)$ be the basis of X_n. Assume that there are numbers C_1, C_2, and C_3, so that

(4.5.11a) $\quad |\Phi_{j,n}(x)| \leqslant C_1 \quad$ for all $x \in D$, $1 \leqslant j \leqslant n$,

(4.5.11b) $\quad$ for each $x \in D$ there are at most C_2 indices j with $\Phi_{j,n}(x) \neq 0$,

(4.5.11c) $\quad \mu(\mathrm{supp}(\Phi_{j,n})) \leqslant C_3/n \quad$ for all $1 \leqslant j \leqslant n$,

where μ denotes the Lebesgue measure (i.e., length, area, etc.). Then

(4.5.11d) $\quad \|\sum_{j=1}^{n} |\Phi_{j,n}(\cdot)|\|_\infty \leqslant C_1 C_2,$

(4.5.11e) $\quad \max\{\int_D |\Phi_{j,n}(x)| dx : 1 \leqslant j \leqslant n\} \leqslant C_1 C_3 / n.$

(b) Assume that the coefficients α_{jk} of the matrix A satisfy

(4.5.11f) $\quad \sum_{\substack{k=1 \\ k \neq j}}^{n} |\alpha_{jk}| \leqslant q\, \alpha_{jj} \quad$ for all $1 \leqslant j \leqslant n$ with $q < 1$,

(4.5.11g) $\quad \alpha_{jj} \geqslant 1/(C_4 n) \quad$ for all $1 \leqslant j \leqslant n$.

Then

(4.5.11h) $\quad \|A^{-1}\|_\infty \leqslant C_4 n / (1 - q).$

(c) If C_1, C_2, C_3, C_4, and q are independent of n, then the conditions (11a,b,c,f,g) imply the uniform boundedness (10b) of $C_n(\Phi)$:

(4.5.11i) $\quad C_n(\Phi) \leqslant C_1^2 C_2 C_3 C_4 / (1 - q).$

Proof. Part (a) is obvious. To prove (11h), split A into $D-B$, where $D=\mathrm{diag}(\alpha_{11},\ldots,\alpha_{nn})$ and $B:=D-A$. Note that $A=D(I-C)$ holds with $C:=D^{-1}B$. The row-sum norm of C is $\|C\|_\infty \leqslant q<1$ (cf. (11f)); hence, $\|(I-C)^{-1}\|_\infty \leqslant 1/(1-q)$ (cf. Lemma 1.3.14 with $S=I-C$, $T=I$). Since $\|D^{-1}\|_\infty \leqslant C_4 n$ due to (11g), one concludes (11h) from the inequalities $\|A^{-1}\|_\infty \leqslant \|(I-C)^{-1}D^{-1}\|_\infty \leqslant \|(I-C)^{-1}\|_\infty \|D^{-1}\|_\infty$. QED

Criterion 9 will be applied to two examples. The orthogonal projection onto the space of the piecewise constant or linear functions will be discussed in §4.5.6 and §4.5.7, respectively.

4.5.4 Error Estimates

Since the Galerkin method is a special projection method, the estimate (3.12b) is valid (cf. Theorem 3.12):

(4.5.12a) $\quad \|f-f_n\| \leqslant |\lambda|\,\|(\lambda I-K_n)^{-1}\|\,\|f-\Pi_n f\|.$

Theorem 3.12 does not specify the choice of X. In (12a) we may insert the norm $\|\cdot\|=\|\cdot\|_{L^2(D)}$ as well as $\|\cdot\|=\|\cdot\|_\infty$. Stability with respect to $\|\cdot\|$ guarantees

(4.5.12b) $\quad |\lambda|\,\|(\lambda I-K_n)^{-1}\| \leqslant \text{const}.$

Here, stability with respect to $\|\cdot\|_{L^2(D)}$ is ensured under the assumptions of Theorem 6. Conditions sufficient for stability with respect to $\|\cdot\|_\infty$ can be found in Lemma 8e.

The problem remains to derive estimates of $\|f-\Pi_n f\|$ for concrete subspaces X_n. For this purpose, it is necessary to describe the asymptotic error behaviour quantitatively in terms of n. Let $h=h_n$ be the discretisation parameter (in general, $h=O(n^{-1})$ for one- and $h=O(n^{-2})$ for two-dimensional problems; cf. (9c)). If the *projection error* of the exact solution f is of the order $\varkappa$, i.e.,

(4.5.12c) $\quad \|f-\Pi_n f\| \leqslant \text{const}\; h_n^{\varkappa},$

then (12b) and (12c) lead to the *error estimate*

(4.5.12d) $\quad \|f_n-f\| \leqslant \text{const}\; h_n^{\varkappa}.$

Thus, the *Galerkin solution converges with order* $\varkappa$.

For the space X_n of *piecewise constant* functions, we shall prove convergence of the order $\varkappa=1$ with respect to $\|\cdot\|_{L^2(I)}$ and $\|\cdot\|_\infty$ (cf. §4.5.6). However, the numerical results from §4.5.10 will indicate higher order convergence. This so-called «superconvergence» will be discussed in §4.6.2 and §4.6.4. It results in a *second order convergence*: (12d) holds with respect to a suitable norm with $\varkappa=2$ (cf. Remark 6.5); furthermore, certain pointwise error norms are also second order (cf. Lemma 6.18).

Error estimates for orthogonal projection onto *piecewise linear* functions are stated in §4.5.7. There, second order convergence with respect to $\|\cdot\|_{L^2(I)}$ and $\|\cdot\|_\infty$ will be proved, i.e., $\varkappa=2$ in (12d). Even *convergence of fourth order* can be achieved with respect to weaker norms (cf. Remark 6.8 with $k=2$).

In general, for an partition of step size h, piecewise polynomial interpolation of degree $\leqslant p$ leads to an error $O(h^{p+1})$ in (12d) with respect to $L^2(I)$. The weaker norm $\|\cdot\|_{-p-1}$ even yields convergence order $O(h^{2p+2})$ (cf. §4.6.2). By suitable modifications, one can obtain approximations $\tilde{f}_n$ with the error norm $O(h^{2p+2})$ in $L^2(I)$ or $C(I)$.

4.5.5 Condition Number of the System of Equations

The coefficients a of the solution f_n are computed from system (8), i.e., $(\lambda A-B)a=b$. In the following theorem, the condition number of the matrix $\lambda A-B$ is estimated with respect to the Euclidean norm $\|\cdot\|_2$.

Theorem 4.5.10. Let $\lambda A-B$ be non-singular. Then estimate (13) holds:

(4.5.13) $$\operatorname{cond}_2(\lambda A-B) \leqslant \operatorname{cond}_2(A)\operatorname{cond}_{L^2(D)}(\lambda I-K_n).$$

The proof immediately follows from

Lemma 4.5.11. The estimates (14a,b) are valid:

(4.5.14a) $$\|\lambda A-B\|_2 \leqslant \|A\|_2\,\|\lambda I-K_n\|, \qquad (\|\cdot\|=\|\cdot\|_{L^2(D)\leftarrow L^2(D)})$$

(4.5.14b) $$\|(\lambda A-B)^{-1}\|_2 \leqslant \|A^{-1}\|_2\,\|(\lambda I-K_n)^{-1}\|.$$

Proof. (i) First, we remind the reader of a result from linear algebra. A matrix M is *positive definite*, if and only if it is symmetric with only positive eigenvalues. Let $\lambda_{\min}$ be the smallest and $\lambda_{\max}$ the largest eigenvalue of M. For example, by transforming M onto diagonal form, one proves the following properties for such positive definite matrices:

(4.5.15a) $$\lambda_{\min}\|v\|_2^2 \leqslant (Mv,v) \leqslant \lambda_{\max}\|v\|_2^2,$$

(4.5.15b) $$\frac{1}{\lambda_{\max}}\|v\|_2^2 \leqslant (M^{-1}v,v) \leqslant \frac{1}{\lambda_{\min}}\|v\|_2^2,$$

(4.5.16) $$\|M\|_2 = \lambda_{\max}, \qquad 1/\|M^{-1}\|_2 = \lambda_{\min}.$$

Here, $(\cdot,\cdot)$ is the Euclidean scalar product of $\mathbb{R}^n$ with $(v,v)=\|v\|_2^2$.

(ii) An arbitrary $\varphi\in X_n$ has a representation $\varphi=\sum\gamma_j\Phi_j$. The $L^2(D)$ norm of φ and the vector $c:=(\gamma_j)\in\mathbb{R}^n$ are connected by

(4.5.17) $$\|\varphi\|_{L^2(D)}^2 = (Ac,c),$$

as can be seen from $\|\varphi\|^2=\langle\varphi,\varphi\rangle=\langle\sum\gamma_j\Phi_j,\sum\gamma_k\Phi_k\rangle=\sum_{j,k}\gamma_j\langle\Phi_j,\Phi_k\rangle\gamma_k$ and the definition of A.

(iii) Let a vector $a=(\alpha_j)\in\mathbb{R}^n$ be given. Choose $f\in X:=L^2(D)$ by $f:=f_n:=\sum\alpha_k\Phi_k$. f solves the integral equation with the inhomogeneity $g:=(\lambda I-K)f\in X$. Its projection onto X_n has a representation $g_n:=\Pi_n g=\sum\beta_j'\Phi_j$ and satisfies the Galerkin equation $g_n=(\lambda I-K_n)f_n$. The coefficients form the vector $b':=(\beta_j')$, while $b=(\beta_j)$ is determined by $\beta_j:=\langle g_n,\Phi_j\rangle=\langle g,\Phi_j\rangle$. As exercised in part (i) of the proof of Lemma 8, b and b' are connected by $b=Ab'$. Since A is positive definite (cf. Exercise 4), (15b), (16), and (17) can be applied:

$$\|g_n\|^2 = (Ab',b') = (b,A^{-1}b) \geqslant \frac{1}{\|A\|_2}\|b\|_2^2,$$
$$\|f_n\|^2 = \|f\|^2 = (Aa,a).$$

From $\|b\|_2 \leqslant \|A\|_2^{1/2}\|g_n\| = \|A\|_2^{1/2}\|(\lambda I-K_n)f_n\| \leqslant \|A\|_2^{1/2}\|\lambda I-K_n\|\|f_n\|$ and

$$\|f_n\| = \|f\| = \sqrt{(Aa,a)} \leqslant \|A\|_2^{1/2}\|a\|_2$$

we conclude that $\|b\|_2 \leqslant \|A\|_2\|\lambda I-K_n\|\|a\|_2$. Since the coefficient vectors a of f_n and b of g satisfy (8) (i.e., $(\lambda A-B)a=b$), we have $\|(\lambda A-B)a\|_2 \leqslant C\|a\|_2$ with $C:=\|A\|_2\|\lambda I-K_n\|$ for all a and therefore (14a) follows.
(iv) Let $b=(\beta_j)\in\mathbb{R}^n$ be given and set $b'=(\beta_j'):=A^{-1}b$. We choose $g:=g_n:=\sum\beta_j'\Phi_j$. The coefficients of b can be retained as $\beta_j=\langle g,\Phi_j\rangle$. The Galerkin solution $f_n\in X_n$ has the coefficients $a\in\mathbb{R}^n$ satisfying system (8) (i.e., $(\lambda A-B)a=b$). To estimate $a=(\lambda A-B)^{-1}b$, we again use (15a,b), (16), and (17):

$$\|a\|_2^2 \leqslant \|A^{-1}\|_2(Aa,a) = \|A^{-1}\|_2\|f_n\|^2,$$
$$\|f_n\| = \|(\lambda I-K_n)^{-1}g_n\| \leqslant \|(\lambda I-K_n)^{-1}\|\|g_n\|,$$
$$\|g_n\|^2=(Ab',b') = (AA^{-1}b,A^{-1}b) = (b,A^{-1}b) \leqslant \|A^{-1}\|_2\|b\|_2^2.$$

Together: $\|a\|_2=\|(\lambda A-B)^{-1}b\|_2 \leqslant \|(\lambda I-K_n)^{-1}\|\|A^{-1}\|_2\|b\|_2$. QED

Because of consistency and stability of $\{K_n\}$, the factor $\mathrm{cond}_{L^2(D)}(\lambda I-K_n)$ in (13) is uniformly bounded (cf. Theorem 1.15) and, obviously, cannot be influenced by the choice of the basis $\{\Phi_{1,n},\dots,\Phi_{n,n}\}$. Hence, the condition number of system (8) depends on $\mathrm{cond}_2(A)$. This leads to the following suggestion.

Remark 4.5.12. The basis $\{\Phi_{1,n},\dots,\Phi_{n,n}\}$ should be chosen so that the spectral condition number $\mathrm{cond}_2(A_n)$ is as small as possible and, in particular, remains bounded as $n\to\infty$.

The minimal condition number is $\mathrm{cond}_2(A)=1$. This value is attained, e.g., for $A=\alpha I$ with $\alpha\in\mathbb{R}$. Remark 5b gives a condition sufficient for $A=\alpha I$ and proves

Remark 4.5.13. The optimal value $\mathrm{cond}_2(A)=1$ holds for an orthonormal basis $\{\Phi_{1,n},\Phi_{2,n},\dots,\Phi_{n,n}\}$ or an ortho*gonal* basis with $\langle\Phi_{j,n},\Phi_{j,n}\rangle=c_n$ for all $1\leqslant j\leqslant n$. In both cases, $\mathrm{cond}_2(A)\leqslant C_n(\Phi)$ is valid with $C_n(\Phi)$ from (9).

The requirement that the basis should be orthonormal is very restrictive. Either we admit large supports of the $\Phi_{j,n}$. Then several systems of orthonormal functions are available, e.g., trigonometric functions. Or one insists on the advantage of small supports. Then the only simple choice are the piecewise constant functions (wavelets are an interesting but more complicated alternative).

For piecewise constant or linear functions, we shall prove uniform boundedness of the spectral condition number $\mathrm{cond}_2(A)$ (cf. Remark 20 and Exercise 27c). For a suitable scaling of the piecewise constant basis functions, Remark 13 is applicable and proves the optimal result $\mathrm{cond}_2(A)=1$.

Besides the spectral condition number, also the *condition number* of the matrices $\lambda A_n - B_n$ *with respect to the row-sum norm* $\|\cdot\|_\infty$ is of practical interest. The uniform boundedness of $\mathrm{cond}_2(\lambda A_n - B_n)$ implies by no means the uniform boundedness of $\mathrm{cond}_\infty(\lambda A_n - B_n)$, since general $n\times n$ matrices C only satisfy the inequality $\mathrm{cond}_\infty(C) \leqslant n\,\mathrm{cond}_2(C)$. We next prove a result analogous to Theorem 10. Therein, $\mathrm{cond}_2(A)$ is replaced by the quantity $C_n(\Phi)$ from (9).

Theorem 4.5.14. Let $X_n \subset L^1(D) \cap L^\infty(D)$ and $C_n(\Phi)$ be defined as in (9). $\lambda A_n - B_n$ is assumed to be non-singular. Then, the following condition number estimate holds with respect to the row-sum norm $\|\cdot\|_\infty$:

(4.5.18a) $$\mathrm{cond}_\infty(\lambda A_n - B_n) \leqslant \left(C_n(\Phi)\right)^2 \mathrm{cond}_{L^\infty(D)}(\lambda I - K_n).$$

We recall that, following Lemma 8, convergence, consistency, and stability of $\{K_n\}$ with respect to the supremum norm $\|\cdot\|_\infty$ can be concluded mainly from the uniform boundedness property

(4.5.18b) $$\sup_{n\in\mathbb{N}} C_n(\Phi) < \infty .$$

By Theorem 1.15, consistency and stability of $\{K_n\}$ with respect to $\|\cdot\|_\infty$ prove uniform boundedness of $\mathrm{cond}_{L^\infty(D)}(\lambda I - K_n)$. Hence, estimate (18b) is the essential requirement for the uniform boundedness of the condition number of the *semi*discrete as well as the discrete problem. In Criterion 9, a sufficient condition for (18b) was given.

Proof of Theorem 14. We give only a sketch of the proof, since it is similar to that of Theorem 10. $\|\lambda A_n - B_n\|_\infty$ will be estimated in (i), $\|(\lambda A_n - B_n)^{-1}\|_\infty$ in (ii).
(i) Let the basis representation of $g_n \in X_n$ be $g_n = \sum \beta_j' \Phi_j$. This defines the vector $b' := (\beta_j')$. The components β_j of $b := A_n b'$ are $\beta_j = \langle g_n, \Phi_j\rangle = \langle g, \Phi_j\rangle$. $|\langle g_n, \Phi_j\rangle| \leqslant |\int g_n \Phi_j dx| \leqslant \|g_n\|_\infty \int |\Phi_j| dx$ yields

$$\|b\|_\infty \leqslant \|g_n\|_\infty \max\{\|\Phi_j\|_{L^1(D)} : 1 \leqslant j \leqslant n\}.$$

Let the solution f_n of $\lambda f_n = g_n + K_n f_n$ have the representation $f_n = \sum \alpha_k \Phi_k$ and set $a := (\alpha_k)$. Then

$$\|f_n\|_\infty \leqslant \left\|\textstyle\sum_k |\Phi_k|\right\|_\infty \|a\|_\infty .$$

Combining both estimates by means of $\|g_n\|_\infty \leqslant \|\lambda I - K_n\|\,\|f_n\|_\infty$, where $\|\cdot\| = \|\cdot\|_{L^\infty(D)\leftarrow L^\infty(D)}$, one obtains the intermediate result

(4.5.18c) $$\|\lambda A_n - B_n\|_\infty \leqslant \max_j \|\Phi_j\|_{L^1(D)} \left\|\sum_k |\Phi_k|\right\|_\infty \|\lambda I - K_n\|.$$

(ii) Given the vector $b = (\beta_j)$, define $(\beta_j') := b' := A^{-1}b$ and $g_n = g := \sum \beta_j' \Phi_j$. Then we have $\|g_n\|_\infty \leqslant \|b'\|_\infty \left\|\sum_k |\Phi_k|\right\|_\infty \leqslant \|A^{-1}\|_\infty \|b\|_\infty \left\|\sum_k |\Phi_k|\right\|_\infty$. The solution f_n has the representation $f_n = \sum \alpha_k \Phi_k$ with $(\alpha_k) = a := A^{-1}a'$, $a' = (\alpha_k')$,

$\alpha'_k := \langle f_n, \Phi_k \rangle$. This shows $\|a\|_\infty \leq \|A^{-1}\|_\infty \|a'\|_\infty \leq \|A^{-1}\|_\infty \|f_n\|_\infty \max \|\Phi_j\|_{L^1(D)} \leq \leq \|A^{-1}\|_\infty \max \|\Phi_j\|_{L^1(D)} \|(\lambda I - K_n)^{-1}\| \|g_n\|_\infty$. Together with the estimate of $\|g_n\|_\infty$, we obtain the inequality

(4.5.18d) $\quad \|(\lambda A_n - B_n)^{-1}\|_\infty \leq \|A^{-1}\|_\infty^2 \, \|\sum_k |\Phi_k| \, \|_\infty \, \max_j \|\Phi_j\|_{L^1(D)} \, \|(\lambda I - K_n)^{-1}\|.$

(iii) (18c,d) and definition (9) of $C_n(\Phi)$ yield (18a). QED

4.5.6 Example: Piecewise Constant Functions

In detail, we investigate the following example. Here, we choose X to be $L^2(I)$ equipped with the scalar product (1.3.16).

Partition the finite interval $D = I = [a, b]$ by

(4.5.19a) $\quad a = x_{0,n} < x_{1,n} < \dots < x_{n,n} = b \qquad$ (short notation: $x_j = x_{j,n}$)

into the subintervals (19b) with step size h from (19c):

(4.5.19b) $\quad I_k := [x_{k-1}, x_k) \qquad$ for $1 \leq k < n$, $I_n := [x_{n-1}, x_n]$,

(4.5.19c) $\quad h := h_n := \max\{ |x_{k,n} - x_{k-1,n}| : k = 1, \dots, n \}$.

The sequence of the interval partitions must be *quasi-uniform*, i.e., there exists a constant C_Z such that

(4.5.19d) $\quad |x_{k,n} - x_{k-1,n}| \geq h_n / C_Z \qquad$ for all $n \in \mathbb{N}$ and all $k = 1, \dots, n$.

The midpoints of the subintervals I_k are denoted by ξ_k:

(4.5.19e) $\quad \xi_k := \xi_{k,n} := (x_{k-1,n} + x_{k,n})/2 \qquad$ for $k = 1, \dots, n$.

The subspace $X_n \subset L^2(I)$ defining the orthogonal projection Π_n is

(4.5.19f) $\quad X_n := \{ f \in L^2(I) : f \text{ constant on } I_k \text{ for } k = 1, \dots, n \}$.

The special case of an *equidistant partition* is characterised by

(4.5.19g) $\quad x_{j,n} = a + jh, \qquad h = h_n = (b-a)/n, \qquad \xi_j = a + (j - \tfrac{1}{2})h, \qquad C_Z = 1.$

Exercise 4.5.15. From (19c) and (19d) derive $h_n = O(n^{-1})$, more precisely show

(4.5.19h) $\quad h_n \leq C_Z \frac{b-a}{n}, \qquad x_{k,n} - x_{k-1,n} \geq \frac{1}{C_Z} \frac{b-a}{n} \qquad$ for all k.

The Lagrange functions from Remark 1.4.13c are convenient basis elements of X_n:

(4.5.20a) $\quad \Phi_j(x) := \Phi_{j,n}(x) := \begin{Bmatrix} 1 & \text{for } x \in I_j \\ 0 & \text{otherwise} \end{Bmatrix} \qquad$ for $j = 1, \dots, n$.

A scaling of the functions recommended by Remark 20 below is

(4.5.20b) $\quad \Phi_j(x) := \Phi_{j,n}(x) := \begin{Bmatrix} \sqrt{h/[x_j - x_{j-1}]} & \text{for } x \in I_j \\ 0 & \text{otherwise} \end{Bmatrix} \qquad$ for $j = 1, \dots, n$.

For the equidistant case, (20a,b) coincide. Both bases are orthogonal, as it is to be proved in the following exercise.

Exercise 4.5.16. **(a)** The Gram matrix A defined in (7b) is

(4.5.21a) $$A = A_n = \begin{cases} \operatorname{diag}\{x_1 - x_0, x_2 - x_1, \dots, x_n - x_{n-1}\} & \text{for basis (20a)}, \\ hI & \text{for basis (20b)}. \end{cases}$$

(b) For the basis (20a), the matrix entries of B from (7c) are

(4.5.21b) $$\beta_{jk} = \int_{I_j} [\int_{I_k} k(x,y)dy]dx \qquad \text{for } j,\ k=1,\dots,n.$$

For further analysis, the concrete representation of the image $\Pi_n\varphi$ of a function $\varphi \in L^2(I)$ is needed. For $x \in I_j$, $(\Pi_n\varphi)(x)$ turns out to be the mean value of the function φ over the subinterval I_j, as we shall now show.

Lemma 4.5.17. Let X_n be defined by (19f). The orthogonal projection (with respect to $L^2(I)$) onto X_n has the pointwise representation

(4.5.22) $$(\Pi_n\varphi)(x) = \frac{1}{x_j - x_{j-1}} \int_{I_j} \varphi(\xi)d\xi \qquad \text{for } x \in I_j.$$

Proof. $\psi := \Pi_n\varphi$ has a representation of the form $\psi = \sum \alpha_j \Phi_j$. From $\beta_k := \langle \varphi, \Phi_k \rangle = \langle \varphi, \Pi_n \Phi_k \rangle = \langle \Pi_n\varphi, \Phi_k \rangle = \langle \psi, \Phi_k \rangle = \langle \sum_j \alpha_j \Phi_j, \Phi_k \rangle = \sum_j \alpha_{kj}\alpha_j$ one obtains the equation $Aa = b$ for $a = (\alpha_j)$, $b = (\beta_k)$, and A from (7b). (21a) leads to the representation (22). ∎

Since the image of Π_n is discontinuous, Π_n cannot belong to $L(C(I), C(I))$; however, as a projection in $L^\infty(I)$ it is stable. From (22) and $|\int_{I_k} \varphi(\xi)d\xi| \le |x_k - x_{k-1}| \, \|\varphi\|_\infty$ one concludes the following theorem.

Theorem 4.5.18. Let X_n be defined by (19a,b,f). The orthogonal projection with respect to $L^2(I)$ is *stable* not only in $L^2(I)$ but also in $L^\infty(I)$:

(4.5.23) $$\|\Pi_n\|_{L^\infty(D) \leftarrow L^\infty(D)} = 1 \qquad \text{for all } n \in \mathbb{N}.$$

The choice of the nodal points x_k in (19a) does not influences the L^∞ stability; in particular, it is independent of the constants C_Z from (19d). But note that Π_n is *not* convergent in $L^\infty(I)$. Convergence $\Pi_n\varphi \to \varphi$ with respect to the $L^\infty(I)$ norm holds only for $\varphi \in C(I)$.

Exercise 4.5.19. Assume (19a-f) and prove the following: For the respective bases (20a) or (20b), Criterion 9 yields the constants

$$C_1 = 1 \text{ or } \sqrt{C_Z},\ C_2 = 1,\ C_3 = C_Z(b-a),\ q = 0,\ C_4 = \frac{1}{b-a}C_Z \text{ or } \frac{1}{b-a}, \text{ resp.}$$

Conclude from (10c) and (11i) that

(4.5.24) $$C_n(\Phi) \le C_Z^2 \qquad \text{and} \qquad C_n(\Phi) = 1 \text{ in the equidistant case.}$$

Remark 4.5.20. Assume (19a-f). The value of $\operatorname{cond}_2(A)$ is the deciding quantity for the behaviour of the linear system with respect to the Euclidean vector norm (cf. Remark 12). One has

(4.5.25a) $$\operatorname{cond}_2(A) = 1 \text{ for basis (20b)}, \qquad \operatorname{cond}_2(A) \le C_Z \text{ for basis (20a)}.$$

The corresponding amplification factor in the $\|\cdot\|_\infty$ condition number estimate (18a) is uniformly bounded:

(4.5.25b) $$C_n(\Phi)^2 \le C_Z^4.$$

Proof. (25a) follows from (21a) and (16). (25b) is a consequence of (24). ∎

The *estimate of the projection error* $\varphi - \Pi_n\varphi$ depends on the norm used and on the regularity of φ. In the following, quasi-uniformity (19d) is not necessary. Inequality (19h) is needed if $O(h)$ is to be estimated by $O(n^{-1})$.

Lemma 4.5.21. Assume (19a-c,f). **(a)** If $\varphi \in C_L(I)$ has Lipschitz constant L_φ (an upper bound is $L_\varphi \leqslant \|\varphi\|_{C_L(I)}$), then the projection errors are bounded by

(4.5.26a) $$\|\varphi - \Pi_n\varphi\|_\infty \leqslant \tfrac{1}{2} h\, L_\varphi .$$

(b) If $\varphi' \in L^2(I)$, the following estimates hold:

(4.5.26b) $$\|\varphi - \Pi_n\varphi\|_{L^2(I)} \leqslant \tfrac{2}{3} h \,\|\varphi'\|_{L^2(I)},$$

(4.5.26c) $$\|\varphi - \Pi_n\varphi\|_\infty \leqslant \tfrac{2}{3}\sqrt{h}\max\{\|\varphi'\|_{L^2(I_k)} : 1 \leqslant k \leqslant n\} \leqslant \tfrac{2}{3}\sqrt{h}\,\|\varphi'\|_{L^2(I)}.$$

Proof. (i) Consider a subinterval I_k of length $\delta := \delta_k := x_k - x_{k-1}$. Using the representation (22) of the projection, we have

(4.5.27a) $$\varphi(x) - (\Pi_n\varphi)(x) = \frac{1}{\delta}\int_{I_k}[\varphi(x) - \varphi(\xi)]\,d\xi \qquad \text{for } x \in I_k.$$

The Lipschitz estimate $|\varphi(x) - \varphi(\xi)| \leqslant L_\varphi |x - \xi|$ leads to

(4.5.27b) $$|\varphi(x) - (\Pi_n\varphi)(x)| \leqslant \frac{1}{\delta}\int_{I_k}|\varphi(x) - \varphi(\xi)|\,d\xi \leqslant L_\varphi \frac{1}{\delta}\int_{I_k}|x-\xi|\,d\xi =$$
$$= L_\varphi \frac{1}{2\delta}[(x_k - x)^2 + (x - x_{k-1})^2] \leqslant \frac{1}{2\delta} L_\varphi\, \delta^2 = \tfrac{1}{2}\delta\, L_\varphi$$

for $x \in I_k$. Since k is arbitrary, (26a) is proved.
(ii) To estimate the L^2 norm, we represent the integrand $|\varphi(x) - \varphi(\xi)|$ in (27b) by

$$|\varphi(x) - \varphi(\xi)| = \left|\int_\xi^x \varphi'(t)\,dt\right| \qquad (x, \xi \in I_k)$$

and use the *Schwarz inequality* $[\int fg\,dx]^2 \leqslant \int f^2 dx \int g^2 dx$, which is identical to the inequality $|\langle f, g\rangle| \leqslant \|f\|\,\|g\|$ with the L^2 scalar product. Taking $f = \varphi'$ and $g = 1$, we find

(4.5.27c) $$|\varphi(x) - \varphi(\xi)| \leqslant |x - \xi|^{1/2}\left|\int_\xi^x (\varphi'(t))^2 dt\right|^{1/2} \leqslant$$
$$\leqslant |x - \xi|^{1/2}\left|\int_{I_k}(\varphi'(t))^2 dt\right|^{1/2} = |x - \xi|^{1/2}\,\|\varphi'\|_{L^2(I_k)}.$$

Since $\int_{I_k}|x - \xi|^{1/2} d\xi = \frac{2}{3}[(x_k - x)^{3/2} + (x - x_{k-1})^{3/2}] \leqslant \frac{2}{3}\delta^{3/2}$, the previous inequality inserted into (27b) yields

(4.5.27d) $$|\varphi(x) - (\Pi_n\varphi)(x)| \leqslant \tfrac{2}{3}\delta^{1/2}\,\|\varphi'\|_{L^2(I_k)} \qquad \text{for all } x \in I_k.$$

The L^2 norm of $\varphi - \Pi_n\varphi$ over the interval I_k reads

(4.5.27e) $$\|\varphi - \Pi_n\varphi\|_{L^2(I_k)} = \left[\int_{I_k}|\varphi(x) - (\Pi_n\varphi)(x)|^2 dx\right]^{1/2} \leqslant \tfrac{2}{3} h\,\|\varphi'\|_{L^2(I_k)},$$

since $\delta = \delta_k \leqslant h$. This implies (26b) as shown in the next Exercise 22. Inequality (26c) immediately follows from (27d). ∎

Exercise 4.5.22. Let an partition (19a,b) be given. Prove $\|\varphi\|_{L^2(I)} \leq c\,\|\psi\|_{L^2(I)}$ under the assumption that $\|\varphi\|_{L^2(I_k)} \leq c\,\|\psi\|_{L^2(I_k)}$ for all k.

To simplify the notation for L^2 estimates of derivatives, we introduce the H^k norms.

Definition 4.5.23. Let $k \in \mathbb{N}_0$. For functions $\varphi \in L^2(D)$ admitting (generalised or weak) derivatives $\varphi^{(\nu)}$ up to order k, we define

$$\text{(4.5.28)} \qquad \|\varphi\|_{H^k(D)} := \left[\sum_{\nu=0}^{k} \|\varphi^{(\nu)}\|^2_{L^2(D)}\right]^{1/2}.$$

The term «generalised derivative» can be ignored as long as $\varphi \in C^k(D)$. If D has dimension ≥ 2, the sum in (28) must be taken over all *partial* derivatives of order $\leq k$ (i.e., ν is a multi-index $|\nu| \leq k$).

One can verify that $\|\cdot\|_{H^k(D)}$ is a norm. $C^k(D)$ equipped with this norm yields a *non*-complete normed space. The subspace $(C^k(D), \|\cdot\|_{H^k(D)})$ has a completion in $L^2(D)$, which defines the Banach subspace $H^k(D) \subset L^2(D)$ known as a *Sobolev space*. For $k=0$, $H^0(D)$ and $L^2(D)$ coincide. Because of the completion, for $k>0$, the *Sobolev space* $H^k(D)$ contains functions admitting no classical kth derivatives but only generalised (or weak) derivatives (cf. Yosida [1, §I.8], Hackbusch [2, §6.2.1]). A special case of the *Sobolev embedding theorem* is contained in

Lemma 4.5.24. Let $D = I \subset \mathbb{R}$ be an interval. For $k \in \mathbb{N}$, $H^k(I)$ is contained in the space $C^{k-1/2}(I)$ of Hölder continuously differentiable functions. In particular, $\|\varphi\|_{C^{k-1/2}(I)} \leq C_I \|\varphi\|_{H^k(I)}$ holds. This statement is also valid, if D is a one-dimensional (sufficiently smooth) curve in $\mathbb{R}^d$.

Proof. (i) Let $I = [a, b]$ and assume $k=1$. Schwarz' inequality can be applied to the integral in

$$\varphi(x) - \varphi(y) = \int_x^y \varphi'(\xi)\,d\xi = \int_x^y 1 \cdot \varphi'(\xi)\,d\xi \qquad (a \leq x \leq y \leq b)$$

and yields $|\varphi(x) - \varphi(y)|^2 \leq \int_x^y 1^2 d\xi \int_x^y \varphi'(\xi)^2 d\xi \leq |y-x| \int_I \varphi'(\xi)^2 d\xi = |y-x|\,\|\varphi\|^2_{H^1(I)}$. This proves the Hölder continuity with the constant $\|\varphi\|_{H^1(I)}$.
(ii) Analogously, one concludes from $\varphi(x)^2 - \varphi(y)^2 = \frac{1}{2}\int_x^y \varphi(\xi)\varphi'(\xi)\,d\xi$ that $\varphi(x)^2 - \varphi(y)^2 = r(x,y)$ with $|r(x,y)| \leq \frac{1}{2}\|\varphi\|\,\|\varphi'\|$, where $\|\cdot\| = \|\cdot\|_{L^2(I)}$. Integration over x for fixed y leads to $\|\varphi\|^2 = (b-a)\varphi(y)^2 + R$ with $|R| \leq \frac{1}{2}(b-a)\|\varphi\|\,\|\varphi'\|$. Hence, $\varphi^2(y)$ is bounded by $\frac{1}{2}\|\varphi\|\,\|\varphi'\| + \|\varphi\|^2/(b-a) \leq C_I'\|\varphi\|^2_{H^1(I)}$, proving $\|\varphi\|_\infty \leq C_I'^{\,1/2}\|\varphi\|_{H^1(I)}$.
(iii) Together, parts (i) and (ii) imply the assertion with $C_I := \max\{1, C_I'^{\,1/2}\}$. For $b-a \geq 2$, the constant has the value $C_I = 1$ independently of I. It is easy to generalise this result to unbounded intervals.
(iv) One can reduce the analysis for $k>1$ to $k=1$, since $\varphi \in H^k(I)$ has a $(k-1)$st derivative in $H^1(I)$: $\varphi^{(k-1)} \in H^1(I)$. QED

Since $\|\varphi'\|_{L^2(I)} \leq \|\varphi\|_{H^1(I)}$, estimate (26b) can also be written as

$$\text{(4.5.26b')} \qquad \|\varphi - \Pi_n\varphi\|_{L^2(I)} \leq \tfrac{2}{3} h\,\|\varphi\|_{H^1(I)}.$$

Let K be compact in $L^2(I)$. The inequalities (12a-d) and (26b') prove that the Galerkin solution for X_n with piecewise constant functions from (19a-f) converges of first order: $\|f_n - f\|_{L^2(I)} \leqslant C\,h\,\|f\|_{H^1(I)}$.

4.5.7 Example: Piecewise Linear Functions

To retain $\dim(X_n)=n$ in the case of piecewise linear functions, we number the nodal points $1,\dots,n$ instead of $0, 1, \dots, n$ as in (19a). Therefore, the notation reads as follows.

Partition the finite interval $D=I=[a,b]$ by

(4.5.29a) $\quad a = x_{1,n} < \dots < x_{n,n} = b \qquad$ (short notation: $x_j = x_{j,n}$)

into the subintervals (29b) with step size h defined in (29c):

(4.5.29b) $\quad I_k := [x_{k-1}, x_k) \qquad$ for $2 \leqslant k < n$, $I_n := [x_{n-1}, x_n]$,

(4.5.29c) $\quad h := h_n := \max\{ |x_{k,n} - x_{k-1,n}| : k = 2,\dots,n\}$.

Let the partition be quasi-uniform, i.e., there is a number C_Z with

(4.5.29d) $\quad |x_{k,n} - x_{k-1,n}| \geqslant h_n / C_Z \qquad$ for all $n \in \mathbb{N}$ and $k=2,\dots,n$.

The subspace $X_n \subset L^2(I)$ defining the orthogonal projection Π_n is

(4.5.29e) $\quad X_n := \{ f \in C(I) : f \text{ linear on } I_k \text{ for } k=2,\dots,n\}$.

Remark 4.5.25. In the case of periodic functions on I, we have to numerate as in (19a): $a = x_0 < x_1 < \dots < x_n = b$ in order to obtain $\dim(X_n)=n$.

An obvious choice of the basis is (30a) (cf. Fig. 1.4.1):

(4.5.30a) $\quad \Phi_k = L_k \qquad (1 \leqslant k \leqslant n) \qquad$ with $L_k = L_{k,n}$ from (1.4.6).

In extremely non-equidistant cases, the condition number can be improved by a suitable scaling, e.g., by

(4.5.30b) $\quad \Phi_k = \sqrt{3/(\delta_k + \delta_{k+1})}\, L_k \qquad (1 \leqslant k \leqslant n) \qquad$ with L_k as in (30a)

(cf. Remark 27c). δ_k denotes the length of the subinterval I_k:

(4.5.30c) $\quad \delta_k := x_k - x_{k-1} \qquad$ for $2 \leqslant k \leqslant n$, $\delta_1 := \delta_{n+1} := 0$.

Remark 4.5.26. Assume (29a-c,e) and (30c) and choose the basis (30a). In the equidistant ($\delta_k = h$ for all k) or non-equidistant case, one obtains the respective tridiagonal matrices

(4.5.31a)
$$A = \frac{h}{3} \begin{bmatrix} 1 & \frac{1}{2} & & & O \\ \frac{1}{2} & 2 & \frac{1}{2} & & \\ & \frac{1}{2} & \ddots & \ddots & \\ & & \ddots & 2 & \frac{1}{2} \\ O & & & \frac{1}{2} & 1 \end{bmatrix} \quad \text{or} \quad A = \frac{1}{6} \begin{bmatrix} 2(\delta_1+\delta_2) & \delta_2 & & O \\ \delta_2 & 2(\delta_2+\delta_3) & \delta_3 & \\ & \ddots & \ddots & \delta_n \\ O & & \delta_n & 2(\delta_n+\delta_{n+1}) \end{bmatrix}.$$

The scaled basis (30b) leads to the matrix entries

(4.5.31b) $\quad \alpha_{jj} = 1, \qquad \alpha_{j,j-1} = \alpha_{j-1,j} = \frac{1}{2}\delta_j / \sqrt{(\delta_{j-1}+\delta_j)(\delta_j+\delta_{j+1})}.$

Exercise 4.5.27. **(a)** Assume (29a-e). Prove that the basis (30a) satisfies Criterion 9 (with $n-1$ instead of n) with the constants

(4.5.32a) $\quad C_1 = 1, \quad C_2 = 2, \quad C_3 = 2C_Z(b-a), \quad C_4 = \frac{3C_Z}{b-a}, \quad q = \frac{1}{2}.$

Deduce that

(4.5.32b) $\quad C_n(\Phi) \leqslant 24\, C_Z^2$ $\qquad$ (where $C_Z = 1$ in the equidistant case).

The norms of the (symmetric) matrix A are bounded by

(4.5.32c) $\quad \|A\|_2 \leqslant \|A\|_\infty \leqslant h, \qquad \|A^{-1}\|_2 \leqslant \|A^{-1}\|_\infty \leqslant \frac{6C_Z}{b-a}(n-1).$

(b) Assume (29a-c,e) and let the local mesh ratio be bounded by $\delta_j/\delta_{j+1} \in [1/\varkappa, \varkappa]$ with $1 \leqslant \varkappa < 4$ for all $2 \leqslant j \leqslant n$. The choice of the basis (30b) yields

(4.5.32d) $\quad q = \frac{1}{2}\sqrt{\varkappa}, \quad C_4 = 1/(n-1)$ $\qquad$ in (11f,g),

(4.5.32e) $\quad \|A\|_2 \leqslant \|A\|_\infty \leqslant 1 + \frac{1}{2}\sqrt{\varkappa}, \qquad \|A^{-1}\|_2 \leqslant \|A^{-1}\|_\infty \leqslant 2/(2-\sqrt{\varkappa}).$

(c) The spectral condition number of the matrix A is bounded by

(4.5.33) $\quad \operatorname{cond}_2(A) \leqslant \begin{cases} 6\,C_Z^2 & \text{for basis (30a)}, \\ (2+\sqrt{\varkappa})/(2-\sqrt{\varkappa}) & \text{for basis (30b)}. \end{cases}$

Condition $\delta_j/\delta_{j+1} \in [1/\varkappa, \varkappa]$ in part (b) requires that the interval lengths δ_j may not change abruptly from one interval to the other. However, the quotient $\max\delta_j/\min\delta_j$ may become arbitrarily large as n increases and need not be bounded by C_Z as in (29d). This admits «graded grids». The computations reported in Table 4.5 have used such a non-equidistant partition. To construct appropriate graded meshes automatically, one needs error estimators. Concerning this subject, we refer to Faermann [1].

Stability and convergence of the projection Π_n with respect to the supremum norm $\|\cdot\|_\infty$ in $C(I)$ follow from Lemma 8 and (32b).

The analysis of *error in the projection method* is prepared in following subsection and continued in §4.5.9.

4.5.8 General Analysis of Projection Errors

In the case of piecewise linear functions, the representation and estimation of projection error $\varphi - \Pi_n\varphi$ is more involved than in §4.5.6. The reason is that a simple explicit representation of the error as in Lemma 17 is not available. Instead we exploit the *interpolation error*. We suppose the following:

(4.5.34a) $\quad X_n \subset X \cap C(D)$ $\qquad$ (X: Hilbert space).

The Lagrange functions $\Phi_j = L_j$ are chosen as basis in X_n:

(4.5.34b) $\quad X_n = \operatorname{span}\{\Phi_1, \ldots, \Phi_n\}, \qquad \Phi_j(x_k) = \delta_{jk},\ x_k \in D.$

The interpolation in X_n with the support abscissae x_k is denoted by $\hat{\Pi}_n$. It has the following representation:

(4.5.34c) $\quad \hat{\Pi}_n \varphi = \sum \hat{\alpha}_j \Phi_j \quad$ with $\hat{\alpha}_j := \varphi(x_j) \qquad$ for $\varphi \in C(D)$

(cf. (1.4.7)). The *interpolation error* is denoted by

(4.5.34d) $\quad \delta\varphi := \varphi - \hat{\Pi}_n \varphi.$

We suppose that error estimates for $\delta\varphi$, e.g., with respect to $\|\cdot\|_{L^2(D)}$ and $\|\cdot\|_\infty$, are known. In the following, we want to transfer these estimates to the (orthogonal) projection error $\varphi - \Pi_n\varphi$. The relationship between $\delta\varphi$ and $\varphi - \Pi_n\varphi$ is described in Lemmata 28 and 29.

Lemma 4.5.28. Let Π_n be the orthogonal projection onto $X_n \subset X \cap C(D)$ with respect to the scalar product in X. For $\varphi \in X \cap C(D)$, we have

(4.5.35) $\quad \|\varphi - \Pi_n\varphi\|_X \leqslant \|\delta\varphi\|_X \qquad$ with $\delta\varphi$ from (34d).

Proof. Remark 1.3.33c shows $\|\varphi - \Pi_n\varphi\|_X \leqslant \|\varphi - \psi\|_X$ for all $\psi \in X_n$ and in particular for $\psi = \hat{\Pi}_n \varphi$. QED

Estimate (35) has two disadvantages. The inequality may yield a pessimistic estimate of $\|\varphi - \Pi_n\varphi\|_X$. Further, it is restricted to the norm $\|\cdot\|_X$ of the Hilbert space, in particular, to the L^2 norm in the usual applications. (35) yields no $\|\cdot\|_\infty$ estimate. To obtain a more precise analysis, we prove the following exact representation of the projection error.

Lemma 4.5.29. Let (34a-c) hold. Define the vector $\hat{a} = (\hat{\alpha}_j) \in \mathbb{R}^n$ by $\hat{\alpha}_j$ from (34c). The matrix $A = A_n$ results from (7b) with the Lagrange basis (34b). Then, the projection error $e = e_n = \varphi - \Pi_n\varphi$ at the abscissae x_k takes the values

(4.5.36a) $\quad \varepsilon_k = e(x_k) = (\varphi - \Pi_n\varphi)(x_k) \qquad (1 \leqslant k \leqslant n),$

where the vector $\varepsilon = (\varepsilon_k)$ is the solution of equation (36b):

(4.5.36b) $\quad A\varepsilon = \delta$ with $\delta = (\delta_j), \qquad \delta_j := -\langle \Phi_j, \delta\varphi \rangle, \qquad \delta\varphi$ from (34d).

Proof. As in the proof of Lemma 17, one shows the representation

(4.5.36c) $\quad \Pi_n\varphi = \sum \alpha_j \Phi_j \quad$ with $a = (\alpha_j), \qquad Aa = b = (\beta_j), \qquad \beta_j = \langle \varphi, \Phi_j \rangle.$

Hence, $\varepsilon_k = e(x_k) = \varphi(x_k) - \sum \alpha_j \Phi_j(x_k) = \varphi(x_k) - \alpha_k$ (cf. (34b)) and $\delta := A\varepsilon = Ac - Aa = Ac - b$ with $c = (\varphi(x_k))_{k \in \mathbb{N}}$. From $\hat{\Pi}_n \varphi = \sum \varphi(x_k)\Phi_k$ one concludes equation (36b):

$$\delta_j = \sum \alpha_{jk}\varphi(x_k) - \beta_j = \sum \langle \Phi_k, \Phi_j \rangle \varphi(x_k) - \langle \varphi, \Phi_j \rangle =$$
$$= \langle \sum \varphi(x_k)\Phi_k - \varphi, \Phi_j \rangle = \langle \hat{\Pi}_n \varphi - \varphi, \Phi_j \rangle = -\langle \delta\varphi, \Phi_j \rangle.$$ QED

Corollary 4.5.30. The projection error $\varphi - \Pi_n\varphi$ has the representation

(4.5.36d) $\quad \varphi - \Pi_n\varphi = \delta\varphi + \sum \varepsilon_j \Phi_j \qquad$ with ε_j from (36a,b).

Any norm $\|\cdot\|$ defined on X_n and φ satisfies the estimate

(4.5.36e) $\quad \|\varphi - \Pi_n\varphi\| \leqslant \|\delta\varphi\| + \|\sum \varepsilon_j \Phi_j\|$.

In the following obvious estimates of $\|\sum\varepsilon_j\Phi_j\|$, the quantity $C_n(\Phi)$ is defined in (9):

(4.5.36f) $\quad \|\sum\varepsilon_j\Phi_j\|_\infty \leqslant \|\varepsilon\|_\infty \|\sum|\Phi_j|\|_\infty$,

(4.5.36g) $\quad \|\sum\varepsilon_j\Phi_j\|_{L^2(D)} \leqslant \|A\|_2^{1/2}\|\varepsilon\|_2 \qquad (\|\varepsilon\|_2^2 = \sum\varepsilon_j^2)$.

The representation (36b) implies the following bounds for $\|\varepsilon\|$:

(4.5.36h) $\quad \|\varepsilon\|_\infty \leqslant \|A^{-1}\|_\infty \|\delta\|_\infty$,

(4.5.36i) $\quad \|\varepsilon\|_2 \leqslant \|A^{-1}\|_2\|\delta\|_2 \leqslant \|A^{-1}\|_2^{1/2}\sqrt{C_n(\Phi)}\,\|\delta\varphi\|_{L^2(D)}$.

Altogether, (36f-i) yield

(4.5.36j) $\quad \|\sum\varepsilon_j\Phi_j\|_\infty \leqslant C_n(\Phi)\|\delta\varphi\|_\infty$,

(4.5.36k) $\quad \|\sum\varepsilon_j\Phi_j\|_{L^2(D)} \leqslant [\operatorname{cond}_2(A)\,C_n(\Phi)]^{1/2}\|\delta\varphi\|_{L^2(D)}$.

Under condition (10b) (i.e., $C_n(\Phi)\leqslant C$), the results (36e,j,k) imply the following estimate for $\|\cdot\|_2$ and $\|\cdot\|_\infty$:

(4.5.36l) $\quad \|\varphi - \Pi_n\varphi\| \leqslant \text{const}\,\|\delta\varphi\|$.

Proof. The nodal values of $(\varphi-\Pi_n\varphi)-\delta\varphi = (\varphi-\Pi_n\varphi)-(\varphi-\hat{\Pi}_n\varphi) = (\hat{\Pi}_n-\Pi_n)\varphi$ are ε_k (cf. (36a)). Hence, $(\hat{\Pi}_n-\Pi_n)\varphi = \sum\varepsilon_k\Phi_k$ holds and proves (36d). Concerning the further estimates, we shall prove only (36i) will be demonstrated. By definition, $|\delta_j| = |\langle\delta\varphi,\Phi_j\rangle| \leqslant \|\delta\varphi|\Phi_j|^{1/2}\|_{L^2(D)}\,\||\Phi_j|^{1/2}\|_{L^2(D)}$ holds and implies

$$\begin{aligned}\|\delta\|_2^2 &= \sum\delta_j^2 \leqslant \textstyle\sum_j\left[\int\delta\varphi^2|\Phi_j(x)|\,dx\right]\left[\int|\Phi_j(x)|\,dx\right] \leqslant \\ &\leqslant \max_j\|\Phi_j\|_{L^1(D)}\textstyle\sum_j\left[\int\delta\varphi^2|\Phi_j(x)|\,dx\right] = \\ &= \max_j\|\Phi_j\|_{L^1(D)}\left[\int\delta\varphi^2\textstyle\sum_j|\Phi_j(x)|\,dx\right] \leqslant \\ &\leqslant \max_j\|\Phi_j\|_{L^1(D)}\|\sum|\Phi_j|\|_\infty\int\delta\varphi^2\,dx.\end{aligned}$$

Since A is symmetric, $\|A^{-1}\|_2^{1/2}$ can be estimated by $\|A^{-1}\|_\infty^{1/2}$. QED

4.5.9 Revisited: Piecewise Linear Functions

According to §4.5.8, we have to start with the investigation of the interpolation error $\delta\varphi := \varphi - \hat{\Pi}_n\varphi$.

Lemma 4.5.31. Partition $I=[a,b]$ as in (29a-c). For the piecewise linear interpolation $\hat{\Pi}_n$ at the support abscissae x_k $(1\leqslant k\leqslant n)$, the following error estimates hold:

(4.5.37a) $\quad \|\varphi - \hat{\Pi}_n\varphi\|_\infty \leqslant \frac{1}{8}L_{\varphi'}\,h_n^2 \qquad$ for $\varphi\in C_L^1(I)$,

where $L_{\varphi'} \leqslant \|\varphi\|_{C_L^1(I)}$ is the Lipschitz constant of φ';

(4.5.37b) $\quad \|\varphi - \hat{\Pi}_n\varphi\|_{L^2(I)} \leqslant 90^{-1/2}h_n^2\|\varphi''\|_{L^2(I)} \qquad$ for $\varphi\in H^2(I)$.

Proof. (i) Repeat the proof of Remark 1.4.11 for $\lambda=2$ and improve the constant C_I from $1/4$ to $1/8$. This shows (37a).
(ii) Show that the error $\delta\varphi := \varphi - \hat{\Pi}_n\varphi$ satisfies $\delta\varphi(x_{k-1}) = \delta\varphi(x_k) = 0$ and $\delta\varphi'' = \varphi''$ and deduce the following «Peano kernel representation»:

(4.5.37c$_1$) $\quad (\varphi - \hat{\Pi}_n\varphi)(x) = \int_{I_k} \varkappa(x,\xi)\,\varphi''(\xi)\,d\xi \qquad$ for all $x \in I_k$ with

(4.5.37c$_2$) $\quad \varkappa(x,\xi) = \max\{0, x-\xi\} - (x - x_{k-1})(x_k - \xi)/(x_k - x_{k-1})$.

The Schwarz inequality shows $|\delta\varphi(x)| \leq \|\varkappa(x,\cdot)\|_{L^2(I_k)} \|\varphi''\|_{L^2(I_k)}$. Evaluation of $\int_{I_k} \varkappa(x,\xi)^2 d\xi$ yields $\|\varkappa(x,\cdot)\|_{L^2(I_k)} = (x-x_{k-1})(x_k-x)/\sqrt{3(x_k-x_{k-1})}$ for all $x \in I_k$. Integration over I_k proves

$$\|\varphi - \hat{\Pi}_n\varphi\|^2_{L^2(I_k)} \leq \int_{I_k} \frac{(x-x_{k-1})^2(x_k-x)^2}{3(x_k-x_{k-1})}\,dx\, \|\varphi''\|^2_{L^2(I_k)} =$$
$$= \tfrac{1}{90}(x_k - x_{k-1})^4 \|\varphi''\|^2_{L^2(I_k)} \leq \tfrac{1}{90} h_n^4 \|\varphi''\|^2_{L^2(I_k)}.$$

From Exercise 22 one obtains the assertion. QED

We remark that (37c) can also be used for the proof of (37a). The derivative of $\varphi \in C^1_L(I)$ is of bounded variation; hence, (37c) can be written as Stieltjes integral: $\int_{I_k} \varkappa(x,\xi)\,d\varphi'(\xi)$. The estimate $|d\varphi'(\xi)| \leq L_{\varphi'}d\xi$ of the total variation proves $\left|\int_{I_k} \varkappa(x,\xi)\,d\varphi'(\xi)\right| \leq L_{\varphi'} \int_{I_k} |\varkappa(x,\xi)|\,d\xi = \frac{1}{2} L_{\varphi'}(x-x_{k-1})(x_k-x) \leq \frac{1}{8} L_{\varphi'} h_n^2$.

The results of the previous section leads us to

Theorem 4.5.33. **(a)** Partition $I=[a,b]$ by (29a-c). The orthogonal projection Π_n onto the space X_n of piecewise linear functions (cf. (29e)) satisfies the error estimates

(4.5.38a) $\quad \|\varphi - \Pi_n\varphi\|_{L^2(I)} \leq 90^{-1/2} h_n^2 \|\varphi''\|_{L^2(I)} \quad$ for $\varphi \in H^2(I)$,

(4.5.38b) $\quad \|\varphi - \Pi_n\varphi\|_\infty \leq (1 + C_n(\Phi)) \|\varphi - \hat{\Pi}_n\varphi\|_\infty \leq (1 + C_n(\Phi)) \frac{1}{8} L_{\varphi'} h_n^2$

for $\varphi \in C^1_L(I)$ with the Lipschitz constant $L_{\varphi'}$ of φ'. Assuming in addition quasi-uniformity (29d), we can derive the estimate

(4.5.38c) $\quad \|\varphi - \Pi_n\varphi\|_\infty \leq (\frac{1}{8} + 3C_Z^2) L_{\varphi'} h_n^2 \leq (\frac{1}{8} + 3C_Z^2) h_n^2 \|\varphi\|_{C^1_L(I)}$.

(b) If λ is a regular value of $K \in K(X,X)$ for $X = L^2(I)$ or $C(I)$ and if the solution f of the integral equation belongs to $H^2(I)$ or $C^1_L(I)$, respectively, the respective inequalities (38c) or (38a) imply the error estimate (38d) for the Galerkin solution f_n:

(4.5.38d) $\quad \|f_n - f\|_{L^2(I)} \leq C h_n^2 \|f\|_{H^2(I)} \quad$ or $\quad \|f_n - f\|_\infty \leq C h_n^2 \|f\|_{C^1_L(I)}$.

Proof. (i) Because of Lemma 28, (38a) is an immediate consequence of (37b). Inequality (38b) can be concluded from (36e,j) and (37a). For estimating $C_n(\Phi)$, one uses (32b), i.e., $C_n(\Phi) \leq 24\,C_Z^2$, and obtains (38c).
(ii) Part (b) follows by applying Theorem 6. QED

Exercise 4.5.34. Repeat the considerations of Corollary 30 and Lemma 31 for $\|\cdot\| = \|\cdot\|_{L^1(D)}$ and prove

(4.5.38e) $\quad \| \varphi - \Pi_n \varphi \|_{L^1(D)} \le (1 + C_n(\Phi)) \| \varphi - \hat{\Pi}_n \varphi \|_{L^1(D)}$, where

(4.5.38f) $\quad \| \varphi - \hat{\Pi}_n \varphi \|_{L^1(D)} \le \frac{1}{8} h_n^2 \| \varphi'' \|_{L^1(D)}$.

Hint. The corresponding vector norm is $\|\varepsilon\|_1 = \sum |\varepsilon_j|$. The associated matrix norm satisfies $\|A^{-1}\|_1 = \|A^{-1}\|_\infty$ for symmetric A^{-1}.

4.5.10 Numerical Examples

As in §4.2.8 and §4.4.5, the first example is given by (2.19a), (2.21a,b) with

$$k(x,y) = \cos(\pi x y), \qquad f(x) = e^x \cos(7x), \qquad I = [0,1], \qquad \lambda = 1.$$

For the discretisation we choose the *piecewise constant functions* for an equidistant partition of the interval $I = [0,1]$. The integral in $K\Phi_k$ can still be evaluated explicitly:

(4.5.39) $$(K\Phi_k)(x) = \int_{(k-1)h}^{kh} \cos(\pi x y)\, dy = h \frac{\sin x\pi h/2}{x\pi h/2} \cos \pi x \xi_k,$$

where $\xi_k := (k - \frac{1}{2})h \;(1 \le k \le n)$ is the midpoint (19e) of I_k. To compute $\beta_{jk} = \langle K\Phi_k, \Phi_j \rangle$, we have to integrate the function (39) a second time over $x \in I_j$. Since this integral is not representable elementarily, we approximate β_{jk} by the (non-composed) Simpson rule (1.4.18d) with step size $\frac{h}{2}$ and support abscissae $x_{j-1} = (j-1)h$, $\xi_j = (j - \frac{1}{2})h$, $x_j = jh$:

(4.5.40a) $$\tilde{\beta}_{jk} = \frac{h}{6}[(K\Phi_k)(x_{j-1}) + 4(K\Phi_k)(\xi_j) + (K\Phi_k)(x_j)],$$

(4.5.40b) $$\tilde{b}_j = \frac{h}{6}[g(x_{j-1}) + 4g(\xi_j) + g(x_j)] \qquad \text{with } g \text{ from (2.21b).}$$

h	$e_{h,\infty}$		$e_{h,2}$	
1/2	0.240 00		0.175 97	
		1.02		1.02
1/4	0.234 53		0.173 16	
		3.14		3.92
1/8	0.074 67		0.044 19	
		3.88		3.97
1/16	0.019 23		0.011 12	
		4.00		3.99
1/32	0.004 81		0.002 78	
		3.99		4.00
1/64	0.001 21		0.000 70	

Table 4.5.1 Galerkin method for example (2.19a), (2.21a,b), piecewise *constant* functions, numerical quadrature (Simpson rule)

h	$e_{h,\infty}$		$e_{h,2}$	
1/2	0.744 50		0.486 39	
		1.00		1.18
1/4	0.741 13		0.413 31	
		4.21		4.34
1/8	0.175 93		0.095 34	
		4.19		4.27
1/16	0.042 00		0.022 81	
		4.03		4.07
1/32	0.010 43		0.005 60	
		3.97		4.04
1/64	0.002 62		0.001 39	

Table 4.5.2 Galerkin method for example (2.19a), (2.21a,b), piecewise *linear* functions, numerical quadrature (Simpson rule)

Since the relative error of Simpson's rule is $O(h^4)$, the quadrature errors of $\tilde{\beta}_{jk}$ and $\tilde{b}_j$ are negligible compared to the discretisation error. Due to (21a), $A = hI$. The solution of equation $(A-B)a = b$ produces the values $f_n(\xi_j) = \sum \alpha_k \Phi_k(\xi_j) = \alpha_j$ at the midpoints ξ_j. In Table 1, the error norms

$$e_{h,\infty} := \max\{|f_n(\xi_j) - f(\xi_j)| : 1 \le j \le n\},$$

$$e_{h,2} := \{h \textstyle\sum_k |f_n(\xi_j) - f(\xi_j)|^2\}^{1/2}$$

are given. The ratios lying very close to 4 indicate second order convergence (superconvergence, cf. Lemma 6.18) and are superior to the linear convergence predicted in §4.5.6.

Since Simpson's rule is more accurate than is necessary, one could integrate $K\Phi_k$ and g over I_j by means of the tangential trapezoidal rule (1.4.18e):

$$(4.5.40c) \qquad \tilde{\beta}_{jk} = h\,(K\Phi_k)(\xi_j), \qquad \tilde{b}_j = h\,g(\xi_j).$$

The respective results are already given in Table 4.2 (numerical results for *collocation*), since the following holds:

Exercise 4.5.35. Consider the piecewise constant functions defined above. Prove that, up to an irrelevant scaling factor h, the tangential trapezoidal rule (40c) produces the same matrix B and the same vector b as the collocation method for piecewise constant interpolation at the interval midpoints ξ_k. Since also A coincides for both discretisations, the solutions f_n are identical.

For piecewise linear functions, the integral $(K\Phi_k)(x)$ is of the form $h\cos(\pi x\, x_k)\left[\frac{\sin x\pi h/2}{x\pi h/2}\right]^2$ when $x_k \in (0,1)$ is an inner nodal point. Here again, the use of numerical quadrature for the scalar products $\langle\Phi_j, g\rangle$ and $\langle\Phi_j, K\Phi_k\rangle$ cannot be avoided. Simpson's rule with step size $h/2$ in $[x_{j-1}, x_j]$ and $[x_j, x_{j+1}]$ yields the approximation

$$(4.5.40d) \qquad \langle\Phi_j, g\rangle \approx \tfrac{h}{3}\left[g(x_j - \tfrac{h}{2}) + g(x_j) + g(x_j + \tfrac{h}{2})\right] \quad \text{at inner nodal points } x_j.$$

Exercise 4.5.36. (a) What is the formula corresponding to (40d) at the boundary points $x_j = 0$ and $x_j = 1$?
(b) Why is Simpson's rule with step size h (instead of $\frac{h}{2}$) unsatisfactory?

As second example, we use (4.22a-c) with the kernel $|x-y|^{-1/2}$. For *piecewise constant* functions, the double integration involved in $\beta_{jk} = \langle K\Phi_k, \Phi_j\rangle$ can be performed analytically, to obtain

$$\beta_{jk} = \tfrac{4}{3} h^{3/2}\left[\,\big||k-j|+1\big|^{3/2} + \big||k-j|-1\big|^{3/2} - 2\,|k-j|^{3/2}\right].$$

Since function g from (4.22c) is too complicated to allow an exact integration, Simpson's rule with step size $h/2$ is applied. The results are shown in Table 3. The errors are very similar to the collocation results from Table 4.5. Since f lies «almost» in $H^1(I)$ (more precisely: in $H^s(I)$ for $s<1$, so that $\|f - \Pi_n f\|_{L^2} \leq C_s h^s$) the considerations of this section yield an error $\approx O(h)$. Also §4.6.4 will predict more pessimistic convergence orders than observed in Table 3.

h	$e_{h,\infty}$		$e_{h,2}$	
1/2	0.038 92	0.49	0.038 92	0.49
1/4	0.079 62	1.73	0.072 63	2.21
1/8	0.045 94	2.63	0.032 80	3.19
1/16	0.017 45	3.16	0.010 29	3.67
1/32	0.005 51	3.55	0.002 80	3.61
1/64	0.001 55	3.98	0.000 78	3.25
1/128	0.000 39		0.000 24	

Table 4.5.3 Galerkin method for Example (4.22a-c) with piecewise constant functions, numerical quadrature of $\langle\Phi_j, g\rangle$ by Simpson's rule

4.6 Additional Comments Concerning Projection Methods

Sections 4.6.1 and 4.6.3 will combine the discretisations introduced above with other algorithms. §4.6.2 and §4.6.4 concern the convergence analysis. §4.6.5 generalises the formulation of the projection method. Finally, §§4.6.6-7 discusses numerical quadrature and its influence on the accuracy of the solution.

4.6.1 Regularisation Method

The error representations (3.12a,c) connect the error in f_n with the projection error $f - \Pi_n f$. However, the latter is sufficiently small only if f is smooth. This is again illustrated by

Exercise 4.6.1. Consider $f(x) = \sqrt{x}$ in $[0,1]$. Define the projection Π_n by piecewise linear interpolation at the support abscissae $\{0, h, 2h, \ldots\}$. Although the interpolation is of second order, the interpolation error equals $\| f - \Pi_n f \|_\infty = \sqrt{h}/4$. The exponent $\frac{1}{2}$ of $\sqrt{h}$ corresponds to the smoothness property of $f \in C^{1/2}(I)$.

As a consequence, the numerical solution by the projection method above is satisfactory only if the solution is smooth. Because of the regularity theorem (Theorem 3.5.1), the solution f is smooth, provided that the inhomogeneity g in the equation $\lambda f = g + Kf$ is sufficiently smooth. But what can be done, if g is not smooth?

Assume that the integral operator has a smoothing effect: $K \in L(C(I), C^\varkappa(I))$. If $g \in C^\varkappa(I)$, Theorem 3.5.1 yields $f \in C^\varkappa(I)$. In the following, nothing more than $f \in C(I)$ is assumed. The solution has the representation $f = \frac{1}{\lambda}(g + Kf)$. On the right-hand side, Kf/λ is unknown, but the assumption on K ensures $Kf/\lambda \in C^\varkappa(I)$. Therefore, we may make the ansatz

$$f = \frac{1}{\lambda} g + \varphi \tag{4.6.1a}$$

with a function $\varphi \in C^\varkappa(I)$ still to be determined. Inserting (1a) into the integral equation, we obtain

$$g + \lambda\varphi = \lambda f = g + Kf = g + \frac{1}{\lambda} K g + K\varphi.$$

Hence, we have derived the following integral equation for φ:

$$\lambda\varphi = \psi + K\varphi \qquad \text{with } \psi = \frac{1}{\lambda} K g. \tag{4.6.1b}$$

The assumptions $g \in C(I)$ and $K \in L(C(I), C^\varkappa(I))$ guarantee $\psi \in C^\varkappa(I)$. Applying a projection method of order $\varkappa$ to (1b), one obtains the full convergence order $\| \varphi - \varphi_n \| = O(h^\varkappa)$. (1a) leads to the approximation

$$f_n := \frac{1}{\lambda} g + \varphi_n \qquad (\varphi_n\text{: projection solution of (1b)}) \tag{4.6.1c}$$

of f. Since $f - f_n = \varphi - \varphi_n$, f_n also has the optimal convergence order. The name «regularisation» for this method is due to the fact that the original problem is reformulated into a new problem (1b) possessing a sufficiently regular (smooth) solution.

The regularisation method is applicable even if the solution f does not belong to basic space $C(I)$ or $L^2(I)$. As an extreme example we take the equation $\lambda f = g + Kf$ with the Dirac delta function $g = \delta_z$ for an $z \in I$. The distribution (generalised function) $\delta_z \in C(I)'$ is the functional with $\delta_z(f) = f(z)$ for any $f \in C(I)$. In this case, the collocation method fails, since δ_z cannot be interpolated reasonably. The Galerkin method can be defined, but all the previous error estimates are not applicable, since δ_z does not belong to $L^2(I)$. In this case, the function $\psi = \frac{1}{\lambda} Kg$ becomes $\psi(x) = \frac{1}{\lambda} k(x,z)$ and is sufficiently smooth, if the kernel k is smooth with respect to the first argument. Therefore, the discretisation of equation (1b) is easy. According to (1c), the solution reads $f_n = \frac{1}{\lambda}\delta_z + \varphi_n$. In an indirect form this method is described in Sloan [4].

The described method can be iterated. If the inhomogeneity ψ of (1b) is still not smooth enough, one can regularise equation (1b) a second time, provided that $K\psi$ is smoother than ψ.

4.6.2 Estimates with Respect to Weaker Norms

In Remark 1.3.3, two norms $\|\cdot\|$ and $|||\cdot|||$ are called *equivalent*, if one can be estimated in terms of the other and vice versa. If, however, $|||\cdot||| \le C\|\cdot\|$ holds but not $\|\cdot\| \le C'|||\cdot|||$, then $|||\cdot|||$ represents a *weaker norm* than $\|\cdot\|$. Obviously, an error estimate $\|f_n - f\| = O(h^\alpha)$ also implies $|||f_n - f||| = O(h^\alpha)$ with respect to the weaker norm, whereas, in general, the converse is false. For example, $\|\cdot\|_{L^2(I)}$ is weaker than the maximum norm $\|\cdot\|_{\infty,I}$ for bounded I. In (5.26c), $\|\Pi_n g - g\|_\infty = O(h^{1/2})$ for $g \in H^1(I)$ is the optimal convergence order. This leads to the inequality $\|\Pi_n g - g\|_{L^2(I)} = O(h^{1/2})$ which is no longer optimal, since in (5.26b) we have proved $\|\Pi_n g - g\|_{L^2(I)} = O(h)$.

The last example gives rise to the hope that one can achieve higher (better) convergence orders by means of weaker norms. But notice that the better convergence order is compensated by a weaker measuring of the error. Nevertheless, there are cases where an estimate by a weaker norm is completely sufficient. A norm even weaker than the $L^2(I)$ norm is the $(H^1(I))'$ dual norm.

Remark 4.6.2. For all $f \in L^2(D)$, the $H^1(D)'$ norm is defined by (2) as the dual norm of the Sobolev norm $\|\cdot\|_{H^1(D)}$ (cf. end of §1.3.6):

(4.6.2) $$\|f\|_{-1} := \|f\|_{H^1(D)'} := \sup\{|(f,\varphi)_{L^2(D)}| / \|\varphi\|_{H^1(D)} : 0 \ne \varphi \in H^1(D)\},$$

where $(\cdot,\cdot)_{L^2(D)}$ is the L^2 scalar product (1.3.16). $\|\cdot\|_{-1}$ represents a norm on $L^2(D)$. The completion of $L^2(D)$ with respect to the $\|\cdot\|_{-1}$ norm yields the Hilbert space $H^1(D)'$, which is the dual space of $H^1(D)$. The embedding $L^2(D) \subset H^1(D)'$ is continuous, i.e., $\|f\|_{-1} \le \|f\|_{L^2(D)}$. The $L^2(D)$ scalar product can be extended continuously onto $H^1(D) \times H^1(D)'$ (cf. Lemma 4c), so that

(4.6.3) $$|(f,\varphi)_{L^2(D)}| \le \|f\|_{H^1(D)} \|\varphi\|_{-1} \qquad \text{for all } f \in H^1(D),\ \varphi \in H^1(D)'.$$

The reason for introducing the $\|\cdot\|_{-1}$ norm is given in the following result.

Lemma 4.6.3. Let Π_n be the orthogonal projection onto $X_n \subset L^2(D)$ and assume

(4.6.4a) $$\|\Pi_n\varphi - \varphi\|_{L^2(D)} \le Ch\|\varphi\|_{H^1(D)}.$$

Then the following inequalities hold with the same constant C:

(4.6.4b) $\qquad \|\Pi_n\varphi - \varphi\|_{-1} \leqslant Ch\|\varphi\|_{L^2(D)},$

(4.6.4c) $\qquad \|\Pi_n\varphi - \varphi\|_{-1} \leqslant C^2h^2\|\varphi\|_{H^1(D)}.$

The operator norms of $\Pi_n - I$ can be estimated by

(4.6.4d) $\qquad \|\Pi_n - I\|_{H^1(D)' \leftarrow L^2(D)} = \|\Pi_n - I\|_{L^2(D) \leftarrow H^1(D)} \leqslant Ch,$

(4.6.4e) $\qquad \|\Pi_n - I\|_{H^1(D)' \leftarrow H^1(D)} \leqslant C^2h^2.$

Proof. (i) Let $\varphi \in L^2(D)$. Since Π_n is selfadjoint (cf. Definition 1.3.32b), for each $\psi \in H^1(D)$ we have

$$|(\Pi_n\varphi - \varphi, \psi)_{L^2(D)}| = |(\varphi, \Pi_n\psi - \psi)_{L^2(D)}| \leqslant$$
$$\leqslant \|\varphi\|_{L^2(D)}\|\Pi_n\psi - \psi\|_{L^2(D)} \leqslant \|\varphi\|_{L^2(D)} Ch\|\psi\|_{H^1(D)}.$$

Definition (2) implies the inequality (4b), i.e., $\|\Pi_n\varphi - \varphi\|_{-1} \leqslant Ch\|\varphi\|_{L^2(D)}$. Then (4b) proves the second inequality in (4d). The first equality is a consequence of the next result, Lemma 4a.
(ii) As Π_n, also $I - \Pi_n$ is a projection: $(I - \Pi_n)^2 = I - \Pi_n$. The estimate

$$\|I - \Pi_n\|_{H^1(D)' \leftarrow H^1(D)} = \|(I - \Pi_n)(I - \Pi_n)\|_{H^1(D)' \leftarrow H^1(D)} \leqslant$$
$$\leqslant \|I - \Pi_n\|_{H^1(D)' \leftarrow L^2(D)}\|I - \Pi_n\|_{L^2(D) \leftarrow H^1(D)} \leqslant (Ch)(Ch)$$

implies (4e). Hence, (4c) follows. QED

Lemma 4.6.4. **(a)** The operator T' dual to $T \in L(X,Y)$ (cf. end of §1.3.6) belongs to $L(Y', X')$ and satisfies $\|T'\|_{X' \leftarrow Y'} = \|T\|_{Y \leftarrow X}$. In the case that X, Y are Hilbert spaces, the dual operator is called the *adjoint operator* $T^* \in L(Y', X')$.
(b) If X is a Hilbert space continuously embedded in $L^2(D)$, X' can be understood as a Hilbert space continuously embedded into $L^2(D)$:

$$X \subset L^2(D) \subset X'.$$

Here, $L^2(D)$ is identified with its dual space, i.e., $f \in L^2(D)$ is identified with the functional $F\colon \varphi \in L^2(D) \mapsto F(\varphi) = \int_D \overline{f}(x)\varphi(x)dx$.
(c) Let the Hilbert spaces X, Y be continuously embedded in $L^2(D)$, or $L^2(D)$ be continuously embedded in X or Y. Then the adjoint operator $T^* \in L(Y', X')$ can be defined by means of the L^2 scalar product:

$$(T^*y, x) = (y, Tx) \qquad \text{for all } y \in Y' \quad \text{and } x \in X.$$

Here, we use the fact that (by the identification described in (b)) the L^2 scalar product $(\cdot,\cdot)_{L^2(D)}$ can be extended continuously into the *dual form* onto $X \times X'$: $x'(x) = (x, x')_{L^2(D)}$ for all $x' \in X'$ and $x \in X$.
(d) If k is the kernel of the integral operator K, then the adjoint operator K^* is the integral operator with the kernel k^* given by (5):

(4.6.5) $$k^*(x,y) = \overline{k(y,x)} \qquad \text{for all } x, y \in D.$$

Proof. For the parts (a–c) we refer to Hackbusch [2,§6.3]. For the proof of part (d) note that $(K\varphi, \psi) = \int_D\int_D k(x,y)\varphi(y)\overline{\psi(x)}dxdy$. QED

Estimate (4a) applies to the Galerkin method with piecewise constant basis functions. For $f \in H^1(D)$, (4c) proves an $O(h^2)$ estimate of the projection errors. Using Theorem 3.13 with $X = H^1(D)'$, one is led to the following result.

Remark 4.6.5. Suppose $(\lambda I - K)^{-1} \in L(X,X)$ for $X = H^1(D)'$. Assume also that the orthogonal projection defining the Galerkin method satisfies (4a). If the solution f of $\lambda f = g + Kf$ belongs $H^1(D)$, then the error estimate

(4.6.6a) $$\|f - f_n\|_{-1} \leqslant |\lambda| \, \|(\lambda I - K_n)^{-1}\|_{H^1(D)' \leftarrow H^1(D)'} \, C^2 h^2 \|f\|_{H^1(D)}$$

holds for the Galerkin solution f_n. Define the Galerkin method by the *piecewise constant* functions given by (5.19a-f). Let λ be a regular value of K. Under the assumption $K \in L(H^1(I)', L^2(I))$ discussed in Lemma 6e, $\{K_n\}$ is stable and convergent in $H^1(I)'$:

(4.6.6b) $$\|f - f_n\|_{-1} \leqslant \text{const}\, h^2 \|f\|_{H^1(D)}.$$

Proof. For the second part use the *compact* embedding $H^1(I)$ into $L^2(I)$ {To see this, observe that the second embedding in $H^1(I) \subset C^{1/2}(I) \subset C(I) \subset$ $\subset L^2(I)$ is compact (cf. Theorem 1.3.27), the others are continuous (cf. Lemma 5.24). By Theorem 1.3.23a, the product of the embeddings is compact. Cf. Hackbusch [2,§6.4] }. Then, according to Theorem 1.3.23d, $L^2(I) \subset H^1(I)'$ is compactly embedded. Analogously to Remark 3.4.13, K is a compact operator in $H^1(I)'$, so that Theorem 3.11 is applicable with $X = H^1(I)'$. QED

To check the assumptions of Remark 5 {i.e., $K \in L(H^1(D)', H^1(D)')$ and $(\lambda I - K)^{-1} \in L(H^1(D)', H^1(D)')$}, we state the following sufficient conditions.

Lemma 4.6.6. **(a)** The statement $K \in L(H^1(D)', H^1(D)')$ is equivalent to $K^* \in L(H^1(D), H^1(D))$. In particular, all conditions of the form $K^* \in L(X,Y)$ with continuous embeddings $Y \subset H^1(D) \subset X$ are sufficient, e.g., $K^* \in L(L^2(D), H^1(D))$, $K^* \in L(L^2(D), C^1(D))$ and, if $D \subset \mathbb{R}^1$ or $D = \partial\Omega$ with $\Omega \subset \mathbb{R}^2$, also $K^* \in L(C(D), C^1(D))$.
(b) Let $\lambda \neq 0$. The inverse $(\lambda I - K)^{-1} \in L(H^1(D)', H^1(D)')$ exists in this space, if $(\lambda I - K)^{-1} \in L(L^2(D), L^2(D))$ and $K \in L(H^1(D)', L^2(D))$ or $K^* \in L(L^2(D), H^1(D))$.
(c) The stability $\|(\lambda I - K_n)^{-1}\|_{H^1(D)' \leftarrow H^1(D)'} \leqslant \text{const}$ with respect to $H^1(D)'$ is a consequence of the stability with respect to $L^2(D)$ and the regularity condition $\|K_n^*\|_{H^1(D) \leftarrow L^2(D)} \leqslant \text{const}$.
(d) In the case of projection methods, the latter regularity condition $\|K_n^*\|_{H^1(D) \leftarrow L^2(D)} \leqslant \text{const}$ follows from $K \in L(H^1(D)', L^2(D))$ or $K^* \in L(L^2(D), H^1(D))$ (cf. §3.4) and the stability of the projection with respect to $L^2(D)$ (cf. (3.3.c)).
(e) The conditions k, $k_y \in L^2(D \times D)$ are sufficient to ensure the property $K \in L(H^1(D)', L^2(D))$ and hence imply $K^* \in L(L^2(D), H^1(D))$.
(f) The condition $\sup\{\int_D [k(x,y)^2 + k_y(x,y)^2]\,dy : x \in D\} < \infty$ is sufficient for $K \in L(H^1(D)', L^\infty(D))$.

Proof. (i) Part (a) results from Lemma 4a. For the given examples note that the embeddings $C^1(D) \subset H^1(D)$ {as well as $H^1(D) \subset C(D)$ for one-dimensional domains D} are continuous (cf. Lemma 5.24).
(ii) According to Lemma 4b, $L^2(D) = L^2(D)'$. Hence, by Lemma 4a,

$(\lambda I-K)^{-1}\in L(L^2(D),L^2(D))$ is equivalent to $(\lambda I-K^*)^{-1}\in L(L^2(D),L^2(D))$. From Theorem 3.5.1 it follows that $(\lambda I-K^*)^{-1}\in L(H^1(D),H^1(D))$. This together with Lemma 4a leads to the assertion of part (b). The proof of part (c) is similar.
(iii) For part (d) use $K_n=\Pi_n K$, $K_n^*=K^*\Pi_n^*$ and $\|\Pi_n^*\|_{L^2(D)\leftarrow L^2(D)}=\|\Pi_n\|_{L^2(D)\leftarrow L^2(D)}\leqslant \text{const}$.
(iv) Let $\varphi\in H^1(D)'$. Due to inequality (3), $\Phi:=K\varphi$ can be estimated by

$$|(K\varphi)(x)|=|\int_D k(x,y)\varphi(y)dy|=|(k(x,\cdot),\varphi)|\leqslant \|k(x,\cdot)\|_{H^1(D)}\|\varphi\|_{-1}.$$

Integration over $x\in D$ together with the definition of the norm $\int_D\|k(x,\cdot)\|^2_{H^1(D)}dx=\int_D\int_D[k^2+k_y^2]\,dy\,dx$ yields the assertion (e). Part (f) follows from the estimate of $|(K\varphi)(x)|$ given above. QED

Exercise 4.6.7. **(a)** Define k on $I=[a,b]$ as convolution kernel $k(x-y)$ with $k\in C([a-b,b-a])$. Prove that $K,K^*\in L(H^1(I),C^1(I))$.
(b) Let $k(x,y)=|x-y|^\gamma$ with $\gamma\in\mathbb{R}$. Prove that $K\in L(H^1(D)',L^2(D))$ for $\gamma\geqslant 0$ and $K\in L(H^1(D)',L^\infty(D))$ for $\gamma>\frac{1}{2}$. *Hint for (a)*: Theorem 3.4.5b.

The previous considerations demonstrate the use of the weaker norm $\|\cdot\|_{-1}$ for a special situation. Forming the dual norm $\|\cdot\|_{-k}$ of $H^k(D)$ (cf. end of §1.3.6 and Definition 5.23), we can prove the following generalisation of (4a).

Remark 4.6.8. **(a)** An error estimate corresponding to (4a) is

(4.6.7a) $\quad \|\Pi_n\varphi-\varphi\|_{L^2(D)}\leqslant Ch^k\|\varphi\|_{H^k(D)}$ for $\varphi\in H^k(D)$.

Moreover, (4a) implies (7b) or, more generally, (7c):

(4.6.7b) $\quad \|\Pi_n\varphi-\varphi\|_{-k}\leqslant C^2h^{2k}\|\varphi\|_{H^k(D)}$ for $\varphi\in H^k(D)$

(4.6.7c) $\quad \|\Pi_n\varphi-\varphi\|_{\alpha}\leqslant C'h^{\beta-\alpha}\|\varphi\|_{H^\beta(D)}$ for $-k\leqslant\alpha\leqslant 0\leqslant\beta\leqslant k$, $\varphi\in H^\beta(D)$.

(b) The Galerkin method with the subspace of *piecewise linear functions* allows the error estimates (7a-c) with $k=2$. With respect to the H^{-2} norm, one obtains $O(h_n^4)$ as maximal convergence order.

Inequality (7c) is extendable to non-integers α and β. The Sobolev space $H^s(D)$ for general $s\in\mathbb{R}$ is explained, e.g., in Hackbusch [2,§6.2.4].

The proof of Lemma 3 exploits the selfadjointness of the orthogonal projection Π_n. Therefore, the results cannot be generalised to a collocation method with of a non-orthogonal projection Π_n. Nevertheless, one obtains better convergence orders in weaker norms also for some collocation methods as shown in the following example. The following inequality (8b) is an example of a *non-optimal error estimate*, since the difference $2-(-1)=3$ is larger than the h-exponent 2.

Example 4.6.9. Let Π_n be the *interpolation by piecewise constant* functions in $I=[a,b]$ as in Remark 4.8. The estimate of the interpolation error in $L^2(I)$ corresponding to (4a) is

(4.6.8a) $\quad \|\Pi_n\varphi-\varphi\|_{L^2(I)}\leqslant Ch\|\varphi\|_{H^1(I)}$ for all $\varphi\in H^1(I)$.

In addition, it holds that

(4.6.8b) $\quad \|\Pi_n\varphi-\varphi\|_{-1}\leqslant Ch^2\|\varphi\|_{H^2(I)}$ for all $\varphi\in H^2(I)$.

Proof. Use Taylor expansion as in the proof of Lemma 5.21. In the equidistant case, one can make use of Fourier expansions. QED

The use of weaker norms can even lead to stability of a projection: $\|\Pi_n\|_{Y\leftarrow X}\leqslant$ const can hold for $Y\subset X$, although $\sup\|\Pi_n\|_{X\leftarrow X}=\infty$. An example is *polynomial interpolation* in $I=[a,b]$. As mentioned at the beginning of §4.4.3, the interpolation Π_n by polynomials is unstable in the space $X=C(I)$. However, the following *Theorem of Erdös and Turan* holds; its proof is to be found in Szegö [1, p. 331]. The orthogonal polynomials mentioned in the theorem are the same as those of the Gauß quadrature with a weight function w (cf. Stoer [1,§3.5]).

Theorem 4.6.10. Let $w(x)\geqslant\varepsilon>0$ in $I=[a,b]$. Define the collocation method by polynomial interpolation (i.e., $X_n=\{f:$ polynomial of degree $\leqslant n-1\}$) at the collocation points $\Xi_n=\{\xi_{j,n}: 1\leqslant j\leqslant n$, zeros of the n th orthogonal polynomial with respect to the weight function $w\}$. Then $\|f-\Pi_n f\|_{L^2(I)}\to 0$ for all $f\in C(I)$.

Corollary 4.6.11. (a) Under the assumptions of Theorem 10, we have the stability estimate $\|\Pi_n\|_{L^2(I)\leftarrow C(I)}\leqslant$ const.
(b) If K satisfies the regularity property $K\in L(L^2(I),C(I))$, consistency of the collocation method follows, i.e., $K_n\varphi\to K\varphi$ for all $\varphi\in L^2(I)$.
(c) Under the stronger assumption $K\in K(L^2(I),C(I))$, $K_n\to K$ with respect to the operator norm $\|\cdot\|_{L^2(I)\leftarrow L^2(I)}$.

Proof. (a): By Theorem 1.3.15. (b) is a direct consequence of Theorem 10. (c) follows from Lemma 3.8, where $A_n\in L(X,X)$ and $K\in K(X,X)$ can be replaced by $A_n\in L(Y,X)$ and $K\in K(X,Y)$. QED

In general, we cannot expect such good negative norm estimates for collocation methods, but Berthold-Silbermann [1] have described an interesting combination of a collocation with a lower dimensional Galerkin method which satisfies this property.

4.6.3 The Iterated Approximation

Let f_n be an approximate solution obtained by some method. Then the «iterated approximation» $\tilde{f}_n$ can be defined by

$$\tilde{f}_n := \tfrac{1}{\lambda}(g+Kf_n). \tag{4.6.9}$$

The error of $\tilde{f}_n$ can immediately be expressed by the error of f_n:

$$f-\tilde{f}_n = \tfrac{1}{\lambda}K(f-f_n). \tag{4.6.10}$$

This procedure makes sense, since often one obtains better convergence orders for $K(f-f_n)$ than for $f-f_n$. The general situation is discussed below.

Lemma 4.6.12. Let $Y\subset X\subset Z$. An error estimate $\|f-f_n\|_Z\leqslant Ch_n^{\varkappa}\|f\|_Y$ together with the regularity assumption $K\in L(Z,X)$ yields the following estimate for the iterated approximation:

(4.6.11) $\quad \|f - \tilde{f}_n\|_X \leqslant C' h_n^{\varkappa} \|f\|_Y$.

If the Z-estimate is quasi-optimal with respect to the subspace X_n of the projection, i.e., $\|f - f_n\|_Z \leqslant C \inf\{\|f - \varphi\|_Z : \varphi \in X_n\}$, then (11) is valid in the form

(4.6.11') $\quad \|f - \tilde{f}_n\|_X \leqslant C' \inf\{\|f - \varphi\|_Z : \varphi \in X_n\}$.

In the application, X is the basic space, while $\|\cdot\|_Z$ is chosen as a weaker norm which, according to §4.6.2, may lead to a better convergence order than $\|\cdot\|_X$. Inserting the results of the previous subsections, we obtain the following application:

Example 4.6.13. **(a)** The iterated solution $\tilde{f}_n$ of the *Galerkin method with piecewise constant functions* has second order convergence in $L^2(I)$: $\|f - \tilde{f}_n\|_{L^2(I)} \leqslant C h_n^2 \|f\|_{H^1(I)}$, if $K^* \in L(L^2(I), H^1(I))$.
(b) The *collocation method with piecewise constant interpolation* leads to the 2nd order error estimate $\|f - \tilde{f}_n\|_{L^2(I)} \leqslant C h_n^2 \|f\|_{H^2(I)}$, if $K^* \in L(L^2(I), H^1(I))$.
(c) Under the stronger condition $K \in L(H^1(I)', L^\infty(I))$ (cf. Lemma 6f), the methods from (a) and (b) converge uniformly: $\|f - \tilde{f}_n\|_\infty \leqslant C h_n^2 \|f\|_{H^\beta(I)}$ with $\beta = 1$ for case (a) and $\beta = 2$ for case (b).
(d) In the case of the *Galerkin method with piecewise linear functions*, the iterated solution $\tilde{f}_n$ is of fourth order in $L^2(I)$: $\|f - \tilde{f}_n\|_{L^2(I)} \leqslant C h_n^4 \|f\|_{H^2(I)}$, if $K^* \in L(L^2(I), H^2(I))$.

Proof. (i) The condition $K^* \in L(L^2(I), H^1(I))$ of part (a) is needed in Lemma 6d as well as in Lemma 11 (there in the equivalent form $K \in L(Z, X)$ with $Z = H^1(I)'$ and $X = L^2(I)$).
(ii) For (b) and (d) use Example 9 or Remark 8b, respectively. QED

A further application of Lemma 12 is based on the possibility of estimating the error $f - \tilde{f}_n$ with respect to a *stronger norm*. If, e.g., one is not primarily interested in f but in its derivative f', one needs an error estimate in $H^1(I)$ or $C^1(I)$. This one obtains, e.g., from $\|f - f_n\|_{L^2(I)}$, if $K \in L(L^2(I), H^1(I))$ or $K \in L(L^2(I), C^1(I))$, respectively.

Another possibility to obtain the error estimates mentioned above is the representation of the error in $\tilde{f}_n$ given by (12c) below.

Theorem 4.6.14. Let f_n be the solution of the projection method characterised by Π_n: $\lambda f_n = \Pi_n g + \Pi_n K f_n$. The iterated solution $\tilde{f}_n$ defined by (9) is a solution of the semidiscrete equation (12a), which has already been classified as a modified projection equation (3.8):

(4.6.12a) $\quad \lambda \tilde{f}_n = g + K \Pi_n \tilde{f}_n$.

The projection applied to $\tilde{f}_n$ reproduces the projection solution f_n:

(4.6.12b) $\quad \Pi_n \tilde{f}_n = f_n$.

$\tilde{f}_n$ has the error representation (12c), where the inverse matrix $(\lambda I - K \Pi_n)^{-1} \in L(X, X)$ which arises exists as well as $(\lambda I - \Pi_n K)^{-1}$:

(4.6.12c) $\quad f - \tilde{f}_n = (\lambda I - K \Pi_n)^{-1} K (I - \Pi_n)(f - f_n)$.

Proof. (i) Multiplication of (9) by Π_n yields $\lambda \Pi_n \tilde{f}_n = \Pi_n g + \Pi_n K f_n$. The comparison with the projection equation $\lambda f_n = \Pi_n g + \Pi_n K f_n$ proves (12b). Inserting the representation $f_n = \Pi_n \tilde{f}_n$ into (9), one obtains (12a).
(ii) From $\lambda(f - \tilde{f}_n) = K(f - f_n)$ and (12b) (i.e., $\Pi_n \tilde{f}_n = f_n$) one concludes that

$$(\lambda I - K\Pi_n)(f - \tilde{f}_n) = \lambda(f - \tilde{f}_n) - K(\Pi_n f - \Pi_n \tilde{f}_n) =$$
$$= K(f - f_n) - K(\Pi_n f - f_n) = K(f - \Pi_n f) = K(I - \Pi_n)(f - f_n).$$

Concerning the inverse $(\lambda I - K\Pi_n)^{-1}$ use Remark 3.10 and (3.10a). QED

The method described above can be iterated by computing the second iterated approximation $\tilde{\tilde{f}}_n := \frac{1}{\lambda}(g + K\tilde{f}_n)$. Since $f - \tilde{\tilde{f}}_n = \lambda^{-2} K^2 (f - f_n)$, this procedure can be justified in the following case. Let $\| f - f_n \|_{-2} = O(h_n^4)$, while an error estimate in $L^2(D)$ is desired. An operator may happen to satisfy $K \in L(H^2(D)', H^1(D)')$ and $K \in L(H^1(D)', L^2(D))$ but not $K \in L(H^2(D)', L^2(D))$. In this case, only the second iterate satisfies the estimate $\| f - \tilde{\tilde{f}}_n \|_{L^2(D)} = O(h_n^4)$.

Finally, we recall that the computation of $\tilde{f}_n$ by (9) requires an (exact) integration of $\int k(x,y) f_n(y) dy$. If we approximate this by quadrature, then we have to analyse the additional error.

Apparently, the iterated approximation is restricted to equations of the second kind. However, in some case one succeeds to reformulate a first kind problem as a second kind problem. For an analysis see Graham-Atkinson [1]. Details of the reformulation can also be found in §9.3.2.

4.6.4 Superconvergence

The term «superconvergence» expresses that a better convergence order is observed than expected by the general estimate. If, e.g., $\| f - f_n \|_\infty = O(h)$ is the optimal error estimate, there may be special points $\xi \in D$ (superconvergence points) with $| f(\xi) - f_n(\xi) | = O(h^2)$. In the case of the collocation method discussed below, these superconvergence points will be the collocation points.

The error $f - f_n$ can be split into the projection error $f - \Pi_n f$ of f and the projection $\Pi_n f - f_n = \Pi_n (f - f_n)$ of the total error $f - f_n$:

(4.6.13a) $$f - f_n = (f - \Pi_n f) + (\Pi_n f - f_n).$$

In (3.12c), the second term $\Pi_n f - f_n$ is represented by

$$\Pi_n f - f_n = (\lambda I - K_n)^{-1} K_n (I - \Pi_n) f,$$

leading to the estimate (3.12d) repeated below:

(4.6.13b) $$\| \Pi_n f - f_n \| \leqslant \| (\lambda I - K_n)^{-1} \| \, \| \Pi_n \| \, \| K(I - \Pi_n) f \|.$$

A similar decomposition is stated in

Lemma 4.6.15. Let $\lambda \neq 0$. The projection solution admits the representation

(4.6.14) $$f - f_n = f - \Pi_n f + \tfrac{1}{\lambda} \Pi_n K (f - f_n).$$

Proof. Application of Π_n to $\lambda f = g + Kf$ yields $\lambda \Pi_n f = \Pi_n g + \Pi_n Kf$. Subtracting the projection equation $\lambda f_n = \Pi_n g + \Pi_n K f_n$, one obtains $\Pi_n f - f_n = \frac{1}{\lambda} \Pi_n K(f - f_n)$. Together with (13a), this proves (14). QED

Remark 4.6.16. The estimation technique discussed in §§4.6.2-3 can be applied to $K(I - \Pi_n)f$ from (13b) as well as to $K(f - f_n)$ from (14). One obtains error norms $\| K(I - \Pi_n) f \|$, $\| K(f - f_n) \|$ of higher order than $\| (I - \Pi_n) f \|$ and $\| f - f_n \|$. Therefore, the projection error $f - \Pi_n f$ represents the *dominant* part of the error $f - f_n$.

By definition of the projection Π_n of the collocation method, $f(\xi) = (\Pi_n f)(\xi)$ holds for all collocation points $\xi \in \Xi_n$. Hence, the usually dominant error $(f - \Pi_n f)(\xi)$ (cf. Remark 16) vanishes. There remains the error

(4.6.15) $$f(\xi) - f_n(\xi) = (\Pi_n f - f_n)(\xi) = \tfrac{1}{\lambda} [\Pi_n K(f - f_n)](\xi) \qquad \text{for all } \xi \in \Xi_n.$$

In the case of a Lagrange function basis (cf. Remark 4.3 and Theorem 4.13), only the nodal values $\{f_n(\xi): \xi \in \Xi_n\}$ from (15) are to be computed. Hence, the examples above are based on the discrete maximum norm $\max\{ |f(\xi) - f_n(\xi)|: \xi \in \Xi_n \}$ or the corresponding Euclidean norm $\left(h \sum_{\xi \in \Xi_n} |f(\xi) - f_n(\xi)|^2\right)^{1/2}$, using only the values from (15).

Two examples of superconvergence statements for collocation solutions are presented in

Theorem 4.6.17. Assume that λ is not an eigenvalue of the integral operator $K \in K(C(I), C(I))$. **(a)** Define the collocation method by *piecewise constant interpolation* as in Remark 4.8 or (5.19a-c,e). Then

(4.6.16a) $$\max\{ |f(\xi) - f_n(\xi)|: \xi \in \Xi_n \} \le Ch \| f \|_{H^1(I)} \qquad \text{for } f \in H^1(I),$$

provided $K \in L(L^2(I), L^\infty(I))$. A sufficient condition for this is $\int_I k(x,y)^2 dy \le C_I$ for all $x \in I$. Moreover,

(4.6.16b) $$\max\{ |f(\xi) - f_n(\xi)|: \xi \in \Xi_n \} \le Ch^2 \| f \|_{H^2(I)} \qquad \text{for } f \in H^2(I),$$

if $K \in L(H^1(I)', L^\infty(I))$. A sufficient condition for this is given in Lemma 6f.

(b) For collocation by *piecewise linear interpolation* as in (5.29a-d), the second order estimate (16b) holds under the assumption $K \in L(L^2(I), L^\infty(I))$.

Proof. Use (14), (15), and Example 9 and exploit the stability of Π_n with respect to $\| \cdot \|_\infty$. QED

In the following, we want to prove an analogous result for the *Galerkin method.* It is based on the following lemma.

Lemma 4.6.18. Let Π_n be the orthogonal projection onto the space X_n of *piecewise constant functions* based on the partition of $I = [a,b]$ satisfying (5.19a-d). The projection error is of *second* order at the midpoints (5.19e): $\xi_i := \frac{1}{2}(x_i + x_{i-1})$ $(i = 1, 2, \ldots, n)$ of the subintervals, i.e.,

(4.6.17a) $$|(\Pi_n g)(\xi_i) - g(\xi_i)| \le Ch^2 \| g'' \|_{\infty, [x_{i-1}, x_i]} \le Ch^2 \| g \|_{C^2(I)}$$

for all $g \in C^2(I)$, whereas the error converges with $O(h)$ at other points. Analogously, for all $g \in C_L^1(I)$ with Lipschitz constant $L_{g'}$ of g' we have

(4.6.17b) $$|(\Pi_n g)(\xi_i) - g(\xi_i)| \leq C h^2 L_{g'}.$$

If the second derivatives are only square-integrable, i.e., $g \in H^2(I)$, one still obtains the order 1.5 at the points from (5.19e), in fact

(4.6.17c) $$|(\Pi_n g)(\xi_i) - g(\xi_i)| \leq C h^{3/2} \|g''\|_{L^2(I)}.$$

The Euclidean norm of the vector $\big((\Pi_n g)(\xi_i) - g(\xi_i)\big)_{i=1,\ldots,n}$ weighted by the interval length can be estimated by

(4.6.17d) $$\Big(\sum_{i=1}^{n} |x_i - x_{i-1}|\,|(\Pi_n g)(\xi_i) - g(\xi_i)|^2\Big)^{1/2} \leq C h^2 \|g''\|_{L^2(I)}.$$

If, in addition, λ is a regular value of the operator $K \in L(H^{-1}(I), C(I))$ or $K \in L(H^{-1}(I), L^2(I))$ and if the solution f of the integral equation belongs to $C_L^1(I)$ or $H^2(I)$, respectively, then the following estimates respectively hold for the Galerkin solution f_n:

(4.6.17e) $$\max\{|f_n(\xi_i) - f(\xi_i)| \leq C h^2 \|f\|_{C_L^1(I)},$$

(4.6.17f) $$\Big(\sum_{i=1}^{n} |x_i - x_{i-1}|\,|f_n(\xi_i) - f(\xi_i)|^2\Big)^{1/2} \leq C h^2 \|f\|_{H^2(I)}.$$

Proof. (i) Integration by parts proves the representation

(4.6.17g) $$(g - \Pi_n g)(\xi_k) = \int_{x_{k-1}}^{\xi_k} (\xi_k - x) g''(x)\,dx - \frac{1}{2}\frac{1}{x_k - x_{k-1}} \int_{x_{k-1}}^{x_k} (x_k - x)^2 g''(x)\,dx.$$

This immediately implies (17a,c). Concerning (17b), we refer to the paragraph following the proof of Lemma 5.31.
(ii) $K \in L(H^{-1}(I), C(I))$ implies $K \in L(X, X)$ for $X = C(I)$ and $X = L^2(I)$. Since the embeddings $C(I) \subset H^{-1}(I)$ and $L^2(I) \subset H^{-1}(I)$ are compact, K is even compact in $X = C(I)$ and $X = L^2(I)$ (cf. proof of Remark 5). According to Lemma 15, we have to estimate $f - \Pi_n f$ and $\frac{1}{\lambda}\Pi_n K(f - f_n)$ at ξ_i. (17b) yields the estimate for $f - \Pi_n f$. Lemma 6b,c is applicable, since $K \in L(H^{-1}(I), C(I))$ also implies $K \in L(H^{-1}(I), L^2(I))$. Hence, the assumptions of Remark 5 are satisfied and yield (6a): $\|f - f_n\|_{-1} \leq C h^2 \|f\|_{H^1(I)} \leq C' h^2 \|f\|_{C_L^1(I)}$. By $K \in L(H^{-1}(I), C(I))$ and L^∞ stability of Π_n, one obtains (17e).
(iii) The proof of (17f) is almost the same. Instead of (17b), use (17d). As in part (ii), $K \in L(H^{-1}(I), L^2(I))$ guarantees the estimate $\|\varphi\|_{L^2(I)} \leq C h^2 \|f\|_{H^1(I)}$ for $\varphi := K(f - f_n)$. Note that the application of the Schwarz inequality ($[\int_{I_k} \varphi\,dx]^2 \leq \mu(I_k) \int \varphi^2 dx$) to the right-hand side in (5.22) leads to

$$\sqrt{x_k - x_{k-1}}\; |(\Pi_n \varphi)(x)| \leq \|\varphi\|_{L^2(I_k)} \qquad \text{for all } x \in I_k.$$

Summation of the squared inequalities yields $O(h^2)\|f\|_{H^1(I)}$ for the term $\frac{1}{\lambda}\Pi_n K(f - f_n)$ in the representation (14) of $f - f_n$ and completes the proof of (17f).

QED

The proof in part (i) exploits the fact that the projection error $g - \Pi_n g$ of a linear function g vanishes at all «superconvergence points» $\xi_i := \frac{1}{2}(x_i + x_{i-1})$. In general, the following hint concerning the existence and construction of superconvergence points can be given.

Remark 4.6.19. Let Π_n be exact for all polynomials of degree $\leqslant m$, i.e., $\Pi_n x^\nu = x^\nu$ for all $x \in I$ and $0 \leqslant \nu \leqslant m$. Under the assumptions (S.19a-d), $\| g - \Pi_n g \|_\infty \leqslant C h^{m+1} \| g \|_{C_L^m(I)}$ or $\| g - \Pi_n g \|_{L^2(I)} \leqslant C h^{m+1} \| g \|_{H^{m+1}(I)}$ holds. An error estimate which is one order higher is obtainable at the roots $\xi \in I$ of $(\Pi_n x^{m+1})(\xi) = \xi^{m+1}$.

4.6.5 More General Formulations of the Projection Method

The Chapters §4.4 and §4.5 describe two kinds of projections Π_n: interpolation and the orthogonal projection. Of course, there are a lot of other possibilities for selecting a projection.

The projections introduced in §4.3 fulfil *two* tasks. First, they define the subspace X_n, in which the solution f_n is desired. Therefore, the semidiscrete equation for the defect $d_n := \lambda f_n - K f_n - g$ is obtained from the condition $\Pi_n d_n = 0$ (cf. (3.5a)). A further, appearantly more general condition is $\hat{\Pi}_n d_n = 0$ instead of $\Pi_n d_n = 0$ with a projection $\hat{\Pi}_n$ different from Π_n. If $\hat{\Pi}_n$ is a projection acting in X, then the dual operator $\hat{\pi}_n := \hat{\Pi}_n'$ represents a projection in the dual space X'. $\hat{\Pi}_n d_n = 0$ is equivalent to each of the following conditions:

(i) $\varphi(\hat{\Pi}_n d_n) = 0$ for all $\varphi \in X'$,

(ii) $(\hat{\pi}_n \varphi) d_n = 0$ for all $\varphi \in X'$,

(iii) $\varphi d_n = 0$ for all $\varphi \in X_n' := \mathrm{range}(\hat{\pi}_n)$.

We start from the characterisation (iii) and in addition to X_n we introduce another n-dimensional subspace X_n' of the dual space X'. Let $\{\Phi_1', \ldots, \Phi_n'\}$ be a basis of X_n', i.e.,

(4.6.18a) $$X_n' = \mathrm{span}\{\Phi_1', \ldots, \Phi_n'\} \subset X'.$$

The new projection $\hat{\Pi}_n d_n = 0$ for $d_n := \lambda f_n - K f_n - g$ is rewritten as

(4.6.18b) $$\varphi(\lambda f_n - K f_n - g) = 0 \qquad \text{with } f_n \in X_n \text{ for all } \varphi \in X_n'.$$

By the choice of the basis (18a), (18b) is equivalent to the n equations

(4.6.18b') $$\Phi_j'(\lambda f_n - K f_n - g) = 0 \qquad \text{with } f_n \in X_n \text{ for all } j = 1, \ldots, n.$$

Define the *dual product* $\langle \varphi, f \rangle := \varphi(f)$ on $X' \times X$ for $\varphi \in X'$ and $f \in X$. Then, (18b') becomes

(4.6.18b") $$\langle \Phi_j', \lambda f_n - K f_n - g \rangle = 0 \qquad \text{with } f_n \in X_n \text{ for all } j = 1, \ldots, n.$$

Since f_n is constructed as a function in X_n, one calls

(4.6.18c) $$X_n = \mathrm{span}\{\Phi_1, \ldots, \Phi_n\} \subset X$$

the *ansatz space (trial space)* spanned by the *ansatz functions* (basis functions) Φ_j. On the other hand, the functionals Φ_j' involved in (18d") are the *test functions*. X_n' is named the *space of the test functions* or *test space*.

Having chosen concrete basis functions in X_n and X_n', one obtains the discrete system of equations after inserting the ansatz $f_n = \sum \alpha_j \Phi_j$ into (18b").

Remark 4.6.20. Problem (18b") with the ansatz $f_n=\sum\alpha_j\Phi_j$ is equivalent to the system $(\lambda A-B)a=b$, where the coefficients of A, B, and b are defined by

$$\alpha_{jk}=\langle\Phi_j',\Phi_k\rangle,\qquad \beta_{jk}=\langle\Phi_j',K\Phi_k\rangle,\qquad \beta_j=\langle\Phi_j',g\rangle.$$

Exercise 4.6.21. Prove the following: If the matrix A from Remark 20 is non-singular, then there are functions $\Phi_k^*\in X_n$ so that $\langle\Phi_j',\Phi_k^*\rangle=\delta_{jk}$ (Kronecker symbol, $1\leqslant j,k\leqslant n$). $\{\Phi_1^*,\dots,\Phi_n^*\}$ forms a basis in X_n.

To obtain a semidiscrete representation of the discretisation, we choose Φ_k^* ($1\leqslant k\leqslant n$) according to Exercise 21 and introduce the projection

(4.6.19a) $\quad \pi_n: X'\to X_n'\subset X' \qquad$ with $\pi_n: \varphi\mapsto\sum_k\langle\varphi,\Phi_k^*\rangle\,\Phi_k'$.

It is easily verified that $\pi_n\in L(X',X')$, $\mathrm{range}(\pi_n)=X_n'$, and $\pi_n\Phi_i'=\Phi_i'$. Hence, π_n is a projection of X' onto X_n'. One gains the representation (19b) of the projection dual to π_n from the equation $\langle\pi_n\varphi,f\rangle=\sum\langle\varphi,\Phi_k^*\rangle\langle\Phi_k',f\rangle=\langle\varphi,\sum\langle\Phi_k',f\rangle\,\Phi_k^*\rangle$. Denote the projection by Π_n:

(4.6.19b) $\quad \Pi_n f:=\sum_k\langle\Phi_k',f\rangle\,\Phi_k^*$.

Since $\{\Phi_1^*,\dots,\Phi_n^*\}$ forms a basis in X_n, Π_n is a projection onto X_n.

The equivalent characterisations (i) - (iii) of $\hat{\Pi}_n d_n=0$ stated in the beginning of §4.6.5 can be continued by the following equivalent statements:

(iv) $\quad \langle\pi_n\varphi,\,d_n\rangle=0 \qquad$ for all $\varphi\in X'$,

since $X_n'=\mathrm{range}(\pi_n)$;

(v) $\quad \langle\varphi,\,\Pi_n d_n\rangle=0 \qquad$ for all $\varphi\in X'$;

(vi) $\quad \Pi_n d_n=0$.

Hence, we have rewritten $\hat{\Pi}_n d_n=0$ as a projection method by means of a projection Π_n onto the ansatz space X_n - as described in §4.3.

If $\chi_n\in T_n$ are suitable test functions ($\dim T_n=\dim X_n$) and Q_n a quadrature formula, $\hat{\Pi}_n d_n:=Q_n(\chi_n d_n)$ describes a projection, which leads to the very interesting *qualocation method* (cf. Sloan [3], Chandler-Sloan [1]) achieving higher convergence orders with respect to suitable norms.

The following examples formulate the Galerkin and collocation method by means of (18a,b").

Example 4.6.22 (Galerkin method). Identify the Hilbert space with its dual space: $X=X'$. The dual product $\langle\cdot,\cdot\rangle$ is identical to the scalar product. Choosing $X_n=X_n'$ (ansatz functions = test functions), one obtains the Galerkin method.

Example 4.6.23 (collocation method). Let $X=C(D)$ and Ξ_n the set of the collocation points $\xi_1,\dots,\xi_n$. As test functions we take the Dirac delta functions $\Phi_j'=\delta_{\xi_j}$ at $\xi_j\in\Xi_n$, which belong to the dual space $C(D)'$ and are defined by $\langle\delta_{\xi_j},f\rangle:=f(\xi_j)$. Then, (18b") defines the collocation method. The «dual basis functions» Φ_k^* used above are the Lagrange functions L_k.

The situation dual to the previous choice of Φ_j and Φ_j' is described in

Example 4.6.24 (cf. Routsalainen - Saranen [1,2]). Choose the delta functions $\Phi_j = \delta_{\xi_j}$ for $\xi_j \in \Xi_n$ as ansatz functions: $X_n \subset X := C(D)'$. The test functions must be at least continuous functions $\Phi_j' \in X_n' \subset C(D)$. According to Remark 20, the matrix entries of A and B are $\alpha_{jk} = \Phi_j'(\xi_k)$ and $\beta_{jk} = \int \Phi_j'(x) k(x, \xi_k) dx$ (cf. (4.9b,c)). The semidiscrete solution $f_n = \sum \alpha_k \delta_{\xi_k}$ belongs to $C(D)'$, but not to $L^\infty(D)$ or $L^2(D)$. This means that an error estimate with respect to $\|\cdot\|_\infty$ or $\|\cdot\|_{L^2(D)}$ cannot be found. In the one-dimensional case $D = I \subset \mathbb{R}$, the continuous (even compact) embedding $H^1(I) \subset C(I)$ implies the continuity (compactness) of the embedding $C(I)' \subset H^1(I)'$ (cf. Lemma 5.24). Therefore, besides the dual norm of $C(I)'$, we may choose the dual norm $\|\cdot\|_{-1}$ defined in §4.6.2.

Example 4.6.25 («least squares method»). Let $X_n = \operatorname{span}\{\Phi_1, \ldots, \Phi_n\}$ be a subspace of the Hilbert space X. If one chooses $X_n' = \operatorname{span}\{\Phi_1', \ldots, \Phi_n'\} \subset X$ with $\Phi_j' := (\lambda I - K)\Phi_j$, then f_n from (18b") is the solution of the problem of minimising $\|(\lambda I - K)\varphi - g\|_X$ over all $\varphi \in X_n$.

4.6.6 Numerical Quadrature

The numerical examples of the collocation and Galerkin method have shown that often the exact integration needed to determine the coefficients of $A = A_n$, $B = B_n$, and $b = b_n$ is not available. Therefore, one has to use suitable quadrature formulae.

In the following, we reduce ourselves to the discussion of the quadrature in B. Let β_{jk} denote the entries of B, while $\tilde{\beta}_{jk}$ are their approximations by means of a quadrature method. Correspondingly, the matrix $(\tilde{\beta}_{jk})$ is denoted by $\tilde{B}$. The quadrature error of $\tilde{\beta}_{jk}$ is $\delta\beta_{jk} := \tilde{\beta}_{jk} - \beta_{jk}$ and forms the error matrix $\delta B = \tilde{B} - B$. We let f_n denote the semidiscrete solution and $\tilde{f}_n$ the solution with numerical quadrature. Then $\delta f_n = \tilde{f}_n - f_n$ is the effect of the quadrature errors.

The influence of the numerical quadrature can be discussed at two levels. (i) One can start from the deviation δB for deriving error estimates or (ii) one may directly analyse the difference δf_n. Approach (i) has the advantage of ensuring stability without further conditions and yielding error estimates of δf_n, which do *not* depend on the smoothness of the solution f. The considerations based on (ii) can yield better estimates of δf_n. However, a consistent application of this argument leads to the Nyström method (cf. §4.7, §4.8.1.3) and hence away from the projection method.

We start with the description of the error matrix $\delta B = \tilde{B} - B$. In the following we assume that the support of the basis functions Φ_j have a length $O(h)$. Further, let Φ_j be scaled in such a way that $\beta_{jk} = O(h)$. The latter leads to $\|B\|_\infty = O(1)$ and $\|B\|_2 = O(1)$. Note that all integrations have to be taken only over the support of Φ_j. Let a suitable quadrature lead to a quadrature error $\delta\beta_{jk} = O(h^{1+\varkappa})$ with $\varkappa > 0$. Hence, the row-sum and spectral norm of the matrices are

$$\text{(4.6.20)} \qquad \|\delta B\|_\infty = O(h^\varkappa), \quad \|\delta B\|_2 = O(h^\varkappa), \quad \|B\|_\infty^{-1} = O(1), \quad \|B\|_2^{-1} = O(1).$$

Instead of $(\lambda A-B)a=b$, we solve the system perturbed by δB:

(4.6.21) $$(\lambda A-\widetilde{B})\widetilde{a} = b.$$

According to §4.4.4 and §4.5.5, it is reasonable to assume that $\mathrm{cond}(\lambda A-B)=O(1)$. Given that, then Theorem 1.4.30 together with (20) guarantees the respective estimates (22a) or (22b) below of the relative error for sufficiently small h:

(4.6.22a) $$\|\widetilde{a}-a\|_\infty / \|a\|_\infty \leqslant O(h^\varkappa),$$

(4.6.22b) $$\|\widetilde{a}-a\|_2 / \|a\|_2 \leqslant O(h^\varkappa)$$

(cf. (1.4.23e)). Using the respective representations $f_n=\sum\alpha_j\Phi_j$ or $\widetilde{f}_n=\sum\widetilde{\alpha}_j\Phi_j$, we can carry over the estimates (22a,b) to those with respect to the norms of $\widetilde{f}_n-f_n$, since the constant or linear ansatz functions Φ_j satisfy the assumptions of Criterion 5.9 (cf. Exercise 5.19/27).

Remark 4.6.26. (a) Let $C_n(\Phi)$ be the quantity defined in (5.9), which, under the assumptions of Criterion 5.9, is uniformly bounded. Denote the semidiscrete Galerkin solutions by f_n (for exact integration) and $\widetilde{f}_n$ (for numerical quadrature). Then, under assumption (22a), the following estimate of the relative error of $\delta f_n := \widetilde{f}_n - f_n$ holds:

(4.6.23a) $$\|\widetilde{f}_n-f_n\|_\infty / \|f_n\|_\infty \leqslant C_n(\Phi)\,\|\widetilde{a}-a\|_\infty / \|a\|_\infty .$$

(b) Let $A=A_n$ be the matrix from (5.7b). Then

(4.6.23b) $$\|\widetilde{f}_n-f_n\|_{L^2(D)} / \|f_n\|_{L^2(D)} \leqslant \sqrt{\mathrm{cond}_2(A)}\,\|\widetilde{a}-a\|_\infty / \|a\|_\infty .$$

(c) Under the assumptions $C_n(\Phi)\leqslant\mathrm{const}$ or $\mathrm{cond}_2(A)\leqslant\mathrm{const}$, the estimate $\|\widetilde{f}_n-f_n\| / \|f_n\| \leqslant O(h^\varkappa)$ follows from (22a/b) or (23a/b), respectively.

Proof. The proof of Lemma 5.8 showed $\|\delta f_n\|_\infty=\|\Pi_n\delta f_n\|_\infty \leqslant \|\sum|\Phi_j|\|_\infty \|\widetilde{a}-a\|_\infty$ and $\|a\|_\infty \leqslant \|A^{-1}\|_\infty \max_j \|\Phi_j\|_{L^1(D)} \|f_n\|_\infty$. This implies (23a). The analogous inequalities for (23b) can be found in part (iii) of the proof of Lemma 5.11. QED

The assumption $\varkappa>0$ ensures stability [here in the form of the uniform boundedness of $(\lambda A_n-\widetilde{B}_n)^{-1}$]. Concerning stability, the estimate $\|\delta B\|=O(h^\varkappa)$ can be weakened by $\|\delta B\|=o(1)$ or even $\|\delta B\|\leqslant\eta$ (with sufficiently small η).

For the description of concrete quadrature methods for an integral of the form $\int\varphi(\xi)\Phi_j(\xi)d\xi$, which, e.g., involves $\varphi=k(x,\cdot)$ in the case of β_{jk} and $\varphi=g$ in β_j, we assume that we have piecewise smooth basis functions Φ_j on a partition $a=x_0<x_1<\ldots<x_n=b$ (cf. (5.19a-c), (5.29a-c)). If Φ_j is scaled such that

(4.6.24a) $$\|\Phi_j\|_\infty = O(1)$$

(as for Lagrange functions), the νth derivative existing in the open intervals (x_{k-1},x_k) and the Lipschitz constant of $\Phi_j^{(\nu-1)}$ satisfy no better estimate than

(4.6.24b) $\|\Phi_j^{(\nu)}\|_\infty = O(h^{-\nu})$ and $L_{\Phi_j^{(\nu-1)}} = O(h^{-\nu})$.

These can be deduces from the mean-value theorem, (24a), and $\mathrm{supp}(\Phi_j) = O(h)$.

The following example shows that not every numerical quadrature can be analysed in this way. Let Φ_j be the piecewise linear Lagrange functions (1.4.6) corresponding to the equidistant interval partition $x_j = a + jh$. The composite Simpson's rule (1.4.18d) with the same support abscissae x_j applied to $\varphi\Phi_j$ yields $\frac{2}{3}h\varphi(x_j)$ or $\frac{4}{3}h\varphi(x_j)$, depending on whether j is even or odd. Both values deviate by $O(h)$ from $\int\varphi\Phi_j d\xi = h\varphi(x_j) + O(h^2)$. The reason is that the (non-composite) Simpson rule covers the interval $[x_{k-1}, x_{k+1}]$, in which $\varphi\Phi_j$ only belongs to $C_L(I) = \hat{C}^1(I)$. Therefore, the error estimate $|R(\varphi\Phi_j)| \leqslant Ch^{1+\varkappa}\|\varphi\Phi_j\|_{\hat{C}^\varkappa(I)}$ – this is (1.4.17) with the support length $b - a = 2h$ and $n = 3$ – only applies with $\varkappa = 1$. Because $\|\varphi\Phi_j\|_{\hat{C}^1(I)} = O(h^{-1})$ (cf. (24b)), this yields an error of the same order as the integral $\int\varphi\Phi_j d\xi$ itself. From $\delta\beta_{jk} = O(h)$ one concludes that $\delta B = O(1)$, so that stability seems questionable and the inequalities (22a,b) with $\varkappa = 0$ do not lead to a convergence statement. However, the following *second approach* to the error discussion shows that, nevertheless, positive statements are possible.

Let Q be the quadrature used in the collocation method, i.e.,

$$\beta_{jk} = \int_I k(\xi_j, y)\Phi_k(y)\,dy, \qquad \tilde{\beta}_{jk} = Q(k(\xi_j,\cdot)\Phi_k).$$

Let $f_n = \sum\alpha_k\Phi_k$ be the semidiscrete solution. In (4.7b) we replace $(K_nf_n)(\xi_j)$ by

$$\textstyle\sum_k\tilde{\beta}_{jk}\alpha_k = Q(k(\xi_j,\cdot)\sum_k\alpha_k\Phi_k) = Q(k(\xi_j,\cdot)f_n).$$

If the system is solvable, it leads to a solution $\tilde{f}_n$ of $\lambda\tilde{f}_n = g_n + \tilde{K}_n\tilde{f}_n$, where $\tilde{K}_n$ is obtained from K_n by replacing the integration by Q. From this equation and $\lambda f_n = g_n + K_nf_n$, one obtains the representation

$$(\lambda I - \tilde{K}_n)\,\delta f_n = (\tilde{K}_n - K_n)f_n$$

for the difference $\delta f_n := \tilde{f}_n - f_n$. For the right-hand side one finds the following reformulation, where k is a short notation for $k(\xi_j,\cdot)$ and f can be chosen arbitrarily:

$$((\tilde{K}_n - K_n)f_n)(\xi_j) = \int_I(f - f_n)k\,dy - Q((f - f_n)k) + \left\{Q(fk) - \int_I fk\,dy\right\}.$$

Assuming that f is the solution of $\lambda f = g + Kf$, we obtain an estimate for $\int(f - \Pi_nf)k\,dy$ and $Q((f - f_n)k)$ by the projection error $\|f - \Pi_nf\|_\infty$. The term $Q(fk) - \int_I fk\,dy$ describes the quadrature error $R(fk)$ for the integrand $fk = fk(\xi_j,\cdot)$ and depends on the smoothness of the solution f and of the kernel k. Assuming stability $\|(\lambda I - \tilde{K}_n)^{-1}\|_\infty \leqslant C$ (which does not yet follow directly), we gain the error estimate

(4.6.25) $\|\delta f_n\| \leqslant O(\|f - f_n\|_\infty) + O(\max_j R(fk(\xi_j,\cdot)))$.

Estimate (25) yields the following rule: *The numerical quadrature should be of the order of the projection errors.* Also Simpson rule with step size h is admissible and yields a perturbation of the order $O(h^4)$, although $\delta B = O(h^0)$!

In the case of Simpson's rule one can make the δB-error small, if the subintervals, on which the *composite* quadrature formula is based, coincide with the subintervals I_k or are contained in I_k. By Lemma 1.4.23, piecewise smoothness of Φ_j in the *interior* of I_k is sufficient. In the case of Simpson's rule, the step size of the quadrature formula must be $\leqslant h/2$. The weights $\frac{4}{3}(h/2)$ and $\frac{2}{3}(h/2)$ together with the values $\Phi_j(x_j)=1$ and $\Phi_j(x_j \pm h/2)=\frac{1}{2}$ of the piecewise linear Lagrange function (1.4.6) yield the approximation

$$\int \varphi \Phi_j d\xi \approx Q_{j,h}(\varphi) := \tfrac{1}{3} h \,[\varphi(x_j - h/2) + \varphi(x_j) + \varphi(x_j + h/2)].$$

The error estimates are $|Q_{j,h}(\varphi) - \int \varphi \Phi_j d\xi| \leqslant C h^5 \|\varphi\|_{\hat{C}^4(I)}$.

The quadrature above has been used in our first numerical example of the Galerkin method (§4.5.10, Table 5.2) for approximating $\int \varphi(x) \Phi_j(x) dx$ with $\varphi(x) := (K\Phi_k)(x)$ (cf. (5.40d)).

Exercise 4.6.27. Let Φ_j be the piecewise linear Lagrange functions. Prove the following: **(a)** The *composite trapezoidal rule* approximates $\int \varphi \Phi_j dx$ by $Q_j(\varphi) = \frac{1}{2}(x_{j+1} - x_{j-1})\varphi(x_j)$. The resulting error matrix satisfies $\|\delta B\| = O(h)$ for a general mesh and $\|\delta B\| = O(h^2)$ for an equidistant partition.
(b) The trapezoidal rule Q_j from (a) is also the weighted Newton-Cotes formula with one support abscissa x_j and the weight function $w = \Phi_j$.
(c) Show that α, β, γ in formula (26a) are given by (26b):

(4.6.26a) $$Q_{j,h}(\varphi) := \alpha\,\varphi(x_{j-1}) + \beta\,\varphi(x_j) + \gamma\,\varphi(x_{j+1}),$$

(4.6.26b) $$\alpha = \frac{A^3 + 2A^2B - B^3}{12\,A\,(A+B)}, \qquad \beta = \frac{A^3 + 4A^2B + 4AB^2 + B^3}{12\,A\,B}, \qquad \gamma = \frac{B^3 + 2B^2A - A^3}{12\,B\,(A+B)}$$

with $\quad A := x_j - x_{j-1}, \qquad B := x_{j+1} - x_j.$

$Q_{j,h}$ is exact for polynomials up to degree $\leqslant 2$. Formula (26a,b) represents the *Newton-Cotes formula weighted* with the piecewise linear Lagrange function (1.4.6) using the support abscissae x_j, $x_{j\pm 1}$. In the equidistant case, the error is $O(h^5)$; otherwise

(4.6.26c) $$|Q_{j,h}(\varphi) - \int \varphi \Phi_j d\xi| \leqslant C h^4 \|\varphi\|_{\hat{C}^3(I)}.$$

4.6.7 Product Integration

The second numerical example from §4.4.5 contains the weakly singular kernel (4.22a): $k(x-y) = |x-y|^{-1/2}$. Fortunately, the matrix entries $\beta_{jk} = \int k(\xi_j, y)\Phi_k(y)dy$ could be integrated exactly, so that questions concerning the numerical quadrature did not appear in §4.4.5. The difficulties arising for any standard quadrature method are:

(i) For $k=j$, the singular point $y=\xi_j$ lies in the integration domain (in the support of Φ_k) and the quadrature formula fails.

(ii) Even if $y = \xi_j$ is not in the support of Φ_k but in a small neighbourhood of it, large quadrature errors are induced. If, e.g., the distance is $\leqslant \ell h$ and $k(x-y) = |x-y|^{-1/2}$, the norm $\|\varphi\|_{\hat{C}^{\varkappa}(I)}$ for $\varphi = k(\xi_j, \cdot)$ appearing in the error estimate has the order $O([\ell h]^{-\varkappa - 1/2})$.

If the kernel k is not simple enough to integrate $\int k(\xi,y)\Phi_k(y)dy$ exactly, it may possibly be written as a product

(4.6.27) $$k(x,y) = \ell(x,y)\,\varkappa(x,y),$$

where $\ell(x,y)$ is sufficiently smooth and $\varkappa(x,y)$ allows an exact integration of $\int \varkappa(\xi,y)\Phi_k(y)dy$. In this case, the interpolatory quadrature method with the weight function $\omega = \varkappa(x,\cdot)$ as defined in (1.4.16d,e) is applicable:

(4.6.28a) $$Q_n^{\varkappa(x,\cdot)}(\varphi) := \int_D \varkappa(x,y)\,\Pi_n\varphi(y)\,dy,$$

where Π_n defines an interpolation with support abscissae x_i. Let L_i be the Lagrange functions. According to (1.4.16e), the representation (28b) holds:

(4.6.28b) $$Q_n^{\varkappa(x,\cdot)}(\varphi) := \sum w_i(x)\varphi(x_i) \qquad \text{with } w_i(x) := \int_D \varkappa(x,y)L_i(y)\,dy.$$

Inserting $\varphi := \ell(x,\cdot)\Phi_k$, we obtain the approximation

(4.6.28c) $$Q_n^{\varkappa(x,\cdot)}(\ell(x,\cdot)\Phi_k) = \int_D \varkappa(x,y)\,\Pi_n[\ell(x,y)\Phi_k(y)]\,dy = \\ = \sum_i w_i(x)\,\ell(x,x_i)\,\Phi_k(x_i)$$

of $\int_D k(x,y)\Phi_k(y)dy$. If the basis functions Φ_k are chosen as Lagrange functions L_k of the interpolation Π_n involved in (28a), (28c) becomes

(4.6.28d) $$Q_n^{\varkappa(x,\cdot)}(\ell(x,\cdot)\Phi_k) = w_k(x)\ell(x,x_k) = \ell(x,x_k)\int_D \varkappa(x,y)L_k(y)dy.$$

The weighted quadrature method $Q_n^{\varkappa(x,\cdot)}$, which in this connection is called *product integration*, contains weights $w_i(x)$ depending on x! In the case of $k(x-y) = |x-y|^{-1/2}$ and piecewise linear interpolation, the coefficients β_{jk} from (4.23) describe the weights $w_k(x)$ at $x = x_j$.

When factorising $k(x,y) = \ell(x,y)\varkappa(x,y)$ in (27), one has to take care that $\varkappa(x,y)L_k(y)$ is sufficiently easy to integrate. One may choose $\varkappa$ to asymptotically have the same singularity as k. Then the factor ℓ is defined indirectly by $\ell := k/\varkappa$.

Example 4.6.28. At $x = y$, $k(x,y) = |e^x - e^y|^{-1/2}$ behaves like $e^{-x/2}|x-y|^{-1/2}$. Therefore, take $\varkappa(x,y) := |x-y|^{-1/2}$ and $\ell := k/\varkappa$. Prove that k is analytical with respect to x and y and, therefore, all regularity requirements are satisfied.

As in the standard case, the *quadrature error of the product integration* is given by the interpolation error of $\Pi_n[\ell(x,y)\Phi_k(y)]$ appearing in (28c).

Remark 4.6.29. Since the integration and evaluation of the weights $w_i(x)$ in (28b) is more expensive than the usual quadrature methods, the following compromise is helpful: Compute the integral $\int k(x,y)\Phi_k(y)dy$ by means of product integration as long as x has a distance smaller than $\delta = \delta(h)$ from the supp(Φ_k). Otherwise, use usual quadrature methods as proposed in §4.6.6. Here we assume that $k(x,y)$ is singular only at $x = y$.

4.7 Discretisation by Quadrature: The Nyström Method

4.7.1 Description of the Method

In this section, we study a discretisation which is directly obtainable by applying a quadrature formula to the integral. In this respect, Fredholm integral equations offer a simpler situation than Volterra equation (cf. §2.2.1) because of the fixed integration domain.

We consider n *support abscissae*

$$\Xi_n := \{\xi_{1,n}, \xi_{2,n}, \dots, \xi_{n,n}\} \subset D \tag{4.7.1a}$$

with associated *weights*

$$w_{1,n}, w_{2,n}, \dots, w_{n,n} \tag{4.7.1b}$$

and define a *quadrature method* Q_n ($n \in \mathbb{N}$, cf. §1.4.2),

$$Q_n(\varphi) := \sum_{k=1}^{n} w_{k,n}\, \varphi(\xi_{k,n}), \tag{4.7.1c}$$

for integrals $\int_D \varphi(y)dy$ over D. The integrand of Fredholm's integral equation is $\varphi(y) := k(x,y)f(y)$. The *Nyström method* (cf. Nyström [1,2]) approximates the integral by means of Q_n and determines an approximation f_n by

$$\lambda f_n(x) = g(x) + \sum_{k=1}^{n} w_{k,n}\, k(x,\xi_{k,n}) f_n(\xi_{k,n}) \qquad \text{for all } x \in D. \tag{4.7.2}$$

The solution of the semidiscrete problem (2) can be performed in two steps. First, we rewrite equation (2) only at the points $x = \xi_{j,n} \in \Xi$:

$$\lambda f_n(\xi_{j,n}) = g(\xi_{j,n}) + \sum_{k=1}^{n} w_{k,n}\, k(\xi_{j,n},\xi_{k,n}) f_n(\xi_{k,n}) \qquad (1 \le j \le n). \tag{4.7.3}$$

With the abbreviation

$$f_{k,n} := f_n(\xi_{k,n}) \qquad (1 \le k \le n,\ n \in \mathbb{N}), \tag{4.7.4}$$

(3) represents a system of n equations

$$\lambda f_{j,n} = g(\xi_{j,n}) + \sum_{k=1}^{n} w_{k,n}\, k(\xi_{j,n},\xi_{k,n}) f_{k,n} \qquad (1 \le j \le n) \tag{4.7.5}$$

for the n unknowns $\{f_{1,n}, \dots, f_{n,n}\}$. We introduce the vectors

$$a_n := \begin{bmatrix} f_{1,n} \\ \vdots \\ f_{n,n} \end{bmatrix}, \qquad b_n := \begin{bmatrix} g(\xi_{1,n}) \\ \vdots \\ g(\xi_{n,n}) \end{bmatrix}, \tag{4.7.6a}$$

and the matrix

$$B_n := \begin{bmatrix} \beta_{11} & \beta_{12} & \cdots & \beta_{1n} \\ \vdots & \vdots & & \vdots \\ \beta_{n1} & \beta_{n2} & \cdots & \beta_{nn} \end{bmatrix} \quad \text{with } \beta_{jk} := w_{k,n}\, k(\xi_{j,n},\xi_{k,n}). \tag{4.7.6b}$$

Remark 4.7.1. The *Nyström method* leads to the system (5) of n equations for the n unknown function values $f_{k,n}$ from (4). In matrix representation, the equations (5) may be written as the system

(4.7.7) $$(\lambda I - B_n)\, a_n = b_n .$$

If the values $f_{k,n} = f_n(\xi_{k,n})$ are known, also the right-hand side in equation (2) is known for all $x \in D$. Using (4), we rewrite (2) and solve for $f_n(x)$:

(4.7.8) $$f_n(x) = \tfrac{1}{\lambda}\Big[g(x) + \sum_{k=1}^{n} w_{k,n}\, k(x, \xi_{k,n}) f_{k,n}\Big] \qquad (x \in D).$$

This determines $f_n(x)$ for all $x \in D$.

Remark 4.7.2. If $f_{k,n} := f_n(\xi_{k,n})$ solves the system (7), equation (8) can be regarded as an interpolation, the so-called *Nyström interpolation*. The Nyström interpolation is affine but not linear. The Nyström interpolant (8) of the discrete solution $f_{k,n}$ yields the solution of problem (2).

Proof. Because of (7), the values of f_n at the abscissae $\xi_{k,n}$ are the prescribed values $f_{k,n}$. The appearance of the term $g(x)/\lambda$ in (8) contradicts linearity. QED

The right-hand side in (2) defines the operator $K_n \in L(C(D), C(D))$ with

(4.7.9a) $$(K_n \varphi)(x) = \sum_{k=1}^{n} w_{k,n}\, k(x, \xi_{k,n})\, \varphi(\xi_{k,n}) = Q_n(k(x,\cdot)\varphi(\cdot))$$

for all $x \in D$ and $\varphi \in C(D)$. Hence, equation (2) can be written in the form

(4.7.9b) $$\lambda f_n = g + K_n f_n .$$

Different from previous discretisation methods, the discrete system (7) is *directly* related to the discrete operator equation (9b). The system (7) is a consequence of (9b), if the arguments x are restricted to Ξ_n.

Remark 4.7.3. If the kernel is symmetric, i.e., $k(x,y) = k(y,x)$, the matrix B_n is *not* necessarily symmetric. However, if the weights $w_{k,n}$ are positive, one can transform equation (7) into the equivalent system

(4.7.7') $$(\lambda I - B_n')\, a_n' = b_n'$$

with the symmetric matrix $B_n' = D^{1/2} B_n D^{-1/2}$ and vectors $a_n' = D^{1/2} a_n$, $b_n' = D^{1/2} b_n$, where $D = D_n := \mathrm{diag}\{w_{1,n}, \ldots, w_{n,n}\}$. Explicitly, the matrix entries of B_n' are given by $\beta_{jk}' = \sqrt{w_{j,n}\, w_{k,n}}\; k(\xi_{j,n}, \xi_{k,n})$.

Lemma 4.7.4. The operator K_n defined by (9a) has a finite-dimensional range. If the kernel $k(x,y)$ is continuous with respect to $x \in D$, $K_n \in K(C(D), C(D))$ holds. The operator norm $\|\cdot\|_{C(D) \leftarrow C(D)}$ of K_n has the value

(4.7.10) $$\|K_n\| = \sup_{x \in D} \sum_{k=1}^{n} |w_{k,n}\, k(x, \xi_{k,n})|.$$

Proof. If $k(\cdot, y) \in C(D)$, K_n maps into $C(D)$. By Theorem 1.3.23c $\dim(\mathrm{range}(K_n)) \leqslant n$ implies $K_n \in K(C(D), C(D))$. The proof of (10) is similar to the proof of (4.18). QED

4.7.2 Convergence Analysis

Convergence estimates can be formulated in two ways. Firstly, one can restrict oneself to the support abscissae $\xi_{k,n} \in \Xi_n$ and consider the error

(4.7.11a) $\delta f_{\Xi,n} := \max\{|f_{k,n} - f(\xi_{k,n})| : 1 \leqslant k \leqslant n\}.$

Here, f is the exact solution of the Fredholm integral equation (3.1.1). Secondly, $f_{k,n}$ defines the function f_n (cf. (8)) so that the error of f_n with respect to the supremum norm can be studied:

(4.7.11b) $\delta f_{D,n} := \| f_n - f \|_{\infty,D}.$

Both errors are closely connected as explained in the following remark.

Remark 4.7.5. Let the errors $\delta f_{\Xi,n}$, $\delta f_{D,n}$ be defined by (11a,b). **(a)** One direction is given by the inequality

(4.7.12a) $\delta f_{\Xi,n} \leqslant \delta f_{D,n}.$

(b) For the reverse direction, e.g., the estimate

(4.7.12b) $\delta f_{D,n} \leqslant [\| R_n(f) \|_\infty + \| K_n \| \, \delta f_{\Xi,n}] / |\lambda|$

holds, where $R_n(f)$ is the error of the quadrature in (2):

(4.7.12c) $R_n(f)(x) := \int_D k(x,y) f(y) dy - Q_n(k(x,\cdot) f(\cdot)).$

Proof. (12a) is trivial because of $\Xi_n \subset D$. Subtraction of the discrete equation $\lambda f_n = g + K_n f_n$ from the integral equation $\lambda f = g + Kf$ yields

$$\lambda(f - f_n) = Kf - K_n f_n = (K - K_n) f + K_n (f - f_n).$$

By definition of K_n, $\|(K-K_n)f\|_\infty = \| R_n(f) \|_\infty$ holds. The second term $K_n(f - f_n)$ can be estimated not only by $\| K_n \| \, \| f - f_n \|_\infty = \| K_n \| \, \delta f_{D,n}$ but also by $\| K_n \| \, \delta f_{\Xi,n}$. This proves (12b). QED

In the case of the previous discretisation methods, we could prove convergence $K_n \to K$ with respect to the operator norm under suitable assumptions. Then, consistency, stability, and convergence followed (cf. Theorem 1.13). Error estimates could be gained, e.g., from the inequalities (1.12d,e), which for $g_n = g$ take the form

$$\| f - f_n \| \leqslant \| (\lambda I - K_n)^{-1} \| \, \| (K - K_n) f \| \leqslant \| (\lambda I - K_n)^{-1} \| \, \| K - K_n \| \, \| f \|.$$

The following lemma shows that in the case of the Nyström method, norm convergence is *not* valid (at least with $X = C(D)$). Therefore, the bound $\| (\lambda I - K_n)^{-1} \| \, \| K - K_n \| \, \| f \|$ in the latter error estimate is useless.

Lemma 4.7.6. *D is assumed to be compact and $k \in C(D \times D)$. Let the quadrature method Q_n from (1c) be convergent (cf. Definition 1.4.15). Then*

(4.7.13a) $\| K_n \varphi - K \varphi \|_\infty \to 0 \quad (n \to \infty)$ for all $\varphi \in C(D)$, i.e., $\{K_n\}$ is *consistent*. With respect to the operator norm $\|\cdot\|_{C(D)\leftarrow C(D)}$, we have

(4.7.13b) $\qquad \|K_n - K\| \geq \|K\| \qquad$ for all $n \in \mathbb{N}$,

(4.7.13c) $\qquad \|(K_n - K)K\| \to 0, \quad \|(K_n - K)K_n\| \to 0 \qquad (n \to \infty)$.

Proof. (i) For fixed $x \in D$, $\Phi(y) := k(x,y)\varphi(y)$ is continuous. The convergence of Q_n implies $(K_n\varphi)(x) = Q_n\Phi \to \int_D \Phi(y)dy = (K\varphi)(x)$ for all $x \in D$. Theorem 1.4.17 proves stability of Q_n, i.e., $\|Q_n\| = \sum_{k=1}^{n} |w_{k,n}| \leq C$ for all $n \in \mathbb{N}$. This shows the uniform boundedness of K_n (cf. (10)):

$$\|K_n\| \leq C\|k\|_{\infty, D\times D}.$$

Further, the stability $\|Q_n\| \leq C$ leads to the estimate

$$|(K_n\varphi)(x) - (K_n\varphi)(y)| \leq C\|k(x,\cdot) - k(y,\cdot)\|_\infty \|\varphi\|_\infty$$

showing the equicontinuity of $\{K_n\varphi : n \in \mathbb{N}\}$ (note that k is uniformly continuous on $D \times D$). Therefore, the theorem of Arzelà-Ascoli (Theorem 1.3.26) is applicable: There is a subsequence of $K_n\varphi$ converging uniformly. Having already stated pointwise convergence $K_n\varphi \to K\varphi$, according to part (iv) of the proof ,we obtain (13a): the uniform convergence of the full sequence $K_n\varphi$.
(ii) Since D is compact, there is a $\xi \in D$ according to (3.2.6) so that

$$\|K\| = \int_D |k(\xi,y)|dy.$$

In Lemma 3.2.2, this statement was proved by constructing a $\varphi_\varepsilon \in C(D)$ for each $\varepsilon > 0$ fulfilling $\|\varphi_\varepsilon\|_\infty = 1$ and

$$\int_D k(\xi,y)\varphi_\varepsilon(y)dy \geq \int_D |k(\xi,y)|dy - \varepsilon.$$

Changing φ_ε in a sufficiently small neighbourhood of Ξ_n, one can satisfy $\varphi_\varepsilon(\xi_{k,n}) = 0$ at the quadrature points $\xi_{k,n} \in \Xi_n$. This implies $K_n\varphi_\varepsilon = 0$. Then we have

$$\|K - K_n\| \geq \|(K - K_n)\varphi_\varepsilon\|_\infty = \|K\varphi_\varepsilon\|_\infty \geq \|K\| - 2\varepsilon$$

and (13b) follows.
(iii) By Theorem 3.2.5, K is compact. From (13a) and Lemma 3.8 it follows that $\|(K_n - K)K\| \to 0$. For proving the second assertion $\|(K_n - K)K_n\| \to 0$, define

$$\Phi_n(\xi) := \|(K - K_n)k(\cdot,\xi)\|_\infty \qquad \text{for } \xi \in D.$$

The reverse triangle inequality yields

$$|\Phi_n(\xi) - \Phi_n(\zeta)| \leq \|(K - K_n)(k(\cdot,\xi) - k(\cdot,\zeta))\|_\infty$$
$$\leq \|K - K_n\| \, \|k(\cdot,\xi) - k(\cdot,\zeta)\|_\infty.$$

The uniform boundedness of $\|K - K_n\|$ follows from the convergence (13a). Hence, the uniform continuity of k implies the uniform boundedness and equicontinuity of Φ_n. Together with the compactness of D and the pointwise convergence (13a), the uniform convergence $\Phi_n \to 0$ follows (cf. (iv)). From the representation

$$\|(K_n-K)K_n\| = \sum_{k=1}^{n}|w_{k,n}|\Phi_n(\xi_{k,n})$$

(analogous to (10)) and because of the stability of Q_n: $\sum_{k=1}^{n}|w_{k,n}|\leqslant C$, the assertion $\|(K_n-K)K_n\|\to 0$ follows.

(iv) Here, we give the proof that pointwise convergence $\psi_n\to\psi$ and equicontinuity imply uniform convergence (i.e., $\|\psi_n-\psi\|_\infty\to 0$). The negation of $\|\psi_n-\psi\|_\infty\to 0$ reads: There is an $\varepsilon>0$ and $x_n\in D$ with $|\psi_n(x_n)-\psi(x_n)|\geqslant\varepsilon$ for all $n\in\mathbb{N}$. Since D is compact, a subsequence $x_{n_i}\to x^*\in D$ must exist. Because of the equicontinuity, $|\psi_\ell(x_{n_i})-\psi_\ell(x^*)|\leqslant\varepsilon/4$ holds for $n_i\geqslant n'$ (with n' sufficiently large). This inequality is applied to $\ell=n_i$. Furthermore, the pointwise convergence $(\ell\to\infty)$ yields $|\psi(x_{n_i})-\psi(x^*)|\leqslant\varepsilon/4$. Together with $|\psi(x^*)-\psi_{n_i}(x^*)|\leqslant\varepsilon/4$ for sufficiently large n_i, we obtain $|\psi_{n_i}(x_{n_i})-\psi(x_{n_i})|\leqslant 3\varepsilon/4<\varepsilon$ in contradiction to the assumption. QED

The assertions in (13c) can be viewed as replacements for the missing operator norm convergence. The next subsection shows how (13c) can be used to prove the stability of the discretisation, which as we know is necessary for the convergence (cf. Theorem 4.1.11).

4.7.3 Stability

In Theorem 1.4, we proved existence and uniform boundedness of $\|(\lambda I-K_n)^{-1}\|$ by (1.6b): $\|K-K_n\|<1/\|(\lambda I-K)^{-1}\|$. The following Theorem of Brakhage [1] shows that the corresponding estimate of $\|(K_n-K)K\|$ is also sufficient.

Theorem 4.7.7. Let X be a Banach space and $\lambda\neq 0$ a regular value of the operator $S\in L(X,X)$, i.e., $(\lambda I-S)^{-1}\in L(X,X)$. Let T be a compact operator, i.e., $T\in K(X,X)$. If the inequality

(4.7.14a) $\quad\|(T-S)T\| < |\lambda|/\|(\lambda I-S)^{-1}\|$

holds, then $(\lambda I-T)^{-1}\in L(X,X)$ exists and satisfies the estimate

(4.7.14b) $\quad\|(\lambda I-T)^{-1}\| \leqslant \dfrac{1+\|(\lambda I-S)^{-1}\|\,\|T\|}{|\lambda|-\|(\lambda I-S)^{-1}\|\,\|(T-S)T\|}.$

The solutions of the equations $(\lambda I-S)f_S=g$ and $(\lambda I-T)f_T=g$ satisfy

(4.7.14c) $\quad\|f_S-f_T\|_X \leqslant \|(\lambda I-S)^{-1}\|\dfrac{\|(T-S)T\|\,\|f_S\|_X+\|(T-S)g\|_X}{|\lambda|-\|(\lambda I-S)^{-1}\|\,\|(T-S)T\|},$

(4.7.14d) $\quad\|f_S-f_T\|_X \leqslant \|(\lambda I-T)^{-1}\|\,\|(S-T)f_S\|_X.$

Proof. (i) The identity $I=\frac{1}{\lambda}[(\lambda I-T)+T]$ implies the representation

$$(\lambda I-T)^{-1} = \tfrac{1}{\lambda}[I+(\lambda I-T)^{-1}T],$$

provided that $(\lambda I-T)^{-1}$ exists. In the following, we do not need this assumption, since we replace $(\lambda I-T)^{-1}$ from the right-hand side by $(\lambda I-S)^{-1}$ in order to get an approximation

$$A := \tfrac{1}{\lambda}[I+(\lambda I-S)^{-1}T]$$

of $(\lambda I-T)^{-1}$. Hence, $B:=A(\lambda I-T)$ should approximate the identity I. By

$$B = \tfrac{1}{\lambda}[I+(\lambda I-S)^{-1}T](\lambda I-T) = I-\tfrac{1}{\lambda}[T-(\lambda I-S)^{-1}T(\lambda I-T)] =$$
$$= I-\tfrac{1}{\lambda}(\lambda I-S)^{-1}\{(\lambda I-S)T-T(\lambda I-T)\} = I-\tfrac{1}{\lambda}(\lambda I-S)^{-1}(T-S)T$$

we see that B differs from I by $\frac{1}{\lambda}(\lambda I-S)^{-1}(T-S)T$. Thanks to assumption (14a), the operator norm of this expression is

$$\|\tfrac{1}{\lambda}(\lambda I-S)^{-1}(T-S)T\| \leqslant \|(\lambda I-S)^{-1}\|\,\|(T-S)T\|/|\lambda| < 1,$$

so that Lemma 1.3.14 (with B and I instead of S and T) proves the existence of $B^{-1}\in L(X,X)$.

(ii) If λ is an eigenvalue of T, $B=A(\lambda I-T)$ must be singular in contradiction to part (i). Since T is compact, Theorem 1.3.28 proves the existence of the inverse $(\lambda I-T)^{-1}\in L(X,X)$. Multiplying the equation $B=A(\lambda I-T)$ from the left by B^{-1} and from the right by $(\lambda I-T)^{-1}$, we obtain the representation $(\lambda I-T)^{-1}=B^{-1}A$, i.e.,

$$(4.7.14e) \qquad (\lambda I-T)^{-1} = [\lambda I-(\lambda I-S)^{-1}(T-S)T]^{-1}[I+(\lambda I-S)^{-1}T].$$

According to (1.3.12b), the inverse on the right side can be estimated by

$$\|[\lambda I-(\lambda I-S)^{-1}(T-S)T]^{-1}\| \leqslant 1/[|\lambda|-\|(\lambda I-S)^{-1}\|\,\|(T-S)T\|],$$

while the second factor is bounded by $1+\|(\lambda I-S)^{-1}\|\,\|T\|$. Together, inequality (14b) results.

(iii) Subtraction of the equations $(\lambda I-S)f_S=g$ and $(\lambda I-T)f_T=g$ yields

$$\lambda(f_S-f_T) = Sf_S-Tf_T = T(f_S-f_T)+(S-T)f_S$$

and $\qquad f_S-f_T = (\lambda I-T)^{-1}(S-T)f_S.$

This representation proves estimate (14d). Changing the roles of S and T, we gain the analogous representation

$$f_T-f_S = (\lambda I-S)^{-1}(T-S)f_T.$$

Inserting $f_T=\frac{1}{\lambda}(g+Tf_T)$ into the right-hand side, we obtain

$$f_T-f_S = \tfrac{1}{\lambda}(\lambda I-S)^{-1}(T-S)(g+Tf_T) =$$
$$= \tfrac{1}{\lambda}(\lambda I-S)^{-1}(T-S)(g+Tf_S)+\tfrac{1}{\lambda}(\lambda I-S)^{-1}(T-S)T(f_T-f_S);$$

hence,

$$f_T-f_S = [I-\tfrac{1}{\lambda}(\lambda I-S)^{-1}(T-S)T]^{-1}\tfrac{1}{\lambda}(\lambda I-S)^{-1}(T-S)(g+Tf_S).$$

The associated norm estimate leads to (14c). QED

To prove the stability of Nyström method by Theorem 7, we have to apply (14b) with $T:=K_n$ and $S:=K$. To fulfil assumption (14a), we have to show $\|(T-S)T\|=\|(K_n-K)K_n\|\to 0$. In Lemma 6, $\|(K_n-K)K_n\|\to 0$ was proved under the strong condition $k\in C(D\times D)$. We recall that, according to Lemma 3.8, $\|(K_n-K)K\|\to 0$ follows from the pointwise convergence $K_n\to K$ and the compactness of K. In the case of $\|(K_n-K)K_n\|\to 0$, compactness of K_n is *not* sufficient. Reviewing Lemma 3.7 shows that a precompact subset $M \subset X$ exists containing all images $K_n\varphi$ ($\|\varphi\| = 1$, $n \in \mathbb{N}$). This leads to the next definition.

Definition 4.7.8. A set $\{T_n\}_{n\in\mathbb{N}}$ of operators $T_n \in L(X,X)$ is called *collectively compact*, if the set $\{T_n\varphi:\ \varphi\in X \text{ with } \|\varphi\| \leqslant 1,\ n\in\mathbb{N}\}$ is precompact.

Note that the precompactness of $B_n := \{T_n\varphi:\ \varphi\in X,\ \|\varphi\| \leqslant 1\}$ defines the compactness of T_n, while now the *union* $\bigcup_{n\in\mathbb{N}} B_n$ must be precompact.

Remark 4.7.9. **(a)** If the set $\{T_n\}_{n\in\mathbb{N}}$ is collectively compact, then each T_n is compact.
(b) If a sequence of collectively compact T_n converges pointwise to $T\in L(X,X)$, then T is compact.
(c) If the T_n are compact operators converging with respect to the operator norm, then $\{T_n\}_{n\in\mathbb{N}}$ is collectively compact.

Proof. (i) If $B\subset X$ is precompact, also the closure $\overline{B}$ and any subset $A\subset\overline{B}$ are precompact. By definition $B := \{T_n\varphi:\ \varphi\in X \text{ with } \|\varphi\|\leqslant 1,\ n\in\mathbb{N}\}$ is precompact. The precompactness of the subset $B_n := \{T_n\varphi:\ \varphi\in X,\ \|\varphi\|\leqslant 1\}$ implies $T_n\in K(X,X)$.
(ii) Let B and B_n be as above and $A := \{T\varphi:\ \varphi\in X,\ \|\varphi\|\leqslant 1\}$. Each $\psi\in A$ has the representation $\psi = T\varphi$ with some $\varphi\in X$, $\|\varphi\|\leqslant 1$. Because of the pointwise convergence, $\psi = \lim \psi_n$ holds for $\psi_n := T_n\varphi$. Since $\psi_n\in B_n\subset B$, $\psi\in\overline{B}$ follows. Since $\psi\in A$ is arbitrary, one concludes that $A\subset\overline{B}$. According to (i), A is precompact and, therefore, T is compact.
(iii) Let $\{\psi_m\}\subset B$ be a sequence in B with B and B_n defined as above. We have to find a convergent subsequence of the ψ_m. First assume that $\{\psi_m\}\subset\bigcup_{0<n\leqslant k} B_n$ for some $k\in\mathbb{N}$. Since T_n is compact, all B_n are precompact. As a union of finitely many precompact sets, $\bigcup_{0<n\leqslant k} B_n$ is again precompact. Hence, there is a convergent subsequence. Each ψ_m has a representation $\psi_m = T_{n(m)}\varphi_m$ with suitable $n(m)$ and $\|\varphi_m\|\leqslant 1$. If the assumption $\{\psi_m\}\subset\bigcup_{0<n\leqslant k} B_n$ is not valid, there must be a subsequence, denoted again by $\psi_m = T_{n(m)}\varphi_m$, so that $n(m)\to\infty$ for $m\to\infty$. Let $T := \lim T_n$ and $y_m := T\varphi_m$. According to Theorem 1.3.23b, T is compact. Therefore, $y_m = T\varphi_m$ contains a subsequence again denoted by y_m. Because of $\|\psi_m - y_m\| = \|(T_{n(m)} - T)\varphi_m\| \leqslant \|T_{n(m)} - T\| \to 0$, the corresponding subsequence ψ_m has the same limit. So, in all cases we have found a convergent subsequence. QED

Theorem 4.7.10. (Anselone-Moore [1], Anselone [1]) Let X be a Banach space. Assume that the discretisation $\{K_n\in L(X,X)\}$ is consistent with $K\in L(X,X)$ (i.e., $K_n\varphi\to K\varphi$ for all $\varphi\in X$, cf. Definition 1.1) and that $\{K_n\}$ is collectively compact. Then K is compact and the sequence $\{K_n\}$ is uniformly bounded. Furthermore,

(4.7.15a) $\quad \|(K-K_n)K\| \to 0 \qquad$ for $n\to\infty$,

(4.7.15b) $\quad \|(K-K_n)K_n\| \to 0 \qquad$ for $n\to\infty$.

Proof. (i) Because of the consistency, Remark 9b is applicable and proves $K \in K(X,X)$.
(ii) The uniform boundedness of $\{K_n\}$ is a consequence of Theorem 1.3.15 (Theorem of Uniform Boundedness).
(iii) Consistency of $\{K_n\}$ and $K \in K(X,X)$ yields (15a) (cf. Lemma 3.8).
(iv) (15b) follows from Lemma 3.7 with $M := B$, where B is given in the proof of Remark 9. QED

Combining the statements of the Theorems 7 and 10, one obtains the following result.

Theorem 4.7.11. Let X be a Banach space. The discretisation $K_n \in L(X,X)$ is assumed to be consistent with $K \in L(X,X)$ and collectively compact. Let $\lambda \neq 0$ be a regular value of K. *Then the discretisation is stable and convergent.* In particular, the following inequality holds for sufficiently large n_0:

(4.7.16a) $\quad \|(K-K_n)K_n\| < |\lambda| / \|(\lambda I-K)^{-1}\| \qquad$ for all $n \geq n_0$,

so that $(\lambda I-K_n)^{-1}$ exists for $n \geq n_0$ and is uniformly bounded because of

(4.7.16b) $$\|(\lambda I-K_n)^{-1}\| \leq \frac{1+\|(\lambda I-K)^{-1}\|\,\|K_n\|}{|\lambda|-\|(\lambda I-K)^{-1}\|\,\|(K-K_n)K_n\|} \qquad \text{for } n \geq n_0.$$

The solutions of $\lambda f = g + Kf$ and $\lambda f_n = g + K_n f_n$ $(n \geq n_0)$ satisfy the following two *error estimates*:

(4.7.16c) $$\|f-f_n\| \leq \|(\lambda I-K)^{-1}\| \frac{\|(K-K_n)K_n\|\,\|f\| + \|(K-K_n)g\|}{|\lambda|-\|(\lambda I-K)^{-1}\|\,\|(K-K_n)K_n\|}$$

(4.7.16d) $$\|f-f_n\| \leq \|(\lambda I-K_n)^{-1}\|\,\|(K-K_n)f\| \qquad \text{for all } n \geq n_0.$$

Proof. Theorem 10 proves $\|(K-K_n)K_n\| \to 0$, so that (16a) follows. Together with the uniform boundedness of the K_n, inequality (16b), obtained from (14b) with $T := K_n$ and $S := K$, shows the uniform boundedness of $\|(\lambda I-K_n)^{-1}\|$; hence, K_n is stable. Consistency, stability, and $K \in K(X,X)$ imply convergence (cf. Theorem 1.12a)). The convergence can also be derived from the inequalities (16c,d), which follow from (14c,d). QED

Corollary 4.7.12. Choose $X = C(D)$. Because of the previous inequality (12a): $\max\{|f_{k,n} - f(\xi_{k,n})| : 1 \leq k \leq n\} \leq \|f-f_n\|_\infty$, (16c) and (16d) are also error estimates of the computed approximations $\alpha_j = f_{j,n} = f_n(\xi_{j,n})$.

Exercise 4.7.13. Prove the converse of Theorem 11. Let X be a Banach space and $K \in K(X,X)$. Assume also that for some n, $(\lambda I-K_n)^{-1} \in L(X,X)$ and

(4.7.17a) $\quad \|(K-K_n)K\| < |\lambda| / \|(\lambda I-K_n)^{-1}\|.$

Then $(\lambda I-K)^{-1}$ exists and satisfies

(4.7.17b) $$\|(\lambda I-K)^{-1}\| \leq \frac{1+\|(\lambda I-K_n)^{-1}\|\,\|K\|}{|\lambda|-\|(\lambda I-K_n)^{-1}\|\,\|(K-K_n)K\|}.$$

The error estimate reads

(4.7.17c) $$\|f-f_n\| \leq \|(\lambda I-K_n)^{-1}\| \frac{\|(K-K_n)K\|\,\|f_n\| + \|(K-K_n)g\|}{|\lambda|-\|(\lambda I-K_n)^{-1}\|\,\|(K-K_n)K\|}.$$

The previous assumptions still allow an arbitrarily slow convergence $\|(K-K_n)K_n\|\to 0$. For practical use, one has to prove as high a convergence order as possible.

4.7.4 Consistency Order

In the following, assume that the quadrature method Q_n is convergent and has the consistency order $\varkappa>0$:

(4.7.18) $$|Q_n\varphi-\int_D\varphi(x)dx| \le Ch^{\varkappa}\|\varphi\|_{\hat{C}^{\varkappa}(D)} \qquad \text{with } h=\tfrac{1}{n} \text{ for all } \varphi\in\hat{C}^{\varkappa}(D).$$

If the kernel k belongs to $\hat{C}^{\varkappa}(D\times D)$, then, in particular, the estimates (19a,b) are valid:

(4.7.19a) $$\|k(\cdot,y)\|_{\hat{C}^{\varkappa}(D)} \le \|k\|_{\hat{C}^{\varkappa}(D\times D)} \qquad \text{for all } y\in D,$$

(4.7.19b) $$\|k(x,\cdot)\|_{\hat{C}^{\varkappa}(D)} \le \|k\|_{\hat{C}^{\varkappa}(D\times D)} \qquad \text{for all } x\in D,$$

where $\|k(\cdot,y)\|_{\hat{C}^{\varkappa}(D)}$ involves the norm with respect to x, while y is a fixed parameter.

Exercise 4.7.14. Let $\varphi,\psi\in\hat{C}^{\varkappa}(D)$ and prove $\|\varphi\psi\|_{\hat{C}^{\varkappa}(D)}\le C\|\varphi\|_{\hat{C}^{\varkappa}(D)}\|\psi\|_{\hat{C}^{\varkappa}(D)}$ with a constant C depending only on $\varkappa\ge 0$.

Theorem 4.7.15. Let D be compact. Assume that the kernel $k(x,y)$ of the operators K satisfies $k\in\hat{C}^{\varkappa}(D\times D)$ for some $\varkappa>0$, Q_n must be stable with the consistency order $\varkappa>0$ (cf. (18)). Then, the Nyström discretisation $\{K_n\}$ is collectively compact and the estimates (20a-c) hold with suitable C:

(4.7.20a) $$\|(K-K_n)K_n\| \le Ch^{\varkappa}, \qquad (\|\cdot\|=\|\cdot\|_{C(D)\leftarrow C(D)})$$

(4.7.20b) $$\|(K-K_n)K\| \le Ch^{\varkappa}, \qquad (h=1/n)$$

(4.7.20c) $$\|K-K_n\|_{C(D)\leftarrow\hat{C}^{\varkappa}(D)} \le Ch^{\varkappa}.$$

Proof. (i) Let $\varphi\in\hat{C}^{\varkappa}(D)$ be arbitrary. By definition (9a) of K_n,

$$((K-K_n)\varphi)(x) = \int_D k(x,y)\varphi(y)dy - Q_n(k(x,\cdot)\varphi(\cdot))$$

holds, so that (18) leads to $|((K-K_n)\varphi)(x)|\le Ch^{\varkappa}\|k(x,\cdot)\varphi(\cdot)\|_{\hat{C}^{\varkappa}(D)}$. Using Exercise 14 and estimate (19b), one obtains $\|k(x,\cdot)\varphi(\cdot)\|_{\hat{C}^{\varkappa}(D)}\le C\|k(x,\cdot)\|_{\hat{C}^{\varkappa}(D)}\|\varphi\|_{\hat{C}^{\varkappa}(D)}\le C\|k\|_{\hat{C}^{\varkappa}(D\times D)}\|\varphi\|_{\hat{C}^{\varkappa}(D)}$. This proves the inequality $\|(K-K_n)\varphi\|_\infty\le Ch^{\varkappa}\|k\|_{\hat{C}^{\varkappa}(D\times D)}\|\varphi\|_{\hat{C}^{\varkappa}(D)}$ with a *new* constant C. Since $\|\cdot\|_\infty$ is the norm associated with $C(D)$, $\|K-K_n\|_{C(D)\leftarrow\hat{C}^{\varkappa}(D)}\le Ch^{\varkappa}\|k\|_{\hat{C}^{\varkappa}(D\times D)}$ follows (cf. Exercise 1.3.12b). Denoting $C\|k\|_{\hat{C}^{\varkappa}(D\times D)}$ as a new constant C, one obtains (20c).

(ii) Theorem 3.4.1 guarantees $\|K\|_{\hat{C}^{\varkappa}(D)\leftarrow C(D)}\le C$. Together with (20c), one obtains $\|(K-K_n)K\|_{C(D)\leftarrow C(D)}\le\|K-K_n\|_{C(D)\leftarrow\hat{C}^{\varkappa}(D)}\|K\|_{\hat{C}^{\varkappa}(D)\leftarrow C(D)}\le C'h^{\varkappa}$ and proves (20b).

(iii) Let $\varphi\in\hat{C}^{\varkappa}(D)$ be arbitrary. Exercise 14 and (19a) prove

$$\|K_n\varphi\|_{\hat{C}^{\varkappa}(D)} = \|\textstyle\sum_j w_{j,n}k(\cdot,\xi_{j,n})\varphi(\xi_{j,n})\|_{\hat{C}^{\varkappa}(D)} \le$$
$$\le\textstyle\sum_j|w_{j,n}||\varphi(\xi_{j,n})|\|k(\cdot,\xi_{j,n})\|_{\hat{C}^{\varkappa}(D)}\le\|k\|_{\hat{C}^{\varkappa}(D\times D)}\|\varphi\|_\infty\sum_j|w_{j,n}|.$$

Since the Q_n are assumed to be stable, $\sum_j |w_{j,n}| \leqslant \text{const}$ is uniformly bounded with respect to n and one ends up with the inequality

$$\|K_n\|_{\hat{C}^{\varkappa}(D) \leftarrow C(D)} \leqslant \text{const}\, \|k\|_{\hat{C}^{\varkappa}(D\times D)}.$$

As in part (ii), this inequality implies (20a). Any sequence $K_{n(j)}\varphi_j$ with $\|\varphi_j\|_\infty \leqslant 1$ is bounded in $\hat{C}^{\varkappa}(D)$ and, therefore, has a convergent subsequence in $C(D)$. Hence, $\{K_n\}$ is collectively compact. ∎

The following theorem shows that the consistency order of the quadrature formula determines the error estimate of the Nyström solution.

Theorem 4.7.16. Let D be compact. Assume that the Nyström method is constructed with a stable quadrature method Q_n with consistency order $\varkappa > 0$ (cf. (18)). Let $k \in \hat{C}^{\varkappa}(D\times D)$, let λ be a regular value of K, and $g \in \hat{C}^{\varkappa}(D)$. Then the components α_j of the solution vector a_n from (7) as well as the Nyström solution f_n satisfy the *error estimate*

(4.7.21) $$\max\{|\alpha_j - f(\xi_{j,n})|: 1 \leqslant j \leqslant n\} \leqslant \|f_n - f\|_\infty \leqslant C h^{\varkappa} \|g\|_{\hat{C}^{\varkappa}(D)}$$

for $n \geqslant n_0$ (n_0 sufficiently large).

Proof. The left inequality in (21) is identical with (12a). The stability of Q_n together with (18) proves the convergence of the quadrature method (cf. Theorem 1.4.17). Therefore, the consistency of the K_n is ensured by Lemma 6. Theorem 15 guarantees the collective compactness of $\{K_n\}$ and the inequalities (20a-c). Theorem 11 proves stability. In particular, there is some n_0 such that the Nyström solution f_n is well-defined for $n \geqslant n_0$. The second inequality in (21) follows from (16c) by estimating the numerator of the fraction therein as follows:

$$\|(K-K_n)K_n\|\,\|f\|_\infty \underset{(20a)}{\leqslant} C h^{\varkappa} \|(\lambda I - K)^{-1}\|\,\|g\|_\infty \leqslant C' h^{\varkappa} \|g\|_{\hat{C}^{\varkappa}(D)},$$

$$\|(K-K_n)g\|_\infty \leqslant \|K - K_n\|_{C(D)\leftarrow \hat{C}^{\varkappa}(D)} \|g\|_{\hat{C}^{\varkappa}(D)} \underset{(20c)}{\leqslant} C h^{\varkappa} \|g\|_{\hat{C}^{\varkappa}(D)}.$$

The denominator $|\lambda| - \|(\lambda I - K)^{-1}\|\,\|(K-K_n)K_n\|$ in (16c) is positive for $n \geqslant n_0$ and tends to $|\lambda|$ for $n \to \infty$. Hence, the denominator remains $\geqslant \varepsilon > 0$. Altogether, the right-hand side in (16c) can be estimated by $\text{const}\, h^{\varkappa} \|g\|_{\hat{C}^{\varkappa}(D)}$ and proves (21). ∎

4.7.5 Condition Number of the System of Equations

Lemma 4.7.17. Let K_n be the Nyström operator (9a) and let B_n be the matrix (6b). Then the following estimate holds for the row-sum norm of $\lambda I - B_n$:

(4.7.22a) $$\|\lambda I - B_n\|_\infty \leqslant \|\lambda I - K_n\|_{C(D)\leftarrow C(D)}.$$

Proof. Let $a = (\alpha_j) \in \mathbb{R}^n$. Piecewise linear interpolation of the values $\varphi(\xi_{j,n}) = \alpha_j$ yields $\varphi \in C(D)$ with $\|\varphi\|_\infty = \|a\|_\infty$. The statement (22a) follows from

$$\|(\lambda I - B_n)a\|_\infty = \max\{|\alpha_j - \textstyle\sum_k w_{k,n} k(\xi_{j,n}, \xi_{k,n})\alpha_k|: 1 \leqslant j \leqslant n\} \leqslant$$
$$\leqslant \sup\{|\varphi(x) - \textstyle\sum_k w_{k,n} k(x, \xi_{k,n})\varphi(\xi_{k,n})|: x \in D\} =$$

$$= \|(\lambda I - K_n)\varphi\|_\infty \leqslant \|\lambda I - K_n\| \, \|\varphi\|_\infty = \|\lambda I - K_n\| \, \|a\|_\infty .$$ QED

Lemma 4.7.18. Let K_n and B_n be as in Lemma 17. The matrix $\lambda I - B_n$ is non-singular if and only if the operator $\lambda I - K_n$ is invertible. In this case,

(4.7.22b) $$\|(\lambda I - B_n)^{-1}\|_\infty \leqslant \|(\lambda I - K_n)^{-1}\|_{C(D)\leftarrow C(D)}.$$

Proof. The considerations in §4.7.1 show that $\lambda I - B_n$ and $\lambda I - K_n$ are simultaneously non-singular or singular. As in the proof of Lemma 17, choose $g \in C(D)$ with $\|g\|_\infty = \|b\|_\infty$ for a given $b = (\beta_j) \in \mathbb{R}^n$. Let f_n be the solution of $(\lambda I - K_n) f_n = g$. According to Remark 1, the values $\alpha_j = f(\xi_{j,n})$ are the components of the solution vector a of $(\lambda I - B_n) a = b$. Since b is arbitrarily chosen, (22b) follows from

$$\|(\lambda I - B_n)^{-1} b\|_\infty = \|a\|_\infty \leqslant \|f_n\|_\infty = \|(\lambda I - K_n)^{-1} g\|_\infty \leqslant$$
$$\leqslant \|(\lambda I - K_n)^{-1}\| \, \|g\|_\infty = \|(\lambda I - K_n)^{-1}\| \, \|b\|_\infty .$$ QED

Theorem 4.7.19. The condition numbers with respect to the row-sum norm $\|\cdot\|_\infty$ and the supremum norm of $X = C(D)$ satisfy the optimal estimate

(4.7.22c) $$\mathrm{cond}_\infty(\lambda I - B_n) \leqslant \mathrm{cond}_{C(D)}(\lambda I - K_n).$$

Here, the right-hand side is uniformly bounded under the assumption of consistency and stability of K_n.

Proof. (22c) is a direct consequence of (22a,b). The uniform boundedness of $\mathrm{cond}_{C(D)}(\lambda I - K_n)$ is stated in Theorem 1.15. QED

4.7.6 Regularisation

The regularisation method described in §4.6.1 immediately carries over to the Nyström method. Let an inhomogeneity g be given which is not sufficiently smooth (e.g., g continuous, but not Hölder continuous). $Kg \in \hat{C}^\varkappa(D)$ holds under the assumptions $g \in C(D)$ and $K \in L(C(D), \hat{C}^\varkappa(D))$ (cf. Theorem 3.4.1). Solve $\lambda\varphi = Kg + K\varphi$ by means of the Nyström method:

$$\lambda \varphi_n = Kg + K_n \varphi_n .$$

According to Remark 1, one has to solve the system $(\lambda I - B_n) a_n = b_n$. The evaluation of the coefficients β_j of the vector b_n,

$$\beta_j := (Kg)(\xi_{j,n}) = \int_D k(\xi_{j,n}, y) g(y) dy,$$

requires additional work. Since g is assumed not to be smooth, the approximation of this integral requires special care or the integrals are to be evaluated exactly. The *regularised solution* is defined by

$$f_n := \tfrac{1}{\lambda}(\varphi_n + g).$$

The error of f_n satisfies the equation

$$f - f_n = \tfrac{1}{\lambda}(\varphi - \varphi_n).$$

If $Kg \in \hat{C}^{\varkappa}(D)$, inequality (21) yields the estimate $\|\varphi - \varphi_n\|_\infty \leqslant C h^{\varkappa} \|Kg\|_{\hat{C}^{\varkappa}(D)}$, which carries over immediately to $f - f_n$:

$$\|f - f_n\|_\infty \leqslant \frac{1}{|\lambda|} C h^{\varkappa} \|Kg\|_{\hat{C}^{\varkappa}(D)}.$$

Then the solution of the problem is obtained with full consistency order $\varkappa$.

4.7.7 Numerical Examples

The example specified in (2.19a), (2.21a,b) satisfies all assumptions concerning smoothness of the kernel required in the previous theorems. As quadrature method we use the *composite trapezoidal rule* and the *composite Simpson rule* over an equidistant partition of step size $h = 1/n$. The Simpson rule is applicable only for even n. The support abscissae are $\xi_{j,n} = jh = j/n$. The weights $w_{j,n}$ are described in (1.4.18c) or (1.4.18d), respectively. The assembling of the system requires only the evaluation of the kernel function k at $(\xi_{j,n}, \xi_{k,n})$ and of the inhomogeneity g at $\xi_{j,n}$. Since no integrals are to be evaluated, this method is very convenient. In spite of its simplicity, the Nyström method is very effective as demonstrated by the results in Tables 1 and 2. $e_{h,\infty}$ denotes the error (4.24a): the maximum over $|f_{j,n} - f(jh)|$, $0 \leqslant j \leqslant n$. The Euclidean norm of the error (cf. (4.24b)) shows the same behaviour. The composite trapezoidal rule is of second order ($\varkappa = 2$). In coincidence with Theorem 16, the discretisation error converges with second order, as shown by the error quotients (right column in Table 1). The ratios in Table 2 converge to $16 = 2^4$ and hence indicate convergence of the Nyström method of the fourth order. Since the Simpson rule is of fourth order, Theorem 16 is again confirmed.

h	$e_{h,\infty}$	
1/2	0.838 018	5.69
1/4	0.147 195	4.24
1/8	0.034 712	4.06
1/16	0.008 560	4.01
1/32	0.002 133	4.00
1/64	0.000 533	4.00
1/128	0.000 133	

Table 4.7.1 Example (2.19a), (2.21a,b). Nyström method with the composite *trapezoidal rule*

h	$e_{h,}$	
1/2	2.297 217 238	22.41
1/4	0.102 493 748	31.57
1/8	0.003 246 200	18.38
1/16	0.000 179 572	16.47
1/32	0.000 010 900	16.11
1/64	0.000 000 676	16.03
1/128	0.000 000 042	

Table 4.7.2 Example (2.19a), (2.21a,b). Nyström method with the composite *Simpson rule*

4.7.8 Product Integration

As described in §4.6.7, one can introduce a quadrature formula based on the weight $\varkappa(x,y)$:

(4.7.23a) $\quad Q_n^{\varkappa}(\varphi) = \int_D \varkappa(x,y)\, \Pi_n \varphi(y)\, dy = \sum_{i=1}^{n} w_{i,n}(x)\, \varphi(\xi_{i,n})$ with

(4.7.23b) $\quad w_{i,n}(x) := \int_D \varkappa(x,y)\, L_i(y)\, dy,$

which is appropriate for integrals involving a kernel $k(x,y)$ with a product representation

$$k(x,y) = \ell(x,y)\varkappa(x,y) \qquad (\ell \text{ smooth, } \varkappa \text{ possibly singular}).$$

$Q_n^{\varkappa}(\ell(x,\cdot)\varphi)$ serves as approximation of $(K\varphi)(x)=\int k(x,y)\varphi(y)dy$. Here, Π_n is an interpolation at the support abscissae $\xi_{i,n}$ by the Lagrange functions L_i. For practical computation one needs the weights $w_{i,n}(x)$ only at the abscissae $x=\xi_{k,n}\in\Xi_n$. The semidiscrete Nyström operator K_n is

(4.7.24) $$(K_n\varphi)(x) = \sum_{k=1}^{n} w_{k,n}(x)\ell(x,\xi_{k,n})\varphi(\xi_{k,n}) = Q_n^{\varkappa}(\ell(x,\cdot)\varphi).$$

The *matrix* $B_n=(\beta_{jk})$ of the system (7) has the coefficients $\beta_{jk}:=w_{k,n}(\xi_{j,n})\,\ell(\xi_{j,n},\xi_{k,n})$ (cf. (6b)). *Collective compactness* and *consistency* of K_n results from the following lemma.

Lemma 4.7.20. Let $\hat{K}$ be the integral operator with kernel $\varkappa(x,y)$ and D a compact set. The interpolation appearing in (23a) is assumed to be stable in $C(D)$. If $\hat{K}$ is compact in $C(D)$ and $\ell\in C(D\times D)$, then the discretisation K_n from (24) constructed by the product integration is *collectively compact*.

Proof. (i) Theorem 3.2.6 proves $\int|\varkappa(x,y)|dy\le C$ for $x\in D$ and $\Phi(\xi,x):=\int|\varkappa(\xi,y)-\varkappa(x,y)|dy\to 0$ for $\xi\to x\in D$. In part (ii) of the proof of Theorem 3.2.6, we have already shown uniform continuity of Φ in $D\times D$. Let $\varphi\in C(D)$ be arbitrary with $\|\varphi\|_\infty\le 1$. $K_n\varphi$ is uniformly bounded by $C\|\Pi_n[\ell\varphi]\|_\infty\le C'\|\ell\|_\infty\|\varphi\|_\infty\le C'\|\ell\|_\infty$. The representation $(K_n\varphi)(\xi)-(K_n\varphi)(x)=\int[\varkappa(\xi,y)-\varkappa(x,y)]\,\Pi_n[\ell(\xi,\cdot)\varphi](y)\,dy + \int\varkappa(x,y)\Pi_n[(\ell(\xi,\cdot)-\ell(x,\cdot))\varphi](y)\,dy$ together with the uniform continuity of Φ and ℓ proves the equicontinuity of $\{K_n\varphi:\ n\in\mathbb{N},\ \|\varphi\|_\infty\le 1\}$. Hence, $\{K_n\}$ is collectively compact. QED

Exercise 4.7.21. In addition to the assumptions of Theorem 20 assume that the interpolation Π_n is convergent. Then K_n is *consistent* in $C(D)$.

More important than consistency is *consistency of the order* $\varkappa$:

Remark 4.7.22. Let D be compact and let $\int_D|\varkappa(x,y)|dy$ be continuous in D. Suppose the factor ℓ in $k=\ell\varkappa$ belongs to $\hat{C}^{\varkappa}(D\times D)$ [$\|\ell(x,\cdot)\|_{\hat{C}^{\varkappa}(D)}\le$ const for all $x\in D$ is sufficient]. If the interpolation Π_n used in quadrature formula (23a) is of the order $\varkappa$: $\|\Pi_n\varphi-\varphi\|_\infty\le Ch^{\varkappa}\|\varphi\|_{\hat{C}^{\varkappa}(D)}$, then K_n is also consistent of the order $\varkappa$: $\|K_n\varphi-K\varphi\|_\infty\le Ch^{\varkappa}\|\varphi\|_{\hat{C}^{\varkappa}(D)}$.

Proof. For any $\varphi\in\hat{C}^{\varkappa}(D)$ the product $\psi:=\ell\varphi$ satisfies the estimate $\|\psi(x,\cdot)\|_{\hat{C}^{\varkappa}(D)}\le C\|\varphi\|_{\hat{C}^{\varkappa}(D)}$ for all $x\in D$ (cf. Exercise 14). The representation $(K_n\varphi-K\varphi)(x)=\int_D\varkappa(x,y)\{\Pi_n[\ell(x,\cdot)\varphi](y)-[\ell(x,y)\varphi(y)]\}dy$ immediately leads to the desired assertion, since $\int_D|\varkappa(x,y)|dy$ is bounded in D. QED

By Theorem 11, *stability* and *convergence* are consequences of collective compactness and consistency. Via (16d), consistency of the order $\varkappa$ guarantees the *error estimate* $\|f-f_n\|_\infty\le Ch^{\varkappa}\|f\|_{\hat{C}^{\varkappa}(D)}$.

For weakly singular kernels, it is not sufficient to improve the integration by means of product integration. Since the solution u has singular derivatives at the boundary (cf. §4.4.5, (4.22b)), the quadrature error is unsatisfactory close to the boundary (end-points of the interval) in spite of the product integration. A remedy is a *graded mesh*, as already applied in §4.4.5 (Table 4.5). Concerning optimal error estimates and a precise description of the choice of step widths we refer to Graham [2] and Schneider [1].

4.8 Supplements

4.8.1 Connection between the Discretisation Methods

The discretisations by kernel approximation, collocation, Galerkin method and Nyström method are not completely different. In the following, we show some interconnections.

4.8.1.1 The Kernel Approximation and the Galerkin Method

Let $X_n \subset L^2(D)$ be a subspace for which we choose an orthonormal basis $\{\Phi_1, \ldots, \Phi_n\}$. The kernel

(4.8.1a) $$k_n(x,y) = \sum_{j=1}^{n} a_j(x)\, b_j(y) \qquad \text{with } a_j := K\Phi_j \text{ and } b_j := \Phi_j$$

is degenerate (cf. (2.1)) and defines the method of the kernel approximation. The following considerations are based on the following result.

Exercise 4.8.1. Let Π_n be the orthogonal projection onto X_n with respect to $L^2(D)$. The integral operator K_n defined by the kernel k_n from (1a) has the representation

(4.8.1b) $$K_n = K\,\Pi_n .$$

This and Remark 3.10b yield the following remark.

Remark 4.8.2. The solution $f_n^{\cdot}$ of the kernel approximation by k_n from (1a) is the solution of equation (3.8). The orthogonal projection $f_n := \Pi_n f_n^{\cdot}$ of the kernel approximation solution is the solution of the Galerkin method $\lambda f_n = \Pi_n g + \Pi_n K f_n$ on X_n. Vice versa, one obtains the kernel approximation solution with kernel (1a) from the Galerkin solution f_n via $f_n^{\cdot} := \frac{1}{\lambda}(g + K f_n)$. $f_n^{\cdot}$ also represents the iterated Galerkin solution (cf. §4.6.3).

4.8.1.2 From the Galerkin Method to the Collocation and Nyström Method

In practice, the Galerkin method can rarely be applied without additional numerical quadrature (cf. examples in §4.6.6). We give a simple example for an interpretation of the quadrature.

Let X_n denote the space of piecewise constant functions over a (for the sake of simplicity) equidistant partition of $I=[a,b]$ and let Φ_j be the Lagrange functions from Example 1.4.13c. Replacing all integrals in the scalar product as well as in $K\varphi$ by the composite tangential trapezoidal rule (1.4.18e), we approximate the matrices A, B from (5.7b,c) by

$$\widetilde{A}_n = A_n = h\,I, \qquad \widetilde{B}_n = (\widetilde{\beta}_{jk}) \quad \text{with } \widetilde{\beta}_{jk} := h^2 k(\xi_j, \xi_k).$$

The arising system $(\lambda\widetilde{A}_n - \widetilde{B}_n)a_n = b_n$ is equivalent with the system of the *Nyström method* (for the tangential trapezoidal rule).

Replacing only the integrals of the scalar products $\langle\Phi_j,\Phi_k\rangle$, $\langle K\Phi_k,\Phi_j\rangle$ by the tangential trapezoidal rule, one obtains the corresponding *collocation equation* (4.10).

Remark 4.8.3. The interpretation of the collocation method as a Galerkin method with numerical quadrature opens a new approach for error estimates: The error of the collocation solution can be estimated by the sum of the Galerkin error and the quadrature errors discussed in §4.6.6.

In general, if an arbitrary quadrature method $\sum_i w_{ij}\varphi(\xi_{ij}) \approx \int\Phi_j\varphi\,dx$ is used to approximate $\int\Phi_j g\,dx$ and $\int\Phi_j(K\Phi_k)\,dx$, then we obtain a projection method $\lambda f_n = g_n + K_n f_n$ with $g_n = \Pi_n g$ and $K_n = \Pi_n K$, where Π_n is described by the following *weighted interpolation*: $\varphi_n = \Pi_n\varphi \in X_n$ satisfies $\sum_i w_{ij}\varphi_n(\xi_{ij}) = \sum_i w_{ij}\varphi(\xi_{ij})$ for all $1 \le j \le n$ (cf. §4.6.5). If for fixed j, $w_{ij} \ne 0$ holds only for *one* $i = i(j)$, then Π_n describes the usual interpolation at the support abscissae $\xi_i := \xi_{i(j),i}$.

Remark 4.8.4. Modify the Galerkin method in X_n by a numerical quadrature of the scalar products $\langle\Phi_j,g\rangle$ and $\langle\Phi_j,K\Phi_k\rangle$. The quadrature formula should only use the abscissae $\Xi_n = \{\xi_{1,n},\dots,\xi_{n,n}\}$. Then the collocation solution f_n characterised by the subspace X_n and the collocation points Ξ_n is also the Galerkin solution (with numerical quadrature).

Proof. Let $d_n := \lambda f_n - g_n - K_n f_n$. If $d_n(\xi) = 0$ at all $\xi \in \Xi_n$, then the result of applying numerical quadrature to $\int\Phi_j d_n dx$ is also zero. QED

4.8.1.3 From the Collocation to the Nyström Method

The introductory examples of §4.8.1.2 showed the analogous connection: Approximating the integration in the collocation method by the tangential trapezoidal rule, one obtains the Nyström method.

However, even the unmodified collocation method can be interpreted as a Nyström method. Let Φ_j denote the Lagrange functions and x_j the support abscissae of the interpolation involved in the collocation method. For computing the integrals $\int k(x,y)\varphi(y)dy$, choose the product integration from §4.7.8 with $\ell = k$ and $\varkappa = 1$. This yields $\sum w_j(x)\varphi(x_j)$ with the weights $w_j(x) := \int k(x,y)\Phi_j(y)dy$. The computation of these weights may be unpractical, but in any case it defines a Nyström method based on this quadrature. Because of the special choice of the quadrature, the Nyström operator K_n has the property $K_n\Phi_j = K\Phi_j$. This proves the following remark.

Remark 4.8.5. Let f_n be the collocation solution and $\widetilde{f}_n$ the Nyström solution for the special quadrature formula described above. Then the solutions coincide at the support abscissae, i.e., $f_n(x_j) = \widetilde{f}_n(x_j)$. Furthermore, the Nyström solution $\widetilde{f}_n$ is the iterated collocation solution:

$$\widetilde{f}_n = \tfrac{1}{\lambda}(g + Kf_n).$$

4.8.1.4 From the Collocation to the Galerkin Method

For the piecewise linear collocation (more generally: collocation with splines of odd degree) Arnold - Wendland [1] showed the equivalence of the collocation method with a modified Galerkin method. Therefore, in particular, further convergence properties of the Galerkin method (e.g., Remark 6.8) carry over to the collocation method. Concerning a generalisation to the three-dimensional case compare Hsiao-Prößdorf [1].

4.8.2 Method of the Defect Correction

A well-known technique in the field of differential equations is the defect correction (cf. Hackbusch [1,§14.3]). We start from two assumptions:

(i) The matrix $\lambda A_n - B_n$ of the discrete method is available as well as a routine for solving equations with this matrix.

(ii) For each (smooth) φ, we are able to approximate $K\varphi$ by a suitable integration routine very precisely. To simplify the following considerations we assume *exact* integration.

We define the following iterative method:

iterated defect correction:

(4.8.2a) $$f_n^0 := (\lambda I - K_n)^{-1} g_n$$

(4.8.2b) $$f_n^{j+1} := f_n^j - (\lambda I - K_n)^{-1} [(\lambda I - K) f_n^j - g] \qquad \text{for } j = 0,1,2,\ldots$$

Since the expression $[(\lambda I - K) f_n^j - g]$ is called the «defect» of f_n^j, the name «defect correction» results for the iteration step (2b). In the case of the Nyström method or the kernel approximation, the evaluation of $(\lambda I - K_n)^{-1}\ldots$ in (2a) and (2b) requires the solution of a system (Assumption (i)). Concerning the projection method, we refer to Exercise 6. Thanks to Assumption (ii), the evaluation of Kf_n^j in (2b) is possible.

Exercise 4.8.6. In the projection methods which we have considered so far, we have only solved equations $\lambda f_n = g_n + K_n f_n$ with $g_n \in X_n$. Prove that the projection solution f_n of the equation $\lambda f_n = g + K_n f_n$ is $f_n = f^X + f^\perp$ with $f^\perp := (I - \Pi_n) g / \lambda$ and $(\lambda I - K_n) f^X = \varphi_n \in X_n$, where $\varphi_n := \Pi_n g + K_n f^\perp$.

Exercise 4.8.7. Prove that f_n^j from (2a,b) has the error representation (2c), where f is the solution of the integral equation $\lambda f = g + Kf$:

(4.8.2c) $$f_n^j - f = \{(\lambda I - K_n)^{-1}(K - K_n)\}^j (f_n^0 - f) \qquad \text{for } j = 0,1,2,\ldots$$

For a stable projection method of the order $\varkappa$, there are C_1, C_2 with

(4.8.3a) $$\| K - K_n \|_{X \leftarrow X} \leqslant C_1 h_n^\varkappa,$$

(4.8.3b) $$\| (\lambda I - K_n)^{-1} \|_{X \leftarrow X} \leqslant C_2 \qquad \text{for all } n.$$

Because of $K - K_n = (I - \Pi_n) K$, one needs a regularity assumption [e.g., $K \in L(C(D), \hat{C}^\varkappa(D))$] and a corresponding estimate of the projection errors [e.g., $\|(I-\Pi_n)\varphi\|_\infty \leq Ch^\varkappa \|\varphi\|_{\hat{C}^\varkappa(D)}$] for the proof of (3a). Combining of (2c) with (3a,b) proves the following result.

Lemma 4.8.8. Let (3a,b) hold. For the error of f_n^j from (2a,b) one has

(4.8.3c) $$\|f_n^j - f\|_X \leq (C_1C_2)^j h_n^{j\varkappa} \|f_n^0 - f\|_X \qquad \text{for all } j \geq 0.$$

The estimate (3c) can be interpreted in two directions. Firstly, one can study the limit process $j\to\infty$ for fixed h_n. As soon as the step size h_n becomes smaller than $h_{max} := (C_1C_2)^{-1/\varkappa}$, *the iterated defect correction converges* to the solution of the integral equation, i.e., $f_n^j \to f$ as $j\to\infty$. Secondly, one can fix the iteration number j. Then inequality (3c) shows that the j-fold defect correction is a *method of the order* $(j+1)\varkappa$, if one starts from $\|f_n^0 - f\|_X = O(h_n^\varkappa)$.

For the Nyström method, we can apply the previous considerations with $X = \hat{C}^\varkappa(D)$. Since Theorem 7.15 contains a weaker statement than (3a), we need a stronger regularity assumption on the kernel. Then (3a) holds with $X = \hat{C}^\varkappa(D)$ (cf. Remark 5.3.12).

Exercise 4.8.9. Let K_n be defined by the Nyström method.
(a) Assuming stability of K_n with respect to $C(D)$ and $\|K_n\|_{\hat{C}^\varkappa(D)\leftarrow C(D)} \leq C$ for all n, prove estimate (3b) with $X = \hat{C}^\varkappa(D)$ (cf. Theorem 3.5.1).
(b) Prove that the estimate $\|K_n\|_{\hat{C}^\varkappa(D)\leftarrow C(D)} \leq C$ follows from stability of the underlying quadrature and from $\|k(\cdot,y)\|_{\hat{C}^\varkappa(D)} \leq$ const for all $y\in D$.

Exercise 4.8.10. Let $g_n = g$ in (2a). Prove the following: Replacing the starting step (2a) by $f_n^{-1} := 0$ and applying (2b) for $j \geq -1$, we obtain the value f_n^0 from (2a) as the result of (2b) for $j = -1$. Further,

$$\|f_n^0 - f\| \leq C_1C_2h_n^\varkappa \|f\|.$$

4.8.3 Extrapolation Method

Suppose the solution f_n of the semidiscrete equation $\lambda f_n = g_n + K_nf_n$ for step size $h = h_n$ admits an *asymptotic expansion* of the form

(4.8.4a) $$f_n(x) = \sum_{\nu=0}^{\ell-1} h_n^{\gamma_\nu} f_{(\nu)}(x) + h_n^{\gamma_\ell} F_{(\ell)}(x,h_n) \qquad \text{for all } n\in\mathbb{N},\ x\in D$$

with functions $f_{(\nu)}$ independent of h, with exponents

(4.8.4b) $$0 = \gamma_0 < \gamma_1 < \dots < \gamma_\ell$$

and a bounded remainder

(4.8.4c) $$|F_{(\ell)}(x,h_n)| \leq C(x) \qquad \text{for all } h_n \text{ and } x\in D.$$

Then one can determine an extrapolated value

(4.8.5a) $$f_{ex} := \sum_{\mu=1}^{\ell} \alpha_\mu f_{n_\mu}$$

of the order γ_ℓ from the ℓ values $f_{n_1}, f_{n_2}, \dots, f_{n_\ell}$. The coefficients α_μ are the solution of

(4.8.5b) $\quad \sum_{\mu=1}^{\ell} \alpha_\mu = 1, \quad \sum_{\mu=1}^{\ell} \alpha_\mu h_{n_\mu}^{\gamma_\nu} = 0 \qquad$ for $1 \leqslant \nu \leqslant \ell - 1$.

For $\ell = 2$, the extrapolation formula reads

(4.8.5c) $\quad f_{ex} := (h_{n_2}^{\gamma_1} f_{n_1} - h_{n_1}^{\gamma_1} f_{n_2}) / (h_{n_2}^{\gamma_1} - h_{n_1}^{\gamma_1})$.

Under relatively strong assumptions, the exponents (4b) are of the form $\gamma_\nu = \nu$ or even $\gamma_\nu = 2\nu$. Then the extrapolation from ℓ values f_{n_μ} ($1 \leqslant \mu \leqslant \ell$, e.g., with $n_\mu = 2^{\mu-1} n_1$) yields a result of the respective orders $O(h^\ell)$ or $O(h^{2\ell})$ with $h = h_{n_1}$.

Exercise 4.8.11. Let $\gamma_\nu = 2\nu$ and $h_{n_\mu} := h_{n_1}/2^{\mu-1}$. Prove that the coefficients α_μ from (5b) are $\alpha_1 = -1/3$, $\alpha_2 = 4/3$ for $\ell=2$ and $\alpha_1 = 1/45$, $\alpha_2 = -20/45$, $\alpha_3 = 64/45$ for $\ell=3$.

The following example (a Nyström method based on the composite trapezoidal rule (1.4.18c')) shows that the expansions (4a-c) for solutions of integral equations may exist.

The composite trapezoidal rule $Q_h(\varphi)$ on $I = [a,b]$ for step size $h = h_n = (b-a)/n$ admits the following expansion in h^2:

(4.8.6a) $\quad Q_h(\varphi) = \int_I \varphi(x)\,dx + \sum_{\nu=1}^{\ell-1} h^{2\nu} \tau_{2\nu}(\varphi) + h^{2\ell} T_{2\ell}(\varphi; h)$

with the following functionals as coefficients (cf. Stoer [1, §§3.3-4]):

(4.8.6b) $\quad \tau_{2\nu} \in C^{2\nu}(I)' \quad (1 \leqslant \nu < \ell), \qquad |T_{2\ell}(\varphi;h)| \leqslant C_\ell \|\varphi\|_{\hat{C}^{2\ell}(I)} \qquad$ for all h.

The Nyström method uses the quadrature formula $Q_h(k(x,\cdot)\varphi)$. We define the functions $\varkappa_{2\nu}(\varphi)(x)$ for $\varphi \in C^{2\nu}(I)$ and $x \in I$ by

(4.8.6c) $\quad \varkappa_{2\nu}(\varphi)(x) := \tau_{2\nu}(k(x,\cdot)\varphi) \qquad$ for $1 \leqslant \nu < \ell$,

(4.8.6d) $\quad K_{2\ell}(\varphi;h)(x) := T_{2\ell}(k(x,\cdot)\varphi; h)$.

Since $(K_n\varphi)(x)$ is defined by $Q_h(k(x,\cdot)\varphi)$ for $h = h_n$, the expansion

(4.8.6e) $\quad K_n \varphi = K\varphi + \sum_{\nu=1}^{\ell-1} h^{2\nu} \varkappa_{2\nu}(\varphi) + h^{2\ell} K_{2\ell}(\varphi; h) \qquad$ with $|K_{2\ell}(\varphi;h)(x)| \leqslant C(x)$

follows, provided that φ and $k(x,\cdot)$ belong to $\hat{C}^{2\ell}(I)$ for fixed $x \in I$. The precise smoothness properties of the mappings $\varkappa_{2\nu}$ are the following:

(4.8.6f) $\quad \varkappa_{2\nu} \in L(\hat{C}^{\mu}(D), \hat{C}^{\mu-2\nu}(D)) \quad$ for $2\nu \leqslant \mu \leqslant 2\ell$, $\quad K_{2\ell} \in L(\hat{C}^{2\ell}(D), L^\infty(D))$.

Exercise 4.8.12. Prove the following: **(a)** If an expansion (6e) of the order 2ℓ exists, then (6a) also exists for a smaller order $2\ell' < 2\ell$.
(b) The sum in (6e) can be taken from $\nu=0$ to $\ell-1$, if one defines $K\varphi$ as $h^0 \varkappa_0(\varphi)$. For $\ell=0$, the remainder equals $K_0(\varphi; h) := K_n\varphi$.
Hint for (a): $K_{2\ell-2}(\varphi; h) = \varkappa_{2\ell-2}(\varphi) + h^2 K_{2\ell}(\varphi; h)$.

Lemma 4.8.13. Let the integral operator K satisfy the regularity property $(\lambda I-K)^{-1}\in L(\hat{C}^{\mu}(D),\hat{C}^{\mu}(D))$ for $\mu=2,4,...,2\ell$. The discretisation $\{K_n\}$ is assumed to be stable and admitting the expansion (6e,f). Then the solution f_n of the semidiscrete equation $\lambda f_n=g+K_nf_n$ has an asymptotic expansion (4a-c) with exponents $\gamma_\nu=2\nu$, where the leading coefficient $f_{(0)}$ in (4a) represents the solution f of the integral equation $\lambda f=g+Kf$.

Proof. For the sake of clearness, we give the proof for $\ell=2$. We start from an ansatz of the semidiscrete solution f_n in the form

$$f_n(x) = f_{(0)}(x)+h^2 f_{(1)}(x)+h^4 F_{(2)}(x;h)$$

with not yet determined functions $f_{(0)}$ and $f_{(1)}$. An asymptotic expansion is proved if we can show the boundedness (4c) of $F_{(2)}$. According to Exercise 12a, the ansatz induces the representations $f_n(x)=f_{(0)}(x)+h^2F_{(1)}(x;h)=F_{(0)}(x;h)$. Insertion into the expansion (6e) yields:

$$\begin{aligned} K_nf_n &= K_nf_{(0)}+h^2K_nf_{(1)})+h^4K_nF_{(2)} = \\ &= Kf_{(0)} + h^2\varkappa_2(f_{(0)})+h^4K_4(f_{(0)};h)+ \\ &\qquad + h^2Kf_{(1)} + h^4K_2(f_{(1)};h)+ \\ &\qquad\qquad + h^4K_0(F_{(2)};h). \end{aligned}$$

The expression coincides with $\lambda f_n-g=\lambda f_{(0)}-g+h^2\lambda f_{(1)}+h^4\lambda F_{(2)}$. Equating the coefficients in this polynomial equation, we obtain

$$\lambda f_{(0)}=g+Kf_{(0)},\quad \lambda f_{(1)}=\varkappa_2(f_{(0)})+Kf_{(1)},\quad \lambda F_{(2)}=K_4(f_{(0)};h)+K_2(f_{(1)};h)+K_0(F_{(2)};h).$$

The first equation proves $f=f_{(0)}$. The second defines the coefficient function $f_{(1)}$ as a solution of an integral equation. According to Exercise 12b, we have $K_0(F_{(2)};h)=K_nF_{(2)}$. Therefore, the third equation represents a semidiscrete equation $\lambda F_{(2)}=G_n+K_nF_{(2)}$ with $G_n:=K_4(f_{(0)};h)+K_2(f_{(1)};h)$. The stability of $\{K_n\}$ implies the boundedness (4c) of $F_{(2)}$. QED

As numerical example we use again the integral equation (2.19a), (2.21a,b) with $\lambda=0.1$. For the step sizes $h=1$, $1/2$, $1/4$, ..., $1/32$, the Nyström method based on the composite trapezoidal rule yields the error $\max\{|f_n(\nu h)-f(\nu h)|: \nu=0,1,...,n\}$ reported in the first column of Table 1. Because of the second order, the error quotients are about 4. The coefficients needed in the extrapolation (5a) are given in Exercise 11. Extrapolation from $\ell=2$ values yields the order 4, as confirmed by the ratios in the second column of Table 1 approximating $16=2^4$. The values of the third column belong to the extrapolation from h, $2h$, and $4h$. The given ratios are close to $64=2^6$ and confirm the order 6.

The assumptions required in Lemma 13 are not valid for non-smooth kernels. Then, more general sequences (4b) of exponents can appear (cf. Bulirsch [1], Stoer [1,§3.6]). Literature concerning the extrapolation method (also called «Richardson extrapolation» or «deferred approach to the limit») for integral equations of the second kind can be found, e.g., in Baker [1,§4.18], Saranen [2], and Graham-Qun-Ruifeng [1]

step size h	error without extrapolation		error with extrapolation from 2 values		3 values	
1	1.42					
1/2	1.66	0.86	2.69			
1/4	$2.63_{10}-1$	6.31	$4.46_{10}-1$	6.0	$6.55_{10}-1$	
1/8	$5.19_{10}-2$	5.06	$1.84_{10}-2$	24.2	$1.49_{10}-2$	43.8
1/16	$1.22_{10}-2$	4.24	$9.89_{10}-4$	18.6	$2.26_{10}-4$	66.1
1/32	$3.01_{10}-3$	4.06	$5.93_{10}-5$	16.7	$2.97_{10}-6$	76.0

Table 4.8.1 Discretisation error with and without extrapolation

4.8.4 Eigenvalue Problems

Up to now, we considered the integral *equations* $\lambda f = g + Kf$. Now we study the *eigenvalue problem*. A complex number $\lambda \in \mathbb{C}$ is called an *eigenvalue* and f an *eigenfunction* of K, if

(4.8.7) $\lambda f = Kf$ with $0 \neq f \in X$.

For compact operators $K \in K(X,X)$, the Riesz-Schauder Theory (cf. Theorem 1.3.28) states that all singular values of K except possibly $\lambda = 0$ must be eigenvalues. The set of singular values of K was already introduced in §1.3.1 as the *spectrum* $\sigma(K)$:

$$\sigma(K) := \{\lambda \in \mathbb{C} : \lambda I - K \text{ is not bijective}\}.$$

All discretisation methods mentioned so far lead to a sequence of operators K_n. In each case, the semidiscrete analogue of the eigenvalue problem (7) reads

(4.8.8) $\lambda_n f_n = K_n f_n$ $(n \in \mathbb{N})$,

where the representation of f_n depends on the discretisation method chosen. In the case of the kernel approximation, $f_n \in \mathrm{span}\{a_1, \dots, a_n\}$ holds with a_i from (2.1) or (2.11a), respectively. For projection methods, f_n lies in X_n, while f_n results from the Nyström interpolation in the case of the Nyström method.

In the same way as one transforms the semidiscrete integral equation $\lambda f_n = g_n + K_n f_n$ into the discrete system $(\lambda A_n - B_n) a_n = b_n$ [possibly with $A_n = I$], the *eigenvalue problem* (7) *is equivalent to the matrix eigenvalue problem*

(4.8.9) $\lambda A_n a_n = B_n a_n$.

For the kernel approximation, the Nyström method, and special projection methods (e.g., collocation with Lagrange functions as basis elements), $A_n = I$ holds. In this case, (9) is a usual eigenvalue problem, whereas (9) for $A_n \neq I$ is called a *generalised eigenvalue problem*. Since symmetric eigenvalue problems are easier to solve, the following statement is important.

Remark 4.8.14. Let the integral operator K be *symmetric*, i.e., $k(x,y)=k(y,x)$. Then the matrix B_n is symmetric if
(a) the kernel approximation is used with $a_j=b_j$ in (2.1),
(b) the Galerkin method is applied or
(c) the weights of the quadrature method used in the Nyström method are positive and B_n' from (7.7') instead of B_n is used for the representation.

Since (8) and (9) are equivalent, they lead to the same eigenvalues λ_n. More precisely, for any n there are up to n different eigenvalues $\lambda_n^{(1)}, \lambda_n^{(2)},\ldots,$ forming the spectrum $\sigma(K_n)=\sigma(A_n^{-1}B_n)$ of the operator K_n and of the matrix $A_n^{-1}B_n$. In the following, the connection between the discrete eigenvalue spectrum $\sigma(K_n)$ [and the associated eigenfunctions f_n] and the spectrum $\sigma(K)$ [and the eigenfunctions f] is to be studied. The only assumption of the following theorem is that the discretisation $\{K_n\}$ is collectively compact and consistent with K. Since collective compactness follows from the operator norm convergence $K_n \to K$ (cf. Remark 7.9), this assumption holds for all previously mentioned methods.

Theorem 4.8.15. Let the discretisation $\{K_n\}$ be consistent with K and collectively compact. λ_n $(n\in\mathbb{N})$ is a sequence of eigenvalues $\lambda_n\in\sigma(K_n)$ of the semi-discrete problem (8) with associated normalised eigenfunctions $\|f_n\|=1$. Then there exists a subsequence λ_{n_k} either converging to zero or an eigenvalue $\lambda\in\sigma(K)$. If $\lambda=\lim\lambda_{n_k}\neq 0$, the subsequence can be chosen so that the f_{n_k} converge to an eigenfunction f of the problem (7) belonging to λ.

Proof. From $\lambda_n f_n=K_n f_n$ and $|\lambda_n|\,\|f_n\| = \|\lambda_n f_n\| \le \|K_n f_n\| \le \|K_n\|\|f_n\|$ one concludes that $|\lambda_n|\le\|K_n\|$ for all $\lambda_n\in\sigma(K_n)$. The consistency of $\{K_n\}$ together with the uniform boundedness (Theorem 1.3.17) proves the inequality

$$|\lambda_n| \le \text{const} \qquad \text{for all } \lambda_n\in\sigma(K_n) \text{ and all } n\in\mathbb{N}.$$

Hence, there is a convergent subsequence of $\{\lambda_n\}$, which for the sake of simplicity we denote again by $\{\lambda_n\}$. Since $\|f_n\|=1$, the collective compactness implies the existence of a convergent subsequence $K_{n_i}f_{n_i}\to\varphi$. If $\lambda:=\lim\lambda_n\neq 0$, $f_{n_i}=\lambda_{n_i}^{-1}K_{n_i}f_{n_i}$ converges to $f:=\lambda^{-1}\varphi$. The first two terms in $K_{n_i}f_{n_i}=K_{n_i}(f_{n_i}-f)+(K_{n_i}-K)f+Kf$ tend to zero because of uniform boundedness and consistency, so that $K_{n_i}f_{n_i}\to Kf$ follows. Since also $K_{n_i}f_{n_i}=\lambda_{n_i}f_{n_i}\to\lambda f$ and $\|f\|=\lim\|f_{n_i}\|=1$ (hence, $f\neq 0$), λ is an eigenvalue and f an eigenfunction of problem (7). QED

Theorem 15 shows that subsequences of discrete eigenvalues can only converge to eigenvalues of the problem $\lambda f=Kf$. Now we want to discuss the converse statement: Each eigenvalue of $\lambda f=Kf$ is a limit of discrete eigenvalues.

Theorem 4.8.16. Let the discretisation $\{K_n\}$ be consistent and collectively compact. Then for any eigenvalue $\lambda\neq 0$ of K there is a sequence $\{\lambda_n\}$ of discrete eigenvalues of problem (8) with $\lim_{n\to\infty}\lambda_n=\lambda$.

The proof will be prepared by two lemmata.

Lemma 4.8.17. Define $\varkappa_n(\lambda) := \begin{cases} 1/\|(\lambda I-K_n)^{-1}\|, & \text{if } \lambda \notin \sigma(K) \\ 0 & \text{otherwise} \end{cases}$.

(a) If the discretisation $\{K_n\}$ is consistent, then $\lim_{n\to\infty} \varkappa_n(\lambda)=0$ holds for any eigenvalue $\lambda\in\sigma(K)$.

(b) If $\{K_n\}$ is consistent and collectively compact and if $\lim_{n\to\infty} \varkappa_n(\lambda)=0$ for some $\lambda\neq 0$, then λ is an eigenvalue of K.

Proof. (i) Let $\lambda\in\sigma(K)$ be an eigenvalue of K with associated eigenfunction $f\neq 0$. We set $g_n := \lambda f - K_n f$. By consistency, $g_n \to \lambda f - Kf = 0$ holds. If $\varkappa_n(\lambda)$ does not tend to zero as $n\to\infty$, then there is a subsequence with $\varkappa_{n_i}(\lambda)\geqslant\varepsilon>0$ for some $\varepsilon>0$, i.e., $\|(\lambda I-K_{n_i})^{-1}\| \leqslant C := 1/\varepsilon$. Hence, $\|f\| = \|(\lambda I-K_{n_i})^{-1} g_{n_i}\| \leqslant \|(\lambda I-K_{n_i})^{-1}\| \, \|g_{n_i}\| \leqslant C\,\|g_{n_i}\| \to 0$ in contradiction to $f\neq 0$. Hence, $\varkappa_n(\lambda)\to 0$ is proved.

(ii) Assume $\varkappa_n(\lambda)\to 0$ $(n\to\infty)$ for some $\lambda\neq 0$, i.e., $\|(\lambda I-K_n)^{-1}\|\to\infty$, where formally $\|(\lambda I-K_n)^{-1}\|=\infty$ is defined for $\lambda\in\sigma(K_n)$. Therefore, there are sequences $\{f_n\}$ and $\{g_n\}$ with

$$(\lambda I - K_n) f_n = g_n, \qquad \|f_n\| = 1, \qquad g_n \to 0 \text{ in } X.$$

Because of the collective compactness, $K_n f_n$ contains a convergent subsequence $K_{n_i} f_{n_i}$. By $f_n = (g_n + K_n f_n)/\lambda$, f_{n_i} converges to some f. Hence $f = \lim f_{n_i} = \lim (g_{n_i} + K_{n_i} f_{n_i})/\lambda = Kf/\lambda$ (cf. part (i)) and so λ is an eigenvalue of K. QED

In Theorem 4.7.11, where we proved stability (i.e., $\underline{\lim}\, \varkappa_n(\lambda)\geqslant\varepsilon>0$) using collective compactness, we required that λ is a regular value. Lemma 17 shows that this assumption also necessary.

Exercise 4.8.18. Prove that $|\varkappa_n(z)-\varkappa_n(\zeta)| \leqslant |z-\zeta|$ for all $z,\zeta\in\mathbb{C}$. This inequality shows the Lipschitz continuity of $\varkappa_n$ in $\mathbb{C}$.

The function $\varkappa_n(\lambda)\geqslant 0$ defined in Lemma 17 for all $\lambda\in\mathbb{C}$ (and continuous by Exercise 18) satisfies the following *minimum principle*.

Lemma 4.8.19. Let $G\subset\mathbb{C}\setminus\{0\}$ be a compact set. Then the following alternative holds: *Either* a discrete eigenvalue $\lambda_n\in\sigma(K_n)$ lies in G, *or* $\varkappa_n$ attains its minimum on the boundary ∂G.

Proof. Suppose that no discrete eigenvalue $\lambda_n\in\sigma(K_n)$ is in G. We want to prove the second part of the alternative. Since $\varkappa_n$ is continuous (cf. Exercise 18), it has a minimum at a point $\lambda\in G$. Let $\varphi\in X$ be a fixed arbitrary function. According to our assumption, λ is not a discrete eigenvalue, so that $(\lambda I-K_n)^{-1}\varphi$ is well-defined. By the Hahn-Banach Theorem (cf. Heuser [1, Theorem 36.4] or Yosida [1, §IV.4]), there is a functional $\Phi\in X'$ with

$$\|\Phi\|_{X'} = 1, \qquad \Phi\big((\lambda I-K_n)^{-1}\varphi\big) = \|(\lambda I-K_n)^{-1}\varphi\|.$$

Because $\varkappa_n(z)>0$ in G, $(zI-K_n)^{-1}$ exists for all $z\in G$. We set

$$f(z) := \Phi\big((zI-K_n)^{-1}\varphi\big) \qquad \text{for all } z\in G.$$

Since also the derivative $f'(z) = -\Phi\big((zI-K_n)^{-2}\varphi\big)$ exists in G, $f(z)$ is holomorphic in G. The maximum principle for holomorphic functions states that $|f(z)|$ attains its maximum on the boundary ∂G, i.e.,

$$\|(\lambda I-K_n)^{-1}\varphi\| = f(\lambda) = |f(\lambda)| \leqslant \max\{|f(z)|:\ z\in\partial G\}.$$

Because of $|f(z)| \leqslant |\Phi((zI-K_n)^{-1}\varphi)| \leqslant \|\Phi\|_{X'} \|(zI-K_n)^{-1}\varphi\| = \|(zI-K_n)^{-1}\varphi\| \leqslant \|(zI-K_n)^{-1}\|\,\|\varphi\| = \|\varphi\|/\varkappa_n(z)$ for all $z\in G$, one concludes

$$\|(\lambda I-K_n)^{-1}\varphi\| \leqslant \max\{\|\varphi\|/\varkappa_n(z):\ z\in\partial G\} = \|\varphi\|/\min\{\varkappa_n(z):\ z\in\partial G\}.$$

Since φ is arbitrarily chosen, the operator norm is bounded by

$$1/\varkappa_n(\lambda) = \|(\lambda I-K_n)^{-1}\| \leqslant 1/\min\{\varkappa_n(z):\ z\in\partial G\}.$$

The inequality $\varkappa_n(\lambda) \geqslant \min\{\varkappa_n(z):\ z\in\partial G\}$ shows that the minimum of $\varkappa_n$ is taken on ∂G. QED

Proof of Theorem 16. (i) Since $\{K_n\}$ is consistent and collectively compact, $K=\lim K_n$ is compact (cf. Remark 7.9b). Any eigenvalue $\lambda\neq 0$ of K is isolated according to Theorem 1.3.28c. That is, there is a neighbourhood $G\subset\mathbb{C}$ of λ, so that

$$0\notin G,\qquad G\cap\sigma(K)=\{\lambda\},\qquad \lambda\in G\setminus\partial G,\qquad G \text{ compact}.$$

Set $\alpha_n := \min\{\varkappa_n(z):\ z\in\partial G\}$. Now we will show $\alpha_n \geqslant \varepsilon > 0$ $(n\geqslant n_0)$. Suppose this is not true. Then a subsequence $\alpha_{n_i}\to 0$ and $z_i\in\partial G$ exist with $\alpha_{n_i}=\varkappa_{n_i}(z_i)\to 0$. Since ∂G is compact, a further subsequence of the z_i converges to some $\zeta\in\partial G$. $|\varkappa_{n_i}(\zeta)-\varkappa_{n_i}(z_i)| \leqslant |\zeta - z_i| \to 0$ (cf. Exercise 18) implies $\varkappa_{n_i}(\zeta)\to 0$ as $i\to\infty$. By Lemma 17b, ζ must be an eigenvalue of K. $\zeta\in\partial G$ implies $\zeta\in G$ and $\zeta\neq\lambda$ in contradiction to the assumption that λ is the only eigenvalue in G.
(ii) Part (i) shows $\alpha_n \geqslant \varepsilon > 0$ for $n\geqslant n_0$. By Lemma 17a, $\lim_{n\to\infty}\varkappa_n(\lambda)=0$ holds. Hence, one can choose n_0 such that $\varkappa_n(\lambda)<\varepsilon$ for $n\geqslant n_0$. Since α_n is the boundary minimum of $\varkappa_n$, $\varkappa_n$ cannot attain its minimum on the boundary. From Lemma 19 one concludes that for all $n\geqslant n_0$, G contains at least one discrete eigenvalue $\lambda_n\in\sigma(K_n)$. Since this statement holds for any sufficiently small neighbourhood, we can find $\lambda_n\in\sigma(K_n)$ with $\lim_{n\to\infty}\lambda_n=\lambda$. QED

Applying Theorem 15 to the sequence $\lambda_n\in\sigma(K_n)$ constructed above, we obtain the additional result that a subsequence of the eigenfunctions f_n of K_n converges to an eigenfunction f of K.

Having ensured the convergence $\lambda_n\to\lambda$, we want to prove *quantitative* error estimates for the difference $|\lambda_n-\lambda|$ of the eigenvalues and the differences $\|f_n-f\|$ of the eigenfunctions. For this purpose, we need the regularity assumption (10a) and the consistency assumption (10b):

(4.8.10a) $\quad K\in L(X,Y),\quad K\in K(X,X),$

(4.8.10b$_1$) $\quad \|\varphi-\Pi_n\varphi\|_X \leqslant Ch^\varkappa\|\varphi\|_Y$ for all $\varphi\in Y$ (in the case of a projection method)

(4.8.10b$_2$) $\quad \|K-K_n\|_{X\leftarrow Y} \leqslant Ch^\varkappa$ (for kernel approximation or Nyström method)

From $\lambda f = Kf$, $\lambda \neq 0$, and (10a) one concludes immediately that $f \in Y$ holds for the eigenfunction.

We assume that λ is a *simple eigenvalue*, i.e.,

(4.8.11a) $\quad \dim E(\lambda) = 1 \qquad$ for the eigenspace $E(\lambda) := \{f \in X:\ \lambda f = Kf\}$.

Then, the dual operator K' also has a one-dimensional eigenspace:

(4.8.11b) $\quad \dim E'(\lambda) = 1 \qquad$ for the eigenspace $E'(\lambda) := \{f' \in X':\ \lambda f' = K'f'\}$.

Hence, there are fixed eigenfunctions f and f' spanning $E(\lambda)$ and $E'(\lambda)$. The next requirement implies that also the *algebraic multiplicity* of λ equals one (cf. Hackbusch [2,§11.2.3]):

(4.8.11c) $\quad f'(f) \neq 0$.

Let f be scaled such that $f'(f) = 1$. From Theorem 15 one concludes that for sufficiently large $n \geq n_0$ also the discrete eigenvalues λ_n must be simple. The associated eigenfunctions f_n can be scaled by $f'(f_n) = 1$ for sufficiently large $n \geq n_0$.

Theorem 4.8.20. Let (10a,b), (11a,c), $K \in K(X,X)$. The eigenfunction of K corresponding to $0 \neq \lambda \in \sigma(K)$ is scaled by $f'(f) = 1$. The discretisation $\{K_n\}$ is assumed to be consistent and collectively compact. The dual operator K_n' must satisfy $\lim K_n' f' = K'f'$ in X'. Let λ_n $(n \geq n_0)$ be a sequence of discrete eigenvalues converging to λ with eigenfunctions f_n, scaled by $f'(f_n) = 1$. Then the following convergence estimates holds:

(4.8.12) $\quad |\lambda - \lambda_n| \leq C h^{\varkappa}, \qquad \|f - f_n\| \leq C h^{\varkappa}$.

Proof. (i) The representation $\lambda_n(f_n - f) = K_n f_n - (\lambda_n/\lambda) Kf$ yields

(4.8.13a) $\quad \lambda_n(f_n - f) = K_n(f_n - f) + d_n \qquad$ with

(4.8.13b) $\quad d_n := [\lambda_n(K_n - K)f + (\lambda - \lambda_n)K_n f]/\lambda$.

From (11a) one concludes

(4.8.13c) $\quad \|d_n\| \leq C(h^{\varkappa} + |\lambda - \lambda_n|)$.

Because of equation (13a) and $f'(f_n) = f'(f) = 1$, the augmented equation

$$\begin{bmatrix} \lambda_n - K_n & f \\ f' & 0 \end{bmatrix} \begin{bmatrix} f_n - f \\ \omega \end{bmatrix} = \begin{bmatrix} d_n \\ 0 \end{bmatrix}$$

is satisfied with $\omega = 0$. In anticipation of §4.8.5, we use Lemma 8.25 with $\varphi_n = \varphi = f$, $\varphi_n' = \varphi' = f$, which ensures stability and leads to

(4.8.13d) $\quad \|f_n - f\| \leq C\|d_n\| \leq C'(h^{\varkappa} + |\lambda - \lambda_n|)$.

(ii) Applying the functional f' to the representation

(4.8.13e) $\quad (\lambda I - K)(f - f_n) = (\lambda_n - \lambda)f_n - (K_n - K)f - (K_n - K)(f_n - f)$,

and using $\langle f', (\lambda I - K)\varphi\rangle = \langle(\lambda I - K')f', \varphi\rangle = 0$ and $f'(f_n) = f'(f) = 1$, we obtain the estimate

$$\text{(4.8.13f)}\qquad |\lambda_n-\lambda| \leqslant C[\|(K_n-K)f\| + \|(K_n'-K')f'\|_{X'}\|f-f_n\|] \leqslant$$
$$\leqslant C[h^\varkappa + \|(K_n'-K')f'\|_{X'}\|f-f_n\|].$$

Insertion into (13d) yields

$$\|f_n-f\| \leqslant C(h^\varkappa + \|(K_n'-K')f'\|_{X'}\|f-f_n\|).$$

By assumption, $\|(K_n'-K')f'\|_{X'} \leqslant 1/(2C)$ holds for sufficiently large $n \geqslant n_0$. Hence, $\|f_n-f\| \leqslant C'h^\varkappa$ is proved. Insertion into (13f) yields also $|\lambda_n-\lambda| \leqslant Ch^\varkappa$. QED

For *Galerkin's method* the order of accuracy of $|\lambda_n-\lambda|$ can be doubled, as we now show.

Corollary 4.8.21. Under the assumptions of Theorem 20, the Galerkin method leads to $|\lambda-\lambda_n| \leqslant Ch^\varkappa\|(I-\Pi_n)f'\|$. If $f' \in Y$, then $|\lambda-\lambda_n| \leqslant Ch^{2\varkappa}$.

Proof. For f' one finds $\tilde{f}_n' \in X_n$ with $\langle f'-\tilde{f}_n', f\rangle = 0$ and $\|f'-\tilde{f}_n'\| \leqslant 2\|(I-\Pi_n)f'\|$. $(\lambda I-K_n)f_n=0$ is identical to $\langle(\lambda I-K)f_n,\varphi_n\rangle=0$ for all $\varphi_n \in X_n$. The special choice $\varphi_n=\tilde{f}_n'$ shows

$$0 = \langle(\lambda I-K)f_n, f'\rangle = \langle(\lambda_n I-K)f_n, f'\rangle + (\lambda-\lambda_n)\langle f_n, f'\rangle =$$
$$= \langle(\lambda_n I-K)f_n, f'-\tilde{f}_n'\rangle + (\lambda-\lambda_n)\langle f_n, f'\rangle =$$
$$= \langle(\lambda_n I-K)(f_n-f), f'-\tilde{f}_n'\rangle + (\lambda-\lambda_n)\langle f_n, f'\rangle.$$

Because of the scaling $\langle f_n, f'\rangle = 1$ and the choice of $\tilde{f}_n'$, one obtains $|\lambda-\lambda_n| \leqslant \|\lambda_n I-K\|\ \|f_n-f\|\ \|f'-\tilde{f}_n'\| \leqslant Ch^\varkappa\|f'-\tilde{f}_n'\| \leqslant 2Ch^\varkappa\|(I-\Pi_n)f'\|$. QED

Concerning the numerical solution of the discrete eigenvalue problems, we refer to Hackbusch [1] and [6].

4.8.5 Complementary Integral Equations

By definition of a singular value, the equation

$$\text{(4.8.14a)}\qquad \lambda f = g + Kf \qquad\qquad \text{with } \lambda \in \sigma(K)$$

not uniquely solvable for all $g \in X$. The Riesz-Schauder theory tells us that equation (14a) has (at least) one solution if and only if $\langle\varphi, g\rangle = 0$ holds for all eigenfunctions $\varphi \in E'(\lambda)$ of the dual operator K'. We restrict our considerations to the case that λ is a *simple eigenvalue*. Choosing eigenfunctions $0 \neq e \in E(\lambda)$ and $0 \neq e' \in E'(\lambda)$, we can represent the eigenspaces $E(\lambda)$ and $E'(\lambda)$ by

$$\text{(4.8.14b)}\qquad E(\lambda) = \operatorname{span}\{e\}, \qquad E'(\lambda) = \operatorname{span}\{e'\}, \qquad e \neq 0, \qquad e' \neq 0.$$

The solvability statement for equation (14a) reads as follows.

Lemma 4.8.22. Let $K \in K(X,X)$, $0 \neq \lambda \in \sigma(K)$, and suppose (14b) holds. The equation (14a) has a solution $f \in X$ if and only if

$$\text{(4.8.14c)} \qquad \langle e', g \rangle = 0 \qquad\qquad (e' \text{ from (14b)}).$$

If f is a solution, then $\{f + \alpha e\colon \alpha \in \mathbb{C}\}$ describes the set of all solutions.

A direct discretisation of equation (14a) can lead to very large complications. A simple remedy is the *augmentation of the singular equation* into a system for $f \in X$ and $\omega \in \mathbb{R}$ (or $\omega \in \mathbb{C}$):

(4.8.15a)	$\lambda f - Kf - \omega\varphi = g,$
(4.8.15b)	$\lambda\omega - \langle \varphi', f \rangle = \gamma.$

Here, $\varphi, g \in X$, $\varphi' \in X'$, and $\gamma \in \mathbb{R}$ are given. The unknowns are $f \in X$, $\omega \in \mathbb{R}$. The system (15a,b) can be rewritten as

$$\text{(4.8.15')} \qquad (\lambda I - \hat{K})\hat{f} = \hat{g} \quad \text{with } \hat{f} = \begin{bmatrix} f \\ \omega \end{bmatrix} \in \hat{X} := X \oplus \mathbb{R}, \quad \hat{g} = \begin{bmatrix} g \\ \gamma \end{bmatrix} \in \hat{X}, \quad \hat{K} = \begin{bmatrix} K & \varphi \\ \varphi' & 0 \end{bmatrix}.$$

Exercise 4.8.23. If K is compact in X, then $\hat{K}$ is compact in $\hat{X}$.

Equation (15') is a well-posed problem as shown in the following lemma.

Lemma 4.8.24. Let λ be a simple eigenvalue of $K \in K(X,X)$ with eigenfunctions according to (14b). The functions $\varphi \in X$ and $\varphi' \in X'$ are to be chosen such that

$$\text{(4.8.15c)} \qquad \langle e', \varphi \rangle \neq 0, \qquad \langle \varphi', e \rangle \neq 0.$$

Then λ is a regular value of $\hat{K}$. If g satisfies condition (14c), then the component f of the solution $\hat{f}$ of (15') is a solution of the original equation (14a). Varying γ in (15b), one obtains the set of all solutions of (14a).

The non-singular equation (15') can be discretised by (16):

$$\text{(4.8.16)} \qquad (\lambda I - \hat{K}_n)\hat{f}_n = \hat{g}_n.$$

In the case of a projection method defined by Π_n, the projection $\hat{\Pi}_n$ in $\hat{X} = X \oplus \mathbb{R}$ is defined by $\hat{\Pi}_n \hat{f} = \begin{bmatrix} \Pi_n f \\ \omega \end{bmatrix}$ for $\hat{f} = \begin{bmatrix} f \\ \omega \end{bmatrix}$. Hence, equation (16) represents the system

$$\text{(4.8.17a)} \qquad \lambda f_n - K_n f_n - \omega\varphi_n = g_n \qquad \text{with } K_n = \Pi_n K,\ \varphi_n = \Pi_n\varphi,\ g_n = \Pi_n g,$$

$$\text{(4.8.17b)} \qquad \lambda\omega - \langle \varphi_n', f_n \rangle = \gamma \qquad \text{with } \varphi_n' = \varphi'.$$

By analogy with Exercise 23 one proves

Lemma 4.8.25. Suppose $\varphi_n \to \varphi$ in X and $\varphi_n' \to \varphi'$ in X' for φ_n and φ_n' from (17a,b). Then, consistency and collective compactness of $\{K_n\}$ carry over to $\{\hat{K}_n\}$.

Lemma 24 and Theorem 7.11 prove the following result.

Theorem 4.8.26. Assume that $K \in K(X,X)$ has a simple eigenvalue $\lambda \in \sigma(K)$, suppose (14b) and (15c) hold, and suppose $\{K_n\}$ is consistent and collectively compact. Let $\varphi_n \to \varphi$ in X and $\varphi_n' \to \varphi'$ in X' hold for φ_n and φ_n' from (17a,b). Then the discretisation $\{\hat{K}_n\}$ from (17a,b) is consistent, collectively compact, stable, and convergent.

Theorem 26 ensures that, thanks to the augmentation (17a,b), one can solve the singular problem (14a) without numerical difficulties. It is easy to see that one obtains the usual error estimates for solutions of the discretisation (16).

4.8.6 Supplement: Perturbation Theorem for Stability

In the first subsection §4.1 we discussed general properties of discretisations, in particular, stability. Since the concept of collective compactness was introduced with the Nyström method, Theorem 28 concerning perturbation of a stable discretisation could not be formulated earlier.

In the following, $\{K_n\}$ is assumed to be stable (with respect to X and λ). Let T_n be a sequence of perturbations. $K_n' := K_n + T_n$ is the perturbed discretisation whose stability has to be investigated.

The first statement is based on the same considerations as Theorem 1.4, so that the proof can be omitted.

Lemma 4.8.27. $\{K_n\}$ is assumed to be stable with the constant C_S: $\|(\lambda I - K_n)^{-1}\| \leq C_S$. Suppose also $\{T_n\}$ satisfies the estimate $\|T_n\| \leq \zeta / C_S$ with $\zeta < 1$ for all $n \geq n_0$. Then the sequence $\{K_n + T_n\}$ is also stable.

Theorem 4.8.28. Let $\lambda \neq 0$. $\{K_n\}$ is assumed to be convergent to and consistent with $K \in L(X,X)$. Let $\{T_n\}$ be collectively compact, converging pointwise to T. If λ is a regular value of $K+T$, then the sequence $\{K_n + T_n\}$ is stable, consistent, and convergent.

Proof. (i) First, we will show that also $(\lambda I - K_n)^{-1} T_n$ $(n \geq n_0)$ is collectively compact. Let $\Phi_j := (\lambda I - K_{n(j)})^{-1} T_{n(j)} \varphi_j$ with $\|\varphi_j\| \leq 1$. Since $\{T_n\}$ is collectively compact, there is a subsequence with $\psi_k := T_{n(j_k)} \varphi_{j_k} \to \psi$, where without loss of generality $n(j_{k+1}) \geq n(j_k)$ can be ensured. Either $n(j_k) = m$ holds for $k \geq k_0$ or $n(j_k) \to \infty$. In the former case, $\Phi_{j_k} = (\lambda I - K_m)^{-1} \psi_k \to (\lambda I - K_m)^{-1} \psi$ follows. In the latter one, $\Phi_{j_k} \to (\lambda I - K)^{-1} \psi$ holds thanks to Lemma 1.8 and Theorem 1.12b.

(ii) Applying again Theorem 1.12b (with $g_n := T_n \varphi \to T\varphi$), it follows that $A_n := (\lambda I - K_n)^{-1} T_n$ converges pointwise to $A := (\lambda I - K)^{-1} T$.

(iii) By Lemmata 1.8 and 1.10, λ is a regular value of K. Since $\lambda I - K - T = (\lambda I - K)(I - A)$, 1 is also a regular value of A. Since $\{A_n\}$ is collectively compact and consistent with A, Theorem 7.11 yields stability, i.e., $\|(I - A_n)^{-1}\| \leq C$. Because $(\lambda I - K_n - T_n)^{-1} = (I - A_n)^{-1} (\lambda I - K_n)^{-1}$, the sum $\{K_n + T_n\}$ is also stable.

(iv) The consistency of $\{K_n + T_n\}$ is trivial. Since $\lambda I - K - T = (\lambda I - K)(I - A)$ is bijective (cf. (iii)), the convergence follows from Theorem 1.12a. QED

5 Multi-Grid Methods for Solving Systems Arising from Integral Equations of the Second Kind

The discretisations derived in Chapter 4 change the Fredholm integral equation into a system of equations. Since this system may consist of a large number of equations involving a full matrix, its solution is not a trivial task. In this chapter, we mainly investigate the solution by the *multi-grid method*. The multi-grid treatment of equations of the first kind is discussed in §7.3.6 and §9.3.

5.1 Preliminaries

5.1.1 Notation

The discretisation of the integral equation

(5.1.1) $\lambda f = g + Kf \qquad (\lambda \neq 0)$

of the second kind led to systems of the form

(5.1.2a) $(\lambda I - B_n)\, a_n = b_n$

(cf. (4.2.9a), (4.4.11), (4.7.7)) or

(5.1.2b) $(\lambda A_n - B_n)\, a_n = b_n$

(cf. (4.4.10), (4.5.8)). Here, n is a discretisation parameter which in the former examples is more or less directly related to the dimension of the system.

Only in the case of the projection method, may the matrix A_n be different from the identity matrix I (cf. §4.4 and §4.5). In that case, A_n is always non-singular since, otherwise, the interpolation is not unique (in the collocation case) or the ansatz functions are linearly dependent (in the Galerkin case). Therefore, equation (2b) can always be written in the form

(5.1.2b') $(\lambda I - A_n^{-1} B_n)\, a_n = A_n^{-1} b_n$

corresponding to (2a) with $A_n^{-1}B_n$ and $A_n^{-1}b_n$ instead of B_n and b_n. To express the fact that the *discrete* equation (2b') corresponds to the *continuous* equation (1), we write (2b') as

(5.1.3) $\lambda \boldsymbol{f}_n = \boldsymbol{g}_n + \boldsymbol{K}_n \boldsymbol{f}_n,$

with

(5.1.4) $\quad \boldsymbol{f}_n := a_n, \qquad \boldsymbol{g}_n := A_n^{-1} b_n, \qquad \boldsymbol{K}_n := A_n^{-1} B_n$

where these quantities now have a *different* meaning than in §4. There, f_n and g_n were *functions* (from a finite-dimensional subspace) defined on D, whereas now $\boldsymbol{f}_n$ and $\boldsymbol{g}_n$ are *vectors*. $\boldsymbol{K}_n$ is not an *operator* in $L(X,X)$ as previously, but a *matrix*. To avoid confusion, the new quantities are written in bold face.

5.1.2 Direct Solution of the System of Equations

System (3) can be solved directly, e.g., by Gauß elimination (with pivoting). However, the amount of computational work is considerable.

Remark 5.1.1. If n is the number of equations and unknowns of system (3), then Gauß elimination requires about $\frac{2}{3}n^3$ arithmetical operations.

In contrast to the discretisation of partial differential equations, $\lambda I - \boldsymbol{K}_n$ is a full matrix and so the amount of work mentioned in Remark 1 can be reduced only by other matrix representations (cf. §9.7). The same amount of work is needed for a triangular decomposition or, in the positive definite case, for a Cholesky decomposition.

5.1.3 Picard Iteration

As alternative to the direct solution, one can solve the system *iteratively*. The simplest iteration is the Picard iteration: The $(i+1)$st iterate $\boldsymbol{f}_n^{i+1}$ is obtained by inserting the ith iterate $\boldsymbol{f}_n^i$ into the right-hand side of (3):

(5.1.5) $$\boldsymbol{f}_n^{i+1} := \tfrac{1}{\lambda}(\boldsymbol{g}_n + \boldsymbol{K}_n \boldsymbol{f}_n^i) \qquad (i = 0, 1, 2, \ldots).$$

To start the iteration one needs a «starting iterate» $\boldsymbol{f}_n^0$, e.g., $\boldsymbol{f}_n^0 = 0$ or $\boldsymbol{f}_n^0 = \frac{1}{\lambda}\boldsymbol{g}_n$. The second choice results as first iterate from the start $\boldsymbol{f}_n^0 = 0$.

We gather the main results about the *convergence of iterative methods* in the following theorem. Its proof can be found in Hackbusch [3, Theorem 3.2.7], where the interested reader can find further details about iterative methods.

Theorem 5.1.2. Given an iterative method of the form

(5.1.6a) $$x^{i+1} = Mx^i + c \qquad (c \in \mathbb{R}^n,\ M\ n \times n \text{ matrix}),$$

the sequence $\{x^i\}$ converges for any starting iterate x^0 and any c, if and only if the *spectral radius*

(5.1.6b) $$\rho(M) := \max\{|\lambda| : \lambda \text{ is an eigenvalue of } M\}$$

satisfies the inequality

(5.1.6c) $$\rho(M) < 1.$$

The limit of the sequence $\{x^i\}$ is the (unique) fixed point x of

(5.1.6d) $$x = Mx + c.$$

Sufficient (but not necessary) for (6c) - and thus for the convergence - is the estimate of M with respect to some matrix norm:

(5.1.6e) $\|M\| < 1.$

The matrix M in (6a) is named the *iteration matrix*. We recall that the *matrix norm*

$$\|M\| := \sup\{\|Mx\| / \|x\| : 0 \neq x \in \mathbb{R}^n\} \qquad \text{(cf. (1.4.20))}$$

coincides with the operator norm (1.3.11a) for the space $X = (\mathbb{R}^n, \|\cdot\|)$. The vector norm $\|\cdot\|$ can, e.g., be defined by (1.3.3a) or (1.3.3b). In the case of (6e), the convergence follows directly from the Banach fixed point theorem 1.3.10. This also proves the following

Corollary 5.1.3. Under the assumption (6e), the error estimate

(5.1.6f) $\|x^{i+1} - x\| \leqslant \|M\| \|x^i - x\|$

holds for the iterates $\{x^i\}$ from (6a) and the solution x from (6d). In particular, the errors $x^i - x$ have the representation

(5.1.6g) $x^{i+1} - x = M(x^i - x).$

The iteration matrix of the Picard iteration (5) is $M = \frac{1}{\lambda}\boldsymbol{K}_n$. Applying the convergence criterion to the Picard iteration, we obtain the following result.

Theorem 5.1.4. The Picard iteration (5) converges if and only if

(5.1.7a) $\rho(\boldsymbol{K}_n) < |\lambda|.$

A sufficient convergence condition is the matrix norm estimate

(5.1.7b) $\|\boldsymbol{K}_n\| < |\lambda|.$

It is impossible to satisfy the inequalities (7a,b) by a suitable choice of the discretisation. The convergence depends on the problem (more precisely, on λ and K, not on g). The error representation (6g) yields

(5.1.7c) $\boldsymbol{f}_n^{i+1} - \boldsymbol{f}_n = \frac{1}{\lambda}\boldsymbol{K}_n(\boldsymbol{f}_n^i - \boldsymbol{f}_n).$

Assuming the stronger convergence condition

(5.1.7d) $C_N := \sup\{\|\boldsymbol{K}_n\| : n \in \mathbb{N}\} < |\lambda|$

(cf. (7b)), we need $k \geqslant (\log \varepsilon) / \log(C_N / |\lambda|)$ iterations to ensure

(5.1.7e) $\|\boldsymbol{f}_n^k - \boldsymbol{f}_n\| \leqslant \varepsilon \|\boldsymbol{f}_n^0 - \boldsymbol{f}_n\| \qquad (\varepsilon < 1).$

Remark 5.1.5. When convergence occurs, the amount of computational work for Picard iteration to bound the iteration error by $O(\varepsilon)$ is proportional to $n^2 \log \varepsilon$.

Proof. In each iteration (5), the matrix-vector multiplication $\boldsymbol{K}_n \boldsymbol{f}_n^i$ is the most costly part. Since $\boldsymbol{K}_n$ is a full matrix, $2n^2$ arithmetical operations are needed. This amount of work is to be multiplied by the number k of the iterations required to reach (7e). QED

5.1.4 Conjugate Gradient Method

First, we assume that the matrices $\boldsymbol{K}_n$ or A_n, B_n from (2b) are symmetric:

(5.1.8a) $\quad \boldsymbol{K}_n = \boldsymbol{K}_n^T$ or

(5.1.8b) $\quad A_n = A_n^T$ positive definite, $B_n = B_n^T$ (where $\boldsymbol{K}_n = A_n^{-1} B_n$).

Remark 5.1.6. Without loss of generality, let $\lambda > 0$. Assume (8a) or (8b). Then condition (7a) is equivalent to the requirement that the respective matrices $\lambda I - \boldsymbol{K}_n$ or $\lambda A_n - B_n$ be positive definite.

Proof. (i) Assume (8a). Any eigenvalue σ of $\lambda I - \boldsymbol{K}_n$ has the representation $\sigma = \lambda - \varkappa$, where $\varkappa$ is an eigenvalue of $\boldsymbol{K}_n$. Because of (8a), $\varkappa$ is real, while (7a) leads to $|\varkappa| < \lambda$. Hence, $\sigma = \lambda - \varkappa \geq \lambda - |\varkappa| > 0$. The assertion follows, since $\lambda I - \boldsymbol{K}_n$ is positive definite if and only if all eigenvalues of the symmetric matrix $\lambda I - \boldsymbol{K}_n$ are positive.
(ii) In the case of (8b), $\lambda A_n - B_n$ is positive definite if and only if $\lambda I - A_n^{-1/2} B_n A_n^{-1/2}$ is so. Since $A_n^{-1/2} B_n A_n^{-1/2}$ and $\boldsymbol{K}_n = A_n^{-1} B_n$ are similar, they have the same eigenvalues and the assertion follows as in part (i). QED

Under the assumption

(5.1.8c) $\quad C_n := \lambda I - \boldsymbol{K}_n \quad$ is positive definite

or, alternatively,

(5.1.8d) $\quad C_n := \lambda A_n - B_n \quad$ is positive definite,

the *method of the conjugate gradients* (*cg method*) can be applied to the respective equation (3) or (2b):

Method of the conjugate gradients for solving $C_n \boldsymbol{f}_n = \boldsymbol{g}_n$

Start: $\quad \boldsymbol{f}_n^0 \in \boldsymbol{X}_n$ arbitrary,

(5.1.9a) $\quad \pi^0 := \rho^0 := \boldsymbol{g}_n - C_n \boldsymbol{f}_n^0$.

Iteration over $k = 0, 1, 2, \ldots, n-1$; *termination* if $\rho^k = 0$; otherwise:

(5.1.9b) $\quad c_k := \|\pi^k\|_2^2, \quad d_k := (C_n \pi^k, \pi^k), \quad a_k := c_k / d_k,$

(5.1.9c) $\quad \boldsymbol{f}_n^{k+1} := \boldsymbol{f}_n^k + a_k \pi^k,$

(5.1.9d) $\quad \rho^{k+1} := \rho^k - c_k C_n \pi^k,$

(5.1.9e) $\quad \pi^{k+1} := \rho^{k+1} + \frac{1}{c_k} \|\rho^{k+1}\|_2^2 \, \pi^k.$

$(\cdot,\cdot)$ is the Euclidean scalar product and $\|a\|_2 := (a,a)^{1/2}$ the corresponding Euclidean norm.

Theorem 5.1.7. Let C_n be a positive definite $n\times n$ matrix. Then the cg method (9a-e) terminates after n or less steps with the *exact* solution $\boldsymbol{f}_n^k=\boldsymbol{f}_n$. All iterates $\boldsymbol{f}_n^i$, $0\leqslant i\leqslant k$, fulfil the error estimate

(5.1.10a) $$\|\boldsymbol{f}_n^i-\boldsymbol{f}_n\|_C \leqslant \frac{2(\varkappa_n-1)^i}{(\sqrt{\varkappa_n}+1)^{2i}+(\sqrt{\varkappa_n}-1)^{2i}}\|\boldsymbol{f}_n^0-\boldsymbol{f}_n\|_C,$$ where

(5.1.10b) $$\|a\|_C := \sqrt{(C_n a, a)} = \|C_n^{1/2}a\|_2$$

represents a norm and

(5.1.10c) $$\varkappa_n := \mathrm{cond}_2(C_n) = \gamma_{\max}/\gamma_{\min}$$

denotes the spectral condition number (cf. Definition 1.4.29). $\gamma_{\max}$ and $\gamma_{\min}$ are the respective maximal and minimal eigenvalues of the system matrix C_n.

Proof. For example, in Hackbusch [3]. ∎

The cg method is of particular interest since the condition number $\varkappa_n$ determining the convergence speed is uniformly bounded with respect to n, i.e.,

(5.1.11) $$\varkappa := \sup\{\varkappa_n : n\in\mathbb{N}\} < \infty.$$

The following lemma describes conditions under which the cg method is applicable.

Lemma 5.1.8. The eigenvalues of the operator K are assumed to be smaller than λ. Let the semidiscrete operators $K_n\in L(X,X)$ be consistent and collectively compact. Assume that the matrix $\boldsymbol{K}_n$ is *either* (i) symmetric (condition (8a)), *or* (ii) the matrices A_n, B_n satisfy (8b). Then there is some n_0 ensuring the positive definiteness (8c) [case (i)] or (8d) [case (ii)] for all $n\geqslant n_0$. Furthermore, (11) holds (omitting possibly singular C_n for $n<n_0$). The norms $\|\cdot\|_2$ and $\|\cdot\|_C$ are uniformly equivalent:

$$\tfrac{1}{C}\|\cdot\|_C \leqslant \|\cdot\|_2 \leqslant C\|\cdot\|_C \qquad \text{for all } n\geqslant n_0 \text{ with suitable } C.$$

Proof. (i) Assume (8a). The semidiscrete operators K_n have the same eigenvalues as the matrices $\boldsymbol{K}_n$. In particular, these eigenvalues must be real because of the symmetry (8a). The eigenvalues of K are also real, since they are the limits of the discrete eigenvalues (cf. Theorem 4.8.16). The extreme eigenvalues $\lambda_{\min}$ and $\lambda_{\max}$ of K satisfy $\lambda_{\min}<\lambda_{\max}<\lambda$. Set $\varepsilon:=\frac{1}{2}(\lambda-\lambda_{\max})>0$. Let $\lambda_{n,\max}$ be the maximal eigenvalue of K_n (case (8b)). Since $\lim\lambda_{n,\max}=\lambda_{\max}$ according to Theorem 4.8.16, there is an n_0 such that $\lambda_n\leqslant\lambda_{n,\max}\leqslant\lambda-\varepsilon$ for $n\geqslant n_0$. Similarly, $\lambda_{\min}-\varepsilon\leqslant\lambda_{n,\min}\leqslant\lambda_n$ follows for all eigenvalues λ_n of K_n. Therefore, $0<\varepsilon\leqslant\gamma_{\min}\leqslant\gamma_{\max}\leqslant\lambda-\lambda_{n,\min}\leqslant\lambda-\lambda_{\min}+\varepsilon$ holds for the eigenvalues $\gamma_{\min}$ and $\gamma_{\max}$ of C_n (cf. (c,d)) for all $n\leqslant n_0$.
(ii) In the case of (8b), use the fact that $\lambda I-\boldsymbol{K}_n$ is similar to $E:=A_n^{-1/2}C_nA_n^{-1/2}$. As C_n, also E and therefore $\lambda I-\boldsymbol{K}_n$ have only real eigenvalues. ∎

If the matrices $C_n:=\lambda I-\boldsymbol{K}_n$ or λA_n-B_n are not positive definite (e.g., because of lack of symmetry), algorithm (9) is not applicable, but there are variants of the cg method with similar results as in Theorem 7 (cf. Hackbusch [3,§9.5]. For example, one can apply the cg method to the system

(5.1.12) $$C_n^T C_n \boldsymbol{f}_n = C_n^T \boldsymbol{g}_n$$

equivalent to $C_n \boldsymbol{f}_n = \boldsymbol{g}_n$, since $C_n^T C_n$ is positive definite for any non-singular C_n.

Lemma 5.1.9. $\varkappa_n := \mathrm{cond}_2(C_n^T C_n) = \mathrm{cond}_2(C_n)^2$ holds. Further, a symmetric C_n yields $\varkappa_n \leqslant \mathrm{cond}_\infty(C_n)^2$. In §4, we discussed the uniform boundedness of the condition numbers of the matrices $C_n = \lambda A_n - B_n$. Under this condition, inequality (11) holds.

Corollary 5.1.10. Condition (11) guarantees the uniform convergence speed for all $n \geqslant n_0$. In contrast to systems originating from (partial) differential equations, the convergence speed does not slow down as $n \to \infty$. Since either the cg method (9) or one of its variants converge and require $O(\log \varepsilon)$ iteration steps (cf. Remark 5), the *total amount of work* equals $O(n^2 \log \varepsilon)$.

Although uniform convergence speed is a satisfactory result, much better results can be obtained by the multi-grid method in §§5.4-5.

Remark 5.1.11. The mentioned numbers of arithmetical operations are based on the assumption that the *matrix-vector multiplication* $\boldsymbol{f}_n \mapsto \boldsymbol{K}_n \boldsymbol{f}_n$ requires at least $O(n^2)$ operations. In Section 9.7, we shall discuss important cases for which this amount of work can be reduced substantially.

5.2 Stability and Convergence (Discrete Formulation)

To derive conditions needed later for the formulation of the multi-grid method and its convergence proof, we have to introduce some new definitions. In particular, the terms «consistency», «stability», «convergence», and «regularity» will be used in a new setting. The reader may skip this part and start directly with §5.3.

The terms «stability» and «convergence» have already been defined in §4. However, they are based on the operators $K, K_n \in L(X,X)$ and not directly on the system with the matrix $\lambda I - A_n^{-1} B_n$, now renamed $\lambda I - \boldsymbol{K}_n$. In §4, it was shown that even if the semidiscrete problem were stable, the condition numbers $\mathrm{cond}(\lambda I - \boldsymbol{K}_n)$ could be unbounded because of an unsuitable choice of the basis. From now on, we shall use the discrete equations with their matrices $\lambda I - \boldsymbol{K}_n$ to introduce a new definition of stability and convergence.

5.2.1 Prolongations and Restrictions

The Banach space X was introduced in §4. The matrix $\lambda I - \boldsymbol{K}_n$ is a mapping of the Banach space

(5.2.1) $$\boldsymbol{X}_n := (\mathbb{R}^n, \|\cdot\|_{X_n})$$

into itself. Since one wants to associate vectors from $\boldsymbol{X}_n$ with functions from X, one needs mappings in both directions. A linear mapping

(5.2.2a) $$P_n : \boldsymbol{X}_n \to X \qquad (n \in \mathbb{N}).$$

is called a *prolongation*. The *restriction* acts in the reverse direction

(5.2.2b) $R_n: X \to \mathbf{X}_n$ $(n \in \mathbb{N})$.

In Definition 1, we shall describe so-called canonical prolongations P_n and restrictions R_n corresponding to the approximations discussed in §4 and fulfilling the later conditions (3a-c). Since $\dim(\mathbf{X}_n) < \infty$, P_n and R_n are bounded mappings:

$$P_n \in L(\mathbf{X}_n, X), \qquad R_n \in L(X, \mathbf{X}_n).$$

Their operator norms should be *bounded uniformly* with respect to the discretisation parameter n, i.e., there must be bounds C_P and C_R such that

(5.2.3a) $\|P_n\|_{X \leftarrow X_n} \leqslant C_P$ for all $n \in \mathbb{N}$,

(5.2.3b) $\|R_n\|_{X_n \leftarrow X} \leqslant C_R$ for all $n \in \mathbb{N}$.

To express the fact that the «discrete function» $R_n f \in \mathbf{X}_n$ represents an approximation of the function $f \in X$ and, vice versa, that the function $P_n \boldsymbol{f}_n \in X$ is close to $\boldsymbol{f}_n \in \mathbf{X}_n$ in a suitable sense, we require that the product $P_n R_n \in L(X, X)$ converge pointwise to the identity:

(5.2.3c) $P_n R_n f \to f$ for all $f \in X$.

Requirement (3c) will turn out in §5.2.2 to be a consequence of further conditions.

There are «canonical» prolongations and restrictions, which are associated in a natural way with the discretisations described in §4.

Definition 5.2.1 **(a)** (canonical choice in the case of a projection method). Let the *projection method* be defined by the projection Π_n and the basis $\{\Phi_{1,n}, \dots, \Phi_{n,n}\}$ of the corresponding subspace X_n. The associated *canonical prolongation* reads

(5.2.4a) $P_n \boldsymbol{f}_n := \sum_{j=1}^{n} f_{j,n} \Phi_{j,n}$, where $\boldsymbol{f}_n := (f_{j,n})_{j=1,\dots,n} \in \mathbf{X}_n = \mathbb{R}^n$.

The *canonical restriction* is the left inverse of the canonical prolongation:

(5.2.4b) $R_n P_n = I$,

i.e., for all $\varphi \in X$ the vector $R_n \varphi \in \mathbf{X}_n = \mathbb{R}^n$ represents the basis coefficients of $\Pi_n \varphi$. In the case of the collocation method, the vector $R_n \varphi$ is defined by

(5.2.4c) $R_n \varphi := A_n^{-1} \left(\varphi(\xi_{j,n}) \right)_{j=1,\dots,n}$ (A_n from (4.4.9b))

with collocation points $\xi_{j,n} \in \Xi_n$, whereas in the case of the Galerkin method:

(5.2.4d) $R_n g := A_n^{-1} \left(\langle g, \Phi_{j,n} \rangle \right)_{j=1,\dots,n}$ (A_n from (4.5.7b)).

(b) (canonical choice for Nyström method) In the case of a Nyström method, there is not a unique canonical prolongation P_n. We call P_n canonical, if it is a stable interpolation at the support abscissae $\{\xi_{j,n}\}$ of the quadrature method Q_n defining the *Nyström method*. The order of interpolation P_n must coincide with that of Q_n. The interpolation P_n is represented by (4a) with $\Phi_{j,n}$ being the Lagrange functions. The *canonical restriction* is uniquely defined by

(5.2.4e) $\quad \varphi_n = R_n \varphi \quad$ with $\varphi_n = (\varphi_{j,n})_{j=1,\dots,n}$ and $\varphi_{j,n} := \varphi(\xi_{j,n})$,

i.e., $R_n \varphi$ denotes the evaluation of φ at the abscissae $\{\xi_{j,n}\}$.

Exercise 5.2.2. (a) Let P_n and R_n be the canonical choice in the case of a projection method. Prove that the underlying projection is given by the product

(5.2.4f) $\quad \Pi_n = P_n R_n.$

The convergence (3c) is equivalent to the convergence of the projection.
(b) Also in the case of the Nyström method, the canonical mappings P_n and R_n have the property (4b): $R_n P_n = I$. (3c) is equivalent to the convergence of the interpolation $\Pi_n = P_n R_n$.
(c) Let P_n and R_n be the canonical mappings for the Galerkin method. Let the scalar product in $X_n = \mathbb{R}^n$ be defined by $(\varphi_n, \psi_n)_n := (A_n \varphi_n, \psi_n) = \sum_{i,j} \psi_{i,n} \alpha_{ij} \varphi_{j,n}$ with the matrix A_n from (4.5.7b). If we use $(\cdot,\cdot)_n$ in X_n and the L^2 scalar product $\langle\cdot,\cdot\rangle$ in $X = L^2(D)$, the mappings P_n and R_n are adjoint: $R_n = P_n^*$, $P_n = R_n^*$, i.e., $(R_n \varphi, \psi_n)_n = \langle \varphi, P_n \psi_n \rangle$ for all $\varphi \in X$, $\psi_n \in X_n$.

Remark 5.2.3. Let P_n and R_n be the canonical prolongations and restrictions of a projection method. The matrices $\boldsymbol{K}_n$, which have been denoted by $A_n^{-1} B_n$ in the Chapters 4.4 and 4.5 (cf. (1.4)), and the inhomogeneity $\boldsymbol{g}_n$ satisfy

(5.2.5) $\quad \boldsymbol{K}_n = R_n K P_n, \quad \boldsymbol{g}_n = R_n g.$

Proof. Using the notation $A_n^{-1} B_n$ for $\boldsymbol{K}_n$, we rewrite (5) as $B_n = A_n R_n K P_n$. Let $\boldsymbol{f}_n = \delta_j$ be the jth unit vector. $P_n \boldsymbol{f}_n = \Phi_{j,n}$ holds implying $K P_n \boldsymbol{f}_n = K \Phi_{j,n}$. The assertion follows from definition (4c,d) of R_n and definition (4.4.9c)/(4.5.7c) of B_n. The proof of the representation of $\boldsymbol{g}_n$ is left to the reader. QED

If the functions $\Phi_{j,n}$ in (4a) are Lagrange functions, the prolongation P_n describes the interpolation of the support ordinates $f_{j,n}$. In this case, the collocation method leads to $A_n = I$ (cf. Remark 4.4.3), so that, as in (4e), the restriction R_n from (4b) is the evaluation of g at the collocation points $\xi_{j,n}$.

The *kernel approximation* cannot be regarded so directly as an equation of the pattern (1.3): $\lambda \boldsymbol{f}_n = \boldsymbol{g}_n + \boldsymbol{K}_n \boldsymbol{f}_n$. Let $\boldsymbol{f}_n$ be the coefficient vector $a = a_n$ from equation (4.2.9a). The relationship between this quantity and the function f_n is described in (4.2.5). It is not linear but affine. Therefore, there is *no* linear relationship $f_n = P_n \boldsymbol{f}_n$. However, there are two remedies explained in the following remarks.

Remark 5.2.4. Consider the case of the variant (4.2.13a-d) of the *kernel approximation* described in §4.2.6. There is a linear relationship between the vector $\boldsymbol{f}_n := \hat{a}_n$ of system (4.3.14) and the desired function f_n. Accordingly, the *canonical prolongation* reads

(5.2.6) $\quad P_n \boldsymbol{f}_n := \sum_{j=1}^{n} f_{j,n} a_j, \qquad$ where $\boldsymbol{f}_n := (f_{j,n})_{j=1,\dots,n} \in \boldsymbol{X}_n = \mathbb{R}^n.$

The *canonical restriction* is defined by (4b). Instead of the representation (5), we have $\boldsymbol{K}_n = R_n K_n P_n$ and $\boldsymbol{g}_n = R_n g_n$.

Remark 5.2.5. The semidiscrete equation $\lambda f_n = g + K_n f_n$ of the kernel approximation with the representation $f_n = \frac{1}{\lambda} g + \sum \alpha_j a_j$ can be rewritten as follows. Set $\varphi_n := \sum \alpha_j a_j$. The equation $\lambda f_n = g + K_n f_n$ is equivalent to $\lambda \varphi_n = \tilde{g}_n + K_n \varphi_n$, where $\tilde{g}_n := \frac{1}{\lambda} K_n g$. Since the function φ_n results linearly from the vector $\boldsymbol{\varphi}_n := a_n = (\alpha_j)_j$, the canonical prolongation $\boldsymbol{\varphi}_n \mapsto \varphi_n = P_n \boldsymbol{\varphi}_n$ is described by (6). The *canonical restriction* is the left inverse; cf. (4b). As in Remark 4, $\mathbf{K}_n = R_n K_n P_n$ and $\tilde{\mathbf{g}}_n = R_n g_n$ hold instead of (5).

5.2.2 The Banach Space Y and the Discrete Spaces $\mathbf{Y}_n$

In Subsection 3.4, we studied the range of K and stated conditions under which $K \in L(X,Y)$, where $Y \subset X$ is continuously embedded. Furthermore, the following condition was always satisfied:

(5.2.7) $\quad Y \subset X$ continuously embedded, $\quad Y$ dense in X.

The examples from §3.4 were $X = C(D)$ and $Y = C^\lambda(D)$ with $\lambda > 0$. Analogously to the relationship of the discrete space $\mathbf{X}_n$ to the Banach space X, we shall introduce a further space $\mathbf{Y}_n$ representing a discrete version of Y. $\mathbf{X}_n$ and $\mathbf{Y}_n$ coincide in the sense of sets, but their norms differ. One choice for the norm $\|\cdot\|_{Y_n}$ is contained in

Remark 5.2.6. Assume that the restriction $R_n \in L(X, \mathbf{X}_n)$ is surjective and (7) holds. Then the discrete norm corresponding to Y can be defined by

(5.2.8) $\quad \|\mathbf{g}_n\|_{Y_n} := \inf\{\|g\|_Y : g \in Y \text{ and } \mathbf{g}_n = R_n g\}.$

Definition (8) ensures inequality (9a) with $C_Y = 1$:

(5.2.9a) $\quad \|R_n g\|_{Y_n} \leqslant C_Y \|g\|_Y \qquad$ for all $g \in Y$.

Furthermore, for any $\mathbf{g}_n \in Y_n$ and any $C_Y > 1$ there is a $g \in Y$ with

(5.2.9b) $\quad \mathbf{g}_n = R_n g$ and $\|g\|_Y \leqslant C_Y \|\mathbf{g}_n\|_{Y_n} \qquad$ for suitable $g \in Y$.

If range$(P_n) \subset Y$ and $\Pi_n := P_n R_n$ belongs to $L(Y,Y)$, then we have

(5.2.9c) $\quad \|P_n\|_{Y \leftarrow Y_n} = \|\Pi_n\|_{Y \leftarrow Y}.$

In the following, we require a condition equivalent to (9a):

(5.2.9a') $\quad \|R_n\|_{Y_n \leftarrow Y} \leqslant C_Y \qquad$ for all $n \in \mathbb{N}$.

Regarding $\mathbf{g}_n$ as a discrete function defined at the nodal points ξ_j $(1 \leqslant j \leqslant n)$ and choosing Y as the Hölder/Lipschitz space $\hat{C}^\lambda(D)$, $0 < \lambda \leqslant 1$, one can give the following explicit definition of $\|\cdot\|_{Y_n}$:

(5.2.10a) $\quad \|\mathbf{g}_n\|_{Y_n} := \max\{\|\mathbf{g}_n\|_\infty, |g_{j,n} - g_{k,n}|/|\xi_j - \xi_k|^\lambda : 1 \leqslant j,k \leqslant n, j \neq k\}.$

Similarly, in the case of $Y = C_L^1(D)$ with equidistant collocation points or support abscissae $\xi_j = a + jh$ $(0 \leqslant j \leqslant n)$, one finds the discrete norm

(5.2.10b) $$\|\mathbf{g}_n\|_{Y_n} := \max\{\|\mathbf{g}_n\|_\infty, |g_{j,n} - g_{j+1,n}|/h : 0 \leqslant j < n, \\ |-g_{j-1,n} + 2g_{j,n} - g_{j+1,n}|/h^2 : 1 \leqslant j < n\}.$$

For non-equidistant ξ_j one must use the corresponding first and second divided differences.

Since, by definition, Y is the range of K, we have the *regularity condition*

(5.2.11a) $\quad \|K\|_{Y\leftarrow X} \leqslant C_{YX}.$

Thanks to the continuous embedding $Y \subset X$, it follows that

(5.2.11b) $\quad \|K\|_{Y\leftarrow Y} \leqslant C'_{YX}$

with $C'_{YX} := C_{YX}\|I\|_{X\leftarrow Y}$. As in (11a,b), the discrete estimates for $\boldsymbol{K}_n$ read

(5.2.11c) $\quad \|\boldsymbol{K}_n\|_{Y_n\leftarrow X_n} \leqslant C_K \qquad$ for all $n\in\mathbb{N}$,

(5.2.11d) $\quad \|\boldsymbol{K}_n\|_{Y_n\leftarrow Y_n} \leqslant C'_K \qquad$ for all $n\in\mathbb{N}$.

(11c) is the *discrete regularity condition*, which becomes very important in the convergence proofs of the following section on multi-grid iterations.

In the case of *projection methods*, thanks to property (5), the discrete regularity condition (11c) follows immediately from (11a), while in the *Nyström* case, the discrete regularity can be concluded directly from the smoothness of the kernel, as formulated in

Lemma 5.2.7. **(a)** (3a), (5) [which by Remark 3 holds for projection methods], (9a), and (11a) imply the discrete regularity (11c).
(b) The kernel $k\in C(D\times D)$ is assumed to be uniformly smooth with respect to the first argument, i.e., $k(\cdot,y)\in\hat{C}^{\lambda}(D)$ and $\max\{\|k(\cdot,y)\|_{\hat{C}^{\lambda}(D)}: y\in D\}<\infty$. The quadrature used in the Nyström method must be stable (or convergent). P_n and R_n are to be chosen canonically. Let $\|\cdot\|_{X_n}=\|\cdot\|_{\infty}$. The norm $\|\cdot\|_{Y_n}$ is defined by (8) with $Y=\hat{C}^{\lambda}(D)$ or as in (10a,b) by differences up to order λ. Then the discrete regularity condition (11c) holds.

Proof. (i) From the representation (5) (i.e., $\boldsymbol{K}_n = R_n K P_n$) one concludes that $\|\boldsymbol{K}_n\|_{Y_n\leftarrow X_n} = \|R_nKP_n\|_{Y_n\leftarrow X_n} \leqslant \|R_n\|_{Y_n\leftarrow Y}\|K\|_{Y\leftarrow X}\|P_n\|_{X\leftarrow X_n} = C_Y C_{YX} C_P$.
(ii) Let $\boldsymbol{f}_n\in X_n$. The norm $\|\boldsymbol{K}_n\boldsymbol{f}_n\|_{Y_n}$ contains differences (or in the case of (8) also derivatives) of the function $Q_n(k(x,\cdot)P_n\boldsymbol{f}_n)$. Let δ be such a difference or differential operator. $\delta Q_n(k(x,\cdot)P_n\boldsymbol{f}_n) = Q_n(\delta k(x,\cdot)P_n\boldsymbol{f}_n)$ implies $\|\delta Q_n(k(x,\cdot)P_n\boldsymbol{f}_n)\|_\infty \leqslant \|Q_n\|\|\delta k(x,\cdot)P_n\boldsymbol{f}_n\|_\infty \leqslant \|Q_n\|\|\delta k(x,\cdot)\|_\infty\|P_n\boldsymbol{f}_n\|_\infty \leqslant$ $\leqslant \text{const}\,\|\boldsymbol{f}_n\|_{X_n}$. ∎

The second estimate (11d) is a consequence of (11c) as stated in

Exercise 5.2.8. The continuous embedding $Y\subset X$ corresponds to

(5.2.11e) $\quad \|\boldsymbol{g}_n\|_{X_n} \leqslant C\|\boldsymbol{g}_n\|_{Y_n} \qquad$ for all $g_n\in\boldsymbol{X}_n$ and $n\in\mathbb{N}$.

Prove the following: **(a)** The estimate (11d) results from (11c) and (11e).
(b) (3b), (7), and the norm definition (8) imply (11e) with $C:=C_R\|I\|_{X\leftarrow Y}$.
Hint: Use (9b).

If we define $\|\cdot\|_{Y_n}$ by the maxima (10a) or (10b) containing $\|\cdot\|_{X_n}$ - instead of (8) - inequality (11e) holds with $C=1$.

5.2.3 The Interpolation Error or Projection Error

Projection methods satisfy $P_nR_n = \Pi_n$ (cf. (4f)), so that $I - P_nR_n = I - \Pi_n$ describes the projection error. If R_n is the evaluation at the nodal points and P_n an interpolation, then $I - P_nR_n$ represents the interpolation error. In the following we suppose that there is a positive order β such that

(5.2.12a) $\quad \| I - P_nR_n \|_{X \leftarrow Y} \leqslant C_I h_n^{\beta} \qquad (n \in \mathbb{N},\ \beta > 0).$

Here, $h = h_n$ is the typical discretisation parameter, which often has the meaning of a step size (cf. (4.5.19c/29c)).

In the case of a Galerkin method and the canonical choice of P_n and R_n, inequality (12a) coincides with the projection error estimate (4.5.15a): $\| f - \Pi_n f \| \leqslant \text{const} \cdot h_n^{\beta}$. For example, estimate (4.5.26a) in Lemma 4.5.21 provides (12a) with $\beta = 1$, $X = C(D)$, $Y = C_L(D)$.

For a collocation method or the Nyström method, the canonical choice of P_n and R_n leads to the interpolatory projection $\Pi_n = P_nR_n$. The estimate (12a) with $X = C(D)$ and $Y = \hat{C}^{\beta}(D)$ holds if and only if the interpolation Π_n is (at least) of the order β.

In order to obtain a quantitative estimate, a condition like (12a) is unavoidable. On the other hand, the requirement (3c) follows from (12a):

Lemma 5.2.9. Assume (3a,b), (7), and the estimate (12a). Then (3c), i.e., $P_nR_nf \to f$ holds for all $f \in X$.

Proof. According to (3a,b), $P_nR_n \in L(X,X)$ is uniformly bounded and converges on the dense subset Y. Theorem 1.3.17 (Banach-Steinhaus) proves (3c). QED

A weaker statement than (12a) is

(5.2.12b) $\quad \| I - R_nP_n \|_{X_n \leftarrow Y_n} \leqslant C_I' h_n^{\beta} \qquad (n \in \mathbb{N},\ \beta > 0).$

Obviously, (12b) holds with $C_I' = 0$ if (4b): $R_nP_n = I$ is valid. If (4b) does not apply, (12b) can be derived from (12a) and further technical assumptions.

Exercise 5.2.10. Prove that (12b) follows from (3b), (9b), and (12a) with the constant $C_I' := C_R C_I C_Y$.

5.2.4 Consistency

In the following, we define consistency only quantitatively, i.e., as consistency of a certain order $\beta > 0$. As distinct from Definition 4.1.1, we now compare the *matrix* $\mathbf{K}_n$ and the *operator* K. Two different formulations may serve as definition:

(5.2.13a) $\quad \| \mathbf{K}_n R_n - R_n K \|_{X_n \leftarrow Y} \leqslant C_C h_n^{\beta} \qquad (n \in \mathbb{N},\ \beta > 0)$ or

(5.2.13b) $\quad \| P_n \mathbf{K}_n R_n - K \|_{X \leftarrow Y} \leqslant C_C h_n^{\beta} \qquad (n \in \mathbb{N},\ \beta > 0).$

The following lemma shows that both inequalities are "almost" equivalent.

Lemma 5.2.11. **(a)** Assume equation (4f), i.e., $\mathbf{K}_n = R_n K P_n$, which is characteristic for a projection method. Then, (13a) follows from (3b), $K \in L(X,X)$, and (12a), while (13b) can be concluded from (3a,b), (7), (11a), and (12a).
(b) In the general case, suppose (3a), (11b), and (12a). Then, (13b) is a consequence of (13a).
(c) Vice versa, (13a) follows from (13b), if one assumes (3b), (12a), and either (9a) and (11d) or (11c,e).

Proof. (a) (4f) shows that the difference $P_n \mathbf{K}_n R_n - K$ has the form

$$\mathbf{K}_n R_n - R_n K = -R_n K (I - P_n R_n).$$

$\| \mathbf{K}_n R_n - R_n K \|_{X_n \leftarrow Y} \leq \| R_n \|_{X_n \leftarrow X} \| K \|_{X \leftarrow X} \| I - P_n R_n \|_{X_n \leftarrow Y}$ proves (13a) with the constants $C_C := C_R \| K \|_{X \leftarrow X} C_I$. For the proof of (13b) use the representation

$$P_n \mathbf{K}_n R_n - K = -(I - P_n R_n) K P_n R_n - K (I - P_n R_n).$$

(b) Use $P_n \mathbf{K}_n R_n - K = P_n (\mathbf{K}_n R_n - R_n K) - (I - P_n R_n) K$.
(c) $\mathbf{K}_n R_n - R_n K = R_n (P_n \mathbf{K}_n R_n - K) + (I - P_n R_n) K_n R_n$ implies part (c). QED

Remark 5.2.12. **(a)** The relationship of the new consistency concept to the original Definition 4.1.1 of consistency is as follows. For the canonical choice of P_n and R_n in the projection method, the semidiscrete operator $K_n \in L(X,X)$ from Definition 4.1.1 coincides with $P_n \mathbf{K}_n R_n$. Hence, (13b) describes the convergence $K_n \varphi \to K \varphi$ for all $\varphi \in Y$. Since Y is dense in X, the convergence $K_n \varphi \to K \varphi$ for all $\varphi \in X$ - as required in Definition 4.1.1 - follows from (13b), if besides (3a,b) one requires the uniform boundedness

$$\text{(5.2.11f)} \qquad \| \mathbf{K}_n \|_{X_n \leftarrow X_n} \leq C_{XX} \qquad \text{for all } n \in \mathbb{N},$$

which follows from the inequalities (11c,e).
(b) In the Nyström case, the product $P_n \mathbf{K}_n R_n$ and the semidiscrete operator K_n differ with respect to the interpolation of the values $(K_n f)(\xi_{j,n})$ ($\xi_{j,n}$: abscissae of the quadrature Q_n). In $P_n \mathbf{K}_n R_n$, the interpolation $\Pi_n = P_n R_n$ is used, whereas in K_n the Nyström interpolation is applied. Under the assumption of Theorem 4.7.15, the consistency condition (13b) and the semidiscrete consistency (4.7.20c) of the order $\varkappa = \beta$ are equivalent.

5.2.5 Stability

The discretisation given by the matrices $(\mathbf{K}_n)_{n \in \mathbb{N}}$ is called *stable* (in the discrete sense), if there is some $n_0 \in \mathbb{N}$ such that $\lambda I - \mathbf{K}_n$ is non-singular for all $n \geq n_0$ and (14) holds:

$$\text{(5.2.14)} \qquad \| (\lambda I - \mathbf{K}_n)^{-1} \|_{X_n \leftarrow X_n} \leq C_S \qquad \text{for all } n \geq n_0.$$

A comparison with Definition 4.1.3 shows that the choice (14) of the term «stable» is better suited for practical purposes. We recall that, in contrast to the matrix $\mathbf{K}_n$, the operator $K_n \in L(X,X)$ of the kernel approximation or of the projection method does not depend on the choice of the basis.

5.2.6 Convergence

In the following, f and $\boldsymbol{f}_n$ are the respective solutions of the continuous problems (1.1), i.e., $\lambda f = g + Kf$ and of the discretisation (1.3), i.e., $\lambda \boldsymbol{f}_n = \boldsymbol{g}_n + \boldsymbol{K}_n \boldsymbol{f}_n$. The discretisation is said to be *convergent of the order* β, if

(5.2.15) $\quad \| \boldsymbol{f}_n - R_n f \|_{X_n} \leqslant C h_n^\beta \| g \|_Y \qquad$ for all $n \geqslant n_0$ and $g \in Y$.

Theorem 5.2.13 (convergence). Let λ be a regular value of K, i.e., $(\lambda I - K)^{-1} \in L(X,X)$ exists. Assume (7) and (11a): $K \in L(X,Y)$. Then, convergence (15) follows from stability (14), consistency (13a), and estimate (16):

(5.2.16) $\quad \| R_n g - \boldsymbol{g}_n \|_{X_n} \leqslant C h_n^\beta \| g \|_Y \qquad$ for all $g \in Y$.

We mention that inequality (16) holds with $C = 0$, if the representation (4f): $\boldsymbol{g}_n = R_n g$ holds (cf. Remarks 3, 4, and 5), which is true for all methods mentioned here, when the canonical choice of R_n is used.

Proof. $\tilde{\boldsymbol{f}}_n := (\lambda I - \boldsymbol{K}_n)^{-1} R_n g$ is the solution of $\lambda \tilde{\boldsymbol{f}}_n = R_n g + \boldsymbol{K}_n \tilde{\boldsymbol{f}}_n$. Since the continuous solution can be written as $f = (\lambda I - K)^{-1} g$, the difference is

$$\begin{aligned} R_n f - \tilde{\boldsymbol{f}}_n &= [R_n (\lambda I - K)^{-1} - (\lambda I - \boldsymbol{K}_n)^{-1} R_n] g = \\ &= (\lambda I - \boldsymbol{K}_n)^{-1} [(\lambda I - \boldsymbol{K}_n) R_n - R_n (\lambda I - K)] (\lambda I - K)^{-1} g = \\ &= (\lambda I - \boldsymbol{K}_n)^{-1} [R_n K - \boldsymbol{K}_n R_n] (\lambda I - K)^{-1} g. \end{aligned}$$

Theorem 3.5.1 together with (7) and (11a) yields $(\lambda I - K)^{-1} \in L(Y,Y)$. We obtain

$$\begin{aligned} &\| R_n f - \tilde{\boldsymbol{f}}_n \|_{X_n} \leqslant \\ &\leqslant \| (\lambda I - \boldsymbol{K}_n)^{-1} \|_{X_n \leftarrow X_n} \| \boldsymbol{K}_n R_n - R_n K \|_{X_n \leftarrow Y} \| (\lambda I - K)^{-1} \|_{Y \leftarrow Y} \| g \|_Y \\ &\leqslant C_S C_C \| (\lambda I - K)^{-1} \|_{Y \leftarrow Y} h_n^\beta \| g \|_Y . \end{aligned}$$

Finally, by assumption (16), the error

$$\| \tilde{\boldsymbol{f}}_n - \boldsymbol{f}_n \|_{X_n} = \| (\lambda I - \boldsymbol{K}_n)^{-1} (R_n g - \boldsymbol{g}_n) \|_{X_n} \leqslant C_S \| R_n g - \boldsymbol{g}_n \|_{X_n}$$

is bounded by const $h_n^\beta \| g \|_Y$. ∎

Corollary 5.2.14. Under the assumptions of Theorem 13 and the conditions (3a) and (12a), one also obtains the convergence in the form

(5.2.17) $\quad \| P_n \boldsymbol{f}_n - f \|_X \leqslant C h_n^\beta \| g \|_Y \qquad$ for all $n \geqslant n_0$ and $g \in Y$.

Proof. Write $P_n \boldsymbol{f}_n - f$ as $P_n(\boldsymbol{f}_n - R_n f) + (I - P_n R_n) f$. The first term is bounded by const $h_n^\beta \| g \|_Y$ because of (3a) and (15). The bound on the second is $C_I h_n^\beta \| f \|_Y$. By means of $\| f \|_Y \leqslant \| (\lambda I - K)^{-1} \|_{Y \leftarrow Y} \| g \|_Y$ (cf. Theorem 3.5.1)), one obtains the result (17). ∎

5.3 The Hierarchy of Discrete Problems

5.3.1 Levels of Discretisations

In general, the discretisation parameter n determines the dimension of the matrix systems. However, to prove estimates, it is better to write expressions in terms of a *step size* $h = h_n$ {of the basic interpolation (kernel approximation, collocation method) / of the Galerkin subspaces / of the quadrature method (Nyström method)}. Therefore, in the following, h will be the primary discretisation parameter. Halving a given step size h, one obtains a refined interpolation or a refined subspace or an improved quadrature formula for the step size $h' := h/2$. The discrete solutions $\boldsymbol{f}_n$ and $\boldsymbol{f}_{n'}$ corresponding to h and h' are easy to compare.

Example 5.3.1. Let $D = I = [0,1]$. The collocation points Ξ_n are $\xi_{j,n} := jh$ for $j = 0, 1, \ldots, n$ with $h = h_n = 1/n$. The collocation method is based on the piecewise linear interpolation. The components $\boldsymbol{f}_{j,n}$ of the solution vector $\boldsymbol{f}_n$ are approximations of $f(x)$ at $x = \xi_{j,n} := jh$. Halving h, one obtains the solution $\boldsymbol{f}_{2n}$. Each second component $\boldsymbol{f}_{2j,2n}$ is immediately comparable with $\boldsymbol{f}_{j,n}$, since both refer to the same nodal points $x = j/n$.

If the components of $\boldsymbol{f}_n$ are precisely values of a function value at the collocation points $\xi_{j,n}$, we write

$$\boldsymbol{f}_n(\xi_{j,n}) := \boldsymbol{f}_{j,n}$$

and understand $\boldsymbol{f}_n$ as a function defined on a set of discrete points.

Let n_0 be a fixed discretisation parameter and h_0 the corresponding step size. By successive halving we obtain the step sizes

$$h_\ell := h_0/2^\ell \qquad \text{for } \ell = 0, 1, 2, \ldots . \tag{5.3.1a}$$

More generally, one can consider an *arbitrary* step size hierarchy

$$h_0 > h_1 > \ldots > h_{\ell-1} > h_\ell > \ldots \qquad \text{with } \lim h_\ell = 0. \tag{5.3.1b}$$

The index ℓ is called the *level* or *level number*. The step size h_ℓ is associated with the dimension parameter n_ℓ. Hence, the discrete equation at level ℓ is

$$\lambda \boldsymbol{f}_{n_\ell} = \boldsymbol{g}_{n_\ell} + \boldsymbol{K}_{n_\ell} \boldsymbol{f}_{n_\ell}.$$

To avoid a double indexing, we write

$$\lambda \boldsymbol{f}_\ell = \boldsymbol{g}_\ell + \boldsymbol{K}_\ell \boldsymbol{f}_\ell \qquad (\ell = 0, 1, \ldots), \tag{5.3.2}$$

i.e., in the following the level number ℓ is used as primary parameter. Correspondingly, the collocation points, spaces, norms etc. are renamed:

$$\boldsymbol{X}_\ell := \boldsymbol{X}_{n_\ell}, \quad \boldsymbol{Y}_\ell := \boldsymbol{Y}_{n_\ell}, \quad \xi_{j,\ell} := \xi_{j,n_\ell}, \quad \|\cdot\|_{X_\ell} := \|\cdot\|_{X_{n_\ell}} \quad \text{etc.}$$

5.3.2 Prolongations and Restrictions

Up to now, all discretisations have been regarded as approximations of the *continuous* equation. In the following, we consider $\boldsymbol{f}_{\ell-1} \in \boldsymbol{X}_{\ell-1}$ as approximation of $\mathbf{f}_\ell \in \mathbf{X}_\ell$. To compare both grid functions, we introduce a prolongation p mapping from $X_{\ell-1}$ onto X_ℓ:

$$p: \mathbf{X}_{\ell-1} \to \mathbf{X}_\ell \qquad (\text{i.e., } p \in L(\mathbf{X}_{\ell-1}, \mathbf{X}_\ell)). \tag{5.3.3a}$$

The restriction r acts in the reverse direction:

$$r: \mathbf{X}_\ell \to \mathbf{X}_{\ell-1} \qquad (\text{i.e., } r \in L(\mathbf{X}_\ell, \mathbf{X}_{\ell-1})). \tag{5.3.3b}$$

Both mappings will later be used in the multi-grid method. Therefore, p and r must be relatively easy to compute.

In principle, p and r should be indexed by ℓ, since $p = p_\ell$ and $r = r_\ell$ are different mappings for each ℓ. However, we omit these indices since the level number will be uniquely determined by the context.

The prolongations and restrictions must be uniformly bounded for all $\ell \geq 1$:

$$\|p\|_{X_\ell \leftarrow X_{\ell-1}} \leq C_p \qquad \text{for all } \ell = 1, 2, \ldots, \tag{5.3.4a}$$

$$\|r\|_{X_{\ell-1} \leftarrow X_\ell} \leq C_r \qquad \text{for all } \ell = 1, 2, \ldots. \tag{5.3.4b}$$

There is also a *canonical choice* for p and r just as there are for the prolongations and restrictions P_ℓ and R_ℓ from §5.2.1.

Definition 5.3.2. Choose $P_\ell \in L(\mathbf{X}_\ell, X)$ and $R_\ell \in L(X, \mathbf{X}_\ell)$ canonically. Then the canonical prolongation p and canonical restriction r are

$$p := R_\ell P_{\ell-1}, \qquad r := R_{\ell-1} P_\ell. \tag{5.3.5a}$$

Corollary 5.3.3. **(a)** Let the mappings P_ℓ, R_ℓ, p, and r be chosen canonically. If the inclusion $\text{range}(P_{\ell-1}) \subset \text{range}(P_\ell)$ holds, the canonical p and r are uniquely characterised by the implicit condition (5b):

$$P_\ell p = P_{\ell-1}, \qquad R_{\ell-1} = r R_\ell. \tag{5.3.5b}$$

(b) For canonical P_ℓ and R_ℓ, definition (5a) is a consequence of (5b).
(c) The characterisation $P_\ell p = P_{\ell-1}$ expresses the fact that the coefficient vectors $\boldsymbol{\varphi}_{\ell-1} \in \mathbf{X}_{\ell-1}$ and $\boldsymbol{\varphi}_\ell = p\boldsymbol{\varphi}_{\ell-1} \in \mathbf{X}_\ell$ represent the *same* function $P_\ell \boldsymbol{\varphi}_\ell = P_{\ell-1} \boldsymbol{\varphi}_{\ell-1}$ in X. In this sense, p represents the *identity*.

Proof. (i) To prove part (b), multiply the equations in (5b) by R_ℓ from the left or by P_ℓ from the right. By (2.4b), the assertion (5a) follows.
(ii) The canonical restriction R_ℓ is *surjective* because of (2.4b).
(iii) Insertion of (5a) into (5b) yields $\Pi_\ell P_{\ell-1} = P_{\ell-1}$, where thanks to the surjectivity of R_ℓ, $\Pi_\ell = P_\ell R_\ell$ is a projection *onto* $\text{range}(P_\ell)$. Hence, $\Pi_\ell P_{\ell-1} = P_{\ell-1}$ follows from the assumption $\text{range}(P_{\ell-1}) \subset \text{range}(P_\ell)$. Thus, the prolongation p from (5a) satisfies the characterisation (5b). Part (i) has already proved the uniqueness. The characterisation of r by (5b) is shown analogously. QED

Example 5.3.4 (collocation method, Nyström method). **(a)** In the case of the collocation or Nyström method, $\Pi_\ell = P_\ell R_\ell$ describes an interpolation at the abscissae $\xi_{j,\ell} \in \Xi_\ell$. The inclusion $\text{range}(P_{\ell-1}) \subset \text{range}(P_\ell)$ needed in Corollary 3a holds, e.g., if the subintervals $I_{k,\ell}$ of the *piecewise* (constant, linear, etc.) interpolation Π_ℓ are contained in subintervals $I_{k',\ell'-1}$ of the interpolation $\Pi_{\ell-1}$. If the abscissae Ξ_ℓ are the boundary points of these subintervals as in the case of the piecewise linear interpolation, $\mathrm{range}(P_{\ell-1}) \subset \mathrm{range}(P_\ell)$ is equivalent to $\Xi_{\ell-1} \subset \Xi_\ell$.

(b) According to (2.4a,c), the explicit representation of $p\,\boldsymbol{\varphi}_{\ell-1}$ for $\boldsymbol{\varphi}_{\ell-1} \in \mathbf{X}_{\ell-1}$ is:

$$p\,\boldsymbol{\varphi}_{\ell-1} := A_\ell^{-1}\big(\varphi(\xi_{j,\ell})\big)_{j=1,\dots,n_\ell} \text{ with the interpolant } \varphi = P_{\ell-1}\boldsymbol{\varphi}_{\ell-1},$$

where the matrix A_ℓ from (4.4.9b) equals $A_\ell = I$ for the standard choice $\Phi_{j,\ell}$=Lagrange function in (2.4a).

(c) For the *piecewise linear interpolation* with the nodal points Ξ_ℓ or $\Xi_{\ell-1}$,

$$(p\,\boldsymbol{\varphi}_{\ell-1})(\xi) := \boldsymbol{\varphi}_{\ell-1}(\xi) \qquad (\xi \in \Xi_{\ell-1} \subset \Xi_\ell) \tag{5.3.6a}$$

holds with the notation $\boldsymbol{f}_\ell(\xi_{j,\ell}) := \boldsymbol{f}_{j,\ell}$, whereas for points $\xi_{j,\ell} \in \Xi_\ell$ not belonging to $\Xi_{\ell-1}$, one must interpolate piecewise linearly in the subinterval $(\xi_{k-1,\ell-1}, \xi_{k,\ell-1})$ of length $h_{k,\ell-1}$:

$$(p\,\boldsymbol{\varphi}_{\ell-1})(\xi_{j,\ell}) := [(\xi_{k,\ell-1} - \xi_{j,\ell})\,\boldsymbol{\varphi}_{k-1,\ell-1} + (\xi_{j,\ell} - \xi_{k-1,\ell-1})\,\boldsymbol{\varphi}_{k,\ell-1}]/h_{k,\ell-1}.$$

In the *equidistant case* $\xi_{j,\ell} = a + jh_\ell$, $\xi_{j,\ell-1} = a + jh_{\ell-1}$ with halved step size $h_\ell = h_{\ell-1}/2$, the computation of p reduces to

$$\begin{aligned} (p\,\boldsymbol{\varphi}_{\ell-1})(\xi) &:= \boldsymbol{\varphi}_{\ell-1}(\xi) && \text{for } \xi = a + 2\nu h_\ell,\ \nu = 0,1,\dots,n_{\ell-1},\\ (p\,\boldsymbol{\varphi}_{\ell-1})(\xi) &:= \tfrac{1}{2}[\boldsymbol{\varphi}_{\ell-1}(\xi - h_\ell) + \boldsymbol{\varphi}_{\ell-1}(\xi + h_\ell)] && \text{for } \xi = a + (2\nu+1)h_\ell. \end{aligned} \tag{5.3.6b}$$

In the case $\Xi_{\ell-1} \subset \Xi_\ell$, the canonical restriction r reads

$$(r\,\boldsymbol{\varphi}_\ell)(\xi_{k,\ell-1}) = \boldsymbol{\varphi}_\ell(\xi_{k,\ell-1}) \qquad \text{for all } \xi_{k,\ell-1} \in \Xi_{\ell-1}. \tag{5.3.6c}$$

Example 5.3.5 (Galerkin method). **(a)** For the Galerkin method, $\Pi_\ell = P_\ell R_\ell$ is the orthogonal projection onto the subspace $X_\ell \subset X = L^2(D)$. The inclusion $\mathrm{range}(P_{\ell-1}) \subset \mathrm{range}(P_\ell)$ needed in Corollary 3a holds if and only if the Galerkin spaces form a hierarchy $X_{\ell-1} \subset X_\ell$ (cf. Exercise 4.3.1a).

(b) Definition (5a) of the canonical prolongation p becomes

$$p\,\boldsymbol{\varphi}_{\ell-1} := A_\ell^{-1}\big(\langle\varphi, \Phi_{j,\ell}\rangle\big)_{j=1,\dots,n_\ell} \qquad (\boldsymbol{\varphi}_{\ell-1} \in \mathbf{X}_{\ell-1})$$

with $\varphi = P_{\ell-1}\boldsymbol{\varphi}_{\ell-1} = \sum_k \boldsymbol{\varphi}_{k,\ell-1}\Phi_{k,\ell-1}$ and A_ℓ from (4.5.7b). Correspondingly,

$$r\,\boldsymbol{\varphi}_\ell := A_{\ell-1}^{-1}\big(\langle\varphi, \Phi_{j,\ell-1}\rangle\big)_{j=1,\dots,n_\ell} \qquad \text{with } \varphi = P_\ell\boldsymbol{\varphi}_\ell = \textstyle\sum_k \boldsymbol{\varphi}_{k,\ell}\Phi_{k,\ell}$$

is the representation of the canonical restriction r. Define the rectangular matrix $\tilde{p}$ and its transpose $\tilde{r}$ by

$$\tilde{p} := (\tilde{p}_{jk})_{j=1,\dots,n_\ell,\ k=1,\dots,n_{\ell-1}} \qquad \text{with } \tilde{p}_{jk} := \langle\Phi_{k,\ell-1}, \Phi_{j,\ell}\rangle, \tag{5.3.6d}$$

$$\tilde{r} := (\tilde{r}_{jk})_{j=1,\dots,n_{\ell-1},\ k=1,\dots,n_\ell} \qquad \text{with } \tilde{r}_{jk} := \langle\Phi_{k,\ell}, \Phi_{j,\ell-1}\rangle. \tag{5.3.6e}$$

Then the canonical prolongations p and r have the representations

$$p = A_\ell^{-1}\,\tilde{p}, \quad r = A_{\ell-1}^{-1}\,\tilde{r}. \tag{5.3.6f}$$

(c) In the case of $X_{\ell-1} \subset X_\ell$, the coefficients $p := (p_{jk})$ are uniquely determined by

(5.3.6g) $\Phi_{j,\ell} = \sum_k p_{jk}\Phi_{k,\ell-1}$ for all j.

(d) The canonical restriction r depends on p via

(5.3.6h) $r = A_{\ell-1}^{-1}\,\hat{r}\,A_\ell$ with $\hat{r} := p^T$ (transposed matrix of p).

(e) Define the Galerkin method by the subspace X_ℓ of the *piecewise linear* functions over an equidistant partition with step size $h_\ell = h_{\ell-1}/2$ and the Lagrange function basis. Then one finds the representation (6b) for p, where $(p\boldsymbol{\varphi}_{\ell-1})(jh_\ell)$ denotes the jth component of the vector $p\boldsymbol{\varphi}_{\ell-1}$. The restriction r results from (6h) with $\hat{r}$ given by

(5.3.6i) $(\hat{r}\boldsymbol{\varphi}_\ell)(jh_{\ell-1}) = \frac{1}{2}\boldsymbol{\varphi}_{j-1,\ell} + \boldsymbol{\varphi}_{j,\ell} + \frac{1}{2}\boldsymbol{\varphi}_{j+1,\ell}$.

p as well as $\hat{r}$ is characterised by the coefficients $[\frac{1}{2}\ 1\ \frac{1}{2}]$.

Proof. Parts (a) and (b) follow immediately from the definitions. The representation (6g) is obtained from (5b) (applied to an unit vector). Solving the first equation in (6f) for $\tilde{p}$, using $\tilde{r} = \tilde{p}^T$ and $A_\ell = A_\ell^T$, and inserting $\tilde{r} = p^T A_\ell$ into the second equation of (6f), one gains (6h). QED

The uniform boundedness (4a,b) of p and r can be concluded from (2.3a,b) or from the explicit representation of p and r:

Exercise 5.3.6. **(a)** Prove that (2.3a,b) implies (4a,b) with $C_r = C_p = C_R C_P$.
(b) Let $\mathbf{X}_\ell$ be equipped with the maximum norm $\|\cdot\|_{X_\ell} = \|\cdot\|_\infty$. The prolongations mentioned in Examples 4c and 5b satisfy (4a) with $C_p = 1$.
(c) If $\mathrm{range}(P_{\ell-1}) \subset \mathrm{range}(P_\ell)$ holds for the canonical prolongations $P_{\ell-1}$, P_ℓ, the canonically chosen r and p satisfy the equation

(5.3.7a) $rp = I$.

(d) Consider the case of the Galerkin method. Show using (7a) that

(5.3.7b) $A_{\ell-1} = p^T A_\ell p$.

$I - pr$ describes the *interpolation error* of p. We require

(5.3.8) $\|I - pr\|_{X_\ell \leftarrow Y_\ell} \leqslant C_I h_\ell^\beta$ for all $\ell = 1, 2, \ldots$

by analogy with (2.12a) with a positive order β.

Exercise 5.3.7. Prove the estimate (8) with $\beta = 2$ for p and r from (6b,c) (cf. Example 4c), where Y_ℓ is the analogue of $Y = C_L^1(I)$. The norm $\|\cdot\|_{Y_\ell}$ can be defined by (2.8) or (2.10b).

The relationship between the respective estimates (8) of $I - pr$ and (2.12a) of $I - P_\ell R_\ell$ is discussed in

Exercise 5.3.8. Suppose (2.3b), (2.9c), $\|\Pi_\ell\|_{Y\leftarrow Y} \leqslant$ const for $\Pi_\ell = P_\ell R_\ell$, $h_{\ell-1}/h_\ell \leqslant$ const, and (2.12a). Prove (8) for the canonically chosen P_ℓ, R_ℓ, r, p.
Hint: Use $I - pr = R_\ell(I - \Pi_\ell \Pi_{\ell-1})P_\ell$.

5.3.3 Relative Consistency

Since we regard $\boldsymbol{K}_{\ell-1}$ as an approximation of $\boldsymbol{K}_\ell$, we define the *relative consistency* (of the order β) by

(5.3.9) $$\| r\boldsymbol{K}_\ell - \boldsymbol{K}_{\ell-1} r \|_{X_{\ell-1}\leftarrow Y_\ell} \leqslant C_c h_\ell^\beta \qquad \text{for all } \ell = 1,2,\dots .$$

Lemma 5.3.9. (5b) and (2.5), i.e., $\boldsymbol{K}_\ell = R_\ell K P_\ell$, $\boldsymbol{g}_\ell = R_\ell g$ imply the representation

(5.3.10) $$\boldsymbol{K}_{\ell-1} = r\boldsymbol{K}_\ell p, \quad \boldsymbol{g}_{\ell-1} = r\boldsymbol{g}_\ell \qquad \text{for } \ell \geqslant 1.$$

If $\boldsymbol{K}_\ell = A_\ell^{-1} B_\ell$ (cf. (1.4)) and (6h) (i.e., $r = A_{\ell-1}^{-1}\,\hat{r}\,A_\ell$) hold, the characterisation (10) can be expressed as

(5.3.10') $$\boldsymbol{K}_{\ell-1} = A_{\ell-1}^{-1} B_{\ell-1} \text{ with } B_{\ell-1} = \hat{r}\, B_\ell\, p, \qquad b_{\ell-1} = \hat{r}\, b_\ell \text{ with } b_k := A_k \boldsymbol{g}_k.$$

Proof. $r\boldsymbol{K}_\ell p = r(R_\ell K P_\ell)p = (rR_\ell)K(P_\ell p) = R_{\ell-1} K P_{\ell-1} = \boldsymbol{K}_{\ell-1}$. QED

Remark 5.3.10. If the matrix $\boldsymbol{K}_\ell$ is given for a *fixed* level $\ell = \ell_{\max}$, equation (10) (or (10')) provides a constructive description for generating the further matrices $\boldsymbol{K}_{\ell-1}, \boldsymbol{K}_{\ell-2}, \dots, \boldsymbol{K}_0$ at the lower levels.

Under the assumption (10), the relative consistency (9) reduces to the error estimate (8) of $I - pr$, as shown in

Lemma 5.3.11. From (10) one concludes the representation

(5.3.11) $$r\boldsymbol{K}_\ell - \boldsymbol{K}_{\ell-1} r = r\boldsymbol{K}_\ell (I - pr).$$

The relative consistency (9) of order β is ensured under the assumptions (10), $\| r \|_{X_{\ell-1}\leftarrow X_\ell} \leqslant \text{const}$, and (2.11f), i.e., $\| \boldsymbol{K}_\ell \|_{X_\ell \leftarrow X_\ell} \leqslant C_{XX}$.

Since *projection methods* lead to discretisations with the property (2.5) (i.e., $\boldsymbol{K}_\ell = R_\ell K P_\ell$; cf. Remark 2.3), Lemmata 9 and 11 are applicable and guarantee the relative consistency of the same order β as for the projection error $I - pr$, provided that p and r are canonically chosen and $\operatorname{range}(P_{\ell-1}) \subset \operatorname{range}(P_\ell)$.

For the *Nyström method*, equation (2.5) (i.e., $\boldsymbol{K}_\ell = R_\ell K P_\ell$) is not valid.

Remark 5.3.12. Let the Nyström method be based on the quadrature method Q_ℓ. $\beta > 0$ denotes the order of Q_ℓ with respect to the spaces $X = C(D)$ and $Y = \hat{C}^\beta(D)$. The set Ξ_ℓ of the quadrature abscissae should satisfy the inclusion $\Xi_{\ell-1} \subset \Xi_\ell$. The kernel k is assumed to be smooth uniformly with respect to y: $\| k(x,\cdot) \|_{\hat{C}^\beta(D)} \leqslant C_\beta$ for all $x \in D$. Define $\|\cdot\|_{X_\ell}$ by the maximum norm and $\|\cdot\|_{Y_\ell}$ by (2.8). Finally, let $h_{\ell-1}/h_\ell \leqslant \text{const}$. Then relative consistency (9) of the order β holds. If the derivatives $k_\nu := \partial^\nu k / \partial x^\nu$ of order $\nu < \beta$ and their Hölder/Lipschitz differences satisfy the condition $\| k_\nu(x,\cdot) \|_{\hat{C}^\beta(D)} \leqslant C_\beta$ [corresponding to the required condition on k], then (9) holds in the strengthened form $\| r\boldsymbol{K}_\ell - \boldsymbol{K}_{\ell-1} r \|_{Y_{\ell-1}\leftarrow Y_\ell} \leqslant C_c h_\ell^\beta$.

Proof. Choose $x \in D$ and $\boldsymbol{f}_\ell \in Y_\ell$ arbitrarily. Take $f \in Y$ with $\boldsymbol{f}_\ell = R_\ell f$, where R_ℓ is given by (2.4e). Note that the value $Q_\ell(k(x,\cdot)f)$ only depends on the function values $\boldsymbol{f}_\ell = R_\ell f$ of f at the points $\xi_{j,\ell} \in \Xi_\ell$. The quadrature error of $Q_\ell(k(x,\cdot)f)$

can be estimated by $\| k(x,\cdot)f\|_{\hat{C}^{\beta}(D)} \leqslant C\,C_\beta\, h_\ell^\beta \|f\|_{\hat{C}^\beta(D)} = C\,C_\beta h_\ell^\beta \|f\|_Y$. For $x=\xi_{j,\ell}\in\Xi_\ell$, $Q_\ell(k(x,\cdot)f)$ coincides with the jth component of $\mathbf{K}_\ell \mathbf{f}_\ell$. For $x=\xi_{k,\ell-1}\in\Xi_{\ell-1}\subset\Xi_\ell$, the kth component of $r\mathbf{K}_\ell\mathbf{f}_\ell$ is $Q_\ell(k(\xi_{k,\ell-1},\cdot)f)$ and hence

$$|(r\mathbf{K}_\ell\mathbf{f}_\ell)_k - \int_D k(\xi_{k,\ell-1},y)f(y)dy| \leqslant C\,C_\beta h_\ell^\beta\|f\|_Y.$$

Correspondingly, $Q_{\ell-1}(k(\xi_{k,\ell-1},\cdot)f)$ represents the kth component of $\mathbf{K}_{\ell-1}r\mathbf{f}_\ell$, since the values $r\mathbf{f}_\ell$ coincide with $\{f(\xi_{i,\ell-1}):\ \xi_{i,\ell-1}\in\Xi_{\ell-1}\}$. The quadrature error of $Q_{\ell-1}(k(\xi_{k,\ell-1},\cdot)f)$ satisfies the bound

$$|(\mathbf{K}_{\ell-1}r\mathbf{f}_\ell)_k - \int_D k(\xi_{k,\ell-1},y)f(y)dy| \leqslant C\,C_\beta h_{\ell-1}^\beta\|f\|_Y.$$

Using these two bounds, one obtains $|(r\mathbf{K}_\ell\mathbf{f}_\ell-\mathbf{K}_{\ell-1}r\mathbf{f}_\ell)_k| \leqslant C\,C_\beta(h_\ell^\beta+h_{\ell-1}^\beta)\|f\|_Y$. Taking the infimum over all f with $f_\ell=R_\ell f$ (restriction of f to Ξ_ℓ), one attains $|(r\mathbf{K}_\ell\mathbf{f}_\ell-\mathbf{K}_{\ell-1}r\mathbf{f}_\ell)_k| \leqslant C\,C_\beta(h_\ell^\beta+h_{\ell-1}^\beta)\|\mathbf{f}_\ell\|_{Y_\ell}$ (cf. 2.8)). Forming the maximum over $k\in\{1,\dots,n_{\ell-1}\}$ and using the inequality $h_{\ell-1}/h_\ell \leqslant \leqslant$const, one obtains $\|r\mathbf{K}_\ell\mathbf{f}_\ell-\mathbf{K}_{\ell-1}r\mathbf{f}_\ell\|_{X_\ell} \leqslant C\,C_\beta(1+\text{const}^\beta)h_\ell^\beta\|\mathbf{f}_\ell\|_{Y_\ell}$ and hence (9). QED

The inclusion $\Xi_{\ell-1}\subset\Xi_\ell$ required in Remark 12 can be omitted. In this case, $r\mathbf{f}_\ell$ describes an interpolation of the values $\mathbf{f}_\ell$ at the points $\xi\in\Xi_{\ell-1}$. In the canonical case, a sufficiently accurate interpolation is required (cf. Definition 2.1b and (5a)), so that the additional interpolation error is also of order h_ℓ^β and enables the estimate (9).

5.3.4 Convergence

In the following theorem, we compare the *discrete* solutions $\mathbf{f}_\ell$ and $\mathbf{f}_{\ell-1}$.

Theorem 5.3.13. **(a)** The discrete problem is assumed to be stable and relatively consistent, i.e., (2.14) and (9). Then the solutions of (2) satisfy

(5.3.12) $$\|r\mathbf{f}_\ell-\mathbf{f}_{\ell-1}\|_{X_{\ell-1}} \leqslant C_S\|\mathbf{g}_{\ell-1}-r\,\mathbf{g}_\ell\|_{X_{\ell-1}} + C_S C_c h_\ell^\beta\|\mathbf{f}_\ell\|_{Y_\ell}.$$

(b) Under the assumptions (12), (4), and (8), the estimate (13) holds:

(5.3.13) $$\|\mathbf{f}_\ell-p\mathbf{f}_{\ell-1}\|_{X_\ell} \leqslant (C_I+C_pC_SC_c)h_\ell^\beta\|\mathbf{f}_\ell\|_{Y_\ell} + C_pC_S\|\mathbf{g}_{\ell-1}-r\mathbf{g}_\ell\|_{X_{\ell-1}}.$$

(c) Suppose (2.11c), i.e., $\|\mathbf{K}_\ell\|_{Y_\ell\leftarrow X_\ell}\leqslant C_K$, stability (2.14), and the embedding (2.11e), i.e., $\|\cdot\|_{X_\ell}\leqslant C\|\cdot\|_{Y_\ell}$. Then $\|\mathbf{f}_\ell\|_{Y_\ell}$ in (12-13) is bounded by

(5.3.14) $$\|\mathbf{f}_\ell\|_{Y_\ell} \leqslant |\lambda|^{-1}(\|\mathbf{g}_\ell\|_{Y_\ell}+C_K\|\mathbf{f}_\ell\|_{X_\ell}) \leqslant |\lambda|^{-1}(1+C\,C_SC_K)\|\mathbf{g}_\ell\|_{Y_\ell}.$$

(d) Let $\mathbf{g}_{\ell-1}$ and $\mathbf{g}_\ell$ satisfy condition (2.5) (i.e., $\mathbf{g}_\ell=R_\ell g$) and assume (5b) (i.e., $R_{\ell-1}=rR_\ell$). Then the term $\|\mathbf{g}_{\ell-1}-r\,\mathbf{g}_\ell\|_{X_{\ell-1}}$ vanishes because of $\mathbf{g}_{\ell-1}=r\,\mathbf{g}_\ell$.

Proof. (i) By $\mathbf{g}_\ell=(\lambda I-\mathbf{K}_\ell)\mathbf{f}_\ell$, the following identity holds and leads to (12):

$$\begin{aligned}
\mathbf{f}_{\ell-1}-r\mathbf{f}_\ell &= (\lambda I-\mathbf{K}_{\ell-1})^{-1}\mathbf{g}_{\ell-1}-r\mathbf{f}_\ell = \\
&= (\lambda I-\mathbf{K}_{\ell-1})^{-1}(\mathbf{g}_{\ell-1}-r\mathbf{g}_\ell)+(\lambda I-\mathbf{K}_{\ell-1})^{-1}r\mathbf{g}_\ell-r\mathbf{f}_\ell = \\
&= (\lambda I-\mathbf{K}_{\ell-1})^{-1}(\mathbf{g}_{\ell-1}-r\mathbf{g}_\ell)+(\lambda I-\mathbf{K}_{\ell-1})^{-1}[r(\lambda I-\mathbf{K}_\ell)-(\lambda I-\mathbf{K}_{\ell-1})r]\mathbf{f}_\ell = \\
&= (\lambda I-\mathbf{K}_{\ell-1})^{-1}(\mathbf{g}_{\ell-1}-r\mathbf{g}_\ell)+(\lambda I-\mathbf{K}_{\ell-1})^{-1}[\mathbf{K}_{\ell-1}r-r\mathbf{K}_\ell]\mathbf{f}_\ell.
\end{aligned}$$

(ii) Use $\mathbf{f}_\ell-p\mathbf{f}_{\ell-1}=(I-pr)\mathbf{f}_\ell+p(r\mathbf{f}_\ell-\mathbf{f}_{\ell-1})$ for inequality (13).
(iii) The proof of (14) is analogous to Theorem 3.5.1. QED

5.4 Two-Grid Iteration

5.4.1 The Two-Grid Algorithm

As discussed in §5.1.3, the Picard iteration

(5.4.1) $$f_\ell^0 \mapsto \tilde{f}_\ell := \tfrac{1}{\lambda}(g_\ell + K_\ell f_\ell^0)$$

may diverge. In any case, if it converges it does so more slowly than the multi-grid iteration which we analyse now. Let $\tilde{f}_\ell$ be the result of the Picard iteration given by (1). We compute the error of $\tilde{f}_\ell$ in the form of the defect

(5.4.2) $$\tilde{d}_\ell := \lambda\tilde{f}_\ell - g_\ell - K_\ell\tilde{f}_\ell .$$

Obviously, holds $\tilde{d}_\ell = 0$ if and only if $\tilde{f}_\ell$ coincides with the exact solution f_ℓ. The relationship between the error and the defect is the subject of

Remark 5.4.1. The *defect* of $\tilde{f}_\ell \in X_\ell$ equals

(5.4.3a) $$\tilde{d}_\ell := \lambda\tilde{f}_\ell - g_\ell - K_\ell\tilde{f}_\ell .$$

The solution δ_ℓ of the equation

(5.4.3b) $$\lambda\delta_\ell = \tilde{d}_\ell + K_\ell\delta_\ell$$

represents the error of $\tilde{f}_\ell$:

(5.4.3c) $$\delta_\ell = \tilde{f}_\ell - f_\ell \qquad \text{(where } f_\ell := (\lambda I - K_\ell)^{-1} g_\ell \text{ is the exact solution).}$$

Proof. Subtraction of the equation $\lambda f_\ell = g_\ell + K_\ell f_\ell$ from $\lambda\tilde{f}_\ell = g_\ell + \tilde{d}_\ell + K_\ell\tilde{f}_\ell$ yields $\lambda(\tilde{f}_\ell - f_\ell) = \tilde{d}_\ell + K_\ell(\tilde{f}_\ell - f_\ell)$. QED

Following the last remark, one obtains an *exact correction* of the approximation $\tilde{f}_\ell$ by solving equation (3b) with $\tilde{d}_\ell$ from (2): $f_\ell = \tilde{f}_\ell - \delta_\ell$. However, since (3b) is a system of the same kind as the original problem $\lambda f_\ell = g_\ell + K_\ell f_\ell$, the use of δ_ℓ does not simplify the solution of the problem. Nevertheless, studying the structure of δ_ℓ more carefully, we shall recognise equation (3b) as easier to treat approximately than $\lambda f_\ell = g_\ell + K_\ell f_\ell$. By (1.7c) and (3c),

(5.4.4) $$\delta_\ell = \tilde{f}_\ell - f_\ell = \tfrac{1}{\lambda} K_\ell (f_\ell^0 - f_\ell)$$

lies in the range of K_ℓ. Thanks to property (2.11c) (i.e., $\|K_\ell\|_{Y_\ell \leftarrow X_\ell} \le C_K$), one can suppose that δ_ℓ is a smooth (though only discretely defined) function. By Theorem 3.13, δ_ℓ can be approximated up to an error $O(h_\ell^\beta)$ by the solution $\delta_{\ell-1}$ of the problem

(5.4.5) $$\lambda\delta_{\ell-1} = d_{\ell-1} + K_{\ell-1}\delta_{\ell-1} \qquad \text{with } d_{\ell-1} := r\tilde{d}_\ell .$$

Instead of the desired exact solution $f_\ell = \tilde{f}_\ell - \delta_\ell$, one computes the approximation $f_\ell^1 = \tilde{f}_\ell - p\,\delta_{\ell-1}$ defining the next iterate of the two-grid method.

The sketched method consists of two steps: the Picard step (the so-called *smoothing step*, since the error is smoothed) and the *coarse-grid correction* $\tilde{f}_\ell \mapsto \tilde{f}_\ell - p\,\delta_{\ell-1}$.

(5.4.6) **two-grid algorithm** for solving $\lambda \boldsymbol{f}_\ell = \boldsymbol{g}_\ell + \boldsymbol{K}_\ell \boldsymbol{f}_\ell$, $\ell \geq 1$

start: $\boldsymbol{f}_\ell^0 \in \boldsymbol{X}_\ell$ arbitrary.

iteration $\boldsymbol{f}_\ell^i \mapsto \boldsymbol{f}_\ell^{i+1}$:

(5.4.6a) $\tilde{\boldsymbol{f}}_\ell := \frac{1}{\lambda}(\boldsymbol{g}_\ell + \boldsymbol{K}_\ell \boldsymbol{f}_\ell^i)$ (smoothing)

(5.4.6b) $\boldsymbol{f}_\ell^{i+1} := \tilde{\boldsymbol{f}}_\ell - p(\lambda I - \boldsymbol{K}_{\ell-1})^{-1} r(\lambda \tilde{\boldsymbol{f}}_\ell - \boldsymbol{g}_\ell - \boldsymbol{K}_\ell \tilde{\boldsymbol{f}}_\ell)$ (coarse-grid correction)

An ALGOL-like description of algorithm (6) is given in (7). The second parameter $\boldsymbol{f}$ represents $\boldsymbol{f}_\ell^i$ as *input* and $\boldsymbol{f}_\ell^{i+1}$ as *output*. $\boldsymbol{g}$ replaces $\boldsymbol{g}_\ell$. The call of the following procedure *ZGM* executes one step of the two-grid iteration. *ZGM* abbreviates the German word «Zweigittermethode».

(5.4.7) **two-grid procedure** *ZGM* for solving $\lambda \boldsymbol{f}_\ell = \boldsymbol{g}_\ell + \boldsymbol{K}_\ell \boldsymbol{f}_\ell$, $\ell \geq 0$

```
(5.4.7a)  procedure ZGM(ℓ, f, g); integer ℓ; array f, g;
(5.4.7b)  if ℓ=0 then f:=(λI-K_0)^(-1) g else
          begin array d, δ;
(5.4.7c)      f := (1/λ)(g+K_ℓ*f);          (Picard iteration)
(5.4.7d)      d := r*(λf-g-K_ℓ*f);          (computing the defect)
(5.4.7e)      δ := (λI-K_(ℓ-1))^(-1) d;     (coarse-grid solution)
(5.4.7f)      f := f-p*δ                    (coarse-grid correction)
          end;
```

In contrast to (6), the case $\ell = 0$ is admitted in (7). In this case, there is no predecessor in the hierarchy, so that a coarse-grid correction is impossible. Therefore, the system at level 0 is solved exactly (cf. (7b)). Since h_0 represents the coarsest step size, it corresponds to the smallest dimension n_0 of the system and the exact solution by direct methods yields no difficulty.

A comparable method was first presented in 1960 by Brakhage [1]. However, the algorithm was not extended to the multi-grid algorithm.

5.4.2 Convergence Analysis

Since the mapping $\boldsymbol{f}_\ell^i \mapsto \boldsymbol{f}_\ell^{i+1}$ of the two-grid algorithm (6) is affine, it has a representation

(5.4.8) $$\boldsymbol{f}_\ell^{i+1} = M_\ell^{ZGM} \boldsymbol{f}_\ell^i + c_\ell, \qquad (M_\ell^{ZGM}\text{: two-grid iteration matrix})$$

corresponding to (1.6a) of Theorem 1.2. Instead of the sharp convergence condition $\rho(M_\ell^{ZGM}) < 1$, we shall try to prove the sufficient condition $\|M_\ell^{ZGM}\|_{X_\ell \leftarrow X_\ell} < 1$. Insertion of (6a) into (6b) proves

Lemma 5.4.2. The two-grid iteration matrix M_ℓ^{ZGM} is of the form

(5.4.9) $$M_\ell^{ZGM} = \frac{1}{\lambda}[I - p(\lambda I - K_{\ell-1})^{-1} r(\lambda I - K_\ell)] K_\ell \qquad \text{for all } \ell \geq 1.$$

Theorem 5.4.3. Assume stability (2.14), discrete regularity (2.11c), uniform boundedness (3.4a) of the prolongation p, relative consistency (3.9), and the interpolation error estimate (3.8) with some $\beta>0$. Then

(5.4.10a) $\|M_\ell^{ZGM}\|_{X_\ell \leftarrow X_\ell} \leqslant C_{ZGM} h_\ell^\beta$ for all $\ell \geqslant 1$

The constant C_{ZGM} is given by

(5.4.10b) $C_{ZGM} := \frac{1}{|\lambda|}(C_I + C_p C_S C_c) C_K$.

Proof. We split the iteration matrix M_ℓ^{ZGM} from (9) as follows:

$$M_\ell^{ZGM} = \tfrac{1}{\lambda}\{(I-pr)+p(\lambda I-K_{\ell-1})^{-1}[(\lambda I-K_{\ell-1})r-r(\lambda I-K_\ell)]\}K_\ell =$$
$$= \tfrac{1}{\lambda}\{(I-pr)+p(\lambda I-K_{\ell-1})^{-1}[rK_\ell-K_{\ell-1}r]\}K_\ell,$$

so that

$$\|M_\ell^{ZGM}\|_{X_\ell\leftarrow X_\ell} \leqslant \tfrac{1}{|\lambda|}\{\|I-pr\|_{X_\ell\leftarrow Y_\ell} +$$
$$+\|p\|_{X_\ell\leftarrow X_\ell}\|(I-K_{\ell-1})^{-1}\|_{X_\ell\leftarrow X_\ell}\|rK_\ell-K_{\ell-1}r\|_{X_\ell\leftarrow Y_\ell}\}\|K_\ell\|_{Y_\ell\leftarrow X_\ell}.$$

Insertion of the estimates from the assumptions yields (10a,b). ∎

The estimate (10a) can be interpreted as follows.

Remark 5.4.4. **(a)** There is an ℓ_0 such that the two-grid iteration converges at all levels $\ell\geqslant\ell_0$.
(b) Cancelling levels $\ell=0,1,\dots,\ell_0-1$ from the hierarchy of the discretisations ensures convergence at all levels.
(c) The smaller the step size, i.e., the higher the dimension of the system, the faster the convergence of the two-grid method.
(d) For the standard case $h_{\ell-1}=2h_\ell$, the convergence estimate reads $\|M_\ell^{ZGM}\|_{X_\ell\leftarrow X_\ell}\leqslant \text{const } 2^{-\ell\beta}$.

Remark 5.4.5. One can apply the two-grid iteration with two step sizes h_ℓ and $h_{\ell-1}$, whose ratio is much smaller than $\frac{1}{2}$. In this case, the assumptions (3.8-9) are unrealistic. The factor h_ℓ^β must be replaced by $h_{\ell-1}^\beta$ and therefore, the convergence speed is given by $\|M_\ell^{ZGM}\|_{X_\ell\leftarrow X_\ell}\leqslant C\,h_{\ell-1}^\beta$. Only if the ratio $h_{\ell-1}/h_\ell$ is uniformly bounded, can $h_{\ell-1}^\beta$ be replaced by $\text{const}\cdot h_\ell^\beta$.

5.4.3 Amount of Computational Work

In the following we suppose that the matrix $\boldsymbol{K}_\ell$ is given explicitly. The case of $\boldsymbol{K}_\ell=A_\ell^{-1}B_\ell$ will be discussed in the following Subsection §5.4.4. The expensive parts of the two-grid method are:

(5.4.11a) matrix-vector multiplication $\boldsymbol{f}_\ell \mapsto \boldsymbol{K}_\ell\boldsymbol{f}_\ell$: $2n_\ell^2-n_\ell$ operations,

(5.4.11b) coarse-grid solution $\boldsymbol{g}_{\ell-1}\mapsto(\lambda I-\boldsymbol{K}_{\ell-1})^{-1}\boldsymbol{g}_{\ell-1}$:
$\leqslant\frac{2}{3}n_{\ell-1}^3+O(n_{\ell-1})$ operations.

By «operations» we mean all arithmetical operations +,−,*,/. Here, n_ℓ is defined as the dimension of the system of equations. The vector addition and subtraction is negligible because their work count of $O(n_\ell)$ is much smaller than (11a). The same holds for the evaluation of the prolongation p and the restriction r (or $\hat{r}$, cf. §5.4.4), if p or r are not pathological cases. The restriction r from (3.6c) requires no computational work at all. On the other hand, when evaluating the restricted defect $r(\tilde{\boldsymbol{f}}_\ell-\boldsymbol{g}_\ell-\boldsymbol{K}_\ell\tilde{\boldsymbol{f}}_\ell)$, one can save computational work by the fact that only those components in the vector $\tilde{\boldsymbol{f}}_\ell-\boldsymbol{g}_\ell-\boldsymbol{K}_\ell\tilde{\boldsymbol{f}}_\ell$ corresponding to abscissae from $\Xi_{\ell-1}$ must be evaluated. Hence,

(5.4.11c) $\quad \tilde{\boldsymbol{f}}_\ell \mapsto r\boldsymbol{K}_\ell\tilde{\boldsymbol{f}}_\ell$ in the case of (3.6c) requires $\leqslant 2n_\ell n_{\ell-1}$ operations.

Remark 5.4.6. **(a)** In each iteration, the two-grid algorithm requires

(5.4.12a) $\quad \frac{2}{3}n_{\ell-1}^3+4n_\ell^2+O(n_{\ell-1})$ operations.

(b) For the standard case $n_\ell=2n_{\ell-1}+O(1)$, the number from (12a) becomes

(5.4.12b) $\quad \frac{1}{12}n_\ell^3+4n_\ell^2+O(n_\ell^2)$.

(c) In the case of (11c), the numbers reduce to

(5.4.12a') $\quad \frac{2}{3}n_{\ell-1}^3+2n_\ell^2+2n_\ell n_{\ell-1}+O(n_{\ell-1})$,

(5.4.12b') $\quad \frac{1}{12}n_\ell^3+3n_\ell^2+O(n_\ell^2)$.

(d) Starting the iteration with $\boldsymbol{f}_\ell^0:=0$, we can omit the first matrix multiplication in (6a) or (7c). Hence, the computation of $\boldsymbol{f}_\ell^1$ costs

(5.4.12c) $\quad \frac{2}{3}n_{\ell-1}^3+2n_\ell^2+O(n_{\ell-1}) \qquad \{=\frac{1}{12}n_\ell^3+2n_\ell^2+O(n_\ell)$ for $n_\ell=2n_{\ell-1}\}$,

(5.4.12c') $\quad \frac{2}{3}n_{\ell-1}^3+2n_\ell n_{\ell-1}+O(n_{\ell-1}) \qquad \{=\frac{1}{12}n_\ell^3+n_\ell^2+O(n_\ell)$ for $n_\ell=2n_{\ell-1}\}$

operations. There (12c') relates to case (11c).

To estimate the number of iteration steps, one needs a suitable *stopping criterion*. Following (2.15), the discrete solutions $\boldsymbol{f}_\ell$ and $\boldsymbol{f}_{\ell-1}$ have a discretisation error of the orders $O(h_\ell^\beta)$ and $O(h_{\ell-1}^\beta)$, respectively. Apart from a few exceptions (e.g., extrapolation techniques, cf. §4.8.3), it makes no sense to compute approximations $\tilde{\boldsymbol{f}}_\ell$ with an iteration error $\|\boldsymbol{f}_\ell-\tilde{\boldsymbol{f}}_\ell\|_{X_\ell}$ substantially smaller than the discretisation error $O(h_\ell^\beta)$.

Remark 5.4.7. **(a)** Starting with $\boldsymbol{f}_\ell^0:=0$, as recommended in Remark 6d, one needs exactly *one* two-grid iteration leading to $\boldsymbol{f}_\ell^1$ with an iteration error $\|\boldsymbol{f}_\ell-\boldsymbol{f}_\ell^1\|_{X_n}=O(h_\ell^\beta)$ being of the same order as the discretisation error.
(b) The coincidence of the iteration and discretisation error in part (a) only concerns the asymptotic behaviour with respect to $h_\ell\to 0$; the constants in $O(h_\ell^\beta)$ may differ. To make the iteration error significantly smaller than the discretisation error, one should perform *two* iteration steps. Then the iteration error reads $\|\boldsymbol{f}_\ell-\boldsymbol{f}_\ell^2\|_{X_n}=O(h_\ell^{2\beta})$ and becomes smaller than the discretisation error for sufficiently small h_ℓ.
(c) The work count in case (a) is given by (12c) or (12c'), respectively. In case (b) the number of operations equals (12d), or (12d') if (11c) applies:

(5.4.12d) $\frac{4}{3}n_{\ell-1}^3+6n_\ell^2+O(n_{\ell-1})$ $\{=\frac{1}{6}n_\ell^3+6n_\ell^2+O(n_\ell)$ for $n_\ell=2n_{\ell-1}\}$

(5.4.12d') $\frac{4}{3}n_{\ell-1}^3+2n_\ell^2+4n_\ell n_{\ell-1}+O(n_{\ell-1})$ $\{=\frac{1}{6}n_\ell^3+4n_\ell^2+O(n_\ell)\}$.

Since the direct solution of the system at the level ℓ requires $\frac{2}{3}n_\ell^3$ operations (cf. Remark 1.1), the iterative solution leads to a *reduction of the amount of work* by 87.5% or 75%, respectively, if $n_\ell=2n_{\ell-1}$ (cf. Remark 7a/b).

The use of the two-grid iteration with $n_\ell=2n_{\ell-1}$ was proposed in the paper of Brakhage [1] mentioned above. Atkinson [1] recommends the choice $n_{\ell-1}=O(n_\ell^{2/3})$, since then $n_{\ell-1}^3=O(n_\ell^2)$. In total, the amount of work of the two-grid iteration reduces to $O(n_\ell^2)$ instead of $O(n_\ell^3)$. However, this variant has two disadvantages. First, the prolongation between grids of the step sizes h_ℓ and $h_{\ell-1}=O(h_\ell^{2/3})$ is more involved from the programming point of view than if $h_\ell=2h_{\ell-1}$. Second, the convergence speed is worse. According to Remark 5, the speed equals $O(h_{\ell-1}^\beta)=O(h_\ell^{2\beta/3})$. Nevertheless, *two* iteration steps - according to Remark 7b - would lead to an iteration error $O(h_\ell^{4\beta/3})$, which is also asymptotically smaller than the discretisation error. In §5.5.2, we shall see that the multi-grid method can also be performed in $O(n_\ell^2)$ operations maintaining the advantages of the regular refinement $h_\ell=2h_{\ell-1}$ and the fast convergence.

5.4.4 Variant for $A_\ell \neq I$

In the two-grid algorithm (6) or in the procedure (7) we used the matrix $\boldsymbol{K}_\ell$, although this matrix is not always *directly* available. By definition, the quantities $\boldsymbol{g}_\ell$ and $\boldsymbol{K}_\ell$ are given by

(5.4.13a) $\boldsymbol{g}_\ell := A_\ell^{-1}b_\ell, \quad \boldsymbol{K}_\ell := A_\ell^{-1}B_\ell$

(cf. (1.4)). In the collocation case, $A_\ell=I$ is always attainable by choosing a Lagrange basis (cf. Remark 4.4.3). On the other hand, it is rather difficult to avoid $A_\ell \neq I$ for the Galerkin method. In this case, one should *on no account* compute the matrix $\boldsymbol{K}_\ell=A_\ell^{-1}B_\ell$ explicitly. Instead, the solution methods should be modified so that only few systems of the form $A_\ell x_\ell=y_\ell$ have to be solved for x_ℓ. The notation $x_\ell=A_\ell^{-1}y_\ell$ must always be interpreted in this sense. The explicit computation of A_ℓ^{-1} is completely unnecessary! The discrete integral equation is written again in the form (1.2b) with $a_\ell=\boldsymbol{f}_\ell$:

(5.4.13b) $(\lambda A_\ell-B_\ell)\boldsymbol{f}_\ell = b_\ell.$

In the following, we use b_ℓ instead of $\boldsymbol{g}_\ell=A_\ell^{-1}b_\ell$.

For the case of the Galerkin method mentioned above, the canonical restriction reads

(5.4.13c) $r = A_{\ell-1}^{-1}\hat{r}A_\ell$

(cf. (3.6h)), where $\hat{r}:=p^T$ (transposed matrix of p) in contrast to r being a *simply executable* mapping (cf. (3.6i)). In the two-grid algorithm (7), we replace $\boldsymbol{d}$ by the transformed quantity $\hat{\boldsymbol{d}}:=A_{\ell-1}\boldsymbol{d}$ (renamed in (14) again by $\boldsymbol{d}$) and use

$$\boldsymbol{\delta} := (\lambda I-\boldsymbol{K}_{\ell-1})^{-1}\boldsymbol{d} = (\lambda A_{\ell-1}-B_{\ell-1})^{-1}A_{\ell-1}\boldsymbol{d} = (\lambda A_{\ell-1}-B_{\ell-1})^{-1}\hat{\boldsymbol{d}}.$$

Then we obtain the following *equivalent* formulation of the two-grid method (7), which requires no explicit knowledge of $\boldsymbol{K}_\ell$ or $\boldsymbol{K}_{\ell-1}$, but uses the simply executable restriction $\hat{r}=p^{\mathsf{T}}$.

(5.4.14)	**two-grid procedure** *ZGM'* for solving $(\lambda A_\ell - B_\ell)\boldsymbol{f}_\ell = b_\ell$
(5.4.14a)	procedure $ZGM'(\ell,\boldsymbol{f},\boldsymbol{b})$; integer ℓ; array $\boldsymbol{f}$, $\boldsymbol{b}$;
(5.4.14b)	if $\ell=0$ then $\boldsymbol{f} := (\lambda A_0 - B_0)^{-1}\boldsymbol{b}$ else
	begin array $\boldsymbol{d},\boldsymbol{\delta}$;
(5.4.14c)	$\boldsymbol{f} := \frac{1}{\lambda}A_\ell^{-1}(\boldsymbol{b}+B_\ell*\boldsymbol{f})$; (Picard iteration)
(5.4.14d)	$\boldsymbol{d} := \hat{r}*(\lambda A_\ell\boldsymbol{f}-\boldsymbol{b}-B_\ell*\boldsymbol{f})$; (computing the defect)
(5.4.14e)	$\boldsymbol{\delta} := (\lambda A_{\ell-1}-B_{\ell-1})^{-1}\boldsymbol{d}$; (coarse-grid solution)
(5.4.14f)	$\boldsymbol{f} := \boldsymbol{f}-p*\boldsymbol{\delta}$ (coarse-grid correction)
	end;

5.4.5 Numerical Examples

As example we choose the integral equation (4.2.19a) with $\lambda=1$:

(5.4.15) $$\lambda f(x) = g(x) + \int_0^1 \cos(\pi x y)\, f(y)\, dy \qquad (g \text{ from } (4.2.21b)).$$

We apply the *Nyström method* with the composite trapezoidal rule using step sizes

(5.4.16) $$h_0=1,\ h_1=\tfrac{1}{2},\ \ldots,\ h_\ell=2^{-\ell}.$$

The results of this discretisation are already presented in §4.7.7. Since the quadrature method is of second order, one can choose an interpolation of second order, e.g., use piecewise linear interpolation as prolongation P_ℓ (cf. Definition 2.1b), while the trivial restriction to the abscissae $\{0,h_\ell,2h_\ell,\ldots,1-h_\ell,1\}$ defines the restriction R_ℓ. Therefore, the *canonical prolongation* p is the piecewise linear interpolation described in (3.6b). The *canonical restriction* r is the trivial mapping from (3.6c).

Remark 5.4.8. Choose the kernel k, the discretisation, and p and r as above. Then the assumptions of Theorem 3 are satisfied with the order $\beta=2$.

The following Table 1 contains the iteration error $\delta_\ell^i := \|\boldsymbol{f}_\ell^i-\boldsymbol{f}_\ell\|_\infty$ of the iterate $\boldsymbol{f}_\ell^i$ $(i\geq 1)$ computed by the two-grid algorithm (7), where the starting iterate is $\boldsymbol{f}_\ell^0=0$. $\boldsymbol{f}_\ell$ is the exact discrete solution of the Nyström method for the step size h_ℓ from (16). Note that the *iteration error* δ_ℓ^i is to be distinguished from the *discretisation error* $\|\boldsymbol{f}_\ell-R_\ell f\|_\infty$ reported in Table 4.7.1.

To test the convergence speed predicted by Theorem 3, Table 1 also contains the *ratios* $\varepsilon_\ell^i := \delta_\ell^i/\delta_\ell^{i-1}$. Asymptotically, ε_ℓ^i converges to the convergence rate ε_ℓ of the two-grid method. According to Theorem 3, the ratios ε_ℓ^i

should behave like $O(h_\ell^\beta)=O(h_\ell^2)$. The assumption $\varepsilon_\ell=Ch_\ell^\beta+o(h_\ell^\beta)$ implies that the ratio $\eta_\ell:=\varepsilon_{\ell-1}/\varepsilon_\ell$ satisfies

$$\eta_\ell := [Ch_{\ell-1}^\beta+o(h_{\ell-1}^\beta)]/[Ch_\ell^\beta+o(h_\ell^\beta)] = h_{\ell-1}^\beta/h_\ell^\beta+o(h_\ell^\beta) = 2^\beta+o(h_\ell^\beta).$$

i	$h_1=1/2$		$h_2=1/4$		$h_3=1/8$	
0	2.21	$1.95_{10}{-1}$	2.04	$4.33_{10}{-2}$	2.37	$1.20_{10}{-2}$
1	$4.33_{10}{-1}$	$9.87_{10}{-2}$	$8.94_{10}{-2}$	$3.65_{10}{-2}$	$2.84_{10}{-2}$	$9.42_{10}{-3}$
2	$4.27_{10}{-2}$	$1.72_{10}{-1}$	$3.26_{10}{-3}$	$1.98_{10}{-2}$	$2.67_{10}{-6}$	$8.08_{10}{-3}$
3	$7.35_{10}{-3}$	$1.34_{10}{-1}$	$6.44_{10}{-5}$	$5.16_{10}{-2}$	$2.16_{10}{-6}$	$1.10_{10}{-2}$
4	$9.81_{10}{-4}$	$1.38_{10}{-1}$	$3.33_{10}{-6}$	$1.86_{10}{-2}$	$2.38_{10}{-8}$	$8.39_{10}{-3}$
5	$1.36_{10}{-4}$		$6.18_{10}{-8}$		$2.00_{10}{-10}$	

i	$h_4=1/16$		$h_5=1/32$		$h_6=1/64$	
0	2.46	$3.09_{10}{-3}$	2.47	$7.81_{10}{-4}$	2.48	$1.95_{10}{-4}$
1	$7.59_{10}{-3}$	$2.40_{10}{-3}$	$1.93_{10}{-3}$	$6.02_{10}{-4}$	$4.84_{10}{-4}$	$1.51_{10}{-4}$
2	$1.82_{10}{-5}$	$2.19_{10}{-3}$	$1.16_{10}{-6}$	$5.60_{10}{-4}$	$7.30_{10}{-8}$	
3	$3.99_{10}{-8}$	$2.71_{10}{-3}$	$6.51_{10}{-10}$		*	
4	$1.08_{10}{-10}$		*		*	

Table 5.4.1. Iteration error δ_ℓ^i of the two-grid iteration (7) for $\lambda=1$

level ℓ	h_ℓ	averaged convergence rate $\tilde{\varepsilon}_\ell$
1	1/2	$1.5_{10}{-1}$
2	1/4	$3.1_{10}{-2}$
3	1/8	$9.6_{10}{-3}$
4	1/16	$2.5_{10}{-3}$
5	1/32	$6.3_{10}{-4}$
6	1/64	$1.6_{10}{-4}$
7	1/128	$4.0_{10}{-5}$

Table 5.4.2. Convergence speeds ($\lambda=1$)

level ℓ	h_ℓ	averaged convergence rate for $\lambda=0.1$	$\lambda=0.01$
1	1/2	0.72	**9.3**
2	1/4	$4.5_{10}{-1}$	**5.0**
3	1/8	$1.2_{10}{-1}$	**1.3**
4	1/16	$3.3_{10}{-2}$	$4.1_{10}{-1}$
5	1/32	$8.3_{10}{-3}$	$1.2_{10}{-1}$
6	1/64	$3.6_{10}{-3}$	$3.2_{10}{-2}$
7	1/128	$5.6_{10}{-4}$	$2.0_{10}{-2}$

Table 5.4.3. Two-grid convergence speed for $\lambda=0.1$, 0.01

Table 2 contains the averaged $\tilde{\varepsilon}_\ell$ obtained as a geometric mean, e.g., the mean of ε_ℓ^i $(1\le i\le 5)$ is $\tilde{\varepsilon}_\ell=(\varepsilon_\ell^5/\varepsilon_\ell^0)^{1/5}$. As can easily confirmed, the ratio $\tilde{\eta}_\ell:=\tilde{\varepsilon}_{\ell-1}/\tilde{\varepsilon}_\ell$ coincides very accurately with the factor $4=2^2$, supporting the order $\beta=2$ stated in Theorem 3.

For the value of $\lambda=1$ used here, one obtains convergence of the two-grid method at all levels (cf. Remark 4a).

To explain the convergence behaviour of the two-grid method more precisely for this particular example, we mention that the Picard method used as «smoother» converges but that its convergence speed is worse than 0.5. The averaged rates are 0.69 ($h_1=\frac{1}{2}$), 0.72 ($h_2=1/4$), 0.51 ($h_4=1/16$), 0.60 ($h_6=1/64$). From Theorem 1.4 one concludes, in particular, that the Picard method *diverges* for $|\lambda| \leq 0.5$. This means that, for such λ, the smoothing step (7c) may amplify the errors. Indeed, in Table 3 there is a decrease in the two-grid convergence rate for decreasing $|\lambda|$. For $\lambda=0.1$, one observes still convergence at all levels. However, for $\lambda=0.01$ and $1 \leq \ell \leq 3$ the values exceed 1 and, hence, indicate *divergence* of the two-grid method. Nonetheless, the values behave like $O(h_\ell^2)$ in accordance with Theorem 3 and Remark 8. For $\lambda=0.01$ the critical level number ℓ_0 from Remark 4a is $\ell_0 = 4$.

5.5 Multi-Grid Iteration

5.5.1 Algorithm (Basic Version)

Although the two-grid iteration reduces the amount of computational work drastically compared with the exact solution (cf. §5.4.3), the solution of the coarse-grid equation in (4.7e) still takes the major part of the work. The problem to be solved in (4.7e) reads

(5.5.1) $$(\lambda I - \mathbf{K}_{\ell-1})\boldsymbol{\delta}_{\ell-1} = \mathbf{d}_{\ell-1} \qquad \text{or} \qquad \lambda\boldsymbol{\delta}_{\ell-1} = \mathbf{d}_{\ell-1} + \mathbf{K}_{\ell-1}\boldsymbol{\delta}_{\ell-1},$$

respectively. Obviously, equation (1) has the same form as the original equation $\lambda \boldsymbol{f}_\ell = \boldsymbol{g}_\ell + \mathbf{K}_\ell \boldsymbol{f}_\ell$, which is to be solved by the two-grid method. Hence, the natural strategy is to solve problem (1) only approximately by the two-grid method at the levels $\ell-1$ and $\ell-2$. Then it becomes necessary to solve an auxiliary equation of the form $(\lambda I - \mathbf{K}_{\ell-2})\boldsymbol{\delta}_{\ell-2} = \mathbf{d}_{\ell-2}$ at the level $\ell-2$. Again, the two-grid algorithm can be applied to the levels $\ell-2$ and $\ell-3$, etc. The resulting algorithm - the multi-grid iteration - uses all discretisation levels $0, 1, \ldots, \ell-1, \ell$. For the exact definition one formulates the method recursively.

(5.5.2) **multi-grid algorithm** for solving $\lambda \boldsymbol{f}_\ell = \boldsymbol{g}_\ell + \mathbf{K}_\ell \boldsymbol{f}_\ell$, $\ell \geq 0$.

(5.5.2a) $\underline{\ell=1}$: the multi-grid algorithm MGM_1 at the level 1 is identical to the two-grid algorithm (4.6).

$\underline{\ell>1}$: the multi-grid algorithm MGM_ℓ at the level ℓ reads:

start: $\boldsymbol{f}_\ell^0 \in \mathbf{X}_\ell$ arbitrary; *iteration* $\boldsymbol{f}_\ell^i \mapsto \boldsymbol{f}_\ell^{i+1}$ defined by (2b-f):

(5.5.2b) $\tilde{\boldsymbol{f}}_\ell := \frac{1}{\lambda}(\boldsymbol{g}_\ell + \mathbf{K}_\ell \boldsymbol{f}_\ell^i)$ (smoothing)

(5.5.2c) $\mathbf{d}_\ell := \lambda \tilde{\boldsymbol{f}}_\ell - \boldsymbol{g}_\ell - \mathbf{K}_\ell \tilde{\boldsymbol{f}}_\ell$ (computing the defect)

(5.5.2d) $\mathbf{d}_{\ell-1} := r\,\mathbf{d}_\ell$ (restriction of the defect)

(5.5.2e) solution of the coarse-grid equation $(\lambda I - \mathbf{K}_{\ell-1})\boldsymbol{\delta}_{\ell-1} = \mathbf{d}_{\ell-1}$ by two iterations of the multi-grid algorithm $MGM_{\ell-1}$ starting with $\boldsymbol{\delta}_{\ell-1}^0 := 0$ and resulting in the approximation $\boldsymbol{\delta}_{\ell-1}^2$ of $\boldsymbol{\delta}_{\ell-1}$.

(5.5.2f) $\boldsymbol{f}_\ell^{i+1} := \tilde{\boldsymbol{f}}_\ell - p\,\boldsymbol{\delta}_{\ell-1}^2$ (coarse-grid correction)

Remark 5.5.1. Another possibility of the recursive definition of the multi-grid algorithms MGM_ℓ ($\ell \geq 0$) is as follows: Define MGM_0 at the level $\ell=0$ as the *exact solution* of the equation $\lambda \boldsymbol{f}_0 = \boldsymbol{g}_0 + \boldsymbol{K}_0 \boldsymbol{f}_0$, while the recursion (2b-f) defines MGM_ℓ for $\ell \geq 1$. Since for $\ell = 1$, (2e) produces the exact solution $\boldsymbol{\delta}_{\ell-1}^2 = \boldsymbol{\delta}_{\ell-1}$, (2b-f) for $\ell = 1$ coincides with the two-grid method (cf. (2a)). Hence, both definitions are equivalent.

The convergence analysis will explain why exactly *two* iteration steps $MGM_{\ell-1}$ are executed in the step (2e) (cf. Corollary 8).

The following ALGOL-like description of the multi-grid algorithm (2) is very similar to that of the two-grid method in (4.7). The exact solution (4.7e) is replaced by ($3e_1$) and ($3e_2$) corresponding to step (2e). The resulting program is *recursive*. One call of $MGM(\ell,\cdot,\cdot)$ generates two calls of $MGM(\ell-1,\cdot,\cdot)$, these involve further calls of $MGM(\ell-2,\cdot,\cdot)$, etc., until level $\ell = 0$ is reached (cf. (3b)). The recursive structure can easily be resolved if wanted (cf. Hackbusch [1,§4.1]).

(5.5.3)	**multi-grid procedure** *MGM* for solving $\lambda \boldsymbol{f}_\ell = \boldsymbol{g}_\ell + \boldsymbol{K}_\ell \boldsymbol{f}_\ell$, $\ell \geq 0$	
(5.5.3a)	procedure $MGM(\ell, \boldsymbol{f}, \boldsymbol{g})$; integer ℓ; array $\boldsymbol{f}, \boldsymbol{g}$;	
(5.5.3b)	if $\ell = 0$ then $\boldsymbol{f} := (\lambda I - \boldsymbol{K}_0)^{-1} \boldsymbol{g}$ else	
	begin array $\boldsymbol{d}, \boldsymbol{\delta}$; integer i;	
(5.5.3c)	$\boldsymbol{f} := \frac{1}{\lambda}(\boldsymbol{g} + \boldsymbol{K}_\ell * \boldsymbol{f})$;	(Picard iteration)
(5.5.3d)	$\boldsymbol{d} := r * (\lambda \boldsymbol{f} - \boldsymbol{g} - \boldsymbol{K}_\ell * \boldsymbol{f})$;	(computing the defect)
($5.5.3e_1$)	$\boldsymbol{\delta} := 0$;	(set starting iterate)
($5.5.3e_2$)	for $i=1,2$ do $MGM(\ell-1, \boldsymbol{\delta}, \boldsymbol{d})$;	(two multi-grid calls)
(5.5.3f)	$\boldsymbol{f} := \boldsymbol{f} - p * \boldsymbol{\delta}$	(coarse-grid correction)
	end;	

As in (4.7), the second parameter $\boldsymbol{f}$ represents $\boldsymbol{f}_\ell^i$ as *input* and $\boldsymbol{f}_\ell^{i+1}$ as *output*. $\boldsymbol{g}$ abbreviates the inhomogeneity $\boldsymbol{g}_\ell$. One call of *MGM* executes *one* multi-grid iteration step.

Remark 5.5.2. The result remains unchanged if for $\ell=1$ the *two* multi-grid calls $MGM(0,\cdot,\cdot)$ required in ($3e_2$) are replaced by a *single* one, since in both cases the exact coarse-grid solution $\boldsymbol{\delta}$ is produced.

The multi-grid method for integral equations was first presented in 1978 by the author (cf. Hackbusch [5]; however, *two*-grid variants are discussed in §5.4.3). Independently, the algorithm was used in 1979 by Hemker and Schippers (cf. Hemker – Schippers [1]). The method corresponds very closely to the multi-grid algorithm developed for elliptic boundary value problems (Hackbusch [1]). However, note that the convergence properties for the elliptic boundary value problems are different.

There are several modifications of the algorithm. It can be generalised to nonlinear integral equations (cf. Hackbusch [1,§16.7]) and it can be applied to discretisations of equations $f = g + Kf$ of the second kind, even if K is not explicitly given as integral operator. The essential requirement is the smoothing effect, e.g., $K \in L(C^0(D), \hat{C}^\beta(D))$ or $K \in L(L^2(D), H^\beta(D))$.

5.5.2 Amount of Computational Work

Since, by its recursive structure, one multi-grid step generates an avalanche of further calls, it is not trivial to see that the total work is still proportional to n_ℓ^2. The dominant operation is the matrix-vector multiplication $\boldsymbol{f}_\ell \mapsto \boldsymbol{K}_\ell \boldsymbol{f}_\ell$. If the dimension of the problems is $n_\ell \pm O(1)$, then

(5.5.4a) $\quad \boldsymbol{f}_\ell \mapsto \boldsymbol{K}_\ell \boldsymbol{f}_\ell$ requires $2n_\ell^2 + O(n_\ell)$ arithmetical operations.

Concerning the prolongation and restriction we make the assumption

(5.5.4b) $\quad v \mapsto pv$ and $w \mapsto rw$ require $O(n_\ell)$ operations.

If r is the trivial restriction to the $n_{\ell-1}$ components (cf. (3.6c)), (4c) below holds as in (4.11c):

(5.5.4c) $\quad \boldsymbol{f}_\ell \mapsto r\boldsymbol{K}_\ell \boldsymbol{f}_\ell$ requires $2\,n_\ell n_{\ell-1} + O(n_\ell)$ operations.

The dimension at the level ℓ is assumed to be $n_\ell \pm O(1)$. The numbers n_ℓ decrease linearly by a factor $C_N < 1$:

(5.5.4d) $\quad n_{\ell-1} \leqslant C_N n_\ell$.

Remark 5.5.3. **(a)** The standard case uses a halving of the step sizes: $h_\ell = h_{\ell-1}/2$. For (one-dimensional) integral equations on an interval or a curve, inequality (4d) is satisfied with

(5.5.4e) $\quad C_N = \frac{1}{2}$.

(b) In the case of a d-dimensional integration domain, the sequence $h_\ell = h_{\ell-1}/2$ leads to condition (4d) with

(5.5.4f) $\quad C_N = 2^{-d}$.

(c) From (4c) and (4d) one concludes that the amount of work for $\boldsymbol{f}_\ell \mapsto r\boldsymbol{K}_\ell \boldsymbol{f}_\ell$ equals $2C_N n_\ell^2 + O(n_\ell)$ operations.

Theorem 5.5.4. Under the assumptions (4a-d) and

(5.5.4g) $\quad C_N < 1/\sqrt{2}$ $\qquad$ (follows from (4e) or (4f)),

one multi-grid iteration $\boldsymbol{f}_\ell^i \mapsto \boldsymbol{f}_\ell^{i+1}$ at the level ℓ, i.e., one call of $MGM(\ell,\cdot,\cdot)$ requires

(5.5.5a) $\quad C_1 n_\ell^2 + O(n_\ell)$ operations with $C_1 = 2\,\dfrac{1 + C_N - C_N^2}{1 - 2C_N^2}$.

In the particular case $\boldsymbol{f}_\ell^0 = 0$, one multi-grid step $\boldsymbol{f}_\ell^0 \mapsto \boldsymbol{f}_\ell^1$ costs

(5.5.5b) $\quad C_0 n_\ell^2 + O(n_\ell)$ operations with $C_0 = 2C_N \dfrac{1 + C_N}{1 - 2C_N^2}$.

If (4c) is not valid, the numbers increase: $C_0 = 2(1 + C_N^2)/(1 - 2C_N^2)$ or $C_1 = 2(2 - C_N^2)/(1 - 2C_N^2)$, respectively. In the standard case (4e), i.e., $C_N = \frac{1}{2}$, the numbers from (5a,b) are given by

(5.5.5c) $\quad C_1 = 5, \quad C_0 = 3$.

Proof. We want to determine constants C_0, C_1, and C such that one multi-grid iteration with general f_ℓ^i [with $f_\ell^0=0$] requires not more than $C_1 n_\ell^2 + C n_\ell$ [$C_0 n_\ell^2 + C n_\ell$] operations. By a suitable choice of C, this estimate is true for the level $\ell=0$. The amount of work of one call $MGM(\ell,\cdot,\cdot)$ for $\ell \geqslant 1$ consists of $2n_\ell^2 + O(n_\ell)$ operations for the Picard iteration (3c), $2C_N n_\ell^2 + O(n_\ell)$ operations for the computation of the defect (3d) (cf. Remark 3c), and $O(n_\ell)$ operations for the coarse-grid correction (3f). Since the first of the two procedure calls in ($3e_2$) has zero as second parameter, by the induction hypothesis the amount of work for ($3e_2$) is

$$(C_0+C_1)n_{\ell-1}^2 + C n_{\ell-1} \leqslant (C_0+C_1)C_N^2 n_\ell^2 + C\,C_N n_\ell .$$

Hence, in the general case, the number of the operations is bounded by

$$2n_\ell^2 + 2C_N n_\ell^2 + (C_0+C_1)C_N^2 n_\ell^2 + C\,C_N n_\ell + O(n_\ell) \leqslant$$
$$\leqslant [2 + 2C_N + (C_0+C_1)C_N^2]\,n_\ell^2 + (C' + C\,C_N)\,n_\ell .$$

Here, $C' n_\ell$ is defined as a bound of the $O(n_\ell)$ terms. In order to make (5a,b) valid also for the level ℓ, the inequalities

(5.5.5d) $\qquad 2 + 2C_N + (C_0+C_1)C_N^2 \leqslant C_1, \qquad\qquad C \leqslant C' + C\,C_N$

must be satisfied. The corresponding considerations for the particular case $f_\ell^0=0$ lead to the conditions

(5.5.5e) $\qquad 2C_N + (C_0+C_1)C_N^2 \leqslant C_0, \qquad\qquad C \leqslant C' + C\,C_N .$

One checks that C_0 from (5b) and C_1 from (5a) as well as $C \geqslant C'/(1-C_N)$ are well-defined for $C_N < 1/\sqrt{2}$ and that they satisfy (5d,e). ∎

The value $C_1 = 5$ in (5c) expresses the fact that one multi-grid iteration requires the same amount of work as $2\frac{1}{2}$ Picard iterations at the same level. As we shall see later (in §5.6.1), one multi-grid iteration per level is sufficient so that the total work also proves to be very low.

5.5.3 Convergence

By analogy with (4.8), we write one step $f_\ell^i \mapsto f_\ell^{i+1}$ of the multi-grid algorithm (3) in the form

(5.5.6) $\qquad f_\ell^{i+1} = M_\ell^{MGM} f_\ell^i + c_\ell \qquad$ (M_ℓ^{MGM}: multi-grid iteration matrix).

The multi-grid algorithm is defined *recursively* and so also the iteration matrix M_ℓ^{MGM}. The following Lemma shows that M_ℓ^{MGM} can be regarded as a perturbation of the two-grid iteration matrix M_ℓ^{ZGM}.

Lemma 5.5.5. The multi-grid iteration (3) is represented by (6), where M_ℓ^{MGM} is recursively defined by (7a) for $\ell=1$ and (7b) for $\ell>1$:

(5.5.7a) $\qquad M_1^{MGM} = M_1^{ZGM},$

(5.5.7b) $\qquad M_\ell^{MGM} = M_\ell^{ZGM} + \frac{1}{\lambda}\,p\,(M_{\ell-1}^{MGM})^2(\lambda I - \mathbf{K}_{\ell-1})^{-1} r\,(\lambda I - \mathbf{K}_\ell)\mathbf{K}_\ell .$

An alternative representation of (7b) is

$$(5.5.7b')\qquad M_\ell^{MGM} = M_\ell^{ZGM} + \tfrac{1}{\lambda} p\,(M_{\ell-1}^{MGM})^2 [\, r - (\lambda I - \boldsymbol{K}_{\ell-1})^{-1}(r\boldsymbol{K}_\ell - \boldsymbol{K}_{\ell-1} r)\,]\,\boldsymbol{K}_\ell .$$

If $rp = I$ holds (cf. (3.7a)), (7b) simplifies to

$$(5.5.7c)\qquad M_\ell^{MGM} = M_\ell^{ZGM} + \tfrac{1}{\lambda} p\,(M_{\ell-1}^{MGM})^2\, r\,(M_\ell^{ZGM} - \tfrac{1}{\lambda}\boldsymbol{K}_\ell)\,\boldsymbol{K}_\ell .$$

In order to prove this lemma, we need a corollary which enables us to represent the quantity c_ℓ from (6) explicitly.

Lemma 5.5.6. Let the iteration $\boldsymbol{f}_\ell^i \mapsto \boldsymbol{f}_\ell^{i+1}$ for solving $\lambda \boldsymbol{f}_\ell = \boldsymbol{g}_\ell + \boldsymbol{K}_\ell \boldsymbol{f}_\ell$ have the solutions $\boldsymbol{f}_\ell$ as fixed points for all $\boldsymbol{g}_\ell$, i.e., $\boldsymbol{f}_\ell^i = \boldsymbol{f}_\ell$ yields $\boldsymbol{f}_\ell^{i+1} = \boldsymbol{f}_\ell$. Then the iteration representation (6) holds with

$$(5.5.8)\qquad c_\ell = N_\ell\, \boldsymbol{g}_\ell \qquad \text{with} \quad N_\ell = (I - M_\ell^{MGM})(\lambda I - \boldsymbol{K}_\ell)^{-1}.$$

Proof. Inserting the fixed point $\boldsymbol{f}_\ell = (\lambda I - \boldsymbol{K}_\ell)^{-1}\boldsymbol{g}_\ell$ into (6), one obtains $\boldsymbol{f}_\ell = M_\ell^{MGM}\boldsymbol{f}_\ell + c_\ell$, i.e., $c_\ell = (I - M_\ell^{MGM})\boldsymbol{f}_\ell = (I - M_\ell^{MGM})(\lambda I - \boldsymbol{K}_\ell)^{-1}\boldsymbol{g}_\ell$. QED

Proof of Lemma 5. (i) In the case of $\boldsymbol{f}_\ell^i = \boldsymbol{f}_\ell$, the Picard iteration (3c) reproduces the solution $\boldsymbol{f}_\ell$. In this case, $\boldsymbol{d} = 0$ holds in (3d), so that $\boldsymbol{\delta} = 0$ and thus $\boldsymbol{f}_\ell^{i+1} = \boldsymbol{f}_\ell$. Hence, the representation (8) is applicable.
(ii) (7a) is a consequence of definition (2a) realised by procedure (3).
(iii) Since the quantity $\boldsymbol{g}$ (the third parameter in *MGM*) influences c_ℓ from (6), but does not occur in M_ℓ^{MGM}, we may set $\boldsymbol{g} = 0$ without loss of generality. For $\boldsymbol{f}_\ell^i$ we briefly write $\boldsymbol{f}$. (3c) implies $\tilde{\boldsymbol{f}} = \tfrac{1}{\lambda}\boldsymbol{K}_\ell \boldsymbol{f}$. (3d) yields $\boldsymbol{d} = r(\lambda I - \boldsymbol{K}_\ell)\tilde{\boldsymbol{f}}$. By the induction hypothesis, (6) and (8) hold at level $\ell - 1$. The starting iterate $\boldsymbol{\delta}^0 = 0$ from $(3e_1)$ leads to

$$\boldsymbol{\delta}^1 = M_{\ell-1}^{MGM}\boldsymbol{\delta}^0 + c_{\ell-1} = c_{\ell-1} = N_{\ell-1}\boldsymbol{d} = N_{\ell-1}\, r(\lambda I - \boldsymbol{K}_\ell)\tilde{\boldsymbol{f}} .$$

The second multi-grid iteration step in $(3e_2)$ yields

$$\boldsymbol{\delta}^2 = M_{\ell-1}^{MGM}\boldsymbol{\delta}^1 + c_{\ell-1} = (I + M_{\ell-1}^{MGM})N_{\ell-1}\, r(\lambda I - \boldsymbol{K}_\ell)\tilde{\boldsymbol{f}} .$$

From (8) we obtain the representation

$$\boldsymbol{\delta}^2 = \big(I - (M_{\ell-1}^{MGM})^2\big)(\lambda I - \boldsymbol{K}_{\ell-1})^{-1} r(\lambda I - \boldsymbol{K}_\ell)\tilde{\boldsymbol{f}} .$$

Hence, the final result of the correction (3f) reads

$$\begin{aligned}\boldsymbol{f}_\ell^{i+1} &= \tilde{\boldsymbol{f}} - p\big(I - (M_{\ell-1}^{MGM})^2\big)(\lambda I - \boldsymbol{K}_{\ell-1})^{-1} r(\lambda I - \boldsymbol{K}_\ell)\tilde{\boldsymbol{f}} = \\ &= [\, I - p\big(I - (M_{\ell-1}^{MGM})^2\big)(\lambda I - \boldsymbol{K}_{\ell-1})^{-1} r(\lambda I - \boldsymbol{K}_\ell)\,]\,\tfrac{1}{\lambda}\boldsymbol{K}_\ell \boldsymbol{f} = \\ &= \big\{\tfrac{1}{\lambda}[\, I - p(\lambda I - \boldsymbol{K}_{\ell-1})^{-1} r(\lambda I - \boldsymbol{K}_\ell)\,]\,\boldsymbol{K}_\ell + \\ &\qquad + \tfrac{1}{\lambda} p\,(M_{\ell-1}^{MGM})^2 (\lambda I - \boldsymbol{K}_{\ell-1})^{-1} r(\lambda I - \boldsymbol{K}_\ell)\,]\,\boldsymbol{K}_\ell\big\}\boldsymbol{f} .\end{aligned}$$

Since $\boldsymbol{f}_\ell^i = \boldsymbol{f}$ is arbitrary, the bracket {...} must represent the matrix M_ℓ^{MGM} from (6). The first term therein is M_ℓ^{ZGM} (cf. (4.9)), so that (7b) is proved. Obviously, (7b) and (7b') are equivalent.
(iv) (7c) follows immediately from (7b), (4.9), and $rp = I$. QED

The matrix representation (7b') leads to the estimate

$$\| M_\ell^{MGM} \|_{X_\ell \leftarrow X_\ell} \le \| M_\ell^{ZGM} \|_{X_\ell \leftarrow X_\ell} + \tfrac{1}{|\lambda|} \| p \|_{X_\ell \leftarrow X_{\ell-1}} \| M_{\ell-1}^{MGM} \|^2_{X_{\ell-1} \leftarrow X_{\ell-1}} \times$$
$$\times \left[\| r \|_{X_{\ell-1} \leftarrow X_\ell} \| \boldsymbol{K}_\ell \|_{X_\ell \leftarrow X_\ell} + \right.$$
$$\left. + \| (\lambda I - \boldsymbol{K}_{\ell-1})^{-1} \|_{X_{\ell-1} \leftarrow X_{\ell-1}} \| r \boldsymbol{K}_\ell - \boldsymbol{K}_{\ell-1} r \|_{X_{\ell-1} \leftarrow Y_\ell} \| \boldsymbol{K}_\ell \|_{Y_\ell \leftarrow X_\ell} \right].$$

Using the bounds $\| p \|_{X_\ell \leftarrow X_{\ell-1}} \le C_p$, $\| r \|_{X_{\ell-1} \leftarrow X_\ell} \le C_r$ (cf. (3.4a,b)), $\| r \boldsymbol{K}_\ell - \boldsymbol{K}_{\ell-1} r \|_{X_{\ell-1} \leftarrow Y_\ell} \le C_c h_\ell^\beta$ (cf. (3.9)), $\| (\lambda I - \boldsymbol{K}_{\ell-1})^{-1} \|_{X_{\ell-1} \leftarrow X_{\ell-1}} \le C_S$ (cf. (2.14)), $\| \boldsymbol{K}_\ell \|_{Y_\ell \leftarrow X_\ell} \le C_K$ and $\| \boldsymbol{K}_\ell \|_{X_\ell \leftarrow X_\ell} \le C_{XX}$ (cf. (2.11c/f)) and the two-grid convergence statement $\| M_\ell^{ZGM} \|_{X_\ell \leftarrow X_\ell} \le C_{ZGM} h_\ell^\beta$ (cf. (4.10a)), we can write the inequality in the form

(5.5.9)
$$\| M_\ell^{MGM} \|_{X_\ell \leftarrow X_\ell} \le$$
$$\le C_{ZGM} h_\ell^\beta + \tfrac{1}{|\lambda|} C_p \| M_{\ell-1}^{MGM} \|^2_{X_{\ell-1} \leftarrow X_{\ell-1}} [C_r C_{XX} + C_S C_c h_\ell^\beta C_K].$$

Theorem 5.5.7 (convergence of the multi-grid method). Assume stability (2.14), discrete regularity (2.11c) and (2.11f), uniform boundedness (3.4a,b) of p and r, relative consistency (3.9), and the interpolation error estimate (3.8) with $\beta > 0$. The step size h_1 must be sufficiently small: $h_1 \le h_*$, and $h_{\ell-1} \le C_h h_\ell$ must hold. Then there is a constant C_* such that the multi-grid convergence can be estimated by

(5.5.10)
$$\| M_\ell^{MGM} \|_{X_\ell \leftarrow X_\ell} \le C_{ZGM} (1 + C_* h_\ell^\beta)\, h_\ell^\beta < 1$$

with C_{ZGM} given by (4.10a). As a consequence, the two- and multi-grid algorithms have the same asymptotic speed.

Proof. For sufficiently small $h_\ell < h_{\ell-1} \le h_1 \le h_*$ there is a constant C_* with

(5.5.11)
$$\tfrac{1}{|\lambda|} C_p C_{ZGM} (1 + C_* h_{\ell-1}^\beta)^2 [C_r C_{XX} + C_S C_c C_K h_\ell^\beta] C_h^{2\beta} \le C_* .$$

Assertion (10) is proved by induction with respect to the level number ℓ. The assumptions of the two-grid convergence Theorem 4.3 are satisfied, so that $\| M_\ell^{ZGM} \|_{X_\ell \leftarrow X_\ell} \le C_{ZGM} h_\ell^\beta$ holds (cf. (4.10a)). For $\ell = 1$, inequality (10) is trivial because of (7a). If (10) holds for $\ell - 1$, one inserts this inequality into (9) and obtains

$$\| M_\ell^{MGM} \|_{X_\ell \leftarrow X_\ell} \le$$
$$\le C_{ZGM} h_\ell^\beta + \tfrac{1}{|\lambda|} C_p \big(C_{ZGM} (1 + C_* h_{\ell-1}^\beta)\, h_{\ell-1}^\beta \big)^2 [C_r C_{XX} + C_S C_c h_\ell^\beta C_K].$$

Writing the second term as product of $C_{ZGM} h_\ell^{2\beta}$ and the left-hand side of (11), assertion (10) follows for ℓ. QED

Corollary 5.5.8. Executing only *one* multi-grid iteration at the level $\ell-1$ in step (3e$_2$), the iteration matrix is represented by (7b) or (7b') with $M_{\ell-1}^{MGM}$ instead of $(M_{\ell-1}^{MGM})^2$. Inequality (9) holds again if the square sign in the factor $\| M_{\ell-1}^{MGM} \|^2_{X_{\ell-1} \leftarrow X_{\ell-1}}$ is omitted. Under the assumption

$$\tfrac{1}{|\lambda|} C_p C_h^\beta [C_r C_{XX} + C_S C_c h_\ell^\beta C_K] \le C_* < 1 \qquad \text{for all } \ell \ge 2,$$

the convergence estimate $\|M_\ell^{MGM}\|_{X_\ell \leftarrow X_\ell} \le C_{ZGM} h_\ell^\beta / (1 - C_*)$ holds for all $\ell \ge 2$. Using $C_{**} := \frac{1}{|\lambda|} C_p C_r C_{XX}$, one obtains from (9) the estimate

$$\|M_\ell^{MGM}\|_{X_\ell \leftarrow X_\ell} \le C_\ell \qquad \text{with } C_\ell \ge C_{ZGM} \sum_{k=0}^{\ell-1} h_{\ell-k}^\beta C_{**}^k,$$

so that, in general, a convergence speed better than $O(h_1^\beta)$ cannot be expected.

Proof. Induction by ℓ shows the assertions. QED

Exercise 5.5.9. **(a)** What are the costs (in arithmetical operations), if only one multi-grid iteration is performed in step (3e$_2$) instead of two?
(b) If one wants to keep a similar convergence behaviour to that in Theorem 7 with less amount of work, one can perform *one* iteration step in (3e$_2$) for even ℓ and *two* iteration steps for odd ℓ. What is the amount of work involved?
(c) Show for the variant from part (b) that $\|M_\ell^{MGM}\|_{X_\ell \leftarrow X_\ell} = O(h_\ell^\beta)$.

Exercise 5.5.10. Let the numbers α_ℓ $(\ell \ge 1)$ satisfy $\alpha_1 \le \zeta < \frac{1}{2}$ and $\alpha_\ell \le \varkappa_\ell + A\alpha_{\ell-1}^2$ for $\ell \ge 2$ with $4\zeta A \le 1$ and $\varkappa_\ell \le \zeta$. Prove the following:
(a) $\alpha_\ell < 2\zeta < 1$ for all $\ell \ge 1$.
(b) If in addition $\varkappa_\ell \to 0$, then $\alpha_\ell \to 0$.
(c) If $\varkappa_\ell \le \varkappa h_\ell^\beta$ and $h_{\ell-1} \le C_h h_\ell$ $(\ell \ge 1)$ with $\varkappa > 0$ and $C_h < 1$, there is a constant C such that $\alpha_\ell \le \min(C\varkappa h_\ell^\beta, 2\zeta) < 1$.

A concrete specification of the «sufficiently small» step size h_1 is contained in the following corollary, which relates this step size to the two-grid convergence. For this purpose, we use the standard conditions $rp = I$, $C_r = C_p = 1$. The value C_{XX} stems from (2.11f).

Corollary 5.5.11. Besides the assumptions of Theorem 7, assume $rp = I$ and $C_r = C_p = 1$. The step size $h_1 \le h_*$ must be chosen small enough that

$$\text{(5.5.12)} \qquad \|M_\ell^{ZGM}\|_{X_\ell \leftarrow X_\ell} \le \zeta < \tfrac{1}{2} \quad \text{for all } \ell \ge 1, \qquad 4\zeta \tfrac{1}{|\lambda|}(\zeta + \tfrac{1}{|\lambda|} C_{XX}) C_{XX} \le 1$$

hold. Then the convergence statement (10) is true for a suitable C_*.

Proof. The representation (7c) of M_ℓ^{MGM} implies the estimate

$$\text{(5.5.13)} \qquad \|M_\ell^{MGM}\|_{X_\ell \leftarrow X_\ell} \le \|M_\ell^{ZGM}\|_{X_\ell \leftarrow X_\ell} + \tfrac{1}{|\lambda|} \|M_{\ell-1}^{MGM}\|^2_{X_{\ell-1} \leftarrow X_{\ell-1}} (\|M_\ell^{ZGM}\|_{X_\ell \leftarrow X_\ell} + \tfrac{1}{|\lambda|} C_{XX}) C_{XX}.$$

Using $\alpha_\ell := \|M_\ell^{MGM}\|_{X_\ell \leftarrow X_\ell}$, $\varkappa_\ell := \|M_\ell^{ZGM}\|_{X_\ell \leftarrow X_\ell}$, $A := \frac{1}{|\lambda|}(\zeta + \frac{1}{|\lambda|} C_{XX}) C_{XX}$, and $\varkappa := C_{ZGM}$, one can apply Exercise 10. QED

The «convergence» statement (4.10a) for the *two-grid* method holds (asymptotically) at all levels ℓ independently of whether the iteration converges or diverges. This does not hold for the *multi-grid* algorithm. If divergence is caused by a too large h_1, the algorithm diverges dramatically at all levels, as shown in

Remark 5.5.12. If the two-grid method diverges at the level $\ell=1$ [or if the multi-grid method does so at some level $\ell=\ell_0$], then due to (9), the approximation $\|M_\ell^{MGM}\|_{X_\ell \leftarrow X_\ell} \approx \frac{1}{|\lambda|} C_p C_r C_{XX} \|M_{\ell-1}^{MGM}\|^2_{X_{\ell-1} \leftarrow X_{\ell-1}}$ holds for $\ell \geq 1$ [$\ell \geq \ell_0$], if one neglects the h_ℓ^β terms. Then the divergence rate increases like

$$\|M_\ell^{MGM}\|_{X_\ell \leftarrow X_\ell} \approx O(C^{2^\ell - 1}) \qquad \text{with } C := \tfrac{1}{|\lambda|} C_p C_r C_{XX}. \tag{5.5.14}$$

Remark 5.5.13. The smaller the value of $|\lambda|$, the smaller is the bound h_* of the step size. Therefore, the convergence assumption $h_1 \leq h_*$ is the more restrictive the smaller $|\lambda|$ is.

Proof. Follows from the definition (11) of C_* and from (12). ∎

The numerical examples of Tables 1 and 2, will support this statement. Note that a very restrictive step size condition $h_1 \leq h_*$ cannot be fulfilled in practice, if h_* becomes smaller than the step size h of the original discrete problem. In any case, in step (3b) at the level $\ell=0$ one had still to solve a relatively large system of equations. Therefore, in §5.5.5 we shall propose variants of the multi-grid method which prove to be more robust for small $|\lambda|$.

5.5.4 Numerical Examples

We choose again the example (4.15-16), which we already used in §5.4.5 as a test example for the two-grid method. In the case of $\lambda=1$, the differences between the two-grid and multi-grid convergence rates are so small that the convergence speeds of the multi-grid method (3) reported in Table 5.4.2 are still valid. No deviation from of the two-grid rates given in Table 5.4.3 appear before $\lambda=0.1$.

Table 1 contains the averaged convergence speeds for the step sizes

$$h_\ell := 1/n_\ell \qquad \text{with } n_\ell = 2^\ell n_0 \qquad (\ell \geq 0).$$

The number $n=n_\ell$ denotes the parameter of the problem to be solved, while n_0 characterises the step size of the coarsest grid. A multi-grid method with $n=2n_0$ corresponds to the level $\ell=1$ and coincides with a two-grid method. Fixing n and choosing $n_0=2^{-\ell}n$, one obtains a multi-grid method with ℓ auxiliary grids, in fact, an $(\ell+1)$-grid method. In (3b) one has to solve systems of the dimension n_0. The convergence condition $h_1 \leq h_*$ becomes

$$n_0 \geq 1/(2h_*). \tag{5.5.15}$$

For $n_0=1$ and $n_0=2$, divergence is apparent except for the two-grid case $n=2n_0$. The sizes of the divergence rates correspond to the behaviour described in (14). It is evident that the convergence condition (15) is satisfied for $n_0 \geq 4$: Convergence holds for all $n=n_\ell$. Since the diagonal of Table 1 contains the two-grid rates, one has to compare the values of one row. In agreement with Theorem 7, the multi-grid speed tends to the two-grid speed as n increases.

For $\lambda=0.01$ the difficulties in (11), (12), and (14) arising from $\frac{1}{|\lambda|}$ are enlarged. For this example, the condition (15) necessary for the multi-grid convergence is $n_0 \geqslant 16$, as can be seen from the numbers in Table 2. The factor $\frac{1}{|\lambda|} = 100$ causes a larger constant C_{ZGM} in (4.10b), so that also the *two-grid* convergence sets in later ($n_0 \geqslant 8$).

For comparison, we recall that the Picard iteration (1.5) diverges for the values $\lambda = 0.1$ or 0.01 with rates about 5 or 50, respectively. Also the condition number determining the convergence behaviour of the cg method (§5.1.4) increases with decreasing $|\lambda|$.

n	$n_0=1$	$n_0=2$	$n_0=4$	$n_0=8$	$n_0=16$	$n_0=32$	$n_0=64$
2	$7.2_{10}-1$	-	-	-	-	-	-
4	6.7_{10}**+0**	$4.5_{10}-1$	-	-	-	-	-
8	6.8_{10}**+1**	1.2_{10}**+0**	$1.2_{10}-1$	-	-	-	-
16	7.3_{10}**+4**	4.4_{10}**+0**	$1.2_{10}-1$	$3.3_{10}-2$	-	-	-
32	4.7_{10}**+10**	1.6_{10}**+2**	$2.5_{10}-2$	$1.4_{10}-2$	$8.3_{10}-3$	-	-
64	2.0_{10}**+22**	1.7_{10}**+5**	$9.2_{10}-3$	$2.7_{10}-3$	$2.4_{10}-3$	$3.6_{10}-3$	-
128	2.5_{10}**+45**	4.0_{10}**+11**	$5.1_{10}-4$	$5.7_{10}-4$	$5.5_{10}-4$	$5.5_{10}-4$	$5.6_{10}-4$

Table 5.5.1. Averaged convergence speed for $\lambda = 0.1$

n	$n_0=1$	$n_0=2$	$n_0=4$	$n_0=8$	$n_0=16$	$n_0=32$	$n_0=64$
2	9.3_{10}**+0**	-	-	-	-	-	-
4	7.7_{10}**+3**	5.0_{10}**+0**	-	-	-	-	-
8	1.5_{10}**+7**	1.1_{10}**+3**	1.3_{10}**+0**	-	-	-	-
16	1.7_{10}**+17**	1.0_{10}**+7**	8.5_{10}**+1**	$4.1_{10}-1$	-	-	-
32	1.0_{10}**+36**	1.7_{10}**+16**	1.7_{10}**+4**	5.7_{10}**+0**	$1.2_{10}-1$	-	-
64	1.0_{10}**+74**	8.2_{10}**+33**	4.6_{10}**+10**	2.4_{10}**+2**	$3.8_{10}-1$	$3.2_{10}-2$	-
128	-	8.0_{10}**+69**	8.6_{10}**+22**	9.1_{10}**+6**	$7.3_{10}-1$	$2.9_{10}-2$	$2.0_{10}-2$

Table 5.5.2. Averaged convergence speed for $\lambda = 0.01$

One possibility of compensating for the factor $\frac{1}{|\lambda|} \gg 1$ is to use a quadrature method of higher order, so that the possibly large constants are mitigated by small factors h_ℓ^β. For the example (4.15/16) with $\lambda = -0.001$ one can use, e.g., the *composite Simpson rule* as quadrature method in the Nyström discretisation. The method is of fourth order. To reach $\beta = 4$, one has to choose $Y = \hat{C}^4([0,1])$. According to requirement (3.8), the interpolation error must also be of fourth order. This can be achieved by the *piecewise cubic interpolation*, i.e., $(p\varphi_{\ell-1})(\nu h_{\ell-1}) = \varphi_{\ell-1}(\nu h_{\ell-1})$ at the coarse-grid points $(\nu \in \mathbb{Z})$ and, otherwise, $(p\varphi_{\ell-1})(\nu h_{\ell-1} + h_\ell) = \pi(\nu h_{\ell-1} + h_\ell)$, where π is the cubic polynomial interpolating at the abscissae $(\nu-1)h_{\ell-1}$, $\nu h_{\ell-1}$, $(\nu+1)h_{\ell-1}$, $(\nu+2)h_{\ell-1}$. As before, r is the trivial restriction to the coarse grid. The multi-grid method (3) converges for $n = 128$ and $n_0 \geqslant 16$, but it diverges for $n_0 = 16$ and $n \geqslant 256$. The convergence condition (15) reads $n_0 \geqslant 32$. Note that the Picard method has a divergence rate of about 500. At the positions marked by «+» in Table 3 the convergence is so fast that the machine precision is reached before we can determine the convergence factors ε_ℓ^i

explained in §5.4.5. Therefore the averaged convergence speed cannot be computed.

A further step to ensure multi-grid convergence for small values of λ consists of a modification in the algorithm. This will be discussed in the following subsection.

n	$n_0=8$	$n_0=16$	$n_0=32$	$n_0=64$
64	$1.2_{10}+4$	$2.4_{10}-2$	$2.8_{10}-3$	-
128	$3.5_{10}+10$	$1.3_{10}-1$	$1.1_{10}-4$	$3.0_{10}-4$
256	$1.8_{10}+24$	$1.0_{10}+1$	+	+

Table 5.5.3. Averaged convergence speed for $\lambda=-0.001$ with Simpson's rule in the Nyström method

5.5.5 Variants of the Multi-Grid Methods

By analogy with §5.4.4, we first treat the multi-grid variant for the case $A_\ell \neq I$, in which the equation (4.13b)

$$(\lambda A_\ell - B_\ell)\, \boldsymbol{f}_\ell = b_\ell$$

is to be solved. We recall that the canonical restriction is $r = A_{\ell-1}^{-1}\, \hat{r}\, A_\ell$ (cf. (4.13c)) and that the multiplication by $\hat{r} := p^T$ is simply performed (cf. (3.6i)).

The *equivalent* reformulation of the multi-grid algorithm (3) follows immediately from the two-grid variant (4.14). In only change is that the coarse-grid solution (4.14e) is replaced by a call of two multi-grid steps at level $\ell-1$ corresponding to $(3e_{1,2})$.

(5.5.16)	**multi-grid procedure** *MGM'* for solving $(\lambda A_\ell - B_\ell)\, \boldsymbol{f}_\ell = b_\ell$	
(5.5.16a)	procedure $MGM'(\ell, \boldsymbol{f}, \boldsymbol{b})$; integer ℓ; array $\boldsymbol{f}$, $\boldsymbol{b}$;	
(5.5.16b)	if $\ell=0$ then $\boldsymbol{f} := (\lambda A_0 - B_0)^{-1}\boldsymbol{b}$ else	
	begin array $\boldsymbol{d}, \boldsymbol{\delta}$; integer i;	
(5.5.16c)	$\boldsymbol{f} := \frac{1}{\lambda} A_\ell^{-1}(\boldsymbol{b} + B_\ell * \boldsymbol{f})$;	(Picard iteration)
(5.5.16d)	$\boldsymbol{d} := \hat{r} * (\lambda A_\ell \boldsymbol{f} - \boldsymbol{b} - B_\ell * \boldsymbol{f})$;	(computing the defect)
$(5.5.16e_1)$	$\boldsymbol{\delta} := 0$;	(set starting iterate)
$(5.5.16e_2)$	for $i=1,2$ do $MGM'(\ell-1, \boldsymbol{\delta}, \boldsymbol{d})$;	(two multi-grid calls)
(5.5.16f)	$\boldsymbol{f} := \boldsymbol{f} - p * \boldsymbol{\delta}$	(coarse-grid correction)
	end;	

Since the variant (16) is equivalent to the previous method, a new convergence analysis is superfluous. This is not true for the following modifications (17) and (24), which are modelled especially for problems with small values of λ.

In particular, the factor $\frac{1}{\lambda}$ is detrimental to convergence of the Picard iteration in (3c) or (16c). It will turn out that in the first of the two multi-grid calls $(3e_2)$ the induced Picard step at the level $\ell-1$ is dispensable. Therefore, we introduce a fourth parameter s in the multi-grid variant *MGV*. For s = true, the Picard iteration is performed as «smoother» as before; in the case of s = false it is omitted. The resulting algorithm reads as follows.

(5.5.17)	**multi-grid procedure** *MGV* for solving $\lambda \boldsymbol{f}_\ell = \boldsymbol{g}_\ell + \boldsymbol{K}_\ell \boldsymbol{f}_\ell,\ \ell \geq 0$	
(5.5.17a)	<u>procedure</u> $MGV(\ell, \boldsymbol{f}, \boldsymbol{g}, s)$; <u>integer</u> ℓ; <u>array</u> $\boldsymbol{f}, \boldsymbol{g}$; <u>Boolean</u> s;	
(5.5.17b)	<u>if</u> $\ell = 0$ <u>then</u> $\boldsymbol{f} := (\lambda I - \boldsymbol{K}_0)^{-1} \boldsymbol{g}$ <u>else</u>	
	<u>begin</u> <u>array</u> $\boldsymbol{d}, \boldsymbol{\delta}$;	
(5.5.17c)	<u>if</u> s <u>then</u> $\boldsymbol{f} := \frac{1}{\lambda}(\boldsymbol{g} + \boldsymbol{K}_\ell * \boldsymbol{f})$;	(Picard iteration)
(5.5.17d)	$\boldsymbol{d} := r * (\lambda \boldsymbol{f} - \boldsymbol{g} - \boldsymbol{K}_\ell * \boldsymbol{f})$;	(computing the defect)
(5.5.17e$_1$)	$\boldsymbol{\delta} := 0$;	(set starting iterate)
(5.5.17e$_2$)	$MGV(\ell - 1, \boldsymbol{\delta}, \boldsymbol{d},$ <u>false</u>$)$;	(first multi-grid call)
(5.5.17e$_3$)	$MGV(\ell - 1, \boldsymbol{\delta}, \boldsymbol{d},$ <u>true</u>$)$;	(second multi-grid call)
(5.5.17f)	$\boldsymbol{f} := \boldsymbol{f} - p * \boldsymbol{\delta}$	(coarse-grid correction)
	<u>end</u>;	

This variant *MGV* from (17) proposed by the author in 1983 (called *MGM'''* in Hackbusch [11]) combines two advantages: The amount of computational work is less since some of the Picard steps are omitted and, for small $|\lambda|$, the multi-grid convergence behaves no worse than the two-grid method. An estimate of the amount of computational work is contained in

Exercise 5.5.14. Prove that under the assumptions (4a-d) and (4g) one call $MGV(\ell, \cdot, \cdot,$ <u>true</u>$)$ of the procedure (17) requires

(5.5.18a) $\quad C_1 n_\ell^2 + O(n_\ell)$ operations with $C_1 = 2\dfrac{(1 + C_N)(1 - C_N)^2}{1 - 2C_N^2}$.

For the particular case $\boldsymbol{f} = \boldsymbol{f}_\ell^0 = 0$, one iteration $MGV(\ell, \boldsymbol{f}, \cdot,$ <u>true</u>$)$ costs

(5.5.18b) $\quad C_0 n_\ell^2 + O(n_\ell)$ operations with $C_0 = 2\,C_N \dfrac{1 + C_N - C_N^2}{1 - 2C_N^2}$.

The corresponding numbers for the call $MGV(\ell, \cdot, \cdot,$ <u>false</u>$)$ are

(5.5.18c) $\quad C_1' = C_0, \qquad C_0' = 2\,C_N^2 \dfrac{1 + C_N}{1 - 2C_N^2}$.

For the standard case (4e): $C_N = \frac{1}{2}$ the following values result:

(5.5.18d) $\quad C_1 = 4.5, \qquad C_0 = C_1' = 2.5, \qquad C_0' = 1.5.$

Although the procedure *MGV* is not equivalent to the multi-grid procedure (3) and therefore, generally, yields other results, *MGV* as well as *MGM* from (3) reproduce the *two-grid* method (4.7) for $\ell = 1$, since for any value of s (17e$_2$) as well as (17e$_3$) compute the exact coarse-grid solution $\boldsymbol{\delta}$, when $\ell = 1$. This proves

Remark 5.5.15. Similarly, for variant (17), the algorithm at level $\ell = 1$ is identical to the two-grid method (4.7).

Concerning the convergence analysis, one has first to determine the iteration matrix of the method. If G_ℓ^{MGV} denotes the iteration matrix of the iteration $MGV(\ell,\cdot,\cdot,\underline{false})$ consisting only of the coarse-grid correction (17d-f), then the iteration matrix of the procedure $MGV(\ell,\cdot,\cdot,\underline{true})$ reads

(5.5.19a) $$M_\ell^{MGV} = \tfrac{1}{\lambda} G_\ell^{MGV} K_\ell .$$

In the two-grid case, $M_\ell^{ZGM} = \tfrac{1}{\lambda} G_\ell^{ZGM} K_\ell$ holds with

(5.5.19b) $$G_\ell^{ZGM} = I - pr + p(\lambda I - \boldsymbol{K}_{\ell-1})^{-1}[r\boldsymbol{K}_\ell - \boldsymbol{K}_{\ell-1} r] \qquad \text{(cf. (4.9))}.$$

Repeating the proof idea of Lemma 5 for the method (17), one gains the following representation for G_ℓ^{MGV}.

Lemma 5.5.16. The iteration matrix of MGV is G_ℓ^{MGV} (case s=false) or M_ℓ^{MGV} from (19a) (case s=true), where G_ℓ^{MGV} is defined recursively by

(5.5.19c) $$G_1^{MGV} = G_1^{ZGM} \qquad \text{(for } \ell=1,\text{ cf. (19b))},$$

(5.5.19d) $$G_\ell^{MGV} = G_\ell^{ZGM} + \tfrac{1}{\lambda} p\, G_{\ell-1}^{MGV} \boldsymbol{K}_{\ell-1} G_{\ell-1}^{MGV} (\lambda I - \boldsymbol{K}_{\ell-1})^{-1} r (\lambda I - \boldsymbol{K}_\ell) \quad (\ell \geq 2).$$

(19d) is equivalent to the representation

(5.5.19d') $$G_\ell^{MGV} = G_\ell^{ZGM} + \tfrac{1}{\lambda} p\, G_{\ell-1}^{MGV} \boldsymbol{K}_{\ell-1} G_{\ell-1}^{MGV} [r - (\lambda I - \boldsymbol{K}_{\ell-1})^{-1}(r\boldsymbol{K}_\ell - \boldsymbol{K}_{\ell-1} r)].$$

In the proof of (4.10a,b) we have already estimated G_ℓ^{ZGM} by

(5.5.20a) $$\| G_\ell^{ZGM} \|_{X_\ell \leftarrow Y_\ell} \leq (C_I + C_p C_S C_c)\, h_\ell^\beta .$$

Also the second term in the right-hand side of (19d') must be estimated with respect to the norm $\|\cdot\|_{X_\ell \leftarrow Y_\ell}$. The next condition (20b) is easy to accomplish, e.g., by the trivial restriction (2.4e):

(5.5.20b) $$\| r \|_{Y_{\ell-1} \leftarrow Y_\ell} \leq C_r^* .$$

Furthermore, we require stability in the form

(5.5.20c) $$\| (\lambda I - \boldsymbol{K}_\ell)^{-1} \|_{Y_\ell \leftarrow Y_\ell} \leq C_S^* .$$

This condition follows from the usual stability (2.14), as stated in

Exercise 5.5.17. Conclude (20c) with $C_S^* = \frac{1}{|\lambda|}(1 + C C_S C_K)$ from (2.14), (2.11c), and (2.11e). *Hint:* The proof is analogous to Theorem 3.5.1.

Using (20b,c) and the regularity (2.11c), one proves the inequality

(5.5.20d) $$\| r - (\lambda I - \boldsymbol{K}_{\ell-1})^{-1}(r\boldsymbol{K}_\ell - \boldsymbol{K}_{\ell-1} r) \|_{Y_\ell \leftarrow Y_\ell} \leq C_G := C_r^*(1 + 2 C_S^* C_K).$$

Set

(5.5.20e) $$\alpha_\ell^{ZGM} := \| G_\ell^{ZGM} \|_{X_\ell \leftarrow Y_\ell}, \qquad \alpha_\ell := \| G_\ell^{MGV} \|_{X_\ell \leftarrow Y_\ell}, \qquad \beta_\ell := \tfrac{1}{|\lambda|} C_K \alpha_\ell .$$

According to (19a), β_ℓ is a bound of $\| M_\ell^{MGV} \|_{X_\ell \leftarrow X_\ell}$. By the uniform boundedness (3.4a) of p, regularity (2.11c), and estimate (20d), one concludes from (19d') that

(5.5.20f) $\alpha_\ell \leqslant \alpha_\ell^{ZGM} + \frac{1}{|\lambda|} C_p \alpha_{\ell-1}^2 C_K C_G .$

The constant $C_{ZGM} := C_K (C_I + C_p C_S C_c)/|\lambda|$ from (4.10b) is an upper bound of $\frac{1}{|\lambda|} C_K \alpha_\ell^{ZGM}/h_\ell^\beta$ (cf. (20a/e)). Multiplication of (20f) by $C_K/|\lambda|$ yields (20g), where we formally set $\beta_0 = 0$:

(5.5.20g) $\| M_\ell^{MGV} \|_{X_\ell \leftarrow X_\ell} \leqslant \beta_\ell \leqslant C_{ZGM} h_\ell^\beta + C_p \beta_{\ell-1}^2 C_G$ for $\ell \geqslant 1$.

Remark 5.5.18. (20g) represents a recursive inequality for the upper bound of the norm $\| M_\ell^{MGV} \|_{X_\ell \leftarrow X_\ell}$ for the iteration matrix of the method (17). It requires the assumptions (2.11c,e), (2.14), (3.4a), (3.8), (3.9), and (20b). The inequality (20g) is very similar to the analogous estimate (9) for the standard multi-grid method. *The essential difference is the missing factor* $\frac{1}{|\lambda|}$ *in front of the second term* $C_p \beta_{\ell-1}^2 C_G$ on the right-hand side of (20g). The same argument as in Theorem 7 yields the *convergence result*

(5.5.21) $\| M_\ell^{MGV} \|_{X_\ell \leftarrow X_\ell} \leqslant C_{ZGM} (1 + C_* h_\ell^\beta)\, h_\ell^\beta < 1,$

if $h_\ell \leqslant h_*$. The constant C_* follows indirectly from the condition

(5.5.22) $C_p C_{ZGM} (1 + C_* h_{\ell-1}^\beta)^2 C_G C_h^{2\beta} \leqslant C_*,$

if $h_1 \leqslant h_*$ is sufficiently small. Here, h_* depends only on $C_p C_{ZGM} C_G C_h^{2\beta}$. In the case of the standard multi-grid method (3), the corresponding quadratic inequality (11) contains a further factor $\frac{1}{|\lambda|}$, which diminishes the step size bound h_* by the order $|\lambda|$.

n	$n_0=1$	$n_0=2$	$n_0=4$	$n_0=8$	$n_0=16$	$n_0=32$	$n_0=64$
2	$7.2_{10}-1$	-	-	-	-	-	-
4	$4.2_{10}-1$	$4.5_{10}-1$	-	-	-	-	-
8	$1.9_{10}-1$	$1.5_{10}-1$	$1.2_{10}-1$	-	-	-	-
16	$6.2_{10}-2$	$3.6_{10}-2$	$2.1_{10}-2$	$3.3_{10}-2$	-	-	-
32	$5.8_{10}-3$	$8.8_{10}-3$	$7.9_{10}-3$	$7.5_{10}-3$	$8.3_{10}-3$	-	-
64	$2.1_{10}-3$	$2.0_{10}-3$	$2.0_{10}-3$	$2.0_{10}-3$	$2.1_{10}-3$	$3.6_{10}-3$	-
128	$5.3_{10}-4$	$5.3_{10}-4$	$5.3_{10}-4$	$5.3_{10}-4$	$5.3_{10}-4$	$5.3_{10}-4$	$5.6_{10}-4$

Table 5.5.4. Averaged convergence speed of the modified multi-grid method *MGV* from (17) for $\lambda = 0.1$

We emphasize that the factor $C_p C_{ZGM} C_G C_h^{2\beta}$ is by no means independent of λ. The two-grid constant C_{ZGM} contains a factor $\frac{1}{|\lambda|}$. Furthermore, the same factor appears in C_S^* (cf. Exercise 17) and via (20d) in C_G. Here, we have to remark that the estimate for C_S^* in Exercise 17 is very pessimistic and could be improved. It is even possible to show that a bound C_G on the left-hand side in (20d) exists and behaves like $C_r^* + O(h_\ell^\beta)$. For that purpose, the strengthened consistency condition with respect to the operator norm $\|\cdot\|_{Y_\ell \leftarrow Y_\ell}$ is required:

(5.5.23) $\| r\boldsymbol{K}_\ell - \boldsymbol{K}_{\ell-1} r \|_{Y_{\ell-1} \leftarrow Y_\ell} \leqslant C_c^* h_\ell^\beta,$

as mentioned in Remark 3.12 for the Nyström case. Together with (20c), one obtains the bound $C_G := C_r^{**} + C_S^* C_c^{**} h_\ell^\beta$ for the left-hand side in (20d).

The numerical results reported in Tables 4-5 confirm that the multi-grid convergence of the algorithm *MGV* from (17) does not require any stronger condition than for the two-grid method. *MGV* converges for $\lambda=0.1$ without any restriction. For $\lambda=0.001$, one even observes that the variant *MGV* converges with $n_0=4$ for $n\geqslant 32$, although the two-grid method diverges at level $\ell=1$ $(n_1=8)$.

n	$n_0=1$	$n_0=2$	$n_0=4$	$n_0=8$	$n_0=16$	$n_0=32$	$n_0=64$
2	$9.3_{10}\mathbf{+0}$	–	–	–	–	–	–
4	$1.2_{10}\mathbf{+2}$	$5.0_{10}\mathbf{+0}$	–	–	–	–	–
8	$1.5_{10}\mathbf{+4}$	$1.9_{10}\mathbf{+1}$	$1.3_{10}\mathbf{+0}$	–	–	–	–
16	$2.1_{10}\mathbf{+8}$	$3.1_{10}\mathbf{+2}$	$1.1_{10}\mathbf{+0}$	$4.1_{10}-1$	–	–	–
32	$4.5_{10}\mathbf{+16}$	$8.5_{10}\mathbf{+4}$	$8.1_{10}-1$	$2.3_{10}-1$	$1.2_{10}-1$	–	–
64	$1.9_{10}\mathbf{+33}$	$6.2_{10}\mathbf{+9}$	$8.0_{10}-1$	$7.9_{10}-2$	$4.3_{10}-2$	$3.2_{10}-2$	–
128	$3.8_{10}\mathbf{+66}$	$3.7_{10}\mathbf{+19}$	$5.7_{10}-1$	$1.4_{10}-2$	$9.6_{10}-3$	$8.9_{10}-3$	$2.0_{10}-2$

Table 5.5.5. Averaged convergence speed of the modified multi-grid method *MGV* from (17) for $\lambda=0.01$

Another multi-grid variant has been proposed by Hemker - Schippers [1]. It is the following procedure *MGV'*.

(5.5.24)	**multi-grid procedure** *MGV'* for solving $\lambda \boldsymbol{f}_\ell = \boldsymbol{g}_\ell + \boldsymbol{K}_\ell \boldsymbol{f}_\ell,\ \ell \geqslant 0$	
(5.5.24a)	<u>procedure</u> *MGV'*$(\ell, \boldsymbol{f}, \boldsymbol{g})$; <u>integer</u> ℓ; <u>array</u> $\boldsymbol{f}, \boldsymbol{g}$;	
(5.5.24b)	<u>if</u> $\ell=0$ <u>then</u> $\boldsymbol{f} := (\lambda I - \boldsymbol{K}_0)^{-1}\boldsymbol{g}$ <u>else</u>	
	<u>begin</u> <u>array</u> $\mathbf{d}, \boldsymbol{\delta}$; <u>integer</u> i;	
(5.5.24c)	$\mathbf{d} := \lambda \boldsymbol{f} - \boldsymbol{g} - \boldsymbol{K}_\ell * \boldsymbol{f}$;	(computing the defect)
(5.5.24d)	$\boldsymbol{f} := \boldsymbol{f} - \frac{1}{\lambda}(I - p\,r)\mathbf{d}$;	(extra correction)
(5.5.24e)	$\mathbf{d} := [(\lambda I - \boldsymbol{K}_{\ell-1}) * r - r * \boldsymbol{K}_\ell) * \mathbf{d}$;	(smoothing)
(5.5.24f$_1$)	$\boldsymbol{\delta} := 0$;	(set starting iterate)
(5.5.24f$_2$)	<u>for</u> $i=1,2$ <u>do</u> *MGV'*$(\ell-1, \boldsymbol{\delta}, \mathbf{d})$;	(two multi-grid calls)
(5.5.24g)	$\boldsymbol{f} := \boldsymbol{f} - p * \boldsymbol{\delta}$	(coarse-grid correction)
	<u>end</u>;	

It might seem that the Picard iteration is missing, but it is implicitly contained in (24e). This assertion is the subject of

Exercise 5.5.19. For $\ell=1$ the multi-grid variant (24) reproduces the two-grid method (4.7), i.e., *MGV'*$(1,\cdot,\cdot)$ and *ZGM*$(1,\cdot,\cdot)$ yield identical results.

Concerning the general convergence analysis of the iteration (24), we refer to Hemker–Schippers [1] or Hackbusch [1, §16.2.2.3]. Here, the attention is only drawn to a special result (cf. (7c)).

Theorem 5.5.20. If $rp = I$, the iteration matrix $M_\ell^{MGV'}$ of the method (24) admits the recursive representation $M_1^{MGV'} = M_1^{ZGM}$ for $\ell = 1$ and

(5.5.25) $$M_\ell^{MGV'} = M_\ell^{ZGM} + p\,(M_{\ell-1}^{MGV'})^2\, r\,(M_\ell^{ZGM} - I) \qquad \text{for } \ell \geq 2.$$

If, in addition $C_r = C_p = 1$, the method MGV' converges if

(5.5.26) $$\| M_\ell^{ZGM} \|_{X_\ell \leftarrow X_\ell} \leq 1 / [\,2\,(1 + \sqrt{2})\,] \approx 0.207 \qquad \text{for all } \ell \geq 1.$$

In contrast to the iteration MGV from (17), the method MGV' requires a larger *amount of computational work* than the standard multi-grid iteration (3). The numerical results of MGV' are very similar to those of MGV (cf. Hackbusch [1, p. 318]), but they are not better than the MGV results.

5.6 Nested Iteration

5.6.1 Algorithm

If a discrete integral equation $\lambda \boldsymbol{f}_\ell = \boldsymbol{g}_\ell + \boldsymbol{K}_\ell \boldsymbol{f}_\ell$ is to be solved at a certain level ℓ, one can proceed as for the two-grid case in Table 4.1: Start with an initial value $\boldsymbol{f}_\ell^0$, e.g., with $\boldsymbol{f}_\ell^0 = 0$, and iterate the multi-grid procedure (5.3) or (5.17) a few times. The necessary number of steps is discussed in Remark 4.7, which refers to the two-grid method, but since the multi-grid convergence is asymptotically the same, the considerations remain valid for the multi-grid method. Because of Remark 4.7, one should perform two iterations:

(5.6.1) $$\boldsymbol{f}_\ell^0 = 0 \;\mapsto\; \boldsymbol{f}_\ell^2$$

by means of (5.3) or (5.17), i.e., by means of (1a) or (1b), respectively:

(5.6.1a) $\boldsymbol{f} := 0;\; MGM(\ell, \boldsymbol{f}, \boldsymbol{g}_\ell);\quad MGM(\ell, \boldsymbol{f}, \boldsymbol{g}_\ell);\quad \boldsymbol{f}_\ell^2 := \boldsymbol{f};$

(5.6.1b) $\boldsymbol{f} := 0;\; MGV(\ell, \boldsymbol{f}, \boldsymbol{g}_\ell, \underline{\text{true}});\; MGM(\ell, \boldsymbol{f}, \boldsymbol{g}_\ell, \underline{\text{true}});\; \boldsymbol{f}_\ell^2 := \boldsymbol{f};$

Remark 5.6.1. Let $\boldsymbol{f}_\ell^2$ be computed according to (1). **(a)** The error of $\boldsymbol{f}_\ell^2$ reads

(5.6.2) $$\| \boldsymbol{f}_\ell - \boldsymbol{f}_\ell^2 \|_{X_\ell} \leq [\,C_{ZGM}(1 + C_* h_\ell^\beta)\,]^2\, h_\ell^{2\beta}\, \| \boldsymbol{f}_\ell \|_{X_\ell},$$

where $C_{ZGM}(1 + C_* h_\ell^\beta)$ stems from (5.10) or (5.21), respectively. Then the *relative error* of $\boldsymbol{f}_\ell^2$ is of the order $O(h_\ell^{2\beta})$.
(b) The work to compute (1) corresponds to $(C_0 + C_1) n_\ell^2$ arithmetical operations according to Theorem 5.4 (for (1a)) or according to Exercise 5.14 (for (1b)). For the standard case (5.4e), i.e., $C_N = \frac{1}{2}$, the numbers become

(5.6.3) $$C_0 + C_1 = 8 \quad \text{for (1a)}, \qquad C_0 + C_1 = 7 \quad \text{for (1b)}.$$

However, it will turn out that more can be achieved with less work if one applies the so-called «nested iteration». The idea of the method is the cheap computation of a good starting iterate. The better the starting iterate the fewer iterations required to improve the result. In the following algorithm the starting iterate is gained by the interpolation $\boldsymbol{f}_\ell^0 := \bar{p}\boldsymbol{f}_{\ell-1}^1$ of the approximate value attained at level $\ell-1$. The interpolation $\bar{p}$ may coincide with the interpolation p from the coarse-grid correction step (5.3f), or may differ.

(5.6.4)	**Nested iteration** for solving the equations $\lambda \boldsymbol{f}_\ell = \boldsymbol{g}_\ell + \boldsymbol{K}_\ell \boldsymbol{f}_\ell$ at all levels $\ell = 0, 1, \ldots, \ell_{\max}$	
(5.6.4a)	$\tilde{\boldsymbol{f}}_0 := (\lambda I - \boldsymbol{K}_0)^{-1}\boldsymbol{g}_0$;	(start at level 0)
	<u>for</u> $\ell := 1$ <u>step</u> 1 <u>until</u> $\ell_{\max}$ <u>do</u>	
(5.6.4b)	<u>begin</u> $\tilde{\boldsymbol{f}}_\ell := \bar{p}\tilde{\boldsymbol{f}}_{\ell-1}$; $MGM(\ell, \tilde{\boldsymbol{f}}_\ell, \boldsymbol{g}_\ell)$ <u>end</u>;	

The term «nested iteration» refers to a variant of (4), which in (4b) involves an inner iteration "<u>for</u> $i := 1$ <u>step</u> 1 <u>until</u> i_ℓ <u>do</u> $MGM(\ell, \tilde{\boldsymbol{f}}_\ell, \boldsymbol{g}_\ell)$" instead of only one single step. Then this i-loop is nested in the ℓ-loop. It is essential for the efficiency of the nested iteration (4) that *one* multi-grid iteration step is sufficient per level.

Remark 5.6.2. **(a)** In contrast to procedure (1), the nested iteration (4) produces the approximations $\tilde{\boldsymbol{f}}_0, \tilde{\boldsymbol{f}}_1, \ldots, \tilde{\boldsymbol{f}}_{\ell_{\max}-1}, \tilde{\boldsymbol{f}}_{\ell_{\max}}$ at all levels $0 \leqslant \ell \leqslant \ell_{\max}$. This fact has two advantages. First, for extrapolation techniques, approximations $\tilde{\boldsymbol{f}}_\ell$ of different levels are required (cf. §4.8.3). Second, often one does not know *a priori* the finest step size h_ℓ to be chosen. In this case, the loop in (4b) does not end at a fixed level number $\ell_{\max}$ but is terminated by a suitable stopping criterion (e.g., $\|\tilde{\boldsymbol{f}}_\ell - \bar{p}\tilde{\boldsymbol{f}}_{\ell-1}\|_\infty \leqslant \varepsilon$).
(b) Instead of the standard multi-grid method (5.3), one may also apply the variant (5.17) in (4b). Then (4b) has to be changed to

(5.6.4b') <u>begin</u> $\tilde{\boldsymbol{f}}_\ell := \bar{p}\tilde{\boldsymbol{f}}_{\ell-1}$; $MGV(\ell, \tilde{\boldsymbol{f}}_\ell, \boldsymbol{g}_\ell,$ <u>true</u>$)$ <u>end</u>;

(c) If the solution of a system $\lambda \boldsymbol{f}_\ell = \boldsymbol{g}_\ell + \boldsymbol{K}_\ell \boldsymbol{f}_\ell$ is sought only at the highest level $\ell = \ell_{\max}$, one still needs the vectors $\boldsymbol{g}_\ell$ for $\ell < \ell_{\max}$. Note that the matrices $\boldsymbol{K}_\ell$ $(\ell < \ell_{\max})$ appear in any way in the multi-grid method. The vectors $\boldsymbol{g}_\ell$ can cheaply be computed from the given vector $\boldsymbol{g}_{\ell_{\max}}$ by

(5.6.5) <u>for</u> $\ell := \ell_{\max}$ <u>step</u> -1 <u>until</u> 1 <u>do</u> $\boldsymbol{g}_{\ell-1} := r\boldsymbol{g}_\ell$;

According to (3.10), the relationship $\boldsymbol{g}_{\ell-1} = r\boldsymbol{g}_\ell$ is true for projection methods. It also holds for the Nyström method if $\Xi_{\ell-1} \subset \Xi_\ell$ (cf. (3.6c)).

5.6.2 Amount of Computational Work

The work for the prolongation $\tilde{\boldsymbol{f}}_\ell := \bar{p}\tilde{\boldsymbol{f}}_{\ell-1}$ in (4b) can be neglected compared with the call $MGM(\ell, \tilde{\boldsymbol{f}}_\ell, \boldsymbol{g}_\ell)$, which requires $C_1 n_\ell^2 + O(n_\ell)$ operations (cf. Theorem 5.4 or Exercise 5.14 for MGV instead of MGM). The mentioned statements require the assumption (5.4d): $n_{\ell-1} \leqslant C_N n_\ell$ with $C_N < 1/\sqrt{2}$. Summation of $C_1 n_\ell^2 + O(n_\ell)$ over $1 \leqslant \ell \leqslant \ell_{\max}$ yields $C_1 n_\ell^2 (1 + C_N^2 + C_N^4 + \ldots) + O(n_\ell) < C_1 n_\ell^2 / (1 - C_N^2) + O(n_\ell)$ and proves the following.

Theorem 5.6.3. Assume (5.4d) with $C_N < 1$. According to Theorem 5.4 or Exercise 5.14, let the multi-grid costs be given by C_1. Then the work required by the nested iteration (4) amounts to

(5.6.6) $\quad C_g n_\ell^2 + O(n_\ell)$ operations with $C_g := C_1/(1-C_N^2)$.

The values of C_1 and C_g are given below for the most important cases $C_N = \frac{1}{2}$ (corresponding to $h_{\ell-1} = 2h_\ell$ in the one-dimensional case) and $C_N = \frac{1}{4}$ (corresponding to $h_{\ell-1} = 2h_\ell$ in the two-dimensional case).

	$C_N = \frac{1}{2}$		$C_N = \frac{1}{4}$	
	C_1	C_g	C_1	C_g
MGM	5	$\frac{20}{3} = 6.666...$	$\frac{19}{7} = 2.714...$	$\frac{304}{105} = 2.895...$
MGV	4.5	6	$\frac{75}{28} = 2.678...$	$\frac{20}{7} = 2.857...$

Table 5.6.1. Number of arithmetical operations for the nested iteration

The comparison of the numbers for C_g ($6.\bar{6}$ or 6) with the respective values 8 or 7 from (3) shows that the nested iteration is less costly than the method (1), although it yields not only *one* solution but approximations at all levels. The value $C_g = 6$ expresses the fact that the total nested iteration (4) with (4b') requires the same amount of work as three Picard iterations at the level ℓ_{max}. The effect is even stronger for higher dimensional problems because of the smaller factor $C_N \leq \frac{1}{4}$.

For a modification which will be presented in §5.6.5, the numbers from Table 1 can be further reduced.

5.6.3 Convergence

It remains to discuss the errors of the approximations $\tilde{\boldsymbol{f}}_0, \tilde{\boldsymbol{f}}_1, \dots, \tilde{\boldsymbol{f}}_{\ell_{max}}$, produced by the nested iteration (4). For this purpose, we need the estimate of the *relative discretisation error*

(5.6.7) $\quad \| \boldsymbol{f}_\ell - \bar{p}\boldsymbol{f}_{\ell-1} \|_{X_\ell} \leq C_D h_\ell^\beta \qquad$ for $1 \leq \ell \leq \ell_{max}$.

The prolongation $\bar{p}$ involved in (7) is the one from (4b) or (4b'). A statement similar to (7) has already been proved in (3.13). The essential assumptions were stability and relative consistency. The term $\boldsymbol{g}_{k-1} - r\boldsymbol{g}_k$ appearing in (3.13) vanishes for the projection or Nyström method with canonical r, or if $\boldsymbol{g}_k$ is chosen according to (5) (cf. Remark 2c)).

The boundedness of the prolongations $\bar{p}$ (cf. (3.4a)) together with that of the ratio $h_{\ell-1}/h_\ell$ is formulated in

(5.6.8) $\quad \| \bar{p} \|_{X_\ell \leftarrow X_{\ell-1}} h_{\ell-1}^\beta / h_\ell^\beta \leq \text{const} \qquad$ for $1 \leq \ell \leq \ell_{max}$.

Let M_ℓ denote either the iteration matrix M_ℓ^{MGM} of the multi-grid method (5.3) or the matrix M_ℓ^{MGV} of the variant (5.17). As in the convergence analysis in §5.5.3 (cf. (5.10)), we suppose that

(5.6.9) $$\|M_\ell\|_{X_\ell \leftarrow X_\ell} \leqslant C_{MG}\, h_\ell^\beta \qquad \text{for } 1 \leqslant \ell \leqslant \ell_{\max}.$$

Theorem 5.6.4. Assume (7), (8), and (9). Then the approximations $\tilde{\boldsymbol{f}}_\ell$ satisfy the error estimates (10) at all levels $0 \leqslant \ell \leqslant \ell_{\max}$:

(5.6.10) $$\|\tilde{\boldsymbol{f}}_\ell - \boldsymbol{f}_\ell\|_{X_\ell} \leqslant C_{MG}\, C_D\, h_\ell^{2\beta} + O(h_\ell^{3\beta}) \qquad (\boldsymbol{f}_\ell\text{: exact solution}).$$

Proof. We have to find bounds E_ℓ for the errors $e_\ell := \tilde{\boldsymbol{f}}_\ell - \boldsymbol{f}_\ell$ such that

(5.6.11a) $$\|e_\ell\|_{X_\ell} \leqslant E_\ell\, h_\ell^{2\beta}.$$

Since $\tilde{\boldsymbol{f}}_0 = \boldsymbol{f}_0$ by (4a), (11a) is satisfied for $\ell = 0$ with

(5.6.11b) $$E_0 := 0.$$

Corollary 1.3 can be applied with $\boldsymbol{f}_\ell^0 := \bar{p}\tilde{\boldsymbol{f}}_{\ell-1}$ and $\boldsymbol{f}_\ell^1 := \tilde{\boldsymbol{f}}_\ell$:

$$\|e_\ell\|_{X_\ell} = \|\tilde{\boldsymbol{f}}_\ell - \boldsymbol{f}_\ell\|_{X_\ell} = \|\boldsymbol{f}_\ell^1 - \boldsymbol{f}_\ell\|_{X_\ell} \leqslant \|M_\ell\|_{X_\ell \leftarrow X_\ell} \|\boldsymbol{f}_\ell^0 - \boldsymbol{f}_\ell\|_{X_\ell} \leqslant$$
$$\leqslant C_{MG}\, h_\ell^\beta \|\boldsymbol{f}_\ell^0 - \boldsymbol{f}_\ell\|_{X_\ell} \leqslant C_{MG}\, h_\ell^\beta \|\bar{p}\tilde{\boldsymbol{f}}_{\ell-1} - \boldsymbol{f}_\ell\|_{X_\ell}.$$

Together with

$$\|\bar{p}\tilde{\boldsymbol{f}}_{\ell-1} - \boldsymbol{f}_\ell\|_{X_\ell} = \|\bar{p}(\tilde{\boldsymbol{f}}_{\ell-1} - \boldsymbol{f}_{\ell-1}) + (\bar{p}\boldsymbol{f}_{\ell-1} - \boldsymbol{f}_\ell\|_{X_\ell} \leqslant$$
$$\leqslant \|\bar{p}\|_{X_\ell \leftarrow X_{\ell-1}} \|\tilde{\boldsymbol{f}}_{\ell-1} - \boldsymbol{f}_{\ell-1}\|_{X_{\ell-1}} + \|\bar{p}\boldsymbol{f}_{\ell-1} - \boldsymbol{f}_\ell\|_{X_\ell} \leqslant$$
$$\leqslant \|\bar{p}\|_{X_\ell \leftarrow X_{\ell-1}} \|e_{\ell-1}\|_{X_{\ell-1}} + C_D\, h_\ell^\beta \leqslant$$
$$\leqslant \|\bar{p}\|_{X_\ell \leftarrow X_{\ell-1}} E_{\ell-1}\, h_{\ell-1}^{2\beta} + C_D\, h_\ell^\beta \leqslant$$
$$\leqslant \text{const}\; E_{\ell-1} h_{\ell-1}^\beta\, h_\ell^\beta + C_D\, h_\ell^\beta,$$

we obtain $\|e_\ell\|_{X_\ell} \leqslant C_{MG}\,[\,C_D + \text{const}\; E_{\ell-1} h_{\ell-1}^\beta\,]\, h_\ell^{2\beta}$, so that (11a) with

(5.6.11c) $$E_\ell := C_{MG}\,[\,C_D + \text{const}\; E_{\ell-1} h_{\ell-1}^\beta\,]$$

is satisfied. The equations (11b,c) define a recursion formula leading to $E_\ell = C_{MG} C_D + O(h_\ell^\beta)$. Insertion into (11a) proves (10). QED

Remark 5.6.5. **(a)** Since $C_{MG} = C_{ZGM} + O(h_\ell^\beta)$ (cf. (5.10)), (10) can be rewritten in the following form:

$$\|\tilde{\boldsymbol{f}}_\ell - \boldsymbol{f}_\ell\|_{X_\ell} \leqslant C_{ZGM}\, C_D\, h_\ell^{2\beta} + O(h_\ell^{3\beta}).$$

(b) The leading error term $C_{ZGM}\, C_D\, h_\ell^{2\beta} = [\,C_{ZGM}\, h_\ell^\beta\,][\,C_D\, h_\ell^\beta\,]$ can be interpreted as $[\,C_{ZGM}\, h_\ell^\beta\,] \times$ *relative discretisation error*. That means that the iteration error $\tilde{\boldsymbol{f}}_\ell - \boldsymbol{f}_\ell$ is always smaller than the relative discretisation error (7) by a factor $C_{ZGM}\, h_\ell^\beta$.

(c) The iteration error $\tilde{\boldsymbol{f}}_\ell - \boldsymbol{f}_\ell$ is even small enough to perform extrapolation.

5.6.4 Numerical Examples

Applying the nested iteration (4) with the multi-grid method *MGV* (i.e., (4b')) to the test equation (4.15-16) with $\lambda = 0.1$, one obtains the results reported in Table 2. Here, $\bar{p}$ is constructed by the piecewise linear interpolation p also used in *MGV*. Since for $\lambda = 0.1$ and $h \geqslant 1/4$ the convergence speed is not very fast (cf. Table 5.4, column $n_0 = 1$), the iteration error $\tilde{\boldsymbol{f}}_\ell - \boldsymbol{f}_\ell$ is initially larger than the discretisation error $\boldsymbol{f}_\ell - R_\ell \boldsymbol{f}$. However, as described in Theorem 4, with increasing dimension the iteration error becomes considerably smaller than the discretisation error.

h_ℓ	iteration $\Vert \tilde{\boldsymbol{f}}_\ell - \boldsymbol{f}_\ell \Vert_\infty$	total $\Vert \tilde{\boldsymbol{f}}_\ell - R_\ell f \Vert_\infty$	discretisation error $\Vert \boldsymbol{f}_\ell - R_\ell f \Vert_\infty$
1	0.0	1.423	1.423
1/2	2.261	$9.240_{10}-1$	1.659
1/4	$4.116_{10}-1$	$6.746_{10}-1$	$2.631_{10}-1$
1/8	$8.870_{10}-3$	$6.081_{10}-2$	$5.194_{10}-2$
1/16	$2.756_{10}-3$	$1.325_{10}-2$	$1.224_{10}-2$
1/32	$8.173_{10}-5$	$2.976_{10}-3$	$3.016_{10}-3$

Table 5.6.2. Errors of the nested iteration for $\lambda = 0.1$

In general, the really interesting error is the total error $\tilde{\boldsymbol{f}}_\ell - R_\ell f$ between the approximation $\tilde{\boldsymbol{f}}_\ell$ and the continuous solution f. As soon as $\Vert \tilde{\boldsymbol{f}}_\ell - \boldsymbol{f}_\ell \Vert_\infty \ll \Vert \boldsymbol{f}_\ell - R_\ell f \Vert_\infty$ this error is little different from the discretisation error $\Vert \boldsymbol{f}_\ell - R_\ell f \Vert_\infty$. According to Definition 2.1b, the restriction $R_\ell f$ is the evaluation of the function f at the abscissae $\Xi_\ell = \{\nu h_\ell : 0 \leqslant \nu \leqslant n_\ell = 1/h_\ell\}$.

5.6.5 Nested Iteration with Nyström Interpolation

If a starting iterate $\boldsymbol{f}_\ell^0$ is given, whose error $\Vert \boldsymbol{f}_\ell^0 - \boldsymbol{f}_\ell \Vert_{Y_\ell}$ is small with respect to the Y_ℓ norm, the smoothing in the multi-grid process by the Picard iteration can be omitted. The remaining coarse-grid correction produces $\boldsymbol{f}_\ell^1$ with

$$\Vert \boldsymbol{f}_\ell^1 - \boldsymbol{f}_\ell \Vert_{X_\ell} \leqslant \Vert G_\ell^{MGV} \Vert_{X_\ell \leftarrow Y_\ell} \Vert \boldsymbol{f}_\ell^0 - \boldsymbol{f}_\ell \Vert_{Y_\ell}$$

[cf. (20e); the estimate for β_ℓ proved in Remark 5.18 carries over to α_ℓ because of $\Vert G_\ell^{MGV} \Vert_{X_\ell \leftarrow Y_\ell} = \alpha_\ell = \beta_\ell |\lambda| / C_K$].

Among other terms, the initial approximation $\boldsymbol{f}_\ell^0 := \bar{p} \tilde{\boldsymbol{f}}_{\ell-1}$ contains the interpolation error $\bar{p} \boldsymbol{f}_{\ell-1} - \boldsymbol{f}_\ell = (\bar{p} r - I) \boldsymbol{f}_\ell + \bar{p} (\boldsymbol{f}_{\ell-1} - r \boldsymbol{f}_\ell)$. The term $(\bar{p} r - I) \boldsymbol{f}_\ell$ is estimated in X_ℓ by $\Vert (\bar{p} r - I) \boldsymbol{f}_\ell \Vert_{X_\ell} = O(h_\ell^\beta)$, but the estimate with respect to the Y_ℓ norm only yields $\Vert (\bar{p} r - I) \boldsymbol{f}_\ell \Vert_{Y_\ell} = O(1)$. Therefore, the smoothing by the Picard step cannot be dispensed with.

In the case of the Nyström method, one can replace the interpolation $\tilde{\boldsymbol{f}}_\ell := \bar{p} \tilde{\boldsymbol{f}}_{\ell-1}$ in (4b) or (4b') by the Nyström interpolation (4.7.8). Inserting (4.7.8) for $n = n_{\ell-1}$ and interpolating only at the abscissae $\Xi_\ell = \{\nu h_\ell : 0 \leqslant \nu \leqslant n_\ell = 1/h_\ell\}$, one obtains an equation of the form

$$\hat{\boldsymbol{f}}_\ell = \tfrac{1}{\lambda} (\boldsymbol{g}_\ell + \boldsymbol{K}_{\ell,\ell-1} \boldsymbol{f}_{\ell-1}), \tag{5.6.12}$$

where $\boldsymbol{K}_{\ell,\ell-1}$ is an $n_{\ell-1} \times n_\ell$ rectangular matrix. If $\boldsymbol{f}_{\ell-1}$ is the discrete solution at the level $\ell-1$, $r\hat{f}_\ell = f_{\ell-1}$ holds for the (trivial) restriction r to the support abscissae $\Xi_{\ell-1}$. Interpolating the approximation $\tilde{f}_{\ell-1} = f_{\ell-1} + \delta_{\ell-1}$ by the error $\delta_{\ell-1}$, one gets the result $\tilde{f}_\ell = \hat{f}_\ell + \delta_\ell$ with $\delta_\ell = \frac{1}{\lambda} K_{\ell,\ell-1}\delta_{\ell-1}$. $\| K_{\ell,\ell-1}\|_{Y_\ell \leftarrow X_{\ell-1}} \leqslant C_K^*$ is found in the same way as (2.11c), i.e., $\| K_\ell \|_{Y_\ell \leftarrow X_\ell} \leqslant C_K$ has been shown. This property proves the estimate

$$\| \delta_\ell \|_{Y_\ell} \leqslant C_K^* \| \delta_{\ell-1} \|_{X_{\ell-1}} / |\lambda|.$$

Like the Picard iteration, the Nyström interpolation leads to a smoothing of the iteration errors $\delta_{\ell-1}$. Since one may assume $\| \hat{f}_\ell - f_\ell \|_{Y_\ell} = O(h_\ell^\beta)$, $f_\ell^0 := \tilde{f}_\ell$ is a starting iterate which in the following multi-grid iteration requires no smoothing by any Picard step. In the case of the multi-grid method *MGV*, the omission of the smoothing is accomplished by $MGV(\ell,\cdot,\cdot,\underline{\text{false}})$. The resulting method is the

(5.6.13)	**nested iteration with Nyström interpolation**	
(5.6.13a)	$\tilde{f}_0 := (\lambda I - K_0)^{-1} g_0;$	(start at level 0)
	$\underline{\text{for}}\ \ell := 1\ \underline{\text{step}}\ 1\ \underline{\text{until}}\ \ell_{max}\ \underline{\text{do}}$	
(5.6.13b)	$\underline{\text{begin}}\ \tilde{f}_\ell := \frac{1}{\lambda}(g_\ell + K_{\ell,\ell-1}\tilde{f}_{\ell-1});\ MGV(\ell, \tilde{f}_\ell, g_\ell, \underline{\text{false}})\ \underline{\text{end}};$	

Since one call of $MGV(\ell, \tilde{f}_\ell, g_\ell, \underline{\text{false}})$ [instead of $MGV(\ell, \tilde{f}_\ell, g_\ell, \underline{\text{true}})$ from (4b')] requires only $C_1' n_\ell^2 + O(n_\ell)$ operations with $C_1' < C_1$ from (5.18c), the multi-grid work is reduced. On the other hand, the Nyström interpolation is much more expensive than $\tilde{f}_\ell := \bar{p}\tilde{f}_{\ell-1}$. Since $K_{\ell,\ell-1}$ is a full $n_\ell \times n_{\ell-1}$ matrix, the Nyström interpolation requires $2n_\ell n_{\ell-1} + O(n_\ell)$ operations. Instead of Theorem 3, one obtains

Remark 5.6.6. Assume (5.4d) with $C_N < 1$. Then the nested iteration (13) with the Nyström interpolation requires an amount of work of

(5.6.14) $\quad C_g' n_\ell^2 + O(n_\ell)$ operations with $C_g' := C_1' / (1 - C_N^2)$.

The values of C_g' are $\frac{10}{3} = 3.\bar{3}$ for $C_N = \frac{1}{2}$ and $19/21 = 0.9047...$ for $C_N = \frac{1}{4}$.

The example from Table 2 combined with the Nyström interpolation yields in the values reported in Table 3. They confirm that in spite of less computational work, the approximations show the desired accuracy.

h_ℓ	iteration $\|\tilde{f}_\ell - f_\ell\|_\infty$	total error $\|\tilde{f}_\ell - R_\ell f\|_\infty$	discretisation error $\|f_\ell - R_\ell f\|_\infty$
1	0.0	1.423	1.423
1/2	1.476	$4.291_{10}-1$	1.659
1/4	$5.014_{10}-1$	$7.645_{10}-1$	$2.631_{10}-1$
1/8	$2.078_{10}-1$	$1.559_{10}-1$	$5.194_{10}-2$
1/16	$3.956_{10}-3$	$1.404_{10}-2$	$1.224_{10}-2$
1/32	$2.513_{10}-5$	$3.025_{10}-3$	$3.016_{10}-3$

Table 5.6.3. Errors of the nested iteration. Nyström interpolation for (4.15) with $\lambda = 0.1$

6 Abel's Integral Equation

6.1 Notations and Examples

6.1.1 Abel's Integral Equation and its Generalisations

The following Volterra integral equation of the first kind is due to Abel (1823):

(6.1.1) $$g(x) = \int_a^x \frac{f(y)}{\sqrt{x-y}}\,dy \qquad \text{for } x \geqslant a.$$

Since the denominator $\sqrt{x-y}$ has a zero at $y=x$, the integral in (1) is to be understood in the improper sense (cf. §6.1.3) and Abel's integral equation is an example of a weakly singular equation.

Replacing $\sqrt{x-y}$ by $(x-y)^\lambda$ with some λ from the interval $0<\lambda<1$, we can generalise equation (1) to obtain

(6.1.2) $$g(x) = \int_a^x \frac{f(y)}{(x-y)^\lambda}\,dy \qquad \text{for } x \geqslant a.$$

In this chapter the analysis will be stressed more than the numerics. The solutions of equations (1) and (2) can be represented explicitly. This enables us to study precisely the consequences of the weak singularity of the order λ. It is well-known that integration is the inverse of differentiation. We shall see that in a certain sense the integration in (2) represents the inversion of a derivative of non-integer order.

6.1.2 Examples from Applications

The following problem is due to Huygens (*Horologium oscillatorium*; 1673). Let $y=\varphi(x)$ be a function starting from the origin (i.e., $\varphi(0)=0$) and monotonically increasing up to the ordinate $H>0$. The graph of the function φ has to describe a trajectory along which a point mass is "sliding down". Starting from the height $y\in[0,H]$ (i.e., from a curve point (x,y) with $\varphi(x)=y$) with zero initial speed, the point mass needs a certain time denoted by $t=t(y)$ to arrive at the origin $(0,0)$. Let a function $\tau: [0,H]\to\mathbb{R}$ be given. How must the function φ from above look like so that the time $t(y)$ coincides with the given function $\tau(y)$?

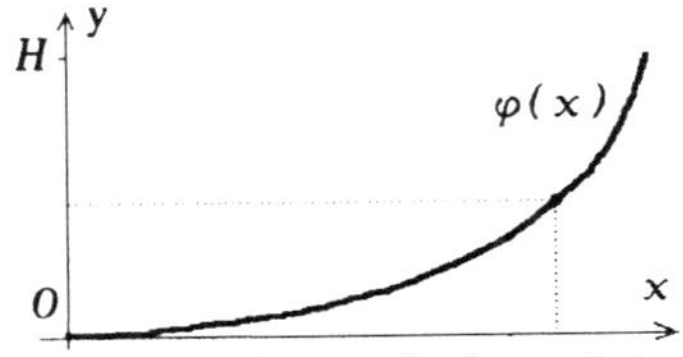

Instead of the curve φ or its inverse φ^{-1}, we shall determine its arc length $\sigma(y)$ between the curve points $(0,0)$ and $(\varphi^{-1}(y),y)$. Using $\varphi'(x)^2+1=\sigma'(\varphi^{-1}(y))$, one retains φ by integration of σ. The necessary connection between σ and the given function τ is obtained as follows. Let $y(t)$ be the height at time t. The starting height at $t=0$ is $y=y_0$. $\sigma(y(t))$ is the arc length of the remaining piece of curve. The absolute value of the speed of the mass point is $v=-d\sigma/dt$, so that $mv^2/2$ is the kinetic energy (m: mass). The potential energy equals $mg(y-y_0)$, where g is the acceleration, y_0 the starting height, y the actual height). The total energy

$$E = \tfrac{1}{2}mv^2+mg(y-y_0)$$

must be constant. The constant equals $E=0$, since $v=0$ holds at the starting point $(\varphi^{-1}(y_0),y_0)$. Solving for v, we obtain $d\sigma/dt=-v=-[2g(y_0-y)]^{1/2}$. The inverse $t=t(\sigma)$ has the derivative

$$\frac{dt}{d\sigma} = -1/[2g(y_0-y)]^{1/2},$$

so that $\frac{dt}{dy}=\frac{dt}{d\sigma}\frac{d\sigma}{dy}=-\sigma'(y)/[2g(y_0-y)]^{1/2}$ describes the fall time as a function of y. The total time has to coincide with τ:

$$\tau(y_0) = t(0)-t(y_0) = -\int_0^{y_0}\frac{dt}{dy}dy = \int_0^{y_0}\frac{\sigma'(y)}{\sqrt{2g(y_0-y)}}dy.$$

Setting $f:=\sigma'/\sqrt{2g}$, $\xi=y$, $x=y_0$, and $a=H$, one obtains Abel's integral equation (1).

The question of Huygens' tautochrone problem was: What is the curve φ with constant fall time: $\tau(y)=1$.

The following *second example* results from an inverse problem appearing, e.g., in the image reconstruction of a centrifuge. In the stationary state, the concentration of a mixture depends only on the radius $r\in(0,1]$. Therefore, the optical absorption coefficient is a function $\alpha=\alpha(r)$ of the radius, too. A ray of intensity I_0 passing the mixture at the height x_0 is reduced to the intensity $I(x_0)=I_0e^{-\beta(x_0)}$, where $\beta=\int_L\alpha$ is the integral of the absorption coefficient along the secant L:

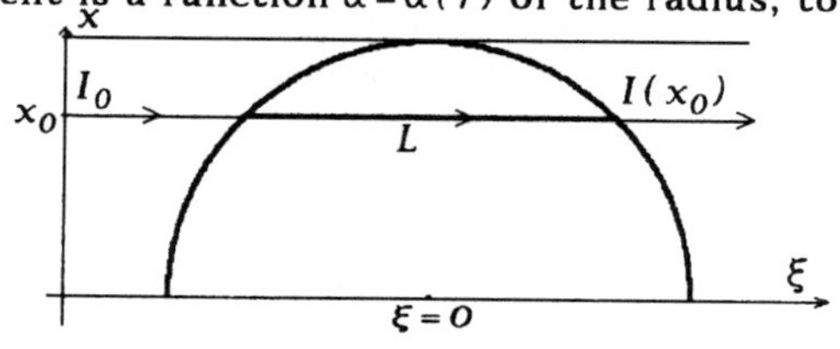

$$\beta(x) = \int_{-\sqrt{1-x^2}}^{\sqrt{1-x^2}}\alpha(\sqrt{x^2+\xi^2})d\xi = 2\int_0^{\sqrt{1-x^2}}\alpha(\sqrt{x^2+\xi^2})d\xi.$$

$\beta(x)=-\log(I(x)/I_0)$ can be measured, so it may be regarded as a known quantity. We are looking for the function α, from which one can draw conclusions about the concentration of the components. The substitution $\rho=1-r^2=1-x^2-\xi^2$ yields

$$\beta(x) = 2\int_0^{\sqrt{1-x^2}}\alpha(\sqrt{x^2+\xi^2})d\xi = -\int_{1-x^2}^{0}\alpha(\sqrt{1-\rho})/\sqrt{1-x^2-\rho}\,d\rho.$$

Setting $g(x) := \beta(\sqrt{1-x})$ and $f(\rho) := \alpha(\sqrt{1-\rho})$, one obtains Abel's integral equation (1).

6.1.3 Improper Integrals

The integrand in (1) has a singularity at $\xi = x$ so that, e.g., the Riemann integral over $[0,x]$ does not exist. The right-hand side in (1) is defined as an improper integral:

$$\int_a^x \frac{f(y)}{\sqrt{x-y}}\,dy := \lim_{\substack{\varepsilon\to 0\\ \varepsilon>0}} \int_a^{x-\varepsilon} \frac{f(y)}{\sqrt{x-y}}\,dy.$$

In the general case, let $S \subset D$ be the set of the singular points of the integral $\int_D f(x)dx$. That means, that for any neighbourhood $U \subset D$ of S, $\int_{D\setminus U} f(x)dx$ is defined as a usual integral. If S intersects ∂D, the neighbourhood U is understood relative to D, i.e., $U = U' \cap D$, where $U' \subset \mathbb{R}^d$ is an open neighbourhood. A singularity can also be caused by the unboundedness of the integration domain D. In this case, ∞ belongs to S. U is called a neighbourhood of ∞ if $D\setminus U$ is bounded. We always assume $\mu(S) = 0$, i.e., S has measure zero. This holds, in particular, if S consists of finitely many points.

Let $U_k \subset U_{k-1}$ be an arbitrary sequence of S-neighbourhoods, which contract to S: $\bigcap_k U_k = S$. We denote this fact by $U_k \to S$. If

$$\lim_{k\to\infty} \int_{D\setminus U_k} f(x)dx$$

exists for any sequence of neighbourhoods $U_k \to S$, the limit is independent of the special sequence U_k and defines the *improper integral* $\int_D f(x)dx$.

The definition given here is adapted to the needs of the chapters which follow and implies the *absolute* integrability, i.e., also the integral $\int_D |f(x)|dx$ exists and is finite. Usually, an integral like, e.g., $\int_0^\infty \frac{\sin x}{x}dx = \pi/2$ is also called an improper integral, since $\int_{[0,\infty)\setminus U} \frac{\sin x}{x}dx \to \pi/2$ holds for all neighbourhoods of the form $U = (R,\infty)$ of ∞ as $R \to \infty$. However, there are other (not connected) neighbourhoods U_k of ∞ such that $\int_{[0,\infty)\setminus U_k} \dots dx$ diverges. Thus, in the sense of the definition above, $\frac{\sin x}{x}$ is not improperly integrable. The functions characterised here as «improperly integrable on D» are ***Lebesgue integrable*** and, hence, belong to $L^1(D)$.

Exercise 6.1.1. Let f be an improperly integrable function on $D = I = [a,b] \subset \mathbb{R}$. Prove the following: **(a)** Let $a \leqslant c \leqslant b$. Even if f is singular at $x = c$, we still have

$$\int_a^b f(x)dx = \int_a^c f(x)dx + \int_c^b f(x)dx.$$

(b) The primitive $\int^x f(\xi)d\xi$ is continuous on I; in particular, it satisfies

$$\lim_{\epsilon\to 0} \int_c^{c+\epsilon} f(x)dx = 0.$$

(c) The following *majorant criterion* holds: If g is integrable on $D\setminus U$ for any neighbourhood U of the singularity and if g satisfies the estimate $|g(x)| \leqslant f(x)$, then improper integrability of f implies that of g. A special application of this criterion reads: Let f be improperly integrable on D and let $g \in L^\infty(D)$. Then also the product fg is improperly integrable on D.
(d) The substitution rule is valid for improper integrals.

(e) Integration by parts is valid.

Remark 6.1.2. **(a)** In $[-1,1]$ the function $f(x)=x^{-\lambda}$ is improperly integrable if and only if $\lambda<1$.
(b) In $D=K_R(0)\subset\mathbb{R}^d$, the function $f(x_1,\ldots,x_d)=r^{-\lambda}$ with $r^2=\sum x_i^2$ is improperly integrable if and only if $\lambda<d$.

Proof. (a) $\int_\varepsilon^1 f(x)dx = x^{1-\lambda}/(1-\lambda)|_\varepsilon^1 \to 1/(1-\lambda)$ as $\varepsilon\to 0$.
(b) Let U be a neighbourhood of the singularity $x=0$. There are $0<\varepsilon<\eta\leqslant R$ with $K_\varepsilon(0)\subset U\subset K_\eta(0)$. Since $f\geqslant 0$, we have $\int_{D\setminus K_\varepsilon(0)} f dx \geqslant \int_{D\setminus U} f dx \geqslant \int_{D\setminus K_\varepsilon(0)} f dx$. Since $U\to\{0\}$ implies $\varepsilon\to 0$, it suffices to show the convergence of $\int_{D\setminus K_\varepsilon(0)} f(x)dx$. Replacing x by polar coordinates ($dx=r^{d-1}drd\Omega$) yields

$$\int_{D\setminus K_\varepsilon(0)} f(x)dx = \int_{D\setminus K_\varepsilon(0)} r^{d-1-\lambda}drd\Omega = \int_\varepsilon^R r^{d-1-\lambda}dr\int_{\partial K_1(0)} d\Omega,$$

where the second integral is taken over the surface of the d-dimensional unit sphere. Hence, the assertion follows from (a). QED

Exercise 6.1.3. Prove that the function $f(x)=1/(x\log x)$ is *not* improperly integrable at $x=0$.

Integral operators involve (possibly improper) integrals with a further *parameter*. For this purpose, we investigate the (improper) integral

$$F(p) := \int_D f(p,x)dx \qquad \text{for } p\in P, \tag{6.1.3}$$

where P is the domain of definition of the parameter p. For $U\subset D$ we set

$$\delta(U) := \sup\{|\int_U f(p,x)dx|:\ p\in P\}. \tag{6.1.4}$$

If $\delta(U_k)\to 0$ holds for any sequence U_k with $\mu(U_k)\to 0$, then «the integrals $\int_D f(p,x)dx$ *exist uniformly*». Note that the measures $\mu(U_k)$ converge to zero if the neighbourhoods $U_k\to S$ contract to the singularity set S.

An situation which often arises is discussed in part (a) of the next exercise.

Exercise 6.1.4. **(a)** Let $f(p,x)=\varphi(x-p)$ for $x,p\in D$. The domain of definition of φ is $E:=\{t=x-p:\ x,p\in D\}$. If φ is improperly integrable on E, the improper integrals $\int_D f(p,x)dx$ exist uniformly.
(b) For $f(p,x)=x^{-1-1/p}$ with $p,x\in D:=[1,\infty)$, the integrals $\int_D f(p,x)dx$ exist, but not uniformly.
(c) Let $F(x)$ be an improperly integrable majorant of $f(p,x)$, i.e., $|f(p,x)|\leqslant F(x)$, then the improper integral of f exists uniformly.

The continuous dependence of the integrals on the parameter p is treated in the following lemma. Note that in general, the set S of singular points may depend on the parameter p. Therefore, it is denoted by S_p.

Lemma 6.1.5. $f(p,x)$ is assumed to be continuous in $P\times D$ except for singularities in $\{(p,x):\ p\in P,\ x\in S_p\}$. Let the integrals $F(p)$ from (3) exist uniformly. Then $F(p)$ is a *continuous* function of $p\in P$.

Proof. Let $\varepsilon > 0$ and $p \in P$ be given. Choose an open neighbourhood $U \supset S_p$ with $\delta(U) \leqslant \varepsilon/3$. By definition (4), $F_U(p) := \int_{D\setminus U} f(p,x)dx$ satisfies

$$|F(p) - F_U(p)| \leqslant \delta(U) \leqslant \varepsilon/3.$$

Because of the compactness of $D\setminus U$ and the continuity of f, there is an η such that $|p-q| \leqslant \delta$ implies

$$|f(p,x) - f(q,x)| \leqslant \delta(U)/\mu(D\setminus U) \quad \text{for all } x\in D\setminus U \quad \text{and } U \supset S_q.$$

Integration over $x \in D\setminus U$ yields $|F_U(p) - F_U(q)| \leqslant \delta(U)$. Since we also have $|F(q) - F_U(q)| \leqslant \delta(U)$, it follows that

$$|F(p) - F(q)| \leqslant 3\,\delta(U) \leqslant \varepsilon \qquad \text{for all } q\in P \text{ with } |p-q| \leqslant \eta.$$ QED

Even if the function $f(p,x)$ itself does not contain singularities, its derivatives may do. The «improperly integrable derivative» f_p is understood in the weak sense that f_p is improperly integrable and has to satisfy

(6.1.5) $$\int_{p_0}^{p_1} f_p(p,x)dp = f(p_1,x) - f(p_0,x) \qquad \text{for } p_1, p_0 \in P,\ x\in D.$$

The following lemma shows when *differentiation under the integral sign* is allowed.

Lemma 6.1.6. Let $P \subset \mathbb{R}$ be a connected interval. The improperly integrable derivative $f_P(p,x)$ must be continuous in $P\times D$ except at its singularities. The integrals $\int_D f_P(p,x)dx$ have to exist uniformly. For at least one $p_0 \in P$, $\int_D f(p_0,x)dx$ must exist. Then the following identity holds:

(6.1.6) $$\frac{d}{dp}\int_D f(p,x)dx = \int_D f_P(p,x)dx \qquad \text{for all } p\in P.$$

Proof. Let p be given. Without loss of generality, let $p \geqslant p_0$ and $P = [p_0, p]$. By Lemma 5, $F(p) := \int_D f_P(p,x)dx$ is continuous. Therefore, the integral $\int_{p_0}^{p}\int_D f_P(p,x)dx\,dp$ exists. The continuity of $F(p)$ implies $f_P \in L^1(P\times D)$. The *Theorem of Fubini* permits an interchanging of the order of the integrations:

(6.1.7) $$\int_P\int_D f_P(p,x)dx\,dp = \int_D\int_P f_P(p,x)\,dp\,dx.$$

Since the double integral on the left-hand side exists, also $\int_P f_P(p,x)dp$ exists almost everywhere and is integrable on D. By (5), the latter integral can be written as $f(p,x) - f(p_0,x)$. Therefore, the right-hand side in (7) becomes $\int_D [f(p,x) - f(p_0,x)]dx$. Since, by assumption, $\int_D f(p_0,x)\,dx$ exists, the existence of $\int_D f(p,x)dx$ follows. Therefore, (7) takes the form

$$\int_{p_0}^{p} F(p')dp' = \int_D f(p,x)dx - \int_D f(p_0,x)dx.$$

Since F is continuous, the left-hand side has the derivative $F(p)$, while $\frac{d}{dp}\int_D f(p,x)dx$ is the derivative of the right-hand side. QED

6.2 A Necessary Condition for a Bounded Solution

We shall see in the next theorem that Abel's generalised integral equation (1.2) cannot have bounded solutions, if the inhomogeneity g does not fulfil certain smoothness and initial conditions.

Theorem 6.2.1. Let $0<\lambda<1$, $a<b$, and $f \in L^{\infty}([a,b])$. Then the integral

(6.2.1) $$g(x) = \int_a^x \frac{f(y)}{(x-y)^{\lambda}}\,dy \qquad \text{for } a \leqslant x \leqslant b$$

exists and satisfies $g \in C^{1-\lambda}([a,b])$ as well as the initial condition

(6.2.2) $$g(a) = 0.$$

Proof. (i) The existence of the integrals (1) follows from the improper integrability of $f(y)/(x-y)^{\lambda}$. Hence, also the assumption (3.4.2a) of Theorem 3.4.2 is satisfied if, formally, we regard Abel's equation as a Fredholm integral equation with kernel k given by

$$k(x,y) = (x-y)^{-\lambda} \text{ for } a \leqslant y \leqslant x \leqslant b \quad \text{and} \quad k(x,y) = 0, \text{ otherwise.}$$

It remains to check condition (3.4.2b). For the left-hand side $\int_a^b |k(\xi,y)-k(x,y)|\,dy$ in (3.4.2b) we may, without loss of generality, assume $a \leqslant x \leqslant \xi \leqslant b$ and obtain

$$\int_a^b |k(\xi,y)-k(x,y)|\,dy = \int_a^x (x-y)^{-\lambda}dy - \int_a^x (\xi-y)^{-\lambda}dy + \int_x^{\xi} (\xi-y)^{-\lambda}dy =$$
$$= \{(x-a)^{1-\lambda} - (\xi-a)^{1-\lambda} + 2(\xi-x)^{1-\lambda}\}/(1-\lambda) \leqslant \frac{3}{1-\lambda}|\xi-x|^{1-\lambda}$$

(cf. Lemma 2). Thereby, Theorem 3.4.2 proves $g = Kf \in C^{1-\lambda}([a,b])$. QED

Lemma 6.2.2. $\varphi(t) = t^{\alpha}$ with $0 \leqslant \alpha \leqslant 1$ is globally Hölder continuous on $[0,\infty)$:

(6.2.3) $$|\xi^{\alpha} - \zeta^{\alpha}| \leqslant |\xi - \zeta|^{\alpha} \qquad \text{for all } \xi,\zeta \geqslant 0.$$

Proof. Without loss of generality, let $0 \leqslant \zeta \leqslant \xi$. The case $\xi = 0$ is trivial; therefore assume $\xi > 0$. After division by ξ^{α} one sees that one can restrict the considerations to the case $0 \leqslant \zeta \leqslant \xi = 1$. Note that (3) is equivalent to $\varphi(\zeta) \leqslant 1$, where $\varphi(\zeta) := (1-\zeta^{\alpha})/(1-\zeta)^{\alpha}$. From $\varphi'(\zeta) \leqslant 0$ one concludes that $\varphi(\zeta) \leqslant \varphi(0) = 1$ for all $\zeta \in [0,1]$. QED

Theorem 1 shows that among others, $g(a) = 0$ is a necessary condition for a *bounded* solution of Abel's integral equation. For Huygens' tautochrone problem from §6.1.2 the constant function $\tau(x) = 1$ takes the place of g. Obviously, condition (2) is not satisfied! At the end of §6.4 we shall have to investigate whether even without condition (2) solutions exist. Of course, by Theorem 1, such solutions must be *unbounded*.

6.3 Euler's Integrals

As tools for further reformulations we introduce the Euler integrals. Since often special functions are not so comprehensively taught in analysis lectures, the following calculations are presented in detail.

Definition 6.3.1. Let $p>0$, $q>0$. *Euler's B-integral (or beta-integral or Euler's integral of the first kind)* is defined by

(6.3.1) $$B(p,q) = \int_0^1 x^{p-1}(1-x)^{q-1}dx.$$

The *second Euler integral* (or *Euler's integral of the second kind*) is the *gamma function*

(6.3.2) $$\Gamma(p) = \int_0^\infty e^{-t}t^{p-1}dt.$$

The integrals are improperly integrable with respect to both ends of the interval as long as the assumptions $p>0$, $q>0$ are valid. The gamma function is well-known because of its relationship with the factorial function.

Exercise 6.3.2. Prove that $\Gamma(p)=(p-1)!$ for $p\in\mathbb{N}$.

In the formulation (2), the substitution $t=x^2$ leads to

(6.3.2′) $$\Gamma(p) = 2\int_0^\infty e^{-x^2}x^{2p-1}dx \qquad \text{for } p>0.$$

For $p>0$ and $q>0$ use (2′) for representing $\Gamma(p)\Gamma(q)$:

$$\Gamma(p)\Gamma(q) = 4\int_0^\infty e^{-x^2}x^{2p-1}dx\int_0^\infty e^{-y^2}y^{2q-1}dy =$$

$$= 4\int_0^\infty\int_0^\infty e^{-(x^2+y^2)}x^{2p-1}y^{2q-1}\,dx\,dy.$$

The latter double integral is taken over the quarter plane $x\geqslant 0$, $y\geqslant 0$. This domain is described in *polar coordinates*

$$x = r\cos\varphi, \quad y = r\sin\varphi$$

with the radius $r\in[0,\infty)$ and the angle φ satisfying $0\leqslant\varphi\leqslant\frac{\pi}{2}$. Since $dx\,dy = r\,dr\,d\varphi$, we have

$$\Gamma(p)\Gamma(q) = 4\int_0^{\pi/2}\int_0^\infty e^{-r^2}r^{2p+2q-2}\cos^{2p-1}\varphi\,\sin^{2q-1}\varphi\,r\,dr\,d\varphi.$$

The integrand can be split into an r- and a φ-dependent factor so that the double integral becomes the following product:

$$\Gamma(p)\Gamma(q) = \underbrace{\Big[2\int_0^{\pi/2}\cos^{2p-1}\varphi\,\sin^{2q-1}\varphi\,d\varphi\Big]}_{=:I}\underbrace{\Big[2\int_0^\infty e^{-r^2}r^{2p+2q-2}dr\Big]}_{=\Gamma(p+q)}.$$

The second integral is recognized as the gamma function evaluated at $p+q$. For further reformulation of the first integral I, substitute $x=\cos^2\varphi$, $dx=2\sin\varphi\cos\varphi\,d\varphi$, and use $\sin^2\varphi=1-\cos^2\varphi$:

$$I = \int_0^{\pi/2} \cos^{2p-2}\varphi \sin^{2q-2}\varphi \; 2\sin\varphi\cos\varphi\,d\varphi =$$

$$= \int_0^{\pi/2} \underbrace{(\cos^2\varphi)^{p-1}}_{x} \underbrace{(1-\cos^2\varphi)^{q-1}}_{1-x} \underbrace{2\sin\varphi\cos\varphi\,d\varphi}_{dx} =$$

$$= \int_0^1 x^{p-1}(1-x)^{q-1}dx = B(p,q).$$

This proves the following relationship between Euler's integrals of the first and second kind:

(6.3.3) $$B(p,q) = \frac{\Gamma(p)\Gamma(q)}{\Gamma(p+q)} \qquad \text{for } p>0,\ q>0.$$

A longer calculation yields the integral value

(6.3.4) $$\int_0^\infty \frac{x^{p-1}}{1+x}dx = \frac{\pi}{\sin\pi p} \qquad \text{for } 0<p<1.$$

For $q=1-p$, we have $\Gamma(p+q)=\Gamma(1)=0!=1$. Therefore, we conclude from (3) that

$$\Gamma(p)\Gamma(1-p) = B(p,1-p) = \int_0^1 x^{p-1}(1-x)^{-p}dx.$$

Substitution of $y=\frac{x}{1-x}$, i.e., $x=\frac{y}{1+y}$, $1-x=\frac{1}{1+y}$, $dx=(1+y)^{-2}dy$, yields

$$\Gamma(p)\Gamma(1-p) = \int_0^\infty \left(\frac{y}{1+y}\right)^{p-1}\left(\frac{1}{1+y}\right)^{p}(1+y)^{-2}dy = \int_0^\infty \frac{y^{p-1}}{1+y}dy.$$

Equation (4) proves the relation

(6.3.5) $$B(p,1-p) = \Gamma(p)\Gamma(1-p) = \frac{\pi}{\sin\pi p} \qquad \text{for } p\notin\mathbb{Z}$$

for $0<p<1$. One obtains the equality for the remaining $p\notin\mathbb{Z}$ by analytical continuation. Another direct proof of (5) exploits the alternative representation $\Gamma(z):=\lim_{n\to\infty} z(z+1)(z+2)\cdot\ldots\cdot(z+n)/(n!n^z)$ of the gamma function and the product representation $\sin x = x\prod_{\nu=1}^{\infty}(1-(\frac{x}{\nu\pi})^2)$. In this case, one obtains integral (4) from equation (5).

Exercise 6.3.3. Let $p>0$ and $q>0$. Show for all $a<b\in\mathbb{R}$ that

(6.3.6) $$\int_a^b (x-a)^{p-1}(b-x)^{q-1}dx = (b-a)^{p+q-1}B(p,q).$$

6.4 Inversion of Abel's Integral Equation

The *uniqueness* of the solution of the generalised Abel integral equation can be shown constructively by a unique representation of a solution, provided there is at least one solution.

Lemma 6.4.1. Let $0<\lambda<1$. **(a)** If the generalised Abel integral equation (1.2) has a solution $f\in C([a,b])$, it can be represented by

$$(6.4.1)\qquad f(t)=\frac{\sin\pi\lambda}{\pi}\frac{d}{dt}\int_a^t\frac{g(x)dx}{(t-x)^{1-\lambda}}\qquad\text{for } a\leqslant t\leqslant b.$$

(b) Assume that the (not necessarily continuous) solution f leads to an improper integral in (1.2) for almost all $a<x\leqslant b$ and that g leads to an improper integral in (1) for almost all $a<t\leqslant b$. Then formula (1) represents the solution f at all points t for which it is continuous.
(c) For all $a<t\leqslant b$, the primitives coincide in both of the cases (a) and (b):

$$(6.4.1')\qquad \int_a^t\frac{g(x)dx}{(t-x)^{1-\lambda}}=\frac{\pi}{\sin\pi\lambda}\int_a^t f(y)dy.$$

Proof. In case (a), f belongs in particular to $L^\infty([a,b])$. By Theorem 2.1, g must be Hölder continuous. Hence, the function

$$\frac{g(x)}{(t-x)^{1-\lambda}}=\frac{1}{(t-x)^{1-\lambda}}\int_a^x\frac{f(y)}{(x-y)^\lambda}dy\qquad(a\leqslant x\leqslant t)$$

is improperly integrable in $[a,t]$ with respect to x. In case (b), this function is assumed to be improperly integrable. In both cases, we have

$$\int_a^t\frac{g(x)dx}{(t-x)^{1-\lambda}}=\int_a^t\Big[\int_a^x\frac{f(y)}{(x-y)^\lambda}dy\Big]\frac{dx}{(t-x)^{1-\lambda}}=\int_a^t\int_a^x\frac{f(y)\,dy\,dx}{(x-y)^\lambda(t-x)^{1-\lambda}}.$$

The latter double integral is taken over the triangle $D:=\{(x,y):0\leqslant y\leqslant x\leqslant t\}$. Interchanging the integrations (Theorem of Fubini) and using (3.6) for $p=1-\lambda$, $q=1-p=\lambda$, $p+q-1=0$, one is led to

$$\int_a^t\frac{g(x)dx}{(t-x)^{1-\lambda}}=\int_a^t\Big[\underbrace{\int_y^t\frac{dx}{(x-y)^\lambda(t-x)^{1-\lambda}}}_{=B(1-\lambda,\lambda)\text{ by (3.6)}}\Big]f(y)dy=B(1-\lambda,\lambda)\int_a^t f(y)dy.$$

Equation (3.5) shows (1'). At any point of continuity of f, the right-hand side is a continuously differentiable function with respect to t. Therefore, at those points, the left-hand side is also differentiable with respect to t. The differentiation yields the desired representation (1). If $f\in C([a,b])$, the representation (1) holds for all $a\leqslant t\leqslant b$. QED

The following question is raised by the representation (1): How can the differentiation of the integral be performed with respect to t? The interchanging of the differentiation d/dt and integration $\int\ldots dx$ is *not admissible*, since then the not (improperly) integrable function $g(x)(\lambda-1)/(1-x)^{2-\lambda}$ would appear under the integral sign. An answer is given by following lemma.

Lemma 6.4.2. Let $0<\lambda<1$ and $g\in C^1([a,b])$. Then the derivative

$$(6.4.2)\qquad \frac{d}{dt}\int_a^t\frac{g(x)dx}{(t-x)^{1-\lambda}}=\frac{g(a)}{(t-a)^{1-\lambda}}+\int_a^t\frac{g'(x)dx}{(t-x)^{1-\lambda}}\qquad\text{for } a<t\leqslant b$$

exists. If in addition $g(a)=0$, the first term vanishes and the derivative also exists for $t=a$.

Proof. Integration by parts yields

$$\int_a^t \frac{g(x)dx}{(t-x)^{1-\lambda}} = -g(x)\,\frac{(t-x)^\lambda}{\lambda}\Big|_{x=a}^{x=t} + \int_a^t g'(x)\frac{(t-x)^\lambda}{\lambda}dx.$$

The integrated part vanishes at the upper boundary $x=t$, so that only $g(a)(t-a)^\lambda/\lambda$ remains. Differentiation of this term with respect to t yields the first term on the right-hand side of (2). To obtain the remaining integral, one has to differentiate with respect to the upper limit of the integration and under the integral sign with respect to t. The differentiation with respect to the upper boundary yields zero, since the integrand vanishes at $x=t$. The derivative under the integral leads to $\int_a^t \frac{g'(x)dx}{(t-x)^{1-\lambda}}$, since the integrand which arise is improperly integrable and, therefore, the differentiation under the integral sign is valid (cf. Lemma 1.6). This finishes the proof of (2). QED

Exercise 6.4.3. The assumption $g\in C^1([a,b])$ in Lemma 2 is not necessary. Prove that (2) holds for $g(x):=(x-a)^{1-\lambda}$, too.

The following first result concerning the existence of a solution requires the continuous differentiability of g.

Theorem 6.4.4. Let $0<\lambda<1$ and $g\in C^1([a,b])$ with $g(a)=0$. Then the generalised Abel integral equation (1.2) has a unique solution $f\in C^\lambda([a,b])$. It is represented by

$$f(t) = \frac{\sin\pi\lambda}{\pi}\int_a^t \frac{g'(x)dx}{(t-x)^{1-\lambda}} \qquad \text{for } a\leqslant t\leqslant b. \tag{6.4.3}$$

Proof. (i) From Lemmata 1 and 2, we know that a continuous solution must have the representation (3), if it exists.

(ii) Applying Theorem 2.1 with $g'\in C([a,b])$ instead of f and $1-\lambda$ instead of λ, one obtains the statement $f\in C^\lambda([a,b])$ for f from (3).

(iii) It remains to show that f given by (3) is a solution. We insert the representation (3) into the integral equation (1.2) and interchange the integrations:

$$\int_a^x \frac{f(y)}{(x-y)^\lambda}dy = \frac{\sin\pi\lambda}{\pi}\int_a^x\int_a^t \frac{g'(x)dx}{(t-x)^{1-\lambda}}\,\frac{dt}{(x-t)^\lambda} =$$

$$= \frac{\sin\pi\lambda}{\pi}\int_a^x \underbrace{\int_t^x \frac{dt}{(t-x)^{1-\lambda}(x-t)^\lambda}}_{=B(1-\lambda,\lambda)=\pi/\sin\pi\lambda \text{ by (3.6)}} g'(x)dx = \int_a^x g'(x)dx = g(x).$$

QED

From Theorem 2.1 we know that $g\in C^{1-\lambda}([a,b])$ and $g(a)=0$ are necessary conditions for bounded solutions. However, the assumption $g\in C^1([a,b])$ from Theorem 4 is stronger than necessary. A weaker condition is used in the following theorem.

Theorem 6.4.5. Let $0<\lambda<1$, $g\in C([a,b])$ with $g(a)=0$ and assume that

(6.4.4) $$G(t) := \int_a^t \frac{g(x)dx}{(t-x)^{1-\lambda}} \in C^1([a,b])$$

is differentiable. Then the generalised Abel integral equation (1.2) admits a unique continuous solution, which is represented by (1).

Proof. The uniqueness is proved in Lemma 1. It remains to check whether f given by (1) is a solution. Insertion of (1) into (1.2) yields

$$\int_a^x \frac{f(y)}{(x-y)^\lambda}dy = \frac{\sin\pi\lambda}{\pi}\int_a^x \frac{G'(y)}{(x-y)^\lambda}dy \underset{\uparrow\ \text{Lemma 2}}{=} \frac{d}{dx}\frac{\sin\pi\lambda}{\pi}\int_a^x \frac{G(y)}{(x-y)^\lambda}dy =$$

$$\underset{\uparrow\ (4)}{=} \frac{d}{dx}\frac{\sin\pi\lambda}{\pi}\int_a^x\int_a^y \frac{g(t)dt}{(y-t)^{1-\lambda}}\frac{dy}{(x-y)^\lambda} \underset{\uparrow\ \text{as in Theorem 4}}{=} \frac{d}{dx}\int_a^x g(t)dt = g(x).$$ ∎

The assumption of Theorem 5 concerning g is indirectly formulated via $G\in C^1([a,b])$. One obtains a sufficient condition for $G\in C^1([a,b])$ from the Hölder continuity of g with exponent $\alpha>1-\lambda$:

Lemma 6.4.6. Let $0<\lambda<1$, $\alpha>1-\lambda$ and $g\in C^\alpha([a,b])$ with $g(a)=0$. Then the function $G(t)$ defined in (4) belongs to $C^1([a,b])$. Its derivative equals

(6.4.5) $$\frac{d}{dt}G(t) := (1-\lambda)\int_a^t \frac{g(t)-g(x)}{(t-x)^{2-\lambda}}dx + (t-a)^{\lambda-1}g(t).$$

Proof. Let $a\leq t'<t$ and set $\delta:=t-t'$. The difference quotient $\frac{1}{\delta}[G(t)-G(t')]$ can be represented as

(6.4.6) $$\frac{1}{\delta}[G(t)-G(t')] = \frac{1}{\delta}\int_{t'}^t [g(x)-g(t)](t-x)^{\lambda-1}dx +$$
$$+\frac{1}{\delta}\int_a^{t'}[g(x)-g(t)][(t-x)^{\lambda-1}-(t'-x)^{\lambda-1}]dx +$$
$$+ g(t)\frac{1}{\delta}\Big[\int_a^t (t-x)^{\lambda-1}dx - \int_a^{t'}(t'-x)^{\lambda-1}dx\Big].$$

Since $|g(x)-g(t)|\leq C|t-x|^\alpha$, the first integral in (6) is of order $O(\delta^{\alpha+\lambda-1})$ and vanishes as $\delta\to 0$. The second integral converges to the first term $\int_a^t[g(x)-g(t)][(\lambda-1)(t-x)^{\lambda-2}]dx$ on the right-hand side of (5), since by $|g(t)-g(x)|/(t-x)^{2-\lambda}=O(|t-x|^{\alpha+\lambda-2})$ and $\alpha+\lambda-2>-1$ the integrand is improperly integrable. After evaluation of the integrals, the third expression in (6) equals $g(t)[(t-a)^\lambda-(t'-a)^\lambda]/(\delta\lambda)$ and converges to $g(t)(t-a)^{\lambda-1}$, which is the second term in (5). ∎

The assumption $g\in C^\alpha([a,b])$ with $\alpha>1-\lambda$ is sufficient for $G(t)\in C^1([a,b])$, but not necessary. Vice versa, $g\in C^{1-\lambda}([a,b])$ is not sufficient as seen below.

Exercise 6.4.7. Prove the following: **(a)** For $g(x):=(x-a)^{1-\lambda}\in C^{1-\lambda}([a,b])$ one obtains the continuously differentiable function $G(t)=(t-a)B(2-\lambda,\lambda)$. **(b)** For $g(x):=(b-x)^{1-\lambda}\in C^{1-\lambda}([a,b])$, the function $G(t)$ is continuously differentiable in $[a,b)$; but its derivative is unbounded at $t=b$.

In the cases of Theorem 4 and Lemma 6, the initial value of the solution is

(6.4.7) $f(a) = 0.$

In contrast, Exercise 7a shows a case for which (7) is not true. Note that $g \in C^{\alpha}([a,b])$ with $\alpha < 1-\lambda$ cannot hold for bounded solutions f. On the other hand, $\alpha > 1-\lambda$ already implies (7).

For another formulation of the previous results, we define the operators D and K_λ by

$$D = \frac{d}{dx}, \quad (K_\lambda \varphi)(x) := \int_a^x \frac{\varphi(y)}{(x-y)^\lambda} dy \qquad (0 \leq \lambda < 1,\ a \leq x \leq b).$$

Abel's integral equation (1.2) takes the form

(6.4.8) $g = K_\lambda f.$

Remark 6.4.8. For all $0<\lambda<1$, the following identity holds:

(6.4.9) $D K_{1-\lambda} K_\lambda = \frac{\pi}{\sin \pi\lambda} I$ on $C([a,b])$.

Proof. Let $f \in C([a,b])$ and $g = K_\lambda f$. Then, f is a solution of (1.2) and, by Lemma 1, has the representation (1), i.e., $f := \frac{\sin \pi\lambda}{\pi} D K_{1-\lambda} g$. QED

Repeating the proof of Lemma 1 step by step, we obtain

(6.4.10) $K_\sigma K_\lambda = B(1-\lambda, 1-\sigma) K_{\sigma+\lambda-1} \qquad (\sigma, \lambda < 1).$

Then (9) follows from (10), since $(K_0 f)(t) = \int_a^t f(\xi) d\xi$ is the primitive of f. For $\lambda = \frac{1}{2}$ one obtains $\frac{\sin \pi\lambda}{\pi} = \frac{1}{\pi}$ and

(6.4.11a) $D K_{1/2}^2 = \pi I$ on $C([a,b])$.

The inverse $K_{1/2}^{-1}: g \mapsto f$ appearing in Abel's integral equation can be viewed as «half differentiation». Introduce the subspace $C_0^1([a,b]) := \{\varphi \in C^1([a,b]) : \varphi(a) = 0\}$. Then, the factors in (11a) can be interchanged:

(6.4.11b) $K_{1/2}^2 D = \pi I$ on $C_0^1([a,b])$.

Proof. Let $g \in C_0^1([a,b])$ be arbitrary. By Theorem 4, $f := \frac{\sin \pi\lambda}{\pi} K_{1-\lambda} D g$ is the solution of $g = K_\lambda f$. The choice $\lambda = \frac{1}{2}$ shows (11b). QED

Now, we concentrate to the case of unbounded solutions, which must arise for $g(a) \neq 0$. The tautochrone problem leads to the inhomogeneity $g(t) = 1$, which does not satisfy condition (2.2). The following theorem describes the singular behaviour of the corresponding solution.

Theorem 6.4.9. Let $0<\lambda<1$ and suppose $g \in C([a,b])$ satisfies (4), i.e. $G \in C^1([a,b])$. Then, the generalised Abel integral equation (1.2) has a unique solution f with the following property:

(6.4.12) $f \in C((a,b]), \quad f(x) = \frac{\sin \pi\lambda}{\pi} g(a)(x-a)^{\lambda-1} + \text{continuous function}.$

Proof. (i) First, we look for a solution for the constant function $g(x) = 1$.

Lemma 1 together with equation (2) from Lemma 2 yields $f(x)=\frac{\sin\pi\lambda}{\pi}(x-a)^{\lambda-1}$. Inserting this function into (1.2) shows that it satisfies the integral equation for all $x>a$.
(ii) Split g into the sum g_0+g_1, where $g_0:=g(a)$ is a constant function and, therefore, $g_1(x):=g(x)-g(a)$ satisfies $g_1(a)=0$. The solution f is the sum f_0+f_1, where the solutions f_i correspond to g_i. By part (i), $f_0(x)=\frac{\sin\pi\lambda}{\pi}g(a)(x-a)^{\lambda-1}$ holds. Theorem 5 is applicable to g_1 and ensures $f_1\in C([a,b])$. ∎

Exercise 6.4.10. Solve the tautochrone problem from §6.1.2.

6.5 Reformulation for Kernels $k(x,y)/(x-y)^{\lambda}$

A somewhat more general Volterra integral equation of the first kind is

(6.5.1) $$g(x) = \int_a^x \frac{k(x,y)}{(x-y)^{\lambda}} f(y)\,dy \qquad \text{for } x\geqslant a,$$

where $k(x,y)$ is a sufficiently smooth function, so that the singular behaviour is described solely by the denominator $(x-y)^{\lambda}$.

We try to imitate the application of the reversion formula (4.1). For that purpose we multiply both sides of equation (1) by $(t-x)^{\lambda-1}$ and integrate over $[a,t]$. Interchanging the integrations, we obtain

(6.5.2) $$\int_a^t \frac{g(x)\,dx}{(t-x)^{1-\lambda}} = \int_a^t\int_a^x \frac{k(x,y)}{(t-x)^{1-\lambda}(x-y)^{\lambda}} f(y)\,dy\,dx = \\ = \int_a^t \Big[\int_y^t \frac{k(x,y)\,dx}{(t-x)^{1-\lambda}(x-y)^{\lambda}}\Big] f(y)\,dy.$$

Since we assume k to be smooth,

(6.5.3) $$\hat{k}(t,y) := \int_y^t \frac{k(x,y)\,dx}{(t-x)^{1-\lambda}(x-y)^{\lambda}} \qquad \text{for } a\leqslant y<t$$

is well-defined. Equation (2) can be rewritten in terms of $\hat{k}$ as

(6.5.4) $$G(t) = \int_a^t \hat{k}(t,y)f(y)\,dy \qquad \text{with } G(t) := \int_a^t \frac{g(x)\,dx}{(t-x)^{1-\lambda}}.$$

Hence, we have succeeded in reducing the generalised Abel integral equation (1) to a Volterra integral equation of the first kind, as it has been treated in §2.5. The solution method is based on differentiation of the integral equation (4) with respect to t, which requires the differentiability of G and $\hat{k}(t,y)$ with respect to t (cf. (2.5.2/3)). Concerning the requirement $G\in C^1(I)$, use Lemmata 4.2 and 4.6. The existence of $\hat{k}_t(t,y)$ is discussed below.

Lemma 6.5.1. Let $k\in C(G)$ with G from (2.1.3). In addition, assume k is Hölder continuous with respect to the first argument: $k(\cdot,y)\in C^{\alpha}(I)$ with $\alpha>1-\lambda$. Then the derivative of $\hat{k}(t,y)$ with respect to t exists and equals

(6.5.5) $$\hat{k}_t(t,y) = (1-\lambda)\int_y^t \frac{k(t,y)-k(x,y)}{(t-x)^{2-\lambda}(x-y)^{\lambda}}\,dx.$$

Proof. Analogous to the proof of Lemma 4.6. ∎

6.6 Numerical Methods for Abel's Integral Equation

The reformulation of the integral equation (5.1) described in §6.5 is more useful for theoretical purposes. Practical solution would require the evaluation of the integrals G and $\hat{k}$. Therefore, in the following we investigate numerical methods, which are directly applicable to equation (5.1).

We consider the problem of solving the integral equation (5.1) for $\lambda=\frac{1}{2}$ and $a=0$, i.e.,

(6.6.1) $$g(x) = \int_0^x \frac{k(x,y)}{\sqrt{x-y}} f(y)\,dy \qquad \text{for } x\geq 0.$$

Let $h>0$ be a fixed step size. We look for approximations

(6.6.2a) $$f_i \approx f(ih) \qquad \text{for } i\geq 0.$$

Linear interpolation of the (unknown) values f_i yields

(6.6.2b) $$\tilde{f}(y) = \tfrac{1}{h}[(y-ih)f_{i+1} + ((i+1)h-y)f_i] \qquad \text{for } ih\leq y\leq (i+1)h.$$

The starting value f_0 in the case of $g\in C^{\alpha}(I)$ with $\alpha>\frac{1}{2}$ and $g(0)=0$ is

$$f_0 := f(0) = 0$$

as in (4.7). If we only know that $g\in C^{1/2}(I)$ and $g(0)=0$, we obtain more generally

(6.6.3) $$f_0 := f(0) = [\lim g(x)/\sqrt{x}]/[2k(0,0)].$$

For the determination of the f_i for $i\geq 1$, we require equation (1) at $x=ih$:

(6.6.4) $$g(ih) = \int_0^{ih} \frac{k(ih,y)}{\sqrt{ih-y}} \tilde{f}(y)\,dy \qquad \text{for } i\geq 1.$$

The integral over $[0,ih]$ is the sum of the integrals over $[jh,(j+1)h]$, $0\leq j<i$, all of which can be transformed onto $[0,1]$:

$$\int_{jh}^{(j+1)h} \frac{k(ih,y)}{\sqrt{ih-y}} \tilde{f}(y)\,dy = \sqrt{h}\int_0^1 \frac{k(ih,(j+\eta)h)}{\sqrt{i-j-\eta}} \tilde{f}((j+\eta)h)\,d\eta,$$

so that the collocation equation takes the form (5):

(6.6.5) $$g(ih) = \sqrt{h}\sum_{j=0}^{i-1}\int_0^1 \frac{k(ih,(j+\eta)h)}{\sqrt{i-j-\eta}} \tilde{f}((j+\eta)h)\,d\eta.$$

The integrals on the right-hand side must be approximated by a suitable quadrature formula. Because of the singularity in the denominator in the case of $j=i-1$, a quadrature (1.6.16c-e) with weight function is unavoidable. But even for $j<i-2$, the denominator is awkward, so that one should choose the denominator for all j as weight function. The simplest quadrature method is the Gauß formula with only one support abscissa $w_k\in(0,1)$ corresponding to the weight function $1/\sqrt{k-\eta}$ $(1\leq k\leq i)$:

(6.6.6a) $$\int_0^1 \frac{\varphi(\eta)}{\sqrt{k-\eta}}\,d\eta = a_k\varphi(w_k) + R_k(\varphi) \qquad \text{with}$$

(6.6.6b) $$a_k = \int_0^1 \frac{d\eta}{\sqrt{k-\eta}}, \qquad w_k = \int_0^1 \frac{\eta\, d\eta}{\sqrt{k-\eta}} \Big/ a_k,$$

where the quadrature error is given by

(6.6.6c) $$R_k(\varphi) = r_k \varphi''(\xi), \qquad \xi \in (0,1), \qquad r_k = \frac{1}{2}\int_0^1 \frac{(t-w_k)^2}{\sqrt{k-\eta}}\, d\eta .$$

Replacing the integrals in (5) by the Gauß approximations, one obtains

(6.6.7) $$g(ih) = \sqrt{h} \sum_{j=0}^{i-1} a_{i-j}\, k(ih,(j+w_{i-j})h)\, \tilde{f}((j+w_{i-j})h).$$

By induction assume that the support ordinates f_j are known for $j \leqslant i-1$. $\tilde{f}$ is defined in $[0, ih-h]$ by (2b), i.e., all terms in (7) with $j<i-1$ are computable. For $j=i-1$, the right-hand side contains

$$\tilde{f}((i-1+w_1)h) = (1-w_1) f_{i-1} + w_1 f_i .$$

Hence, (7) represents a linear equation for the unknown value f_i.

Since the quadrature (6a-c) as well as the piecewise linear interpolation (2b) is of second order, one cannot expect more than second order convergence. In fact, second order is achieved as stated below.

Theorem 6.6.1 (Branca [1]). Let $I=[0,b]$ and $G=\{(x,y)\colon x \in I,\ y \in [0,x]\}$. Suppose k satisfies $k \in C^1(G)$, k_y, $k_{yy} \in C_L(G)$, and $k(x,x) \geqslant \varepsilon > 0$ in I. Assume (1) has a solution $f \in C_L^2(I)$. Then, for sufficiently small h, the equations (3) and (7) define the discrete solution f_i for $ih \in I$, and f_i converges with second order to the solution f:

(6.6.8) $$|f_i - f(ih)| \leqslant Ch^2 \qquad \text{for all } 0 \leqslant ih \leqslant b.$$

Theorem 1 cited from Branca [1] can even be formulated for nonlinear problems,

(6.6.9) $$g(x) = \int_0^x \frac{k(x,y,f(y))}{\sqrt{x-y}}\, dy \qquad \text{for } x \geqslant 0,$$

and proves again convergence of second order.

A discussion of more general methods for the integral equation (5.1) of the first kind can be found in Brunner - van der Houwen [1, §6.4].

In the particular case of Abel's integral equation, the explicit representation formula (4.3) of the solution f can be discretised directly. For (convolution) integrals like

(6.6.10) $$I(x;\varphi,1-\lambda) := \int_0^x (x-y)^{\lambda-1} \varphi(y)\, dy,$$

discrete analogues

(6.6.11) $$I_n(\varphi,1-\lambda) := h^\lambda \sum_{j=0}^{n} w_{n,j}\, \varphi(ih) \approx I(nh;\varphi,1-\lambda)$$

with suitable coefficients $w_{n,j}$ can be defined. Since, according to (4.3), the solution f takes the form (10) with $\varphi(y) := \text{const}\cdot g'(y)$, an approximation can be determined by means of (11). Concerning the definition and analysis of the discrete, «fractional» quadrature formulae (11), we refer to Lubich [1,2] and Brunner - van der Houwen [1,§6.1.2]).

7 Singular Integral Equations

7.1 The Cauchy Principal Value

7.1.1 Definition and Properties

Let the function f be defined on $I=[a,b]$ and, possibly, be singular at an interior point $c\in(a,b)$. Recall that the improper integral was defined by

$$\int_a^b f(x)dx := \lim_{\substack{\varepsilon_1\to 0\\ \varepsilon_1>0}} \int_a^{c-\varepsilon_1} f(x)dx + \lim_{\substack{\varepsilon_2\to 0\\ \varepsilon_2>0}} \int_{c+\varepsilon_2}^b f(x)dx,$$

if both limits exist (cf. §6.1.3). By Remark 6.1.2a, the improper integral exists for $f(x):=|x-c|^s$ with $s>-1$. For $f(x):=\frac{1}{x-c}$ (i.e., $s=-1$) one obtains

$$\text{(7.1.1)}\qquad \int_a^{c-\varepsilon_1} \frac{1}{x-c}dx + \int_{c+\varepsilon_2}^b \frac{1}{x-c}dx = \log\frac{b-c}{c-a} + \log\frac{\varepsilon_1}{\varepsilon_2}.$$

If ε_1 and ε_2 in (1) tend to zero independently, the term $\log\frac{\varepsilon_1}{\varepsilon_2}$ can take any value, so that the limit and therefore the improper integral *does not* exist. However, if $\varepsilon_1=\varepsilon_2$ and both parameters go to zero simultaneously, the term $\log\frac{\varepsilon_1}{\varepsilon_2}$ vanishes and the limit is given by $\log\frac{b-c}{c-a}$. In the general case, *Cauchy's principal value* of an integral over an integrand with a singularity at $c\in(a,b)$ is defined by the limit (2), provided that this limit exists:

$$\text{(7.1.2)}\qquad \oint_a^b f(x)dx := \lim_{\substack{\varepsilon\to 0\\ \varepsilon>0}} \Big\{ \int_a^{c-\varepsilon} f(x)dx + \int_{c+\varepsilon}^b f(x)dx \Big\}.$$

If f has no singularity, we set $\oint f(x)dx = \int f(x)dx$. If f has several singularities at $c_i\in(a,b)$, one has to apply (2) accordingly by extracting an ε_i-neighbourhood around each c_i. For $f(x)=1/(x-c)$, the considerations in (1) prove the principal value

$$\text{(7.1.3)}\qquad \oint_a^b \frac{dx}{x-c} = \log\frac{b-c}{c-a} \qquad \text{for all } a<c<b.$$

If f is improperly integrable at $x=c$, one regains the improper integral also from the special limit for $\varepsilon_1=\varepsilon_2=\varepsilon\to 0$ in (2). This leads to

Remark 7.1.1. If the singularities of f are improperly integrable, Cauchy's principal value $\oint_a^b f(x)dx$ coincides with the improper integral $\int_a^b f(x)dx$.

If the singularity of f at $x=c$ is such that f is improperly integrable at c, f is called *weakly singular* at $x=c$, otherwise, *strongly singular*.

Exercise 7.1.2. Prove the following properties of the Cauchy principal values:
(a) Linearity $⨍_a^b [\alpha f(x)+\beta g(x)]dx = \alpha ⨍_a^b f(x)dx + \beta ⨍_a^b g(x)dx$ holds, provided that the Cauchy principal values of f and g exist in $[a,b]$.
(b) Let $a<c<b$. If $⨍_a^b f(x)dx$ exists and f is not strongly singular at c, then the additive splitting $⨍_a^b f(x)dx = ⨍_a^c f(x)dx + ⨍_c^b f(x)dx$ holds.
(c) If Cauchy's principal value of $|f|$ exists, f has only weak singularities.

Since strongly singular integrands admitting a Cauchy principal value often have a singularity of the form $g(x)/(x-c)$, we replace f in the following by $f(x)/(x-c)$.

A *sufficient* condition for the existence of the Cauchy principal value is given in

Theorem 7.1.3. Let f be Hölder continuous: $f \in C^\alpha([a,b])$ with $\alpha>0$. Then the Cauchy principal value $⨍_a^b \frac{f(x)dx}{x-c}$ exist for $a<c<b$ and can be written as

$$⨍_a^b \frac{f(x)dx}{x-c} = \int_a^b \frac{f(x)-f(c)}{x-c}dx + f(c)\log\frac{b-c}{c-a} \qquad (a<c<b). \tag{7.1.4}$$

Here, the integral on the right-hand side of (4) exists as an improper integral.

Proof. By definition (2), one has to investigate $⨍_a^{c-\varepsilon} \frac{f(x)dx}{x-c}$ and $⨍_{c+\varepsilon}^b \frac{f(x)dx}{x-c}$. Split the integrand according to

$$\frac{f(x)}{x-c} = \frac{f(x)-f(c)}{x-c} + \frac{f(c)}{x-c}.$$

Thanks to the Hölder continuity, the first term $\frac{f(x)-f(c)}{x-c}$ has the improperly integrable majorant $\text{const}\cdot|x-c|^{\alpha-1}$. By Exercise 6.1.1c, also $\frac{f(x)-f(c)}{x-c}$ is improperly integrable. As $\varepsilon\to 0$, one obtains the integral on the right-hand side of (4). The Cauchy principal $⨍\frac{f(c)}{x-c}$ is $f(c)$ times the value in (3). QED

Exercise 7.1.4. **(a)** Assume $f\in C^\alpha([-1,1])$ with $\alpha>0$. Prove that the right-hand side in

$$⨍_{-1}^{+1} \frac{f(x)}{x}dx = \int_0^1 \frac{f(x)-f(-x)}{x}dx \tag{7.1.5}$$

exists as improper integral and represents the Cauchy principal value.
(b) Give a reformulation analogous to (5) for the case that the singularity is not in the middle of the interval.

The Hölder continuity of f in $⨍_a^b \frac{f(x)dx}{x-c}$ is sufficient for the existence of Cauchy's principal value, but not necessary as shown by the following examples, the proof of which is left to the reader.

Example 7.1.5. Prove the following: **(a)** $f(x) = \text{sign}(x)\exp\{-\sqrt{\log(1/|x|)}\}$ is continuous in $[-1,1]$, but not Hölder continuous. Nevertheless, the

Cauchy principal value $\oint_{-1}^{+1} \frac{f(x)}{x} dx$ exists even as improper integral.
(b) $f(x) = \sin\frac{1}{x}$ is not continuous at $x=0$, but $\oint_{-1}^{1} \frac{\sin 1/x}{x} dx$ exists.

In the case of proper and improper integrals, the *substitution formula*

$$\int_I f(x)dx = \int_J f(g(y))g'(y)dy = \int_J f(g(y))dg(y)$$

holds for continuous, monotone functions g mapping $J \subset \mathbb{R}$ uniquely onto I. In particular, this formula holds if g is continuous and *piecewise* differentiable with $g' \geq 0$ in J (or $g' \leq 0$ in J). For Cauchy's principal value, this statement remains valid only under appropriate additional conditions.

Lemma 7.1.6. **(a)** Let x_0 be an inner point of the interval I. f is assumed to be weakly singular on I and bounded in a neighbourhood of x_0. The Cauchy principal value $\oint_I \frac{f(\xi)}{\xi - x_0} d\xi$ is assumed to exist. $g \in C^1(J)$ is an invertible mapping of an interval J onto I. $y_0 \in J$ is the pre-image of $x_0 = g(y_0)$. Then the substitution rule (6a) holds. In particular, the Cauchy principal value on the right-hand side exists:

$$\oint_I \frac{f(\xi)}{\xi - x_0} d\xi = \oint_J \frac{f(g(y))g'(y)}{g(y) - x_0} dy. \tag{7.1.6a}$$

(b) If g is continuous, but only *piecewise* continuously differentiable, we require in addition: The left- and right-sided derivatives $g'(y_0 - 0)$ and $g'(y_0 + 0)$ are different from zero and f is continuous at x_0. Then, the modified substitution rule (6b) holds. The Cauchy principal value on the right-hand side exists:

$$\oint_I \frac{f(\xi)}{\xi - x_0} d\xi = \oint_J \frac{f(g(y))g'(y)}{g(y) - x_0} dy + f(x_0) \log \frac{g'(y_0+0)}{g'(y_0-0)}. \tag{7.1.6b}$$

Proof. For (b): The intervals are $J = [a, b]$ and $I = [\alpha, \beta]$. Without loss of generality let $g' \geq 0$ so that $\alpha = g(a)$, $\beta = g(b)$. Let $\varepsilon > 0$ be sufficiently small. Since $[\alpha, x_0 - \varepsilon]$ and $[x_0 + \varepsilon, \beta]$ contain no singularity, the substitution $\xi = g(y)$ yields

$$I_\varepsilon^- := \int_\alpha^{x_0 - \varepsilon} \frac{f(\xi)}{\xi - x_0} d\xi = \int_a^{\gamma(x_0 - \varepsilon)} \frac{f(g(y))g'(y)}{g(y) - x_0} dy,$$

where $\gamma: I \to J$ denotes the inverse of g. The upper boundary is $y_0 - \eta$ with $\eta := y_0 - \gamma(x_0 - \varepsilon)$. We write the integral over the second interval as

$$I_\varepsilon^+ := \int_{x_0+\varepsilon}^{\beta} \frac{f(\xi)}{\xi - x_0} d\xi = \int_{\gamma(x_0+\varepsilon)}^{b} \frac{f(g(y))g'(y)}{g(y) - x_0} dy = \int_{y_0+\eta}^{b} \dots dy + \int_{\gamma(x_0+\varepsilon)}^{y_0+\eta} \dots dy.$$

To the latter integral we apply the following variant of the mean value theorem: $\int_I \varphi(y)\psi(y)dy = \tilde{\varphi}\int_I \psi(y)dy$ with $\tilde{\varphi} \in [\inf \varphi, \sup \varphi]$, if ψ does not change the sign (φ may be discontinuous!). Hence,

$$I_\varepsilon^\delta := \int_{\gamma(x_0+\varepsilon)}^{y_0+\eta} \dots dy = \varphi_\varepsilon \int_{\gamma(x_0+\varepsilon)}^{y_0+\eta} \frac{g'(y)dy}{g(y) - x_0} = \varphi_\varepsilon \int_{x_0+\varepsilon}^{g(y_0+\eta)} \frac{d\xi}{\xi - x_0} = \varphi_\varepsilon \log \frac{g(y_0+\eta) - x_0}{\varepsilon}$$

with $\varphi_\varepsilon \in \Xi_\varepsilon := [\inf f(g(y)), \sup f(g(y))] = [\inf f(\xi), \sup f(\xi)]$, where in «inf» and «sup» the variable y varies between $\gamma(x_0+\varepsilon)$ and $y_0+\eta$, while ξ is between $x_0+\varepsilon$ and $g(y_0+\eta)$. Together with g, also the inverse γ is continuous. Therefore, $y_0+\eta = -\gamma(x_0-\varepsilon)$ and $\gamma(x_0+\varepsilon)$ tend to y_0 as $\varepsilon\to 0$. From the continuity of f at x_0 one concludes $\varphi_\varepsilon \to f(x_0)$. The right-sided differentiability of g proves $g(y_0+\eta)-x_0 = g'(y_0+0)\eta + o(\eta)$. Solving $\eta = y_0-\gamma(x_0-\varepsilon)$ for $\varepsilon = x_0-g(y_0-\eta)$, one obtains $\varepsilon = g'(y_0-0)\eta + o(\eta)$. This shows that $I_\varepsilon^\delta \to f(x_0)\log[g'(y_0+0)/g'(y_0-0)]$ for $\varepsilon\to 0$. The limit of $I_\varepsilon^- + I_\varepsilon^+$ as $\varepsilon\to 0$ yields (6b).

In case (a), φ_ε is bounded for $\varepsilon\to 0$, while the logarithmic term converges to $\log[g'(y_0)/g'(y_0)] = \log 1 = 0$. Hence, $I_\varepsilon^\delta \to 0$. QED

Theorem 6 is formulated for the integrand $f(\xi)/(\xi-x_0)$ with f being continuous at x_0. Part (b) of the next exercise shows that a corresponding substitution formula (even under the assumption $g\in C^1(J)$) is not valid for general integrands f.

Exercise 7.1.7. **(a)** Prove that the *rule of integration by parts*

$$\int_a^b f'(x)\log|x-x_0|\,dx = -\oint_a^b \frac{f(x)}{x-x_0}dx + \left[f(x)\log|x-x_0|\right]_a^b$$

holds even if f' has a weak singularity of the kind $|f'(x)| = O(|x-x_0|^{-\alpha})$ with $\alpha<1$ at $x=x_0$.
(b) Let $a<0<b$. Evaluate the Cauchy principal value $\oint_a^b x^{-3}dx$ and give a transformation $g\in C^1(J)$, so that the Cauchy principal value does not exist on the right-hand side of (6a).

Definition (2) requires that the singularity $x=c$ lies in the interior of the interval $[a,b]$. One might hope to gain a principal value $\oint_a^b \frac{f(x)}{x-a}dx$ from the limit of $\oint_a^b \frac{f(x)}{x-c}dx$ as $c\to a$. Example (3) for $f=1$ shows that this is *not* possible. More generally, the following exercise can be proved by means of formula (4).

Exercise 7.1.8. Let $f\in C^\alpha([a,b])$ with $\alpha>0$. Then the Cauchy principal value $F(c) := \oint_a^b \frac{f(x)}{x-c}dx$ with $a<c<b$ has the same singularity as $c\to a$ and $c\to b$ as the function

(7.1.7) $$\sigma(c) := f(b)\log(b-c) - f(a)\log(c-a).$$

The difference of the Cauchy principal value $F(c)$ and the function $\sigma(c)$ remains bounded: $\|F-\sigma\|_\infty \le C\|f\|_{C^\alpha([a,b])}$.

In connection with periodic integrands it makes sense to extent the definition of Cauchy's principal value to the case that f is *singular at both ends of the interval*:

(7.1.8) $$\oint_a^b f(x)\,dx := \lim_{\substack{\varepsilon\to 0\\ \varepsilon>0}} \int_{a+\varepsilon}^{b-\varepsilon} f(x)\,dx$$

The connection between (8) and (2) is more closely examined in

Remark 7.1.9. Let $c\in I=[a,b]$, $L := b-a$. Any function defined on $I\setminus\{c\}$ with singularity at $x=c$ can be extended L-periodically onto $\mathbb{R}$ by $f(x)=f(x+\nu L)$ for all $\nu\in\mathbb{Z}$. The additional definition (8) permits an arbitrary shift of the integration domain:

$$(7.1.9)\qquad \oint_a^b f(x)dx = \oint_{a+\delta}^{b+\delta} f(x)dx \qquad \text{for all } \delta\in\mathbb{R}.$$

Vice versa, let f be singular at $c\neq a$ and determine δ from $a+\delta=c$. Then equation (9) defines the principal value on the right-hand side and this value coincides with that in definition (8).

7.1.2 Curvilinear Integrals

In the applications of Chapter 8, the appearing integrals are taken over a curve Γ instead of an interval I.

Definition 7.1.10. **(a)** $\Gamma\subset\mathbb{R}^2$ is called a C^α-*curve* $\{\hat{C}^\alpha,\ C_L^\nu$-curve$\}$, if there is a mapping $\gamma\colon [0,L]\to\Gamma$ defined on $[0,L]$ and uniquely invertible on $[0,L)$ with $\gamma=\binom{x}{y}\in C^\alpha([0,L])$ $\{\gamma\in\hat{C}^\alpha([0,L]),\ \gamma\in C_L^\nu([0,L])$, respectively$\}$, where the *end-point* $\gamma(L)$ may coincide with the *initial point* $\gamma(0)$, but with no *interior* point $\{\gamma(\tau)\colon 0<\tau<L\}$.
(b) Γ is called a *closed* C^α-curve $\{\hat{C}^\alpha,\ C_L^\nu$-curve$\}$, if in addition the L-periodic continuation of γ onto $\mathbb{R}$ leads to $\gamma=\binom{x}{y}\in C^\alpha(\mathbb{R})$ $\{\gamma\in\hat{C}^\alpha(\mathbb{R}),\ \gamma\in C_L^\nu(\mathbb{R})$, respectively$\}$.
(c) Let $\alpha\geq 1$. Γ is called a *piecewise* C^α-curve, if Γ is a C_L-curve and an interval partition $0=\tau_0<\tau_1<\ldots<\tau_e=L$ exists, so that $\gamma\in C^\alpha([\tau_{\nu-1},\tau_\nu])$ for $\nu=1,\ldots,e$. The *closed piecewise* C^α-curve is defined, analogously.

As abbreviation we use following symbols:

$\Gamma\in C^\alpha$, $\Gamma\in\hat{C}^\alpha$, $\Gamma\in C_L^\nu$ for C^α, $\hat{C}^\alpha$, C_L^ν-curves,

$\Gamma\in C_0^\alpha$, $\Gamma\in\hat{C}_0^\alpha$, $\Gamma\in C_{L,0}^\nu$ for closed C^α, $\hat{C}^\alpha$, C_L^ν-curves,

$\Gamma\in C_{\mathrm{stw}}^\alpha$ for piecewise C^α-curves; analogously, $\hat{C}_{\mathrm{stw}}^\alpha$, $C_{L,\mathrm{stw}}^\nu$,

$\Gamma\in C_{0,\mathrm{stw}}^\alpha$ for closed piecewise C^α-curves, etc.

The subscript «stw» stands for «stückweise» (piecewise). By a *parametrisation of* $\Gamma\in C^\alpha$ ($\Gamma\in\hat{C}^\alpha$ etc.) we always mean a mapping γ with the properties mentioned in Definition 10.

Remark 7.1.11. **(a)** Since γ is supposed to be uniquely invertible in $[0,L)$, only *double-point free* curves are admitted, i.e., curves not intersecting or touching themselves. However, the end-points may coincide: $\gamma(0)=\gamma(L)$.
(b) The images $\gamma(\tau_\nu)$ of the partition points τ_ν from Definition 10c are possibly *corners* of the curve. In τ_ν, different one-sided derivatives $x'(\tau_\nu\pm 0)$, $y'(\tau_\nu\pm 0)$ of the coordinates x, y of γ may appear.
(c) The parametrisations $\gamma(\tau)$ and $\hat{\gamma}(\tau):=\gamma(L-\tau)$ describe the two possible *orientations* of the curve Γ. In the case of a closed curve, the *orientation is positive (negative)*, if $\gamma(\tau)$ follows the curve in the anticlockwise (clockwise) sense.
(d) Any continuous and closed curve $\Gamma\in C_0^0$ partitions $\mathbb{R}^2\setminus\Gamma$ into two disjoint

sets: the bounded *interior domain* $\Omega = \Omega_-$, which has Γ as its boundary: $\partial\Omega = \Gamma$, and the unbounded *exterior domain* $\Omega_+ = \mathbb{R}^2 \setminus \overline{\Omega}$.

In order to define the integral of a function $f: \Gamma \to \mathbb{R}$ on Γ, one must suppose at least that γ is a continuous function of bounded variation is (a sufficient condition is $\gamma \in C_L$). In this case, $\int_\Gamma f d\Gamma$ is defined as Stieltjes integral. However, in the following, we always assume

(7.1.10a) $\quad \Gamma \in C^1_{stw} \quad (\Gamma \in C^1_{0,stw}$ respectively).

The integral of a function $f: \Gamma \to \mathbb{R}$ is defined by

(7.1.10b) $$\int_\Gamma f\, d\Gamma := \int_0^L f(\gamma(\tau)) \sqrt{x'(\tau)^2 + y'(\tau)^2}\, d\tau$$

where $\gamma = \binom{x}{y}: [0, L] \to \Gamma$ is a piecewise differentiable parametrisation of Γ and x', y' are the derivatives of the coordinates of γ.

If $f = f(\mathbf{x})$ depends not only on $\mathbf{x} \in \Gamma$, but also on an additional parameter ζ, then the notation $\int_\Gamma f(\mathbf{x}, \zeta)\, d\Gamma_\mathbf{x}$ indicates that $\mathbf{x}$ is the integration variable.

Remark 7.1.12. (a) By the substitution formula one proves that $\int_\Gamma f d\Gamma$ depends neither on the choice of the parametrisation nor on the orientation of Γ.
(b) γ can be chosen such that $x'(\tau)^2 + y'(\tau)^2 = 1$ for all $\tau \in [0, L]$ (parametrisation by the arc length). Then L is the *arc length* of the curve Γ and (10c) holds:

(7.1.10c) $$\left| \int_\Gamma f\, d\Gamma \right| \leqslant L \|f\|_\infty .$$

Finally, we recall some geometric terms. The *tangent* to the curve Γ at the point $\mathbf{x} \in \Gamma$ is described by the normalised vector $\mathbf{t}(\mathbf{x})$. The *normal direction* $\mathbf{n}(\mathbf{x})$ is perpendicular to $\mathbf{t}$. Both are defined only if Γ has a parametrisation γ differentiable at $\mathbf{x}$. Then (11a,b) hold:

(7.1.11a) $$\mathbf{t}(\mathbf{x}) = \left(x'(\tau)^2 + y'(\tau)^2\right)^{-1/2} \binom{x'(\tau)}{y'(\tau)}, \qquad \text{where } \mathbf{x} = \binom{x(\tau)}{y(\tau)} \in \Gamma.$$

(7.1.11b) $$\mathbf{n}(\mathbf{x}) = \left(x'(\tau)^2 + y'(\tau)^2\right)^{-1/2} \binom{y'(\tau)}{-x'(\tau)},$$

Remark 7.1.13. (a) Alternating the orientation of Γ, one has to turn $\mathbf{t}$ and $\mathbf{n}$ into $-\mathbf{t}$ and $-\mathbf{n}$.
(b) For a closed curve with positive orientation, the normal direction points into the exterior domain (cf. Remark 11d).

If γ is twice differentiable, (11c) defines the *curvature* of the curve:

(7.1.11c) $$\varkappa(\mathbf{x}) := (y''x' - x''y') / \left(x'^2 + y'^2\right)^{3/2}.$$

Exercise 7.1.14. For the circle $\Gamma = \{\mathbf{x} = \binom{x}{y} \in \mathbb{R}^2 : x^2 + y^2 = r^2\}$ use the parametrisation $x = r\cos\tau$, $y = r\sin\tau$ over $[0, 2\pi]$ and prove

$$\mathbf{t}(\mathbf{x}) = \binom{-\sin\tau}{\cos\tau}, \quad \mathbf{n}(\mathbf{x}) = \binom{\cos\tau}{\sin\tau}, \quad \varkappa(\mathbf{x}) = 1/r \quad \text{for } \mathbf{x} = \gamma(\tau).$$

Since the $\mathbb{R}^2$ can be identified with the complex numbers $\mathbb{C}$ via $(x,y) \leftrightarrow z = x + iy$, $\Gamma \subset \mathbb{R}^2$ can also be regarded as a complex curve $\Gamma \subset \mathbb{C}$. Then, the parametrisation is described by a scalar complex-valued function $\zeta: [0, L] \to \Gamma$.

Different from (10b), the *complex curvilinear integral* is defined by

(7.1.12) $$\int_\Gamma f(\zeta)\,d\zeta := \int_0^L f(\zeta(\tau))\,\zeta'(\tau)\,d\tau.$$

The right-hand side of (10b) will be denoted by $\int_\Gamma f(\zeta)|d\zeta|$. Remark 12a remains valid for (12) up to one exception: $\int_\Gamma f(\zeta)d\zeta$ changes sign when the *orientation* of Γ is reversed. The *parametrisation by the arc length* is described by the condition $|\zeta'(\tau)|=1$.

7.1.3 Cauchy's Principal Value for Curvilinear Integrals

If the integrand f of the curvilinear integral $\int_\Gamma f d\Gamma$ from (10b) or of the complex integral $\int_\Gamma f(\zeta)d\zeta$ from (12) has singularities, the right-hand side in (10b) or (12), respectively, represents an integral over $[0,L]$ with singular integrand. If, for a certain parameter representation, the respective integrands $f(\gamma(\tau))\sqrt{x'(\tau)^2+y'(\tau)^2}$ or $f(\zeta(\tau))\zeta'(\tau)$ are improperly integrable, then thanks to the substitution rule they do so for any other parametrisation of $\Gamma\in C^1_{stw}$ (cf. Exercise 6.1.1d). This enables the following definition: f is *improperly integrable* on $\Gamma\in C^1_{stw}$, if the integrals of the respective right-hand sides in (10b) or (12) exist as improper integrals for an arbitrary parametrisation of the class C^1_{stw}.

The Cauchy principal value of $\int_\Gamma f d\Gamma$ or $\int_\Gamma f(\zeta)d\zeta$ can also be explained by the definition in §7.1.1, but in the case of C^1_{stw} curves one must take care that the respective parametrisation by $\gamma=\binom{x}{y}$ or ζ fulfils

(7.1.13a) $$x'(\tau)^2+y'(\tau)^2\in C([0,L]) \quad\text{or}\quad |\zeta'|\in C([0,L]),\ \text{respectively,}$$

since in Lemma 6b the substitution rule had to be modified. The condition (13a) is always satisfied if $\Gamma\in C^1$ or if the arc length is chosen as parameter. In the latter case, the functions in (13a) equal the constant value 1.

Definition 7.1.15. Let f have a singularity of first order at $\mathbf{x}_0\in\Gamma$ (and possibly at further points), i.e., $|f(\mathbf{x})|\,\|\mathbf{x}-\mathbf{x}_0\|_2$ is bounded in a neighbourhood of $\mathbf{x}_0$. Then, *Cauchy's principal value* is defined by

(7.1.13b) $$⨏_\Gamma f\,d\Gamma := ⨏_0^L f(\gamma(\tau))\sqrt{x'(\tau)^2+y'(\tau)^2}\,d\tau,$$

(7.1.13c) $$⨏_\Gamma f(\zeta)\,d\zeta := ⨏_0^L f(\zeta(\tau))\zeta'(\tau)\,d\tau,$$

where the parametrisation must satisfy condition (13a).

By Lemma 6a, Definition 15 make sense: Any parametrisation with (13a) leads to the same value in (13b,c).

Remark 7.1.16. By definition (8) and property (9), the initial point $\gamma(0)=\gamma(L)$ of the parametrisation of a closed curve $\Gamma\in C^1_{0,stw}$ may be chosen arbitrarily and may also coincide with a strong singularity point of f.

The following representation explains the Cauchy principal value directly by means of curvilinear integrals without using a parametrisation.

Theorem 7.1.17. **(a)** Let $\Gamma\subset\mathbb{R}^2$, $\Gamma\in C^1_{stw}$ or $\Gamma\in C^1_{0,stw}$. In the first case, $\mathbf{x}_0\in\Gamma$

must not be an end-point of Γ; in the second case, $\mathbf{x}_0 \in \Gamma$ can be arbitrary. The function f defined on Γ is supposed to have a singularity only at $\mathbf{x}_0$. Then

(7.1.13d) $$\oint_\Gamma f\,d\Gamma := \lim_{\substack{\varepsilon\to 0\\ \varepsilon>0}} \int_{\Gamma\setminus K_\varepsilon(\mathbf{x}_0)} f\,d\Gamma.$$

(b) The analogous statement holds for the integral $\oint_\Gamma f(\zeta)\,d\zeta$ over $\Gamma\subset\mathbb{C}$ with a singularity at $\zeta_0\in\Gamma$, where $\Gamma\setminus K_\varepsilon(\mathbf{x}_0)$ must be replaced by $\Gamma\setminus K_\varepsilon(\zeta_0)$.

Proof. (i) Let $\mathbf{x}_0=\gamma(\tau_0)$. In the case of a closed curve, without loss of generality one can restrict to $0<\tau_0<L$ (cf. Remark 16).

(ii) Since $[0,L]$ is compact and γ continuous, the image Γ is also compact. Hereby and from $\gamma\in C_L([0,L])$ and the absence of double points, one concludes the following: For sufficiently small $\varepsilon\in(0,\varepsilon_0]$ the intersection of the circle $K_\varepsilon(\mathbf{x}_0)$ with Γ contains a connected piece of curve, which is the image of some interval $(\tau_0+\eta(-\varepsilon),\tau_0+\eta(\varepsilon))$ under γ. Here, η is monotone with $\eta(-\varepsilon)<\eta(0)=0<\eta(\varepsilon)$. One gains the function η implicitly from the condition

$$\|\gamma(\tau_0+\eta(\varepsilon))-\mathbf{x}_0\|_2 = |\varepsilon| \qquad \text{for } -\varepsilon_0\leqslant\varepsilon\leqslant\varepsilon_0.$$

Differentiation yields the ordinary differential equation

$$d\eta(\varepsilon)/d\varepsilon = \varepsilon/[(x-x_0)x'+(y-y_0)y'](\tau_0+\eta(\varepsilon))$$

with the initial value $\eta(0)=0$, where $\gamma=\binom{x}{y}$, $\mathbf{x}_0=\binom{x_0}{y_0}$. $\tau_0+\eta(\varepsilon)$ is the argument in x and y. If ε_0 is sufficiently small, then τ_0 is the only point in $[\tau_0+\eta(-\varepsilon_0),\tau_0+\eta(\varepsilon_0)]$, at which γ may have a discontinuity in the derivative. From (13a) one obtains $\eta'(0+0)=\eta'(0-0)$, so that η is continuously differentiable in $[-\varepsilon_0,\varepsilon_0]$. We define

$$\Phi(\sigma) := \begin{cases} \tau_0+\eta(\sigma) & \text{for } -\varepsilon_0\leqslant\sigma\leqslant\varepsilon_0,\\ \eta'(-\varepsilon_0)(\sigma+\varepsilon_0)+\tau_0+\eta(-\varepsilon_0) & \text{for } a\leqslant\sigma\leqslant-\varepsilon_0,\\ \eta'(\varepsilon_0)(\sigma-\varepsilon_0)+\tau_0+\eta(\varepsilon_0) & \text{for } \varepsilon_0\leqslant\sigma\leqslant b \end{cases}$$

with $\quad a:=-\varepsilon_0-(\tau_0-\eta(-\varepsilon_0))/\eta'(-\varepsilon_0), \qquad b:=(L-\tau_0-\eta(\varepsilon_0))/\eta'(\varepsilon_0)+\varepsilon_0.$

Obviously, Φ is a continuously differentiable function (also at $\sigma=\pm\varepsilon_0$), mapping $[a,b]$ uniquely onto $[0,L]$. Furthermore, for $|\varepsilon|\leqslant\varepsilon_0$, $[-\varepsilon,\varepsilon]$ is mapped onto $[\tau_0+\eta(-\varepsilon),\tau_0+\eta(\varepsilon)]$. $\hat\gamma(\sigma):=\gamma(\Phi(\sigma))$ is also a parametrisation of the class C^1_{stw}. $\hat x$ and $\hat y$ are the components of $\hat\gamma$. The chain rule yields

$$\hat x'(\sigma)^2+\hat y'(\sigma)^2 = [x'(\Phi(\sigma))^2+y'(\Phi(\sigma))^2]\,\Phi'(\sigma)^2.$$

Since γ satisfies condition (13a), $\hat\gamma$ does so, too. Due to our construction of η and Φ, the integral over $\Gamma\setminus K_\varepsilon(\mathbf{x}_0)$ can be rewritten as

$$\int_{\Gamma\setminus K_\varepsilon(\mathbf{x}_0)} f\,d\Gamma = \int_a^{-\varepsilon}\ldots d\sigma + \int_\varepsilon^b \ldots d\sigma \quad \text{with } \ldots = f(\hat\gamma(\sigma))\sqrt{\hat x'(\sigma)^2+\hat y'(\sigma)^2}$$

with $|\varepsilon|\leqslant\varepsilon_0$. The limit $\varepsilon\to 0$ performed on the right-hand side leads to the Cauchy principal value $\oint_a^b\ldots d\sigma$, which by definition (13b) is the Cauchy principal value $\oint_\Gamma f\,d\Gamma$.

(iii) The proof of part (a) carries over to the complex case (b), since one can identify the real and imaginary part of ζ with the components x, y of γ. QED

7.1.4 The Example $f(\zeta)=1/(\zeta-z)$

By analogy with example (3), we investigate whether Cauchy's principal value of $f(\zeta)=1/(\zeta-z)$ exists over a complex C^1_{stw} curve. For preparation we define the argument function Arg(ζ) on Γ.

Let $\alpha_0\in(0,2\pi]$ be chosen fixed. By $S(\alpha_0)$ we denote the straight line $\{\rho e^{i\alpha_0}: \rho\geqslant 0\}$, running from the origin in the complex direction $e^{i\alpha_0}$. For $z=re^{i\alpha}$ ($r>0$, $\alpha\in\mathbb{R}$), the branch of the logarithm in the complex plane cut along $S(\alpha_0)$ is defined by

(7.1.14a) $$\log z = \log r + i(\alpha+2\nu\pi) \qquad \text{for } z=re^{i\alpha}$$ with $\nu\in\mathbb{Z}$ such that $\alpha+2\nu\pi\in[\alpha_0-2\pi,\alpha_0)$.

The function $\log z$ is holomorphic on $\mathbb{C}\setminus S(\alpha_0)$. The *argument function*

(7.1.14b) $$\arg(z) = Im \log z = \alpha+2\nu\pi \qquad (\alpha,\nu \text{ from (14a)})$$

is defined on $\mathbb{C}\setminus\{0\}$. It jumps by -2π $\{+2\pi\}$, if the argument crosses the straight line $S(\alpha_0)\setminus\{0\}$ in positive {negative} direction. Because of this discontinuity, one introduces the function Arg on the Riemann domain.

Remark 7.1.18 Let $\Gamma\in C^0$ be a continuous and double-point free curve, which does not meet the origin: $0\notin\Gamma$. **(a)** Arg(ζ) is uniquely defined for $\zeta\in\Gamma$ up to a (globally constant) multiple of 2π by the following requirements:

(i) $[\mathrm{Arg}(\zeta)-\arg(\zeta)]/2\pi\in\mathbb{Z}$,

(ii) $\mathrm{Arg}(\zeta(\tau))$ is continuous for $0\leqslant\tau<L$, whenever $\zeta(\tau)$ is a (continuous) parametrisation of Γ.

(b) A parametrisation ζ: $[0,L]\to\Gamma$ defines an «initial point» $\zeta_-=\zeta(0)$ and an «end-point» $\zeta_+=\zeta(L)$ of Γ. Without loss of generality, one can fix the free constant in Arg by $\mathrm{Arg}(\zeta_-)=\arg(\zeta_-)$. Then,

(7.1.14c) $$\mathrm{Arg}(\zeta_+) = \arg(\zeta_+)+(\nu_+-\nu_-)2\pi$$

holds for $\mathrm{Arg}(\zeta_+) := \mathrm{Arg}(\zeta(L-0))$, if Γ crosses the straight line $S(\alpha_0)$ ν_+-times in the anticlockwise direction and ν_--times in the clockwise direction.

(c) Let the closed curve $\Gamma\in C^0_0$ define the interior domain Ω. The following cases can be distinguished:

(i) If $0\notin\Omega$, $\nu_+=\nu_-$ holds, so that $\mathrm{Arg}(\zeta(\tau))$ can be extended L-periodically to a continuous function on $\mathbb{R}$.

(ii) If $0\in\Omega$, $\nu_+-\nu_-=+1$ holds (respectively, $\nu_+-\nu_-=-1$) in the case of a positive (respectively, negative) orientation of Γ. At $\zeta_-=\zeta_+\in\Gamma$, the function Arg(ζ) has different one-sided limits:

(7.1.14c') $$\mathrm{Arg}(\zeta(L-0))-\mathrm{Arg}(\zeta(0+0)) = \pm 2\pi,$$

where the upper (lower) sign holds for a positive (negative) curve orientation.

The fact that Arg can be fixed only up to an arbitrary constant $2\nu\pi$, causes no problems, since in the following only differences of Arg appear. The function $\arg(z)$ is not defined for $z=0$; however, the limit of $\arg(z)$ exists if one moves towards $z=0$ along a fixed direction.

Lemma 7.1.19 (a) Let 0 be the initial point of $\Gamma\in C^0$: $\zeta:\ [0,L]\to\Gamma$, $\zeta(0)=0$, $\zeta(\tau)\neq 0$ for $\tau>0$. ζ is supposed to be one-sided differentiable at 0 with $\zeta'(0):=\zeta'(0+0)\notin S(\alpha_0)$ (α_0 from (14a)). Then (14d) holds:

$$\arg(\zeta(\tau))\ \to\ \arg(\zeta'(0)) \qquad \text{for } \tau\to 0,\ \zeta(\tau)\to 0. \tag{7.1.14d}$$

(b) The condition $\zeta'(0)\notin S(\alpha_0)$ can be satisfied by a suitable choice of the angle α_0. This condition is not needed for the function Arg:

$$\operatorname{Arg}(\zeta(\tau))\ \to\ \operatorname{Arg}(\zeta'(0)) \qquad \text{for } \tau\to 0,\ \zeta(\tau)\to 0. \tag{7.1.14d'}$$

(c) If 0 is not the initial but end-point: $\zeta(L)=0$ with $-\zeta'(L-0)\notin S(\alpha_0)$, then

$$\arg(\zeta(\tau))\ \to\ \arg(-\zeta'(L)) \qquad \text{for } \tau\to L,\ \zeta(\tau)\to 0 \tag{7.1.14d''}$$

holds.

Proof. (14d) follows from $\zeta(\tau)/\tau=[\zeta(\tau)-\zeta(0)]/\tau\to\zeta'(0+0)$ and $\arg(\zeta(\tau))=\arg(\zeta(\tau)/\tau)$ for $\tau>0$. Analogously, one obtains (14d',d''). QED

A C^1_{stw}-curve $\{\zeta(\tau):\ 0\leqslant\tau\leqslant L\}$, which. in contrast to the assumption $0\notin\Gamma$ of Remark 18, meets zero, i.e., $\zeta(\tau_0)=0$, may possibly have a jump of the derivative. In this case, let $\zeta'(\tau_0\pm 0)$ denote both *one-sided tangential directions.* If $\zeta'(\tau_0+0)\neq\zeta'(\tau_0-0)$, the point $\zeta=\zeta(\tau_0)$ is called a *corner*. In the case of a closed curve, $\zeta(0)=\zeta(L)$ is a corner if $\zeta'(0+0)\neq\zeta'(L-0)$.

Corollary 7.1.20. Assume that $\Gamma\in C^1_{\text{stw}}$ with the parametrisation $\zeta:\ [0,L]\to\Gamma$ is a curve with $\zeta(\tau_0)=0$ at $\tau=\tau_0$.
(a) Then $\arg(\zeta(\tau))$ has a jump at $\tau=\tau_0$ with the magnitude given by the angle

$$\omega\ :=\ \arg(\zeta'(\tau_0+0))\ -\ \arg(-\zeta'(\tau_0-0)). \tag{7.1.14e}$$

(b) The geometric interpretation is that ω is the angle between the tangents at the incoming and outgoing pieces of the curve, where the angle must not include the radial line $S(\alpha_0)$. Depending on the direction of the angle, $\omega\geqslant 0$ (cf. Fig. 1a) or $\omega\leqslant 0$ (cf. Fig. 1b) hold.

Proof. (i) (14d) proves $\arg(\zeta(\tau))\to\arg(\zeta'(\tau_0+0))$ for $\tau\to\tau_0$ with $\tau>\tau_0$, while (14d'') leads to $\arg(\zeta(\tau))\to\arg(-\zeta'(\tau_0-0))$ for $\tau\to\tau_0$ with $\tau<\tau_0$. Since the jump is given by $\lim[\arg(\zeta(\tau_1))-\arg(\zeta(\tau_2))]$ as $\tau_1,\tau_2\to\tau_0$, $\tau_2<\tau_0<\tau_1$, (14e) is demonstrated. QED

In the case $0\in\Gamma$, even the argument function Arg cannot be defined continuously on Γ. However, different from ω in (14e), the jump at $\zeta(\tau_0)=0$ should not depend on the arbitrary choice of α_0. Therefore, we require *per definitionem* a jump by ω_+ (see Fig. 1a–c):

$$\operatorname{Arg}(\zeta(\tau_0+0)) := \operatorname{Arg}(\zeta(\tau_0-0))+\omega_+, \qquad \omega_+=\begin{cases}\omega & \text{if } \omega>0\\ \omega+2\pi & \text{if } \omega<0\end{cases} \tag{7.1.14f}$$

with ω from (14e). ω_+ is the angle between the incoming and outgoing tangents in positive (anticlockwise) direction (cf. Fig. 1a). In the exceptional case of $\omega=0$ (i.e., $\vartheta:=\zeta'(\tau_0+0)=-\zeta'(\tau_0-0)$) we choose $\omega_+=0$, if $Im\{[\zeta(\tau_0+\varepsilon)-\zeta(\tau_0-\varepsilon)]/\vartheta\}>0$ for $0<\varepsilon\leqslant\varepsilon_0$ (ε_0 sufficiently small) and

$\omega_+ = 2\pi$, otherwise.

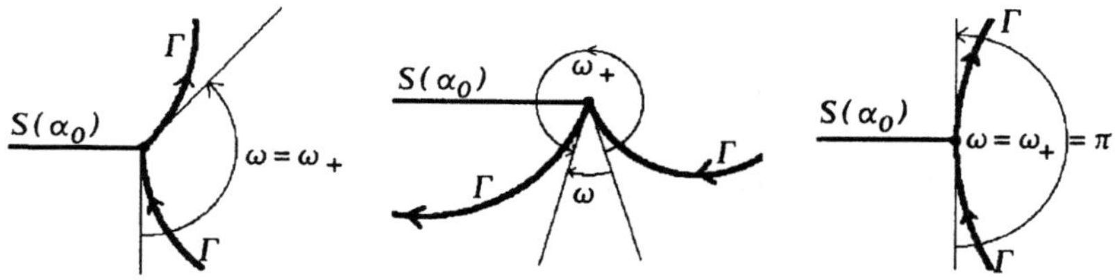

Fig. 7.1.1a $\omega > 0$ **Fig. 7.1.1b** $\omega < 0$ **Fig. 7.1.1c** $\omega = \pi$

Exercise 7.1.21. **(a)** $\omega_+ = \pi + Im \log(\zeta'(\tau_0+0)/\zeta'(\tau_0-0))$ holds if the logarithm is cut by means of the negative half-axis $S(\pi)$.
(b) Changing the orientation of Γ, one changes ω_+ into $2\pi - \omega_+$.

According to Remark 18a, Arg can be defined on $\Gamma \in C^0$ uniquely up to a constant $2\nu\pi$, if $0 \notin \Gamma$. The case $0 \in \Gamma$ is treated in

Remark 7.1.22. **(a)** If $\Gamma \in C^0$ meets the origin at τ_0: $\zeta(\tau_0) = 0$ and the one-sided derivatives $\zeta'(\tau_0 \pm 0)$ exist, then the following conditions (i)-(iii) determine the function Arg uniquely up to a constant $2\nu\pi$ ($\nu \in \mathbb{Z}$):
(i) $[\mathrm{Arg}(\zeta) - \arg(\zeta)]/2\pi \in \mathbb{Z}$,
(ii) $\mathrm{Arg}(\zeta(\tau))$ is piecewise continuous for $0 \le \tau < \tau_0$ and $\tau_0 < \tau < L$,
(iii) the jump condition (14f) holds.
(b) If $\zeta(\tau)$ is differentiable at $\tau = \tau_0$ (i.e., $\zeta'(\tau_0+0)) = \zeta'(\tau_0-0))$, it holds that

(7.1.14g) $\omega_+ = \pi$ in (14f).

Let a C^1_{stw} curve Γ and some $z \in \mathbb{C}$ be given. By $\Gamma - z$ we understand the curve $\{\zeta(\tau) - z : 0 \le \tau \le L\}$ shifted by z. $\mathrm{Arg}(\zeta - z)$ denotes the argument function Arg applied to $\Gamma - z$ instead of Γ. We summarise the previous results:

(i) By Remark 18a, the function $\mathrm{Arg}(\zeta(\cdot) - z)$: $[0, L] \to \mathbb{R}$ is continuous if $z \notin \Gamma$ (i.e., $0 \notin \Gamma - z$).

(ii) If z is a curve point: $\zeta(\tau_0) = z$, $\mathrm{Arg}(\zeta(\cdot) - z)$ jumps at τ_0 by ω_+ (ω_+ from (14f); cf. Remark 22).

(iii) The angle ω needed in (14f) defining ω_+ still has the representation (14e): $\omega = \arg(\zeta'(\tau_0+0)) - \arg(-\zeta'(\tau_0-0))$, since $(\zeta(\tau) - z)' = \zeta'(\tau)$.

Lemma 7.1.23. **(a)** Let Γ be an open C^1_{stw} curve with the parametrisation ζ: $[a, b] \to \Gamma$. $z \in \mathbb{C}$ is a point outside the curve: $z \notin \Gamma$. Then

(7.1.15a) $$\int_\Gamma \frac{d\zeta}{\zeta - z} = \mathrm{Log}(\zeta(b) - z) - \mathrm{Log}(\zeta(a) - z)$$

holds with

(7.1.15b) $$\mathrm{Log}(\zeta(\tau) - z) := \log|\zeta(\tau) - z| + i\,\mathrm{Arg}(\zeta(\tau) - z).$$

(b) If, however, $\Gamma \in C^1_{0,\mathrm{stw}}$ is a closed curve (i.e., $\zeta(b) = \zeta(a)$), the difference $\mathrm{Arg}(\zeta(b) - z) - \mathrm{Arg}(\zeta(a) - z)$ in (15a,b) is to be formulated by means of one-sided limits: $\mathrm{Arg}(\zeta(b-0) - z) - \mathrm{Arg}(\zeta(a+0) - z)$. For the interior domain $\Omega = \Omega_-$ of the closed curve, (15c) holds:

(7.1.15c) $$\int_\Gamma \frac{d\zeta}{\zeta - z} = \begin{cases} 0 & \text{if } z \in \Omega_+ := \mathbb{C}\setminus\overline{\Omega} \quad (z \text{ in the exterior domain}) \\ 2\pi i & \text{if } z \in \Omega \text{ and } \Gamma \text{ has positive orientation,} \\ -2\pi i & \text{if } z \in \Omega \text{ and } \Gamma \text{ has negative orientation.} \end{cases}$$

Proof. (i) Since the functions log and Log differ only by a constant $2\nu\pi$, they have the same derivative $\frac{d}{d\tau}\mathrm{Log}(\zeta(\tau)-z) = \frac{\zeta'(\tau)}{\zeta(\tau)-z}$ in $\mathbb{C}\setminus S(\alpha_0)$. Since α_0 can be chosen arbitrarily, the representation of the derivative holds everywhere. (15a) follows from

$$\int_\Gamma \frac{d\zeta}{\zeta-z} = \int_a^b \frac{\zeta'(\tau)d\tau}{\zeta(\tau)-z} = \int_a^b \frac{d}{d\tau}\mathrm{Log}(\zeta(\tau)-z)d\tau = \left[\mathrm{Log}(\zeta(\tau)-z)\right]_a^b.$$

(ii) Concerning part (b), compare Remark 18c and note that $\log|\zeta(b)-z| = \log|\zeta(a)-z|$ because of $\zeta(a)=\zeta(b)$. We mention that (15c) also follows from the residue theorem. QED

A strong singularity appears as soon as z lies on Γ. In this case, the integral over $1/(\zeta-z)$ must be explained as Cauchy's principal value. By $\partial\Gamma$ we denote the *boundary points of* Γ:

(7.1.16) $$\partial\Gamma = \{\zeta_+, \zeta_-\} \text{ with } \zeta_- = \zeta(0),\ \zeta_+ = \zeta(L), \text{ if } \zeta_- \neq \zeta_+,$$
$\partial\Gamma$ is empty for a *closed* curve.

Theorem 7.1.24. **(a)** Let $\Gamma \in C^1_{\mathrm{stw}}$. Let Γ form the angle ω_+ from (14f) at the curve point $z \in \Gamma\setminus\partial\Gamma$. Then the Cauchy principal value exists and equals

(7.1.17a) $$\oint_\Gamma \frac{d\zeta}{\zeta-z} = \mathrm{Log}(\zeta(L)-z) - \mathrm{Log}(\zeta(0)-z) - i\,\omega_+.$$

(b) If $\Gamma \in C^1_{0,\mathrm{stw}}$ is a closed curve, (17a) simplifies to

(7.1.17b) $$\oint_\Gamma \frac{d\zeta}{\zeta-z} = \begin{Bmatrix} i(2\pi-\omega_+) \\ -i\,\omega_+ \end{Bmatrix} \quad \text{for } \begin{Bmatrix} \text{positive} \\ \text{negative} \end{Bmatrix} \text{ orientation.}$$

(c) If $\Gamma \in C^1_0$, then

(7.1.17c) $$\oint_\Gamma \frac{d\zeta}{\zeta-z} = \pm i\pi \quad \text{for } \begin{Bmatrix} \text{positive} \\ \text{negative} \end{Bmatrix} \text{ orientation.}$$

Proof. (i) There is some $0<\tau_0<L$ with $z=\zeta(\tau_0)$. In the case of an open curve, this is possible because of $z \in \Gamma\setminus\partial\Gamma$. Otherwise, one can choose an appropriate parametrisation (cf. (9)). By Theorem 17b, the limit of the integral over $\Gamma\setminus K_\varepsilon(z)$ is to be determined. For sufficiently small ε, this subset is the union of the two curve pieces $\Gamma_{\varepsilon,1} = \{\zeta(\tau): 0 \le \tau \le \tau_\varepsilon^-\}$, $\Gamma_{\varepsilon,2} = \{\zeta(\tau): \tau_\varepsilon^+ \le \tau \le L\}$, where the parameter values $\tau_\varepsilon^\pm$ are determined such that $|\zeta(\tau_\varepsilon^\pm) - z| = \varepsilon$ (cf. proof of Theorem 17). $\tau_\varepsilon^\pm$ is continuous in ε with $\tau_\varepsilon^\pm \to \tau_0$ as $\varepsilon \to 0$. Since z belongs neither to $\Gamma_{\varepsilon,1}$ nor $\Gamma_{\varepsilon,2}$ (cf. Fig. 2a), Lemma 23 can be applied and yields

$$\int_{\Gamma_{\varepsilon,1}} \frac{d\zeta}{\zeta-z} = \log\left|\frac{\zeta(\tau_\varepsilon^-)-z}{\zeta(0)-z}\right| + i\left[\mathrm{Arg}(\zeta(\tau_\varepsilon^-)-z) - \mathrm{Arg}(\zeta(0)-z)\right],$$

$$\int_{\Gamma_{\varepsilon,2}} \frac{d\zeta}{\zeta-z} = \log\left|\frac{\zeta(L)-z}{\zeta(\tau_\varepsilon^+)-z}\right| + i\left[\mathrm{Arg}(\zeta(L)-z) - \mathrm{Arg}(\zeta(\tau_\varepsilon^+)-z)\right].$$

Since $|\zeta(\tau_\varepsilon^\pm)-z|=\varepsilon$, one can cancel this factor in the sum of log terms:

$$\log\left|\frac{\zeta(\tau_\varepsilon^-)-z}{\zeta(0)-z}\right|+\log\left|\frac{\zeta(L)-z}{\zeta(\tau_\varepsilon^+)-z}\right| = \log\left|\frac{\zeta(L)-z}{\zeta(0)-z}\right|.$$

τ_ε^- [τ_ε^+] tends from below [above] to τ_0, so that

$$\mathrm{Arg}(\zeta(\tau_\varepsilon^\pm)-z) \to \mathrm{Arg}(\zeta(\tau_0\pm 0)-z).$$

The jump relation (14f) leads to

$$\begin{aligned}&\mathrm{Arg}(\zeta(\tau_\varepsilon^-)-z) - \mathrm{Arg}(\zeta(0)-z)+\mathrm{Arg}(\zeta(L)-z) - \mathrm{Arg}(\zeta(\tau_\varepsilon^+)-z) \to\\ &\to \mathrm{Arg}(\zeta(L)-z) - \mathrm{Arg}(\zeta(0)-z)+\mathrm{Arg}(\zeta(\tau_0-0)-z) - \mathrm{Arg}(\zeta(\tau_0+0)-z) =\\ &= \mathrm{Arg}(\zeta(L)-z) - \mathrm{Arg}(\zeta(0)-z)-\omega_+ \qquad \text{for } \varepsilon\to 0.\end{aligned}$$

Altogether, $\int_{\Gamma_{\varepsilon,1}}\frac{d\zeta}{\zeta-z}+\int_{\Gamma_{\varepsilon,2}}\frac{d\zeta}{\zeta-z}$ has the right-hand side of (17a) as limit. By Theorem 17b, this is the Cauchy principal value.

(ii) For a closed curve $\zeta(0)=\zeta(L)$ holds; hence, $\log\left|\frac{\zeta(L)-z}{\zeta(0)-z}\right|=0$. The values of $\mathrm{Arg}(\zeta(0)-z)$ and $\mathrm{Arg}(\zeta(L)-z)$ are obtained from the considerations in Exercise 25. By these, $\mathrm{Arg}(\zeta(L)-z)-\mathrm{Arg}(\zeta(0)-z)$ equals zero or 2π for negative or positive orientation of the curve, respectively. This concludes the proof of part (b). Part (c) follows because of $\omega_+=\pi$, if Γ is differentiable (cf. Fig. 1c). QED

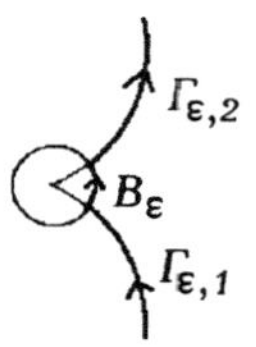

Fig. 7.1.2a

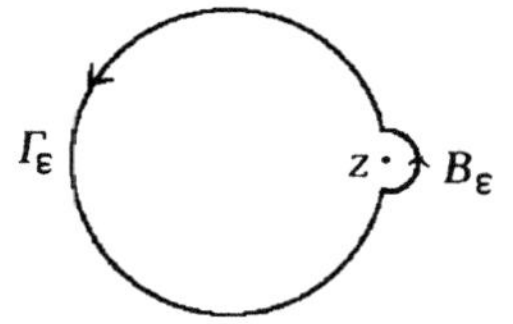

Fig. 7.1.2b

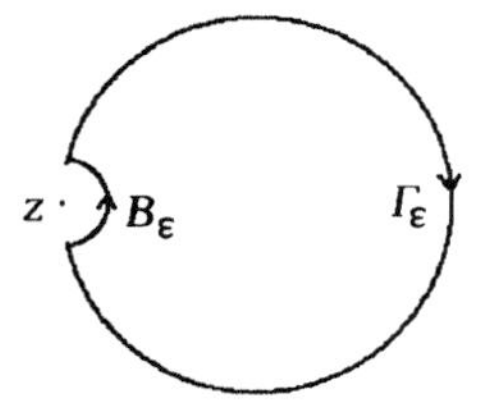

Fig. 7.1.2c

Exercise 7.1.25. Let $z\in\Gamma$. $\Gamma_\varepsilon^* := \Gamma_\varepsilon + B_\varepsilon$ is the union of $\Gamma\setminus K_\varepsilon(z)$ and the circular arc B_ε around z with radius ε oriented in positive direction from the incoming to the outgoing part of Γ_ε (cf. Fig. 2a). Define the argument function $\mathrm{Arg}(\zeta-z)$ on Γ according to (14f). Since z is not a point of Γ_ε^*, one can define $\mathrm{Arg}(\zeta^*-z)$ on Γ_ε^* according to Remark 18. Prove the following:
(a) In the intersection $\Gamma_\varepsilon=\Gamma_\varepsilon^*\cap\Gamma$, both argument functions coincide up to a *global* constant $2\nu\pi$.
(b) If Γ is closed, then Γ_ε^* is closed, too.
(c) If the closed curve Γ has a positive (negative) orientation, then z lies in the interior domain (exterior domain) of Γ_ε^* (cf. Fig. 2b or 2c, respectively).

The previous considerations enables us to define

(7.1.18a) $\quad I(z) := \oint_\Gamma \frac{d\zeta}{\zeta-z} \qquad$ for all $z\in\mathbb{C}\setminus\partial\Gamma$, where $\mathbb{C}\setminus\partial\Gamma=\mathbb{C}$ holds for a closed curve. In the following, we investigate how $I(z)$ behaves as a function of z. First, we restrict ourselves to inner curve points $z\in\Gamma\setminus\partial\Gamma$.

Remark 7.1.26. **(a)** Let $\Gamma\in C^1_{\mathrm{stw}}$ have the angle $\omega_+=\omega_+(z)$ at the corner $z\in\Gamma\setminus\partial\Gamma$, as defined in (14f). Using (17a), we have $I(z)=\mathrm{Log}(\zeta(L)-z)-\mathrm{Log}(\zeta(0)-z)-i\,\omega_+$. While $\log|\frac{\zeta(L)-z}{\zeta(0)-z}|$ is continuous (as long as $z\notin\partial\Gamma$), $\omega_+(z)$ jumps at all corners of Γ; otherwise, away from the corner, $\omega_+(z)$ is the constant function $\omega_+=\pi$. Hence, $I(z)$ has only *removable discontinuities*. The definition

(7.1.18b) $\quad \overline{I}(\zeta(\tau)) := I(\zeta(\tau\pm 0))$

removes the discontinuities and leads to

(7.1.18c) $\quad \overline{I}(z) = \mathrm{Log}(\zeta(L)-z)-\mathrm{Log}(\zeta(0)-z)-i\pi \qquad$ for all $z\in\Gamma\setminus\partial\Gamma$.

(b) $\overline{I}(z)=I(z)$ holds almost everywhere on $\Gamma\setminus\partial\Gamma$ (everywhere, in the case of $\Gamma\in C^1$).
(c) For a closed curve $\Gamma\in C^1_{0,\mathrm{stw}}$, $\overline{I}(z)$ is constant (cf. (17b)):

(7.1.18d) $\quad \overline{I}(z) = \pm i\pi \qquad$ for $z\in\Gamma\in C^1_{0,\mathrm{stw}}$ and $\left\{\begin{matrix}\text{positive}\\ \text{negative}\end{matrix}\right\}$ orientation.

Now we discuss the case $z\notin\Gamma$ and suppose first that the curve is closed.

Remark 7.1.27. The closed curve $\Gamma\in C^1_{0,\mathrm{stw}}$ defines the interior domain $\Omega_-=\Omega$ and the exterior domain $\Omega_+=\mathbb{C}\setminus\overline{\Omega}$ (cf. Remark 11d). The closures of these (open) domains are $\overline{\Omega}_\pm$. According to (15c), $I(z)$ is constant in Ω_+ as well as in Ω_-: It has the values $\pm 2\pi i$ (in the case of positive / negative orientation) in Ω_- and 0 in Ω_+. Hence, $I(z)$ can be extended continuously to $I_+(z)$ in $\overline{\Omega}_+$ and to $I_-(z)$ in $\overline{\Omega}_-$. In the intersection $\Gamma=\overline{\Omega}_+\cap\overline{\Omega}_-$ of the respective domains of definition, the *jump relations* (19a,b) are valid:

(7.1.19a) $\quad I_+(z)-I_-(z) = -2\pi i \qquad (z\in\Gamma)$

for positive orientation (otherwise $I_+(z)-I_-(z)=2\pi i$) and

(7.1.19b) $\quad \frac{1}{2}[I_+(z)+I_-(z)] = \overline{I}(z). \qquad (z\in\Gamma)$

Even in the case of an open curve, $I(z)$ is not only continuous in $\mathbb{C}\setminus\Gamma$, but also differentiable with the derivative

(7.1.20) $\quad \frac{d}{dz}I(z) = \frac{1}{\zeta_- - z} - \frac{1}{\zeta_+ - z} \qquad$ for $z\in\mathbb{C}\setminus\Gamma$ with $\zeta_-=\zeta(0)$, $\zeta_+=\zeta(L)$,

as one sees from (15a). Given a point $\zeta_0\in\Gamma\setminus\partial\Gamma$, we choose an arbitrary compact and simply connected neighbourhood U of ζ_0, which is disjoint to $\partial\Gamma$: $\zeta_0\in\mathring{U}$, $U\cap\partial\Gamma=\emptyset$. For the sake of simplicity, U is supposed to be convex. Γ divides $U\setminus\Gamma$ into two connection components $U_\pm$, where U_- lies on the left and U_+ on the right side of Γ. The choice of $U_\pm$ is such that in the case of a closed curve with positive orientation, U_- lies in the interior domain Ω_- and U_+ in the exterior domain Ω_+. An example for U is the circle $\overline{K_\varepsilon(\zeta_0)}$ with $\varepsilon<\delta:=\mathrm{dist}(\zeta_0,\partial\Gamma)$, where formally $\mathrm{dist}(\zeta_0,\emptyset)=\infty$ is used (cf. Fig. 3).

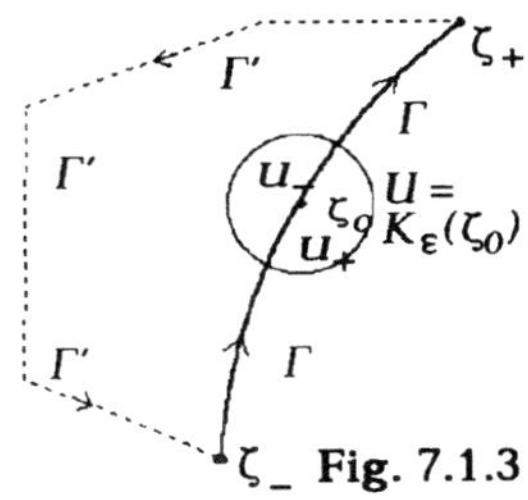

Fig. 7.1.3

The derivative (20) is bounded in $U_\pm$, so that $I(z)$ is uniformly continuous and therefore continuously extendable to I_+ [I_-] on $\overline{U}_+$ [$\overline{U}_-$]. Since $\zeta_0 \in \Gamma \backslash \partial\Gamma$ is arbitrary, I_+ and I_- are well-defined on the interior part $\Gamma \backslash \partial\Gamma$ of the curve.

Remark 7.1.28. **(a)** Choose the neighbourhood U of $\zeta_0 \in \Gamma \in C^1_{stw}$ as above (cf. Fig. 3). $U_\pm \subset K_\varepsilon(\zeta_0)$ are the connection components of $U \backslash \Gamma$. Then, $I_+(z) \in C^\lambda(\overline{U}_+)$ and $I_-(z) \in C^\lambda(\overline{U}_-)$ holds for $0<\lambda<1$ with the Hölder constants $C_\delta = 2\,\delta^{-\lambda} + \text{const}$, where $\delta := \text{dist}(U, \partial\Gamma)$.
(b) In $\mathbb{C} \backslash \Gamma$ all derivatives of $I(z)$ are defined and their continuous extensions to $\Gamma \backslash \partial\Gamma$ are equal.
(c) For $z \in \Gamma \backslash \partial\Gamma$ the *jump relations* (19a,b) are valid.

Proof of (c). (i) Adding the curve Γ' connecting ζ_+ to ζ_- depicted in Fig. 3, we obtain a closed curve $\Gamma+\Gamma'$, which is positively oriented and contains U_- in its interior domain Ω_-. We define $I'(z) := \int_{\Gamma'} \frac{d\zeta}{\zeta - z}$ and $I''(z) := I(z) + I'(z)$. Since I' is continuous in U, $I'_\pm(z) = \overline{I}'(z) = I'(z)$ holds for $z \in \Gamma \backslash \partial\Gamma$. Remark 27 proves $I''_+(z) - I''_-(z) = -2\pi i$ and $\frac{1}{2}[I''_+(z) + I''_-(z)] = \overline{I}''(z) = \overline{I}(z) + \overline{I}'(z)$. Solving with respect to $I_\pm(z) = I''_\pm(z) - I'(z)$, one gains the assertion. QED

Exercise 7.1.29. Let $\Gamma \in C^1_{stw}$ have the angle ω_+ at $z \in \Gamma \backslash \partial\Gamma$. Prove that

$$I(z) = I_+(z) + i(2\pi - \omega_+) = I_-(z) - i\omega_+ = \overline{I}(z) + i(\pi - \omega_+). \tag{7.1.21}$$

7.2 The Cauchy Kernel

7.2.1 Definition and Properties

Let Γ be a C^1_{stw} curve in $\mathbb{C}$, where the case of a closed $C^1_{0,stw}$ curve is included. $d\zeta/(\zeta - z)$ is called the *Cauchy kernel*. The corresponding integral operator Kf is defined by (1):

$$(Kf)(z) := \frac{1}{2\pi i} \oint_\Gamma \frac{f(\zeta)}{\zeta - z}\, d\zeta \tag{7.2.1}$$

A *singular integral equation* connected with K is, e.g.,

$$\lambda f = g + \mu K f \qquad (\lambda, \mu\text{: constants}). \tag{7.2.2}$$

As we shall see in §7.3.1, it makes no sense for equation (2) to distinguish between equations of the first ($\lambda=0$) and second kind ($\lambda \neq 0$). Instead, a relation between λ and μ is essential.

For any $z \in \Gamma$, the integrand in (1) is, in general, strongly singular. To investigate the existence of Cauchy's principal value, we split the integrand into the terms

$$\frac{f(\zeta)}{\zeta - z} = \frac{f(\zeta) - f(z)}{\zeta - z} + f(z)\frac{1}{\zeta - z}. \tag{7.2.3a}$$

Since with respect to the integration, $f(z)$ is a constant, one obtains formally

$$\oint_\Gamma \frac{f(\zeta)}{\zeta - z} d\zeta = \oint_\Gamma \frac{f(\zeta) - f(z)}{\zeta - z} d\zeta + f(z) \oint_\Gamma \frac{d\zeta}{\zeta - z} \tag{7.2.3b}$$

(cf. (1.4)). From Exercise 1.2a we know: If the integrals on the right-hand side of (3b) exist, then the (Cauchy) integral on the left side exists, too. In §7.1.4 we proved not only the existence of $\oint_\Gamma \frac{d\zeta}{\zeta - z}$, but also determined its value. Hence, it remains to investigate the existence and the behaviour of the first term. For this term we introduce the operator M:

$$(7.2.4) \qquad (Mf)(z) := \oint_\Gamma \frac{f(\zeta)-f(z)}{\zeta - z}\, d\zeta.$$

By analogy with Theorem 1.3, we want to show that the integral in (4) exists as an improper one, if f is Hölder continuous. Here, a function f is called Hölder continuous on Γ with exponent $\lambda \in (0,1]$, if

$$(7.2.5a) \qquad |f(\zeta)-f(z)| \leqslant H_f |\zeta - z|^\lambda \qquad \text{for all } \zeta, z \in \Gamma,$$

where in the case of $\lambda = 1$, Lipschitz continuity holds. We write again $f \in C^\lambda(\Gamma) = \hat{C}^\lambda(\Gamma)$ for $0<\lambda<1$ and $f \in C_L(\Gamma) = \hat{C}^1(\Gamma)$ for $\lambda = 1$. As usual, the norm $\|\varphi\|_{C^\lambda(\Gamma)}$ is defined as maximum of the minimal Hölder constants H_f and the maximum norm.

The Hölder continuity on Γ has to be redefined, since also other explanations would be possible, e.g., the reduction to the Hölder continuity of $f(\zeta(\tau))$ on $[0,L]$ for some parametrisation of Γ. For a real curve $\Gamma \subset \mathbb{R}^2$, (5a) is to be replaced by (5b):

$$(7.2.5b) \qquad |f(\mathbf{x})-f(\mathbf{y})| \leqslant H_f \|\mathbf{x}-\mathbf{y}\|_2^\lambda \qquad \text{for all } \mathbf{x}, \mathbf{y} \in \Gamma.$$

Lemma 7.2.1. Let $\Gamma \in C^1_{\text{stw}}$. $K_\varepsilon(z)$ denotes the circle around z with radius ε. Then the following statement holds. **(a)** The sum of the lengths of all curve pieces in the intersection $\Gamma \cap K_\varepsilon(z)$ is bounded by $4\varepsilon + o(\varepsilon)$ as $\varepsilon \to 0$, where the term $o(\varepsilon)$ is uniform in $z \in \mathbb{C}$.
(b) There is a constant $C_\varkappa$ independent of $\varepsilon > 0$ and $z \in \mathbb{C}$ such that

$$(7.2.6a) \qquad \int_{\Gamma \setminus K_\varepsilon(z)} \frac{|d\zeta|}{|\zeta - z|^\varkappa} \leqslant C_\varkappa \varepsilon^{1-\varkappa} \qquad \text{for } \varkappa > 1,\ z \in \mathbb{C},$$

$$(7.2.6b) \qquad \int_{\Gamma \setminus K_\varepsilon(z)} \frac{|d\zeta|}{|\zeta - z'|^{\varkappa-1}|\zeta - z|} \leqslant C_\varkappa \varepsilon^{1-\varkappa} \qquad \text{for } \varkappa > 1,\ z \in \mathbb{C},\ z' \in \overline{K_{\varepsilon/2}(z)},$$

$$(7.2.6c) \qquad \int_\Gamma \frac{|d\zeta|}{|\zeta - z|^\varkappa} \leqslant C_\varkappa \qquad \text{for } \varkappa < 1,\ z \in \mathbb{C},$$

$$(7.2.6d) \qquad \int_{\Gamma \cap K_\varepsilon(z)} \frac{|d\zeta|}{|\zeta - z|^\varkappa} \leqslant C_\varkappa \varepsilon^{1-\varkappa} \qquad \text{for } \varkappa < 1,\ z \in \mathbb{C}.$$

Proof. (i) The curve Γ consists of finitely many smooth pieces $\Gamma_k := \{\zeta(\tau): \tau_{k-1} \leqslant \tau \leqslant \tau_k\}$ (cf. Definition 1.10c). For sufficiently small ε, there is no circle $K_\varepsilon(z)$ intersecting more than *two* pieces Γ_k, where, in addition, the piece $\gamma_k := \Gamma_k \cap K_\varepsilon(z)$ is connected. For each of the at most two curvilinear arcs γ_k we have to estimate their length ℓ_ε by $2\varepsilon + o(\varepsilon)$. Let $\zeta: [0,L] \to \Gamma$ be the arc length parametrisation. ζ' is uniformly continuous on all subintervals $[\tau_{k-1}, \tau_k]$: $|\zeta'(\tau) - \zeta'(\sigma)| \leqslant \delta(|\tau - \sigma|)$ with $\delta(\eta) \to 0$ as $\eta \to 0$. Let the piece $\gamma_k := \Gamma_k \cap K_\varepsilon(z)$ be the image of the parameter interval $[\tau', \tau'']$. Since $\zeta(\tau')$ and $\zeta(\tau'')$ are situated on the boundary of $K_\varepsilon(z)$, we have

$$\begin{aligned}2\varepsilon &\geq |\zeta(\tau')-\zeta(\tau'')| = \left|\int_{\tau'}^{\tau''}\zeta'(\tau)d\tau\right| = \\ &= \left|\int_{\tau'}^{\tau''}[\zeta'(\tau')+(\zeta'(\tau)-\zeta'(\tau'))]\,d\tau\right| \geq \\ &\geq |\tau''-\tau'|\,[\,|\zeta'(\tau')|-\max_{\tau}|\zeta'(\tau)-\zeta'(\tau')|\,] \geq \\ &\geq |\tau''-\tau'|\,[\,1-\delta(|\tau''-\tau'|)\,],\end{aligned}$$

because $|\zeta'(\tau)|=1$ is the characteristic property of the arc length parametrisation. The arc length difference $|\tau''-\tau'|$ represents the length ℓ_ε of the piece γ_k. The supremum of $\delta(|\tau''-\tau'|)$ over all τ',τ'' with $\zeta(\tau'),\ \zeta(\tau'')\in K_\varepsilon(z),\ z\in\Gamma$, tends to zero as $\varepsilon\to 0$, because otherwise, there would be sequences $\{\tau_j''\}$ and $\{\tau_j'\}$ with $|\tau_j''-\tau_j'|\not\to 0$, $|\zeta(\tau_j')-\zeta(\tau_j'')|\to 0$ and for a suitable subsequence a contradiction to the absence of double points would result. From $\delta(|\tau''-\tau'|)=o(1)$ one obtains (a): $\ell_\varepsilon \leq 2\varepsilon/(1-o(1))=2\varepsilon+o(\varepsilon)$.

(ii) We recall the definition (7a) and the estimate (7b):

(7.2.7a) $$\int_\Gamma \varphi(\zeta)|d\zeta| := \int_0^L \varphi(\zeta(\tau))|\zeta'(\tau)|d\tau,$$

(7.2.7b) $$\left|\int_\Gamma \varphi(\zeta)d\zeta\right| \leq \|\varphi\|_\infty\,\mu(\Gamma), \quad \text{where } \mu(\Gamma)=\int_\Gamma|d\zeta| = \text{length of } \Gamma.$$

We choose $\varepsilon_0>0$ small enough so that $o(\varepsilon)$ from part (a) can be estimated by ε. Furthermore, we set

$$\varepsilon_n:=\varepsilon_0 2^{-n},\quad K_n:=K_{\varepsilon_n}(z),\quad R_0:=\Gamma\setminus K_0,\quad R_n:=\Gamma\cap K_{n-1}\setminus K_n,$$
$$I_0:=\int_{R_0}\frac{|d\zeta|}{|\zeta-z|^\varkappa},\quad I_n:=\int_{R_n}\frac{|d\zeta|}{|\zeta-z|^\varkappa}\quad \text{for } n\geq 1.$$

$|\zeta-z|\geq\varepsilon_0$ holds for $\zeta\in R_0$. Hence, the estimate $I_0\leq\mu(\Gamma)/\varepsilon_0^\varkappa$ is *independent* of z. By part (i) and the choice of ε_0, the curve pieces in R_n $(n\geq 1)$ have a length of $|R_n|\leq 5\varepsilon_{n-1}=10\varepsilon_n$. This leads to $I_n\leq 10\varepsilon_n/\varepsilon_n^\varkappa$. Since for $\varepsilon=\varepsilon_n$, $\Gamma\setminus K_\varepsilon(z)$ is the union $R_0\cup\ldots\cup R_n$, one estimates the left-hand side in (6a) by

$$\begin{aligned}&\mu(\Gamma)/\varepsilon_0^\varkappa+10\sum_{k=1}^n\varepsilon_k^{1-\varkappa} = \mu(\Gamma)/\varepsilon_0^\varkappa+10\,\varepsilon_0^{1-\varkappa}\sum_{k=1}^n 2^{k(\varkappa-1)} = \\ &= O(2^{n(\varkappa-1)}) = O(\varepsilon_n^{1-\varkappa}).\end{aligned}$$

Therefore, a bound $C\varepsilon_n^{1-\varkappa}$ exists. Let ε be an arbitrary element of an interval $[\varepsilon_n,\varepsilon_{n-1}]$. The integral in (6a) over $\Gamma\setminus K_\varepsilon(z)$ can be estimated by that over $\Gamma\setminus K_{\varepsilon_n}(z)$, so that the estimate $C\varepsilon_n^{1-\varkappa}\leq 2^{\varkappa-1}C\varepsilon^{1-\varkappa}$ leads to (6a) with $C_\varkappa:=2^{\varkappa-1}C$.

(iii) From $|\zeta-z|\geq\varepsilon$ and $|\zeta-z'|\geq|\zeta-z|-|z-z'|\geq|\zeta-z|-\varepsilon/2\geq\frac{1}{2}|\zeta-z|$, one concludes that $\int_{\Gamma\setminus K_\varepsilon(z)}\frac{|d\zeta|}{|\zeta-z'|^{\varkappa-1}\,|\zeta-z|}\leq 2^{\varkappa-1}\int_{\Gamma\setminus K_\varepsilon(z)}\frac{|d\zeta|}{|\zeta-z|^\varkappa}$, so that (6b) follows from (6a) with enlarged $C_\varkappa$.

(iv) Part (ii) is also applicable with $\varkappa<1$ and proves the improper integrability of $|\zeta-z|^{-\varkappa}$ with a bound independent of z. This proves (6c). Summing the integrals I_k only for $k>n$, one obtains a convergent sum with the value $O(\varepsilon_n^{1-\varkappa})$, which shows (6d) for $\varepsilon=\varepsilon_n$. For general ε, proceed as in (ii). QED

The following lemma states that M maps each Hölder continuous function f to a continuous image Mf.

Lemma 7.2.2. Let $\Gamma\in C^1_{\mathrm{stw}}$. $M\in L(C^\lambda(\Gamma),C(\Gamma))$ holds for all $\lambda>0$. The improper integral $(Mf)(z)$ in (4) exist for all $f\in C^\lambda(\Gamma)$, $z\in\mathbb{C}$.

Proof. (i) Let $f\in C^\lambda(\Gamma)$, $0<\lambda<1$, be a function with the Hölder constant H_f (cf. (5a)). The absolute value of the integrand $(f(\zeta)-f(z))/(\zeta-z)$ has the majorant $H_f/|\zeta-z|^{1-\lambda}$, which by (6c) is improperly integrable. Because of the majorant criterion from Exercise 6.1.1c, $(f(\zeta)-f(z))/(\zeta-z)$ is also improperly integrable. Inequality (6c) yields the pointwise boundedness $\|Mf\|_{\infty,\Gamma}\leqslant C_{1-\lambda}H_f\leqslant C_{1-\lambda}\|f\|_{C^\lambda(\Gamma)}$.
(ii) The estimate (6d) with $\varkappa=1-\lambda$ states the uniform existence of the improper integrals $(Mf)(z)$ for all $z\in\Gamma$. Since the integrand $(f(\zeta)-f(z))/(\zeta-z)$ is continuous on $\{(\zeta,z)\in\Gamma\times\Gamma:\ \zeta\neq z\}$, the integral $(Mf)(z)$ depends continuously on z. This proves $Mf\in C(\Gamma)$ (cf. Lemma 6.1.5). QED

We return again to the integral operator K with the Cauchy kernel. The next theorem discusses the continuity of Kf on Γ. Here, a point $\zeta\in\Gamma$ is called a *corner* of Γ, if two *different* tangential directions $\zeta'(\tau_0\pm 0)$ exist at $\zeta=\zeta(\tau_0)$ (cf. Fig. 1.1 a,b).

Theorem 7.2.3. Define $\partial\Gamma$ for $\Gamma\in C^1_{\mathrm{stw}}$ as in (1.16). We define

$$C^\lambda_{\partial\Gamma}(\Gamma) := \{f\in C^\lambda(\Gamma):\ f(\zeta)=0 \qquad \text{for all } \zeta\in\partial\Gamma\}.$$

(a) $\varphi:=Kf$ is well-defined on Γ for $f\in C^\lambda_{\partial\Gamma}(\Gamma)$, $\lambda>0$. In the case of a smooth curve $\Gamma\in C^1$, φ is continuous, so that $K\in L(C^\lambda_{\partial\Gamma}(\Gamma),C(\Gamma))$. In the general case of $\Gamma\in C^1_{\mathrm{stw}}$, φ is discontinuous at $\zeta_0\in\Gamma$ if and only if ζ_0 is a corner and $f(\zeta_0)\neq 0$. However, the one-sided limits coincide:

$$\varphi(\zeta(\tau_0+0)) = \varphi(\zeta(\tau_0-0)),$$

so that this discontinuity is removable, i.e., $\overline{\varphi}(\zeta(\tau)) := \varphi(\zeta(\tau\pm 0))$ is continuous and coincides almost everywhere with φ. $\overline{K}$ defined by $\overline{\varphi}=\overline{K}f$ satisfies

(7.2.8a) $\quad \overline{K}\in L(C^\lambda_{\partial\Gamma}(\Gamma),C(\Gamma))$,

(7.2.8b) $\quad \overline{K}=K \qquad$ for $\Gamma\in C^1$.

(b) For a closed curve $\Gamma\in C^1_{0,\mathrm{stw}}$, $\overline{K}\in L(C^\lambda(\Gamma),C(\Gamma))$ and, if in addition $\Gamma\in C^1_0$, the we even have $K\in L(C^\lambda(\Gamma),C(\Gamma))$.
(c) Let $\gamma\subset\Gamma$ be an open neighbourhood of a fixed point $\zeta\in\Gamma\setminus\partial\Gamma$, where $\overline{\gamma}$ and $\partial\Gamma$ are disjoint. If $f\in L^1(\Gamma)$ is Hölder continuous in γ, then $(\overline{K}f)(z)$ is defined for $z\in\gamma$ and continuous.

Proof. (i) Split the integral Kf according to (3b). The first term is $Mf/(2\pi i)$ and represents a continuous function provided $f\in C^\lambda(\Gamma)$ (cf. Lemma 2). The second term $f(z)\oint_\Gamma\frac{d\zeta}{\zeta-z}$ has the value $f(z)I(z)$ with $I(z)$ from (1.18a):

(7.2.8c) $\qquad (Kf)(z) = \frac{1}{2\pi i}[(Mf)(z)+f(z)I(z)] \qquad$ for $z\in\Gamma$.

The removable discontinuity of $I(z)$ at the corners of Γ (cf. Remark 1.26) carries over to Kf, provided that $f(z)\neq 0$. The continuous extension at the corners replaces $f(z)I(z)$ by $f(z)\overline{I}(z)$ with $\overline{I}(z)$ from (1.18c). The function

(7.2.8d) $\quad (\overline{K}f)(z) = \frac{1}{2\pi i}[(Mf)(z) + f(z)\overline{I}(z)] \quad$ for $z \in \Gamma$

is continuous in the inner part $\Gamma \backslash \partial\Gamma$ of the curve. For $z \to \zeta_\pm \in \partial\Gamma$, $I(z)$ and $\overline{I}(z)$ have a logarithmic singularity (cf. (1.17a/18c)). However, since $f(\zeta_\pm) = 0$ by definition of $C^\lambda_{\partial\Gamma}(\Gamma)$, $|f(z)I(z)| \le C|\zeta_\pm - z|^\lambda |\log(\zeta_\pm - z)|$ tends to zero as $z \to \zeta_\pm$. By definition of M, the limit $\lim(Kf)(z) = \frac{1}{2\pi i}(Mf)(\zeta_\pm)$ coincides with $(Kf)(\zeta_\pm)$; hence, Kf is continuous at the end-points $\zeta_\pm \in \partial\Gamma$, too. This proves part (a) of the theorem.
(ii) $\partial\Gamma = \emptyset$ implies $C^\lambda_{\partial\Gamma}(\Gamma) = C^\lambda(\Gamma)$, so that part (b) follows from (a).
(iii) For the proof of (c) split Γ into γ and $\Gamma_1 := \Gamma \backslash \gamma$. The integration over γ and Γ_1 defines the respective operators K_0 and K_1. For $z \in \gamma$, the integrand $f(\zeta)/(\zeta - z)$ is uniformly continuous in $\zeta \in \Gamma_1$, so that $(K_1 f)(z)$ is also continuous. Since γ contains no end-point $\zeta_\pm \in \partial\Gamma$, the considerations from (i) applied to γ instead of Γ show that $K_0 f$ is continuous at z. QED

Remark 7.2.4. Theorem 3 and its proof show that for $f \in C^\lambda(\Gamma)$ the image Kf may possess two different kinds of discontinuities:
(a) At the corners $\zeta = \zeta(\tau_0)$ of the curve Γ, $\varphi := Kf$ *jumps* but has identical left- and right-sided limits $\varphi(\zeta(\tau_0 \pm 0))$, which define the «smoothed» function $\overline{\varphi} = \overline{K}f$. In the case of a closed curve with positive orientation, (8e) holds because of $\overline{I}(z) = \pi i$:

(7.2.8e) $\quad \overline{K} = \frac{1}{2\pi i} M + \frac{1}{2} Id \quad$ for $\Gamma \in C^1_{0,\mathrm{stw}}$ with M from (4), Id = identity,

(b) If Γ is not closed, $\varphi = Kf$ may have *logarithmic singularities* at the end-points of Γ.
(c) For a real-valued function f, the discontinuity from (a) only concerns the real part $Re\ \varphi$, whereas the logarithmic singularity from (b) is connected with the imaginary part $Im\ \varphi$.

The mentioned discontinuities really appear, as one sees from the example $f = 1$, which leads to the function Kf given in (1.17a).

7.2.2 Regularity Properties

The statement $K \in L(C^\lambda(\Gamma), C(\Gamma))$ from Theorem 3b can be substantially strengthened. The following *Theorem of Plemelj-Privalov* guarantees the Hölder continuity for the same exponent.

Theorem 7.2.5. Let $\Gamma \in C^1_{\mathrm{stw}}$ and $\lambda \in (0, 1)$. The operators $\overline{K}$ and M satisfy

(7.2.8f) $\quad M, \overline{K} \in L(C^\lambda_{\partial\Gamma}(\Gamma), C^\lambda(\Gamma)) \quad$ ($C^\lambda_{\partial\Gamma}(\Gamma)$ as in Theorem 3).

The assumption $\Gamma \in C^1$ ensures

(7.2.8f') $\quad K \in L(C^\lambda_{\partial\Gamma}(\Gamma), C^\lambda(\Gamma))$

for the Cauchy operator K from (1). For closed curves, one has

(7.2.8g) $\quad M, \overline{K} \in L(C^\lambda(\Gamma), C^\lambda(\Gamma)) \quad$ for $\Gamma \in C^1_{0,\mathrm{stw}}$,

(7.2.8g') $\quad M, K \in L(C^\lambda(\Gamma), C^\lambda(\Gamma)) \quad$ for $\Gamma \in C^1_0$.

Proof. (i) We start with assertion (8g); (8f,f') will be treated in (vi). Because of (8e) and $K=\overline{K}$ for smooth Γ, one has only to show $M\in L(C^\lambda(\Gamma),C^\lambda(\Gamma))$. For given $f\in C^\lambda(\Gamma)$ and $z, z'\in\Gamma$, we set

$$\delta := \frac{1}{2\pi i}\int_\Gamma \varphi(\zeta)d\zeta \qquad \text{with } \varphi(\zeta) := \frac{f(\zeta)-f(z')}{\zeta-z'} - \frac{f(\zeta)-f(z)}{\zeta-z}.$$

The integration of φ over Γ is split into the integrals over $\Gamma\backslash K_\varepsilon(z)$ and $\Gamma\cap K_\varepsilon(z)$, where $\varepsilon := 2|z'-z|$ is chosen as radius. For the proof of the Hölder continuity it suffices to suppose that ε is sufficiently small. In $\Gamma\cap K_\varepsilon(z)$ the Hölder continuity of f leads to $|\varphi(\zeta)|\leqslant H_f[|\zeta-z'|^{\lambda-1}+|\zeta-z|^{\lambda-1}]$ with H_f from (5a). Lemma 1b is applicable with $\varkappa := 1-\lambda$ in (6d) and proves $|\int_{\Gamma\cap K_\varepsilon(z)}|\zeta-z|^{\lambda-1}d\zeta| \leqslant C_{1-\lambda}\varepsilon^\lambda$. Using $K_\varepsilon(z)\subset K_{3\varepsilon/2}(z')$, we conclude

$$\Big|\int_{\Gamma\cap K_\varepsilon(z)}|\zeta-z'|^{\lambda-1}d\zeta\Big| \leqslant \Big|\int_{\Gamma\cap K_{3\varepsilon/2}(z')}|\zeta-z'|^{\lambda-1}d\zeta\Big| \leqslant C_{1-\lambda}(\tfrac{3}{2}\varepsilon)^\lambda$$

again from (6d). By $\varepsilon=2|z'-z|$, we obtain

$$\Big|\int_{\Gamma\cap K_\varepsilon(z)}\varphi(\zeta)d\zeta\Big| \leqslant \text{const}\,|z'-z|^\lambda H_f. \tag{7.2.9a}$$

(ii) For the integration over $\Gamma\backslash K_\varepsilon(z)$ we split φ into $\varphi=\varphi_1+\varphi_2$ with

$$\varphi_1(\zeta) := \frac{f(z)-f(z')}{\zeta-z}, \tag{7.2.9b}$$

$$\varphi_2(\zeta) := (f(\zeta)-f(z'))\Big[\frac{1}{\zeta-z'}-\frac{1}{\zeta-z}\Big] = \frac{(z'-z)[f(\zeta)-f(z')]}{(\zeta-z')(\zeta-z)}. \tag{7.2.9c}$$

The integral over φ_2 will be estimated in (iii), that over φ_1 in (iv).

(iii) The Hölder continuity of f yields the estimate $|\varphi_2(\zeta)|\leqslant$ $\leqslant |z'-z|H_f|\zeta-z'|^{\lambda-1}|\zeta-z|^{-1}$. Lemma 1b is applicable, since $z'\in\overline{K_{\varepsilon/2}(z)}$ by the choice of ε. Inequality (6b) with $\varkappa := 2-\lambda>1$ shows

$$\Big|\int_{\Gamma\backslash K_\varepsilon(z)}\varphi_2(\zeta)d\zeta\Big| \leqslant |z'-z|\,H_f C_{2-\lambda}(2|z'-z|)^{\lambda-1} = \text{const}\,H_f|z'-z|^\lambda. \tag{7.2.9d}$$

(iv) We suppose the standard case: $\Gamma\backslash K_\varepsilon(z)$ is simply connected (for sufficiently small ε and all z, $\Gamma\backslash K_\varepsilon(z)$ consists of at most two connected components. The further proof is analogous). The integral of $\frac{1}{\zeta-z}$ over $\Gamma\backslash K_\varepsilon(z)$ equals $\log|(\zeta_+-z)/(\zeta_--z)|+i[\text{Arg}(\zeta_+)-\text{Arg}(\zeta_-)] = i[\text{Arg}(\zeta_+)-\text{Arg}(\zeta_-)]$ (cf. Lemma 1.23), where $\zeta_\pm$ are the end-points of $\Gamma\backslash K_\varepsilon(z)$: $|(\zeta_\pm-z)|=\varepsilon$. The bracket [...] has the global bound 2π. Therefore, the integral of $\varphi_1=(f(z)-f(z'))/(\zeta-z)$ can be estimated by

$$\Big|\int_{\Gamma\backslash K_\varepsilon(z)}\varphi_1(\zeta)\,d\zeta\Big| = |f(z')-f(z)|\,\Big|\int_{\Gamma\backslash K_\varepsilon(z)}\frac{d\zeta}{\zeta-z}\Big| \leqslant \text{const}\,H_f|z'-z|^\lambda. \tag{7.2.9e}$$

(v) (9a,d,e) prove $|\delta|=|\int_\Gamma\varphi(\zeta)d\zeta|/2\pi\leqslant\text{const}\,H_f|z'-z|^\lambda$. Because of $H_f\leqslant\|f\|_{C^\lambda(\Gamma)}$, we obtain

$$|(Mf)(z')-(Mf)(z)| \leqslant C|z'-z|^\lambda\|f\|_{C^\lambda(\Gamma)}. \tag{7.2.9f}$$

The boundedness of $\|Mf\|_\infty$ (Theorem 3b) and the Hölder bound (9f) imply $M\in L(C^\lambda(\Gamma),C^\lambda(\Gamma))$. This yields the statements for $\overline{K}$, K.

(vi) If $\Gamma\in C^1_{stw}$ is not closed, complete Γ by an additional arc so that a closed curve $\Gamma^*\in C^1_{0,stw}$ results (cf. Fig. 1.3). $f\in C^\lambda_{\partial\Gamma}(\Gamma)$ can be extended by $f(\zeta):=0$ for $\zeta\in\Gamma^*\backslash\Gamma$ so that $f\in C^\lambda(\Gamma^*)$. Kf can be regarded as an integral over Γ^*. Hence, the assertions follows from the previous results. ∎

Corollary 7.2.6. **(a)** In the general case of $\Gamma \in C^1_{stw}$, the weakened statement $M \in L(C^\lambda(\Gamma), C^{\lambda'}(\Gamma))$ still holds for all $0<\lambda'<\lambda<1$.
(b) Let $\gamma \subset \Gamma$ be an open subset of $\Gamma \in C^1_{stw}$ and $f \in L^1(\Gamma)$ be *locally Hölder continuous* in γ with exponent $\lambda \in (0,1)$. Then also Mf and $\overline{K}f$ are locally Hölder continuous in γ with exponent λ. If $\gamma \in C^1$, the same statement holds for K.
(c) Let $\Gamma \in C^1_{0,stw}$. In general, a Lipschitz continuous function $f \in C_L(\Gamma)$ does not lead to $\varphi := \overline{K}f \in C_L(\Gamma)$. A careful estimate of the quantities appearing in the previous proof yields the following inequality as optimal result:

$$|\varphi(z)-\varphi(z')| \leqslant C\,|z-z'|\,\big|\log|z-z'|\big|\,\|f\|_{C_L(\Gamma)}.$$

(d) Let $\gamma \subset \Gamma$ be compact and disjoint with $\partial\Gamma$. Then $M, \overline{K} \in L(C^\lambda(\Gamma), C^\lambda(\gamma))$.

Proof. (i) The analysis of M requires the estimates of φ from the proof of Theorem 5. For the proof of (a) note that because of the integrals $\int_{\Gamma\setminus K_\varepsilon(z)} \frac{d\zeta}{\zeta - z}$ only logarithmic singularities $\log\varepsilon = \log 2|z'-z|$ can appear (cf. Lemma 1.23). Statement (9e) remains valid, when the last inequality is replaced by $\ldots \leqslant \mathrm{const}\, H_f |z'-z|^\lambda \log|z'-z|$. Since this expression remains bounded by $C(\lambda')\, H_f |z'-z|^{\lambda'}$ for $\lambda'<\lambda$, Mf is Hölder continuous with exponent λ'.
(ii) Let $z \in \gamma$ and choose open neighbourhoods U and V with $z \in U \subset \overline{U} \subset V \subset \overline{V} \subset \gamma$. There is a «cut function» $\chi \in C^\infty(\Gamma)$ with $\chi=1$ in U and $\chi=0$ in $\Gamma\setminus V$. Define $f_1 := \chi f$, $f_2 := (1-\chi)f$. f_1 satisfies the assumption of Theorem 5 with Γ replaced by U, so that $\varphi_1 := \overline{K}f_1$ is Hölder continuous in $\overline{U}$ with exponent λ. The support of f_2 is at most $\Gamma\setminus V$ and has a finite distance from z. In Lemma 7 we shall show $\varphi_2 := \overline{K}f_2 \in C^\infty(U)$, so that $\overline{K}f = \varphi_1 + \varphi_2 \in C^\lambda(\overline{U})$. Since $z \in \gamma$ is arbitrary, we have demonstrated $\overline{K}f \in C^\lambda_{local}(\gamma)$ (i.e., assertion (b)). QED

7.2.3 Properties of the Generated Holomorphic Function

So far, we have evaluated and discussed $\varphi(z) := (Kf)(z) = \frac{1}{2\pi i}\oint_\Gamma \frac{f(\zeta)}{\zeta - z}\,d\zeta$ only for $z \in \Gamma$. It is much simpler to define the function

$$\Phi(z) := \frac{1}{2\pi i}\int_\Gamma \frac{f(\zeta)}{\zeta - z}\,d\zeta \qquad \text{for } z \notin \Gamma, \tag{7.2.10}$$

because, for $z \notin \Gamma$, the integrand has no singularity. In particular, the integrand is differentiable with respect to z. The derivative is $f(\zeta)/(\zeta-z)^2$. Since Γ is compact, the differentiation with respect to z may be performed under the integral sign. This proves the first part of

Lemma 7.2.7. Let Φ be defined by (10) with $f \in L^1(\Gamma)$. Then Φ is *holomorphic* on $\mathbb{C}\setminus\Gamma$ with the derivatives

$$\Phi'(z) = \frac{1}{2\pi i}\int_\Gamma \frac{f(\zeta)}{(\zeta - z)^2}\,d\zeta \qquad \text{for } z \notin \Gamma, \tag{7.2.11a}$$

$$\Phi^{(\nu)}(z) = \frac{\nu!}{2\pi i}\int_\Gamma \frac{f(\zeta)}{(\zeta - z)^{\nu+1}}\,d\zeta \qquad \text{for } z \notin \Gamma,\ \nu \in \mathbb{N}_0. \tag{7.2.11b}$$

As $z \to \infty$, $\Phi^{(\nu)}(z)$ tends to zero. More precisely, we have

$$|\Phi^{(\nu)}(z)| \leqslant \frac{\nu!}{2\pi}\|f\|_{L^1(\Gamma)}/(|z|-R)^{\nu+1} \qquad \text{for } |z|>R := \max_{\zeta\in\Gamma}|\zeta|. \tag{7.2.11c}$$

Proof. Since $|\zeta - z| \geq |z| - R$ for all $\zeta \in \Gamma$, $|z| > R$, one obtains $|\Phi^{(\nu)}(z)| \leq \leq \int_\Gamma |f(\zeta)| |d\zeta| \nu! / [2\pi(|z|-R)^{\nu+1}] = \nu! \|f\|_{L^1(\Gamma)} / [2\pi(|z|-R)^{\nu+1}]$. ∎

The further investigations should aim to reveal the behaviour of $\Phi(z)$ as $z \to z' \in \Gamma$. The main result reads as follows.

Theorem 7.2.8. Let $f \in C^\lambda(\Gamma)$, $\lambda \in (0,1)$. **(a)** The closed curve $\Gamma \in C^1_{0,\mathrm{stw}}$ is supposed to be positively oriented. It defines the interior domain Ω_- and the exterior domain Ω_+. Then the function Φ from (10) can be extended (Hölder) continuously onto the closures $\overline{\Omega}_+$, $\overline{\Omega}_-$:

(7.2.12a) $\quad \Phi_+ \in C^\lambda(\overline{\Omega}_+), \quad \Phi_- \in C^\lambda(\overline{\Omega}_-).$

In general, Φ_+ and Φ_- do *not* coincide on their common domain of definition Γ and they are also different from the value $\overline{\Phi} := \overline{K} f$ obtained by evaluation of Φ on Γ. These three quantities are connected as follows:

(7.2.12b) $\quad \Phi_+(z) - \Phi_-(z) = -f(z) \qquad$ for $z \in \Gamma$,

(7.2.12c) $\quad \overline{\Phi}(z) := (\overline{K} f)(z) = \frac{1}{2}[\Phi_+(z) + \Phi_-(z)] \qquad$ for $z \in \Gamma$,

(7.2.12d) $\quad \Phi_\pm(z) = (\overline{K} f)(z) \mp \frac{1}{2} f(z) \qquad$ for $z \in \Gamma$,

where

(7.2.12e) $\quad \overline{\Phi}(z) = \Phi(z) := (Kf)(z), \qquad$ if $z \in \Gamma$ is no corner.

If Γ is negatively oriented, one has to replace $-f(z)$ in (12b,d) by $f(z)$.
(b) For an open curve $\Gamma \in C^1_{\mathrm{stw}}$, the one-sided continuous extensions of Φ exist for any curve point $z \in \Gamma \setminus \partial\Gamma$: Let Φ_- denote the continuation from the left (from the neighbourhood U_- in Fig. 1.3) and Φ_+ the corresponding continuation from the right. The relations (12b-e) are valid for $\Phi_\pm$, where z is restricted to $\Gamma \setminus \partial\Gamma$, since $\Phi(z)$ may have a singularity at $z \in \partial\Gamma$. Let $G \subset \mathbb{C}$ be a closed set not containing $\partial\Gamma$ and touching Γ at most from one side. Then, $\Phi_+ \in C^\lambda(G)$ or $\Phi_- \in C^\lambda(G)$ holds, depending on the location of G.
(c) If $f \in C^\lambda_{\partial\Gamma}(\Gamma)$, the statements of part (b) are also valid for sets G containing $\partial\Gamma$.

Proof. (i) Part (a) follows from (b) with the choice $G = \overline{\Omega}_+$ or $G = \overline{\Omega}_-$. Part (c) is obtained from (a) by supplementing Γ as in part (vi) of the proof of Theorem 5. f is extended to the closed curve $\Gamma^* := \Gamma + \Gamma' \in C^1_{0,\mathrm{stw}}$ by $f = 0$ on Γ'. Hence, it remains to show part (b).
(ii) Let $U \subset \mathbb{C}$ be a compact and convex set disjoint to Γ. Since the derivative Φ' is uniformly bounded in U, Φ is Lipschitz and therefore also Hölder continuous for any exponent $0 < \lambda < 1$.
(iii) $U_\infty := \mathbb{C} \setminus K_R(0)$ is a neighbourhood of $z = \infty$. Choose R so large that $\Gamma \subset K_R(0)$. Since the derivative Φ' (cf. (11c)) remains bounded in U_∞, also the Hölder constant is uniformly bounded in U_∞, so that Φ is *globally* Hölder continuous in U_∞.
(iv) In (vii) we shall prove the global Hölder continuity of Φ in the left-sided neighbourhood U_- (cf. Fig. 1a,b) of $z_0 \in \Gamma \setminus \partial\Gamma$. The exact choice of U_- reads

(7.2.12f) $U_- = \{z \in K_\eta(z_0): z \text{ lies left from } \Gamma\}, \quad z_0 \in \Gamma \setminus \partial\Gamma.$

Here, η is to be chosen sufficiently small, so that (12g,h) apply:

(7.2.12g) $\Gamma \cap K_\eta(z_0)$ is a connected piece of curve,

(7.2.12h) $\partial\Gamma \subset \Gamma \setminus \overline{K_{2\eta}(z_0)}.$

Condition (12h) ensures that the curve point $z' \in \Gamma$ next to some $z \in U_-$ has always a fixed distance $\mathrm{dist}(z', \partial\Gamma) \geq \mathrm{dist}(z_0, \partial\Gamma) - 2\eta > 0$ from the end-points $\zeta_\pm \in \partial\Gamma$. Here, «dist» is the distance function

(7.2.12i) $\mathrm{dist}(z, M) := \inf\{|z - \zeta|: \zeta \in M\}$ for $M \subset \mathbb{C}$, $z \in \mathbb{C}$.

A possible definition of the condition «z *lies left from* Γ» in (12f) may be: For a suitable positively oriented completion of Γ to a closed curve $\Gamma + \Gamma'$, U_- lies in the interior domain Ω_-.

(v) The global Hölder continuity of Φ in an open set V implies the continuous extendability of Φ onto the closure $\overline{V}$. The extended mapping is Hölder continuous on $\overline{V}$ with the same constants. Therefore, it suffices to show global Hölder continuity (13) in U_-. The continuation of Φ onto $\overline{U}_\pm$ is called $\Phi_\pm$.

(vi) Without loss of generality, assume that the set $G \subset \mathbb{C}$ from assertion (b) touches the curve from the left, if at all. By (ii) and (iv), Φ_- is *locally* Hölder continuous in $G \cap \overline{K_R(0)}$. Since $G \cap \overline{K_R(0)}$ is compact, the *global* Hölder continuity in $G \cap \overline{K_R(0)}$ follows (cf. Exercise 1.2.3a). Since $\Phi = \Phi_-$ is globally Hölder continuous in $G \cap U_\infty$ according to (iii), one obtains the global Hölder continuity in $G = (G \cap \overline{K_R(0)}) \cup (G \cap U_\infty)$.

(vii) Let $z_1, z_2 \in U_-$ be points of U_- from (12f-h). We define

$$\delta_i := \mathrm{dist}(z_i, \Gamma) \quad (i = 1, 2), \qquad \delta := \min\{\delta_1, \delta_2\}.$$

Concerning the estimate of $\Phi(z_1) - \Phi(z_2)$, we distinguish the cases $|z_1 - z_2| \leq \delta/2$ and $|z_1 - z_2| > \delta/2$, for which we prepare the Hölder estimate (13) below in Lemmata 10 and 12, respectively:

(7.2.13) $|\Phi_-(z_1) - \Phi_-(z_2)| \leq H_\Phi |z_1 - z_2|^\lambda.$

(viii) Concerning the jump relations (12b-d), one concludes from the representation of Φ in Lemma 9 that $2\pi i \Phi_\pm(z) = (Mf)(z) + f(z) I_\pm(z)$ for $z \in \Gamma$. $2\pi i (\overline{K}f)(z) = (Mf)(z) + f(z)\overline{I}(z)$ from (8d) and the jump relations (1.19a,b) of $\overline{I}(z)$, $I_\pm(z)$ imply (12b-d). QED

Lemma 7.2.9. By analogy with (3b), one can split Φ into

(7.2.14a) $2\pi i\, \Phi(z) = \mu(z', z) + f(z') I(z)$ with $I(z) := \oint_\Gamma \frac{d\zeta}{\zeta - z}$

and

(7.2.14b) $\mu(z', z) := \int_\Gamma \frac{f(\zeta) - f(z')}{\zeta - z} d\zeta$ for $z' \in \Gamma$, $z \in \mathbb{C}$.

Under the assumption $f \in C^\lambda(\Gamma)$ with $\lambda > 0$, the quantity $\mu(z', z)$ from (14b) is defined for all $z' \in \Gamma$, $z \in \mathbb{C}$ except for $z \in \partial\Gamma$, $z' \neq z$. For $z = z' \in \Gamma$, we have

(7.2.14c) $\mu(z', z') = (Mf)(z')$ for $z' \in \Gamma$, M from (4).

Proof. We distinguish three cases: (i) $z \notin \Gamma$, $z' \in \Gamma$. Here, the integrand is regular.
(ii) $z = z' \in \Gamma$: The integral in (14b) coincides with $(Mf)(z)$ from (4), so that Lemma 2 ensures the *improper* integrability.
(iii) $z \in \Gamma \setminus \partial\Gamma$, $z' \in \Gamma$, $z' \neq z$: The integrand is *strongly singular*. The Cauchy principal value, however, exists. QED

Lemma 7.2.10. Let U_- be defined by (12f-h) (cf. Fig. 1a). $z_1, z_2 \in U_-$ are points with $|z_1 - z_2| < \frac{1}{2}\,\mathrm{dist}(z_i, \Gamma)$ for $i = 1, 2$. Then, the Hölder estimate (13) holds with H_Φ independent of z_1 and z_2.

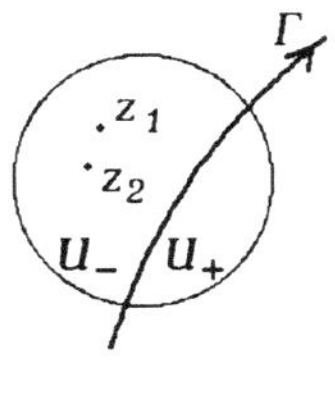

Fig. 7.2.1a

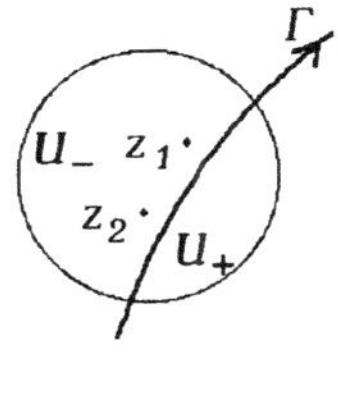

Fig. 7.2.1b

Proof. Without loss of generality, let $0 < \delta := \delta_1 := \mathrm{dist}(z_1, \Gamma) \leq \delta_2 := \mathrm{dist}(z_2, \Gamma)$. We choose $z' \in \Gamma$ with $\mathrm{dist}(z_1, \Gamma) = |z_1 - z'|$. Since $I(z)$ is globally Hölder continuous in U_- (cf. Remark 1.28), only

$$(7.2.14d) \qquad |\mu(z', z_1) - \mu(z', z_2)| \leq C_\mu |z_1 - z_2|^\lambda \qquad \text{for } z' \in \Gamma$$

has to be shown. According to (14b), the difference has the representation

$$(7.2.14e) \qquad \begin{aligned} \mu(z', z_1) - \mu(z', z_2) &= \int_\Gamma \Big[\frac{f(\zeta) - f(z')}{\zeta - z_1} - \frac{f(\zeta) - f(z')}{\zeta - z_2}\Big] d\zeta = \\ &= (z_1 - z_2) \int_\Gamma \frac{f(\zeta) - f(z')}{(\zeta - z_1)(\zeta - z_2)}\, d\zeta . \end{aligned}$$

We split Γ into $\gamma_d := \Gamma \cap K_d(z_1)$, $\Gamma_d := \Gamma \setminus K_d(z_1)$, where $d := 3\delta = 3\delta_1 \leq 3\delta_2$. Using the Hölder continuity (5a) of f, we obtain

$$(7.2.14f) \qquad |\mu(z', z_1) - \mu(z', z_2)| \leq |z_1 - z_2|\, H_f \Big\{ \int_{\gamma_d} \frac{|\zeta - z'|^\lambda |d\zeta|}{|\zeta - z_1||\zeta - z_2|} + \int_{\Gamma_d} \cdots \Big\}.$$

In the first integral, the inequalities $|\zeta - z_1| \geq \mathrm{dist}(z_1, \Gamma) = \delta$ and $|\zeta - z_2| \geq |\zeta - z_1| - |z_1 - z_2| \geq \delta - \delta/2 = \delta/2$ are valid. Application of (6d) to $\int_{\gamma_d} |\zeta - z'|^\lambda |d\zeta|$ yields the bound $C d^{1+\lambda} = C(3\delta)^{1+\lambda}$; hence,

$$\int_{\gamma_d} \frac{|\zeta - z|^\lambda\, |d\zeta|}{|\zeta - z_1||\zeta - z_2|} \leq \delta^{-1} (\delta/2)^{-1} C (3\delta)^{1+\lambda} = 2 \cdot 3^{1+\lambda} C \delta^{\lambda - 1}.$$

By definition, $\delta \leq |\zeta - z_1|/3$ holds for $\zeta \in \Gamma_d$, so that we obtain $|\zeta - z'| \leq |\zeta - z_1| + |z_1 - z'| = |\zeta - z_1| + \delta \leq \frac{4}{3}|\zeta - z_1|$. In the second integral over Γ_d, one estimates the nominator by $[\frac{4}{3}|\zeta - z_1|]^\lambda$ and obtains

$$\int_{\Gamma_d} \frac{|\zeta - z'|^\lambda |d\zeta|}{|\zeta - z_1||\zeta - z_2|} \leq \left(\tfrac{4}{3}\right)^\lambda \int_{\Gamma_d} \frac{|d\zeta|}{|\zeta - z_1|^{1-\lambda}|\zeta - z_2|} \leq \left(\tfrac{4}{3}\right)^\lambda C \delta^{\lambda - 1},$$

by choosing the quantities $\varepsilon := \delta$, $z' := z_1$, $z := z_2$ in (6b). Inserting both integral estimates into (14f) and using $|z_1 - z_2| \leq \delta/2$, one is led to

$$\begin{aligned} |\mu(z', z_1) - \mu(z', z_2)| &\leq |z_1 - z_2|\, C' \delta^{\lambda - 1} = \\ &= C' |z_1 - z_2|^\lambda [|z_1 - z_2|/\delta]^{1-\lambda} \leq C' 2^{\lambda - 1} |z_1 - z_2|^\lambda, \end{aligned}$$

so that the assertion is proved. QED

The next lemma concerns the particular case $z_2 = z' \in \Gamma$, where $z' \in \partial\Gamma$ and any $z \in \mathbb{C} \setminus \Gamma$ is admitted.

Lemma 7.2.11. Let $\Gamma \in C^1_{stw}$. The Hölder constant of $f \in C^\lambda(\Gamma)$ with $\lambda \in (0,1)$ is denoted by H_f. For all $z' \in \Gamma$ and $z \in \mathbb{C} \setminus \Gamma$ we have

$$\text{(7.2.14g)} \qquad |\mu(z',z) - \mu(z',z')| \leqslant C_1 H_f |z'-z|^\lambda \frac{|z'-z|}{\operatorname{dist}(z,\Gamma)}.$$

Proof. (i) The circle $K_\varepsilon(z')$ splits Γ into the pieces $\gamma_\varepsilon := \Gamma \cap K_\varepsilon(z')$ and $\Gamma \setminus K_\varepsilon(z')$. ε is chosen as

$$\text{(7.2.15a)} \qquad \varepsilon := 3|z-z'| > 0.$$

The inequalities $|\zeta - z'| \leqslant |\zeta - z| + |z - z'| \leqslant 2\varepsilon/3 < \varepsilon$ for $|\zeta - z| \leqslant \varepsilon/3$ prove

$$\text{(7.2.15b)} \qquad K_{\varepsilon/3}(z) \subset K_\varepsilon(z'), \quad \Gamma \setminus K_{\varepsilon/3}(z) \supset \Gamma \setminus K_\varepsilon(z').$$

(ii) According to (14e), the difference $\mu(z',z) - \mu(z',z')$ equals

$$\text{(7.2.15c)} \qquad \mu(z',z) - \mu(z',z') = (z-z') \int_\Gamma \frac{f(\zeta) - f(z')}{(\zeta - z)(\zeta - z')} \, d\zeta.$$

Using the Hölder constant H_f of f with respect to the exponent λ, we get

$$|\mu(z',z) - \mu(z',z')| \leqslant |z-z'| H_f \int_\Gamma \frac{|d\zeta|}{|\zeta - z|\,|\zeta - z'|^{1-\lambda}}.$$

(iii) We estimate $|\zeta - z|$ for $\zeta \in \gamma_\varepsilon = \Gamma \cap K_\varepsilon(z')$ from below by $\operatorname{dist}(z,\Gamma)$. Inequality (6d) from Lemma 1 proves that $\int_{\gamma_\varepsilon} |d\zeta| / |\zeta - z'|^{1-\lambda}$ is bounded by $C\varepsilon^\lambda$. By (15a), we obtain (15d) with $C_{11} := 3^\lambda C$:

$$\text{(7.2.15d)} \qquad |z-z'| H_f \int_{\gamma_\varepsilon} \frac{|d\zeta|}{|\zeta - z|\,|\zeta - z'|^{1-\lambda}} \leqslant C_{11} H_f |z-z'|^\lambda \frac{|z'-z|}{\operatorname{dist}(z,\Gamma)}.$$

(iv) For $\zeta \in \Gamma_\varepsilon(z')$, one obtains $|\zeta - z'| \geqslant |\zeta - z| - |z' - z| = |\zeta - z| - \frac{\varepsilon}{3} \geqslant \frac{1}{2}|\zeta - z|$ because of $|\zeta - z| \geqslant |\zeta - z'| - |z' - z| = |\zeta - z'| - \frac{\varepsilon}{3} \geqslant \varepsilon - \frac{\varepsilon}{3} = 2\varepsilon/3$ or $\frac{\varepsilon}{3} \leqslant \frac{1}{2}|\zeta - z|$. Therefore, the integrand can be estimated for $\zeta \in \Gamma_\varepsilon(z')$ by $2^{1-\lambda}/|\zeta - z|^{2-\lambda}$. By (15b), this bound may be integrated over $\Gamma \setminus K_{\varepsilon/3}(z)$ instead of $\Gamma \setminus K_\varepsilon(z')$. Inequality (6a) from Lemma 1 yields the result

$$\int_{\Gamma \setminus K_\varepsilon(z')} \frac{|d\zeta|}{|\zeta - z|\,|\zeta - z'|^{1-\lambda}} \leqslant 2^{1-\lambda} \int_{\Gamma \setminus K_\varepsilon(z')} \frac{|d\zeta|}{|\zeta - z|^{2-\lambda}} \leqslant$$

$$\leqslant 2^{1-\lambda} \int_{\Gamma \setminus K_{\varepsilon/3}(z)} \frac{|d\zeta|}{|\zeta - z|^{2-\lambda}} \leqslant 2^{1-\lambda} C \left(\tfrac{\varepsilon}{3}\right)^{\lambda - 1} \leqslant C_{12} |z - z'|^{\lambda - 1}$$

with $C_{12} := 2^{1-\lambda} C$.
(v) Altogether, the results from (iii) and (iv) produce the estimate

$$\text{(7.2.15e)} \qquad |\mu(z',z) - \mu(z',z')| \leqslant \left(C_{11} \frac{|z'-z|}{\operatorname{dist}(z,\Gamma)} + C_{12}\right) H_f |z - z'|^\lambda.$$

Since the constants from Lemma 1 are globally valid, also C_{11} and C_{12} are independent of z and z'. As $|z'-z| \geqslant \operatorname{dist}(z,\Gamma)$, (15e) implies the assertion (14g) with $C_1 := C_{11} + C_{12}$. QED

Lemma 7.2.12. Let U_- be as in Lemma 10. $z_1, z_2 \in U_-$ are supposed to satisfy $|z_1 - z_2| \geqslant \frac{1}{2}\delta$ with $\delta := \min\{\delta_1, \delta_2\}$, $\delta_i := \operatorname{dist}(z_i, \Gamma)$ (cf. Fig. 1b). $f \in C^\lambda(\Gamma)$ fulfils the Hölder estimate (13) with H_Φ independent of z_1, z_2.

Proof. Without loss of generality, let $\delta=\delta_1\leqslant\delta_2$. Determine $z'_1, z'_2\in\Gamma$ with $|z_i-z'_i|=\delta_i$. Let Φ_- and I_- be the restrictions of Φ and I to U_-:

$$2\pi i\,\Phi_-(z) = \mu(z',z)+f(z')I_-(z).$$

We split $\Phi_-(z_1)-\Phi_-(z_2)$ into

$$\Phi_-(z_1)-\Phi_-(z_2) =$$

$$\text{(7.2.16a)}\qquad = \frac{1}{2\pi i}\{[\mu(z'_1,z_1)+f(z'_1)I_-(z_1)]-[\mu(z'_2,z_2)+f(z'_2)I_-(z_2)]\} =$$

$$= \frac{1}{2\pi i}\{[\mu(z'_1,z_1)-\mu(z'_2,z_2)]+[f(z'_1)I_-(z_1)-f(z'_2)I_-(z_2)]\}.$$

Since f is Hölder continuous in Γ and $I_-(z)$ does so in $\overline{U}_-$ (cf. Remark 1.28a), the second bracket is bounded by $C(|z'_1-z'_2|^\lambda+|z_1-z_2|^\lambda)\leqslant$ $\leqslant(1+6^\lambda)C|z_1-z_2|^\lambda$, as we see from

$$\text{(7.2.16b)}\qquad \delta_1+\delta_2\leqslant 5|z_1-z_2|,\qquad |z'_1-z'_2|\leqslant\delta_1+|z_1-z_2|+\delta_2\leqslant 6|z_1-z_2|.$$

Here we have used the inequalities $\delta_2\leqslant|z_2-z'_1|\leqslant|z_1-z_2|+\delta_1$ and $\delta_1=\delta\leqslant 2|z_1-z_2|$.

It remains to derive a Hölder estimate for the first bracket in (16a). We rewrite the difference in the form

$$\text{(7.2.16c)}\qquad \mu(z'_1,z_1)-\mu(z'_2,z_2) = [\mu(z'_1,z_1)-\mu(z'_1,z'_1)] +$$
$$+[\mu(z'_1,z'_1)-\mu(z'_2,z'_2)]+[\mu(z'_2,z'_2)-\mu(z'_2,z_2)].$$

Lemma 11 applies to the first and third bracket in (16c):

$$\text{(7.2.16d)}\qquad |\mu(z'_i,z_i)-\mu(z'_i,z'_i)|\leqslant C_1H_f|z'_i-z_i|^\lambda = C_1H_f\delta_i^\lambda,$$

since $|z'_i-z_i|/\mathrm{dist}(z_i,\Gamma)=\delta_i/\delta_i=1$ due to the choice of z'_i.

Because of (14c), the second bracket in (16c) coincides with $(Mf)(z'_1)-(Mf)(z'_2)$. By construction (cf. (12h)), z'_1 and z'_2 lie in the set $\gamma=\Gamma\cap\overline{K_{2\eta}(z_0)}$ disjoint to $\partial\Gamma$. Corollary 6d guarantees $Mf\in C^\lambda(\gamma)$, so that

$$\text{(7.2.16e)}\qquad |\mu(z'_1,z'_1)-\mu(z'_2,z'_2)|\leqslant C_2H_f|z'_1-z'_2|^\lambda \underset{(16b)}{\leqslant} 6^\lambda C_2H_f|z_1-z_2|^\lambda$$

follows and concludes the proof. QED

The statement of Theorem 8 ensures that the convergence $\Phi(z)\to\Phi_\pm(z_0)$ is uniform for any approach of $z\in U_\pm$ to $z_0\in\Gamma\setminus\partial\Gamma$. The angle condition described by Muskhelishvili [1, p.46] is need only if the Hölder continuity of f is restricted to an isolated point $z_0\in\Gamma$.

Exercise 7.2.13. Let $\partial\Gamma$ be the boundary of $\Gamma\in C^1_{\mathrm{stw}}$. For $f\in C^\lambda(\Gamma)$ prove the following: Locally (i.e., for a sufficiently small neighbourhood U of z, z'),

$$|\mu(z',z)-\mu(z',z')|\leqslant CH_f|z-z'|^\lambda\log|z-z'|\qquad\text{for } z'\in\partial\Gamma,\ z\in U,$$

holds, i.e., in the case of $f(\zeta_\pm)\neq 0$ for $\zeta_\pm\in\partial\Gamma$, the singularity of Φ is dominated by $f(\zeta_\pm)I(z)$.

At the end of this subchapter we recall some elementary properties of holomorphic functions.

Lemma 7.2.14. Let $\Omega\subset\mathbb{C}$ be open and simply connected and f be holomorphic in Ω. The positively oriented curve $\Gamma\in C^1_{0,\mathrm{stw}}$ with $\Gamma\subset\Omega$ defines an interior domain Ω_- and an exterior domain Ω_+. Then Cauchy's integral formula and integral theorem yield

(7.2.17a) $$\frac{1}{2\pi i}\oint_\Gamma \frac{f(\zeta)}{\zeta-z}\,d\zeta = f(z) \qquad \text{for } z\in\Omega_-,$$

(7.2.17b) $$\frac{1}{2\pi i}\oint_\Gamma \frac{f(\zeta)}{\zeta-z}\,d\zeta = 0 \qquad \text{for } z\in\Omega_+.$$

Exercise 7.2.15. $\Gamma\in C^1_{0,\mathrm{stw}}$ defines the interior domain Ω_-. Let $\Gamma_k\in C^1_{0,\mathrm{stw}}$ with $\Gamma_k\subset\Omega_-$ $(k\in\mathbb{N})$ be curves of uniformly bounded length ($\sup|\Gamma_k|<\infty$), which satisfy $\Gamma_k\to\Gamma$ in the sense of $\sup\{\mathrm{dist}(z,\Gamma): z\in\Gamma_k\}\to 0$ as $k\to\infty$. Prove for $\varphi\in C(\overline{\Omega}_-)$ that

$$\int_{\Gamma_k}\varphi(\zeta)\,d\zeta \to \int_\Gamma \varphi(\zeta)\,d\zeta.$$

In Lemma 14, Γ is contained in the interior of Ω. This condition can be weakened.

Corollary 7.2.16. Assume that the positively oriented curve $\Gamma\in C^1_{0,\mathrm{stw}}$ has Ω as interior domain. Let f be holomorphic in Ω and continuous in $\overline{\Omega}$. Then, (17a) holds for $z\in\Omega$ and (17b) for $z\in\mathbb{C}\setminus\overline{\Omega}$.

Proof. (i) Let $z\in\Omega$ and $\Gamma_k\subset\Omega$ with $\Gamma_k\to\Gamma$ as in Exercise 15. For sufficiently large $k\geq k_0$, z lies in the interior domain Ω^k_- of Γ_k, so that $\int_{\Gamma_k}\frac{f(\zeta)}{\zeta-z}d\zeta = 2\pi i f(z)$ according to Lemma 14. $k\to\infty$ proves (17a) for z.
(ii) $z\in\mathbb{C}\setminus\overline{\Omega}$ lies in the exterior domain of Γ_k, so that $\int_{\Gamma_k}\frac{f(\zeta)}{\zeta-z}d\zeta=0$. QED

Lemma 7.2.17. $\Gamma\in C^1_{0,\mathrm{stw}}$ defines the interior domain Ω_- and the exterior domain Ω_+. Let Ψ be continuous in $\mathbb{C}$ and holomorphic in Ω_+ and Ω_-, i.e., the continuous extensions $\Psi_\pm$ exist on $\overline{\Omega}_\pm$ and coincide on Γ:

(7.2.18) $$\Psi_+ = \Psi_- \qquad \text{on } \Gamma.$$

Then Ψ is holomorphic in the whole complex plane $\mathbb{C}$.

Proof. Let $z_0\in\Gamma$ be arbitrary. Choose the circle $K_\varepsilon(z_0)$ so small that a connected piece $\gamma:=\Gamma\cap K_\varepsilon(z_0)$ is cut out. Let $S:=\partial K_\varepsilon(z_0)$ be the positively oriented surrounding circle. We define

$$S_\pm := S\cap\Omega_\pm,\quad \Gamma_+ := S_+ - \gamma,\quad \Gamma_- := S_- + \gamma$$

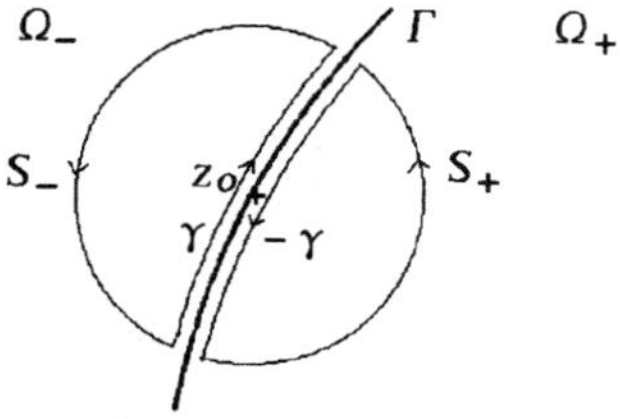

Fig. 7.2.2

(cf. Fig. 2). $\Gamma_\pm$ are the positively oriented boundary curves of the right- and left-sided half-circles $K_\varepsilon(z_0)\cap\Omega_\pm$. The function

$$\psi(z) := \frac{1}{2\pi i}\oint_S \frac{\Psi(\zeta)}{\zeta-z}\,d\zeta \qquad (z\in K_\varepsilon(z_0))$$

is holomorphic in the (open) circle $K_\varepsilon(z_0)$ (cf. Lemma 7). Since γ has different orientations on $\Gamma_\pm$, one concludes from $\int_{\Gamma_\pm}\ldots = \int_{S_\pm}\ldots \mp \int_\gamma \ldots$ the representation

$$\psi(z) = \frac{1}{2\pi i}\oint_{S_+}\frac{\Psi(\zeta)}{\zeta - z}d\zeta + \frac{1}{2\pi i}\oint_{S_-}\frac{\Psi(\zeta)}{\zeta - z}d\zeta = \psi_1(z) + \psi_2(z)$$

with

$$\psi_1(z) := \frac{1}{2\pi i}\oint_{\Gamma_+}\frac{\Psi(\zeta)}{\zeta - z}d\zeta, \qquad \psi_2(z) := \frac{1}{2\pi i}\oint_{\Gamma_-}\frac{\Psi(\zeta)}{\zeta - z}d\zeta \qquad \text{for } z\in K_\varepsilon(z_0)\setminus\gamma.$$

In the interior domain $K_\varepsilon(z_0)\cap\Omega_-\setminus\gamma$ of S_-, $\psi_1=\Psi$ and $\psi_2=0$ holds according to Corollary 16; hence, $\psi=\psi_1+\psi_2=\Psi$. Similarly, one concludes that $\psi_1=0$ and $\psi_2=\Psi$ in the interior domain of Γ_- implying $\psi=\psi_1+\psi_2=\Psi$ there, too. This shows $\psi=\Psi$ in $K_\varepsilon(z_0)\setminus\gamma$. Therefore, the unique and continuous extensions $\Psi_\pm$ must coincide with the holomorphic function ψ. ∎

Lemma 7.2.18. In addition to the assumptions of Lemma 17, suppose $\Psi(\infty)=0$, i.e., $\Psi(z)\to 0$ as $z\to\infty$. Then Ψ vanishes identically, i.e., $\Psi=0$ in $\mathbb{C}$.

Proof. By Lemma 17, Ψ is holomorphic in $\mathbb{C}$. Because of $\Psi(\infty)=0$, Ψ is bounded in $\mathbb{C}$. The theorem of Liouville states that Ψ must be constant:

Lemma 7.2.19. The positively oriented curve $\Gamma\in C^1_{0,\mathrm{stw}}$ defines the interior domain Ω_- and the exterior domain Ω_+. Let $f\in C^\lambda(\Gamma)$. Then the function $\Phi(z) := \frac{1}{2\pi i}\int_\Gamma \frac{f(\zeta)}{\zeta - z}d\zeta$ from (10) is the only one with the properties (a-c):

(a) Φ is holomorphic in Ω_+ and Ω_- with $\Phi(\infty)=0$,

(b) the continuous extensions Φ_+, Φ_- exist on $\overline{\Omega}_+$ and $\overline{\Omega}_-$, respectively.

(c) the following *jump condition* holds:

(7.2.19) $$\Phi_+ - \Phi_- = -f \qquad \text{on } \Gamma.$$

Proof. According to Lemma 7 and Theorem 8, Φ from (10) satisfies the properties (a-c). Let Ψ be a further function fulfilling conditions (a-c). The difference $\varphi := \Phi - \Psi$ satisfies (a-b) and the jump condition (18), i.e., $\varphi_+=\varphi_-$. Lemma 18 proves $\varphi=0$ and thus the uniqueness $\Phi=\Psi$. ∎

As an application of Lemma 19, we will show that $-K$ is an operator adjoint to the Cauchy operator K in the following sense.

Lemma 7.2.20. Let $\Gamma\in C^1_{0,\mathrm{stw}}$ and $f, g\in C^\lambda(\Gamma)$. Then

(7.2.20) $$\int_\Gamma f(\zeta)(Kg)(\zeta)d\zeta = -\int_\Gamma g(\zeta)(Kf)(\zeta)d\zeta.$$

Proof. (i) The functions (10) associated to f and g are denoted by Φ and Ψ. The continuous extensions in $\overline{\Omega}_\pm$ are $\Phi_\pm$, $\Psi_\pm$. We want to show

$$\int_\Gamma \Phi_+\Psi_+ d\zeta = 0, \qquad \int_\Gamma \Phi_-\Psi_- d\zeta = 0.$$

The second equation follows from the Cauchy integral theorem. For the first one use $\int_\Gamma \Phi_+\Psi_+d\zeta = \int_{\partial K_R(0)}\Phi_+\Psi_+d\zeta$ for all sufficiently large R, i.e., R so large that $\Gamma\subset K_R(0)$. (11c) shows $\Phi_+(z)=O(1/|z|)$ and $\Psi_+(z)=O(1/|z|)$;

therefore $\Phi_+(z)\Psi_+(z)=O(1/|z|^2)$. The estimate $|\int_{\partial K_R(0)}\Phi_+\Psi_+d\zeta| \leqslant O(R/R^2)\to 0$ proves $\int_\Gamma \Phi_+\Psi_+ d\zeta = 0$.

(ii) On Γ, $\Phi_+\Psi_+ - \Phi_-\Psi_- = (\Phi_+ - \Phi_-)\frac{1}{2}(\Psi_+ + \Psi_-) + (\Psi_+ - \Psi_-)\frac{1}{2}(\Phi_+ + \Phi_-)$ is valid. The jump relations (12b,c) show $\Phi_+\Psi_+ - \Phi_-\Psi_- = -f\overline{K}g - g\overline{K}f$. By (i), the integration over Γ yields $-\int_\Gamma (f\overline{K}g + g\overline{K}f)d\zeta = 0$; hence (20). Under the integral, $\overline{K}$ can be replaced by K, since the values coincide almost everywhere. ∎

7.2.4 Representation of K^2

As we have seen in the previous chapter, Abel's integral equation (6.4.8), i.e., $g = K_\lambda f$ can be solved explicitly by means of a multiplication by $K_{1-\lambda}$. The singular integral equation $g = Kf$ with the Cauchy operator K from (1) can be similarly solved explicitly using a multiplication by K.

Theorem 7.2.21. Let $\Gamma \in C^1_{0,\mathrm{stw}}$ be a closed curve. For any Hölder continuous function $g \in C^\lambda(\Gamma)$ $(\lambda > 0)$, the equation

(7.2.21 a) $\quad g = \overline{K}f \quad$ on Γ

has the following unique solution in $C^\lambda(\Gamma)$:

(7.2.21 b) $\quad f = 4\overline{K}g \quad$ on Γ.

In the case of no corners ($\Gamma \in C^1_0$), the equations (21a,b) become

(7.2.21a'/b') $\quad g = Kf, \quad f = 4Kg \quad$ on Γ.

In any case,

(7.2.21a''/b'') $\quad g = Kf, \quad f = 4Kg \quad$ *almost everywhere* on Γ.

Lemma 7.2.22 (Poincaré–Bertrand formula). For $\Gamma \in C^1_{0,\mathrm{stw}}$ and $\lambda > 0$,

(7.2.22) $\quad \overline{K}^2 = \frac{1}{4}I \quad$ on $C^\lambda(\Gamma)$ $\quad$ (I: identity).

For a more general formulation for the Poincaré–Bertrand formula compare Dautray – Lions [1, chapter XI,§2.3].

Proof. (i) Let $f \in C^\lambda(\Gamma)$ and $g = \overline{K}f$. Define $F(z) := \frac{1}{2\pi i}\int_\Gamma \frac{f(\zeta)}{\zeta - z}d\zeta$ on $\mathbb{C}\setminus\Gamma$. By Theorem 8, the continuous extensions $F_\pm$ of F onto the interior and exterior domains $\Omega_\pm$ exist and satisfy

(7.2.23a) $\quad F_+ + F_- = 2\overline{K}f = 2g \quad$ on Γ,

(7.2.23b) $\quad F_+ - F_- = -f \quad$ on Γ.

The function $G(z) := \begin{cases} -F(z) & \text{for } z \in \Omega_+ \\ +F(z) & \text{for } z \in \Omega_- \end{cases}$ defined on $\mathbb{C}\setminus\Gamma$ admits the continuous extensions $G_+ = -F_+$ and $G_- = F_-$, so that (23a,b) are equivalent to

(7.2.23a') $\quad G_+ - G_- = -2g \quad$ on Γ,

(7.2.23b') $\quad G_+ + G_- = f \quad$ on Γ.

By Lemma 7 (cf. (11c)), $G(z) = -F(z) \to 0$ holds for $z \to \infty$. Lemma 19 states that G must have the following representation:

(7.2.23c) $$G(z) = \frac{1}{2\pi i}\int_\Gamma \frac{2g(\zeta)}{\zeta - z}d\zeta \qquad (z \notin \Gamma).$$

Theorem 8 applied to G from (23c) shows

$$G_+ + G_- = 2\overline{K}(2g) = 4\overline{K}g \qquad \text{on } \Gamma$$

(cf. (12c)). A comparison with (23b') proves $f = 4\overline{K}g = 4\overline{K}(\overline{K}f) = 4\overline{K}^2 f$. Since f is an arbitrary element from $C^\lambda(\Gamma)$, (22) is proved. QED

Proof of Theorem 21. Lemma 22 states that $\overline{K}$ has the inverse $\overline{K}^{-1} = 4\overline{K}$. $g = \overline{K}f$ is equivalent to $f = \overline{K}^{-1}g = 4\overline{K}g$. QED

The inversion of the Cauchy integrals $g = \overline{K}f$ in the case of an open curve is more complicated and can, e.g., be found in Muskhelishvili [1,§86].

7.2.5 The Cauchy Integral on the Unit Circle

The simplest example of a closed curve is the positively oriented unit circle

(7.2.24) $$\Gamma := \{\zeta : |\zeta| = 1\} \quad \text{with } \zeta(\sigma) = e^{i\sigma} \qquad \text{for } 0 \leqslant \sigma \leqslant 2\pi.$$

Exercise 7.2.23. (a) Let Γ be positively oriented with $z=0$ in the interior domain (as, e.g., for Γ from (24)). Prove for the particular case $f(\zeta) = \zeta^\nu, \nu \in \mathbb{Z}$, that

(7.2.25) $$(\overline{K}\zeta^\nu)(z) = \tfrac{1}{2}z^\nu \quad \text{for } \nu \geqslant 0, \qquad (\overline{K}\zeta^\nu)(z) = -\tfrac{1}{2}z^\nu \quad \text{for } \nu < 0.$$

(b) Verify $\overline{K}^2\zeta^\nu = \frac{1}{4}\zeta^\nu$. *Hint:* (12c) and residue theorem for $\Phi_\pm$.

Let $f(\zeta)$ be given on the unit circle Γ by the parameter representation

(7.2.26) $$f(\zeta(\sigma)) = F(\sigma) \qquad \text{for } 0 \leqslant \sigma < 2\pi.$$

F can be extended by $F(\sigma) = F(\sigma + 2\pi)$ to a 2π-periodic function on $\mathbb{R}$. In the following we will use the representation of F as *Fourier series*

(7.2.27a) $$F(\sigma) = \frac{1}{\sqrt{2\pi}} \sum_{\nu \in \mathbb{Z}} \alpha_\nu e^{i\nu\sigma} \qquad (\sigma \in \mathbb{R}).$$

First assume that F is a *finite* Fourier series in order to avoid any convergence problems (i.e., $\alpha_\nu = 0$ for almost all $\nu \in \mathbb{Z}$ in (27a)). Using the parametrisation $\zeta(\sigma) = e^{i\sigma}$ (cf. (24)) and (26), one obtains the representation of f by

(7.2.27b) $$f(\zeta) = \frac{1}{\sqrt{2\pi}} \sum_{\nu \in \mathbb{Z}} \alpha_\nu \zeta^\nu \qquad (\zeta \in \Gamma,\ \Gamma \text{ as in (24)}).$$

Set $F(\sigma) = f(\zeta(\sigma))$ and $G(\sigma) = g(\zeta(\sigma))$. The $L^2(0, 2\pi)$ scalar product

(7.2.28a) $$\langle F, G\rangle := \int_0^{2\pi} F(\sigma)\overline{G}(\sigma)d\sigma$$

is identical with the scalar product

(7.2.28a') $$\langle f, g\rangle := \int_\Gamma f(\zeta)\overline{g}(\zeta)|d\zeta|$$

defined over Γ. Therefore, one need not distinguish between $\langle F, G\rangle$ and $\langle f, g\rangle$. The corresponding L^2 norms are

(7.2.28b) $\quad \|f\|_{L^2} := \|f\|_{L^2(\Gamma)} := \|F\|_{L^2(0,2\pi)} = \sqrt{\langle f,f\rangle} = \sqrt{\langle F,F\rangle}.$

From the theory of the Fourier series the *Parseval* identity

(7.2.28c) $\quad \|f\|^2_{L^2} = \|F\|^2_{L^2(0,2\pi)} = \sum_{\nu\in\mathbb{Z}} |\alpha_\nu|^2$

is well-known. In addition, one obtains the *Fourier coefficients* α_ν of the respective functions f and F from

(7.2.29) $\quad \alpha_\nu = \frac{1}{\sqrt{2\pi}}\langle f,\zeta^\nu\rangle = \frac{1}{\sqrt{2\pi}}\langle F, e^{i\nu\sigma}\rangle.$

By Π_- and Π_+ we denote the following orthogonal projections:

(7.2.30) $\quad (\Pi_- f)(\zeta) := \frac{1}{\sqrt{2\pi}}\sum_{\nu<0}\alpha_\nu\zeta^\nu, \qquad (\Pi_+ f)(\zeta) := \frac{1}{\sqrt{2\pi}}\sum_{\nu\ge 0}\alpha_\nu\zeta^\nu.$

Using the values $K\zeta^\nu$ evaluated in (25), one proves

(7.2.31) $\quad Kf = \frac{1}{2}\frac{1}{\sqrt{2\pi}}\left(-\sum_{\nu<0}\alpha_\nu\zeta^\nu + \sum_{\nu\ge 0}\alpha_\nu\zeta^\nu\right) = \frac{1}{2}f - \Pi_- f = \Pi_+ f - \frac{1}{2}f,$

since $K=\overline{K}$ because of $\Gamma\in C_0^1$. This shows the identity

(7.2.32) $\quad K = \frac{1}{2}I - \Pi_- = \Pi_+ - \frac{1}{2}I = \frac{1}{2}(\Pi_+ - \Pi_-).$

Exercise 7.2.24. Prove for any finite sum (27b) that

(7.2.33) $\quad \|Kf\|_{L^2} = \frac{1}{2}\|f\|_{L^2}.$

Since the finite trigonometric polynomials f from (27b) are dense in $L^2(\Gamma)$, for any $f\in L^2(\Gamma)$ one can find a sequence of finite Fourier series such that $f_\nu \to f$ in $L^2(\Gamma)$. Because of $\|g_\nu - g_\mu\|_{L^2} = \|Kf_\nu - Kf_\mu\|_{L^2} = \frac{1}{2}\|f_\nu - f_\mu\|_{L^2}$, the sequence $g_\nu := Kf_\nu$ is Cauchy; hence, $g := \lim g_\nu = \lim Kf_\nu$ exists in $L^2(\Gamma)$. One verifies that the definition $Kf := g$ is independent of the choice of the sequence $\{f_\nu\}$. Hereby, the domain of K has been extended. K maps any $f\in L^2(\Gamma)$ into $Kf\in L^2(\Gamma)$. The properties of K are summarised in

Theorem 7.2.25. The operator K belongs to $L(L^2(\Gamma), L^2(\Gamma))$ with the norm

(7.2.33') $\quad \|K\|_{L^2\leftarrow L^2} = \frac{1}{2}.$

The statements (30), (31), and (33) are valid for all $f\in L^2(\Gamma)$.

Note that the L^2 function Kf need not admit a unique value $(Kf)(z)$ for any special $z\in\Gamma$. Also the integral $\frac{1}{2\pi i}\int_\Gamma \frac{f(\zeta)}{\zeta - z}d\zeta$ need not exist for any $z\in\Gamma$ as a Cauchy's principal value.

The kth derivative of a finite Fourier series (27) equals

(7.2.34a) $\quad F^{(k)}(\sigma) = \frac{i^k}{\sqrt{2\pi}}\sum_{\nu\in\mathbb{Z}}\nu^k\alpha_\nu a^\sigma \qquad (\sigma\in\mathbb{R})$

and has the norm $\|F^{(k)}\|_{L^2} = \left[\sum_{\nu\in\mathbb{Z}}\nu^{2k}|\alpha_\nu|^2\right]^{1/2}$. For periodic functions, the H^k-norm defined in Definition 4.5.23 is equivalent to

(7.2.34b) $\quad \|F\|_{H^k} := \|F\|_{H^k(0,2\pi)} := \left[\sum_{\nu\in\mathbb{Z}}(1+\nu^2)^k|\alpha_\nu|^2\right]^{1/2}.$

Another equivalent norm is

(7.2.34c) $\quad \| F \|_{H^k} := [\, |\alpha_0|^2 + \sum_{\nu \neq 0} \nu^{2k} |\alpha_\nu|^2 \,]^{1/2}.$

For f with $F(\sigma)=f(\zeta(\sigma))$ (cf. (26)) we set $\|f\|_{H^k}=\|F\|_{H^k}$. The definitions (34b,c) can be used also for *negative* $k\in\mathbb{Z}$:

Exercise 7.2.26. **(a)** For $k=0$, the norms (34b) and (34c) coincide with $\|F\|_{L^2}$.
(b) The dual norm of $\|\cdot\|_{H^k}$ is $\|\cdot\|_{H^{-k}}$ (i.e., k in (34b,c) replaced by $-k$).

In addition, one defines the Sobolev space $H^k(0,2\pi)$ for all *real* $k\in\mathbb{R}$ as the Banach space of all f with finite $\|f\|_{H^k}$ norm. The representation (31) proves

Theorem 7.2.27. For all $\varkappa\in\mathbb{R}$, $K\in L(H^\varkappa(\Gamma), H^\varkappa(\Gamma))$ holds with $\|K\|_{H^\varkappa \leftarrow H^\varkappa}=\frac{1}{2}$.

7.3 The Singular Integral Equation

7.3.1 The Case of Constant Coefficients

Let Kf be the Cauchy integral $\frac{1}{2\pi i}\oint \frac{f(\zeta)}{\zeta - z}d\zeta$ over Γ (in the following, Γ is always closed) and $\overline{K}f$ the continuous extension (cf. Theorem 2.3). The equation

(7.3.1) $\quad (\alpha + 2\beta\overline{K})f = g \qquad$ on $\Gamma \qquad (\alpha,\beta\in\mathbb{C})$

is an abbreviation for the singular integral equation

(7.3.1') $\quad \alpha f(z) + \frac{\beta}{\pi i}\oint_\Gamma \frac{f(\zeta)}{\zeta - z}d\zeta = g(z) \qquad$ for $z\in\Gamma$, z no corner.

Theorem 7.3.1. Let $\Gamma\in C^1_{0,\mathrm{stw}}$. The integral equation (1) is solvable for all $g\in C^\lambda(\Gamma)$, if and only if

(7.3.2) $\quad \alpha^2 \neq \beta^2.$

In the case of (2), the unique solution $f\in C^\lambda(\Gamma)$ is explicitly represented by

(7.3.3) $\quad f = \frac{1}{\alpha^2 - \beta^2}(\alpha - 2\beta\overline{K})g.$

Proof. (i) We define $T := \alpha + 2\beta\overline{K}$. For $f=1$, one obtains $Tf=\alpha+\beta$. For $f(\zeta)=1/(\zeta-\zeta_0)$ with ζ_0 from the interior of Γ, one has $Tf=(\alpha-\beta)f$ (cf. Exercise 2.23). This shows that for $\alpha+\beta=0$ or $\alpha-\beta=0$, T is *not injective*. The same functions combined with (2.20) yield the respective equations

$$\int_\Gamma f(\zeta)(T\varphi)(\zeta)d\zeta = (\alpha\mp\beta)\int_\Gamma f(\zeta)\varphi(\zeta)d\zeta,$$

proving that T is also *not surjective* for $\alpha+\beta=0$ or $\alpha-\beta=0$.
(ii) Let condition (2) be satisfied. The Poincaré-Bertrand formula $4\overline{K}^2=I$ (cf. (2.22)) shows $(\alpha - 2\beta\overline{K})(\alpha + 2\beta\overline{K}) = (\alpha + 2\beta\overline{K})(\alpha - 2\beta\overline{K}) = (\alpha^2-\beta^2)I$. Hence, $T=\alpha+2\beta\overline{K}$ is bijective and thus f from (3) is the unique solution. QED

7.3.2 The Case of Variable Coefficients

In the variable case, the equation reads as in (1), but α and β are Hölder continuous functions on Γ: $\alpha,\beta\in C^\lambda(\Gamma)$:

(7.3.4) $\quad \alpha(z)f(z) + 2\beta(z)(\overline{K}f)(z) = g(z) \qquad$ for all $z\in\Gamma$.

The obvious generalisation of condition (2) is

(7.3.5) $\alpha(z)^2 \neq \beta(z)^2$ for all $z \in \Gamma$.

Under the assumption (5), $\frac{\alpha(\zeta)-\beta(\zeta)}{\alpha(\zeta)+\beta(\zeta)}$ describes a closed curve, which remains bounded and does not meet the origin. Hence, is defines a winding number $\varkappa \in \mathbb{Z}$. If $\zeta: [0,L] \to \mathbb{C}$ is the parametrisation, this number can be defined by the Arg function from §7.1.4:

(7.3.6) $$\varkappa := \frac{1}{2\pi i} \operatorname{Arg} \frac{\alpha(\zeta)-\beta(\zeta)}{\alpha(\zeta)+\beta(\zeta)} \Big|_{0+0}^{L-0}.$$

$\varkappa$ is called the *index* of the integral equation (4). As, e.g., described by Muskhelishvili [1,§47], the behaviour of the solution of equation (4) can be characterised explicitly. In particular, the following alternative holds:

(7.3.7a) $\dim \operatorname{Kern}(\alpha + 2\beta\overline{K}) = \varkappa$, $\quad \operatorname{range}(\alpha + 2\beta\overline{K}) = C^{\lambda}(\Gamma)$ for $\varkappa \geq 0$,

(7.3.7b) $\ker(\alpha + 2\beta\overline{K}) = \{0\}$, $\quad \dim C^{\lambda}(\Gamma)/\operatorname{range}(\alpha + 2\beta\overline{K}) = -\varkappa$ for $\varkappa \leq 0$.

Statement (7a) expresses the fact that equation (4) is solvable for all right-hand sides $g \in C^{\lambda}(\Gamma)$ and the solutions form a $\varkappa$-dimensional affine space, whereas statement (7b) guarantees uniqueness. The second part in (7b) manifests that the right-hand side must satisfy $\varkappa$ compatibility conditions of the kind $\int \varphi_j g \, d\Gamma = 0$ $(1 \leq j \leq \varkappa)$ in order to ensure existence of a solution. Conditions (7a,b) imply

(7.3.7c) $\alpha + 2\beta\overline{K}$ is bijective on $C^{\lambda}(\Gamma)$ for $\varkappa = 0$,

(7.3.7d) $\operatorname{ind}(\alpha + 2\beta\overline{K}) :=$
$\dim \operatorname{Kern}(\alpha + 2\beta\overline{K}) - \dim C^{\lambda}(\Gamma) \setminus \operatorname{range}(\alpha + 2\beta\overline{K}) = \varkappa$.

The left-hand side in (7d) is called the *index of the operator* $\alpha + 2\beta K$.

Exercise 7.3.2. Constant coefficients $\alpha, \beta \in \mathbb{C}$ lead to $\varkappa = 0$.

7.3.3 General Singular Integral Equations

Instead of the Cauchy kernel, we now admit $k(\zeta, z)/(\zeta - z)$, i.e.,

(7.3.8) $$\alpha(z) f(z) + \frac{1}{\pi i} \oint_{\Gamma} \frac{k(\zeta, z)}{\zeta - z} f(\zeta) d\zeta = g(z) \quad \text{for almost all } z \in \Gamma.$$

The fact that (8) is true «almost everywhere» allows us to ignore the corners of Γ. Concerning the smoothness of k, we suppose Hölder continuity, i.e.,

(7.3.9) $k \in C^{\lambda}(\Gamma \times \Gamma)$ for some $\lambda \in (0,1)$.

Splitting k into the sum

(7.3.10) $k = k_0 + k_1$, $\quad k_0(\zeta, z) := k(z,z)$, $\quad k_1(\zeta, z) := k(\zeta, z) - k(z,z)$,

one can write the integral operator in (8) in the form

$$\frac{1}{\pi i} \oint_{\Gamma} \frac{k(\zeta, z)}{\zeta - z} f(\zeta) d\zeta = (K_0 + K_1) f \quad \text{with}$$

$$K_0 = 2\beta\overline{K}, \qquad \beta(z) := k(z,z), \qquad (K_1 f)(z) = \frac{1}{\pi i} \oint_{\Gamma} \frac{k_1(\zeta, z)}{\zeta - z} f(\zeta) d\zeta.$$

Since $|k_1(\zeta, z)/|\zeta - z|| \leq C|\zeta - z|^{\lambda - 1}$ because of (9), K_1 is compact in

$C^{\lambda}(\Gamma)$. Omitting K_1, one obtains the *principal part* $T := \alpha I + 2\beta\overline{K}$ of the left-hand side of (8). Therefore, the general equation (8) can be regarded as a compact perturbation of equation (4) discussed in §7.3.2. If, in particular, $\varkappa = 0$ (i.e., T bijective), (8) reads $(T+K_1)f = g$ and is equivalent to $(I+\widetilde{K}_1)f = \widetilde{g} := T^{-1}g$, where $\widetilde{K}_1 := T^{-1}K_1$ is again compact (cf. Theorem 1.3.23a). The Riesz-Schauder theory (Theorem 1.3.28) is applicable to $I+\widetilde{K}_1$.

For the further study of singular integral equations we refer, e.g., to Prößdorf-Silbermann [2] or Pipkin [1, §§10-12].

7.3.4 Approximation of the Cauchy Integral on the Unit Circle

The analysis of the Cauchy integral on the unit circle in §7.2.5 suggest the approximation of the functions by *trigonometric polynomials*. For solving

(7.3.11) $$(\alpha + 2\beta K)f = g \qquad (\alpha^2 \neq \beta^2 \text{ constant}),$$

approximate g by the partial sum

(7.3.12) $$g_n(\zeta) = \frac{1}{\sqrt{2\pi}} \sum_{\nu=-n}^{n-1} \beta_\nu \zeta^\nu \qquad \text{with } \beta_\nu = \tfrac{1}{\sqrt{2\pi}} \langle g, \zeta^\nu \rangle$$

(cf. (2.27b), (2.29)). The approximation error is estimated by

(7.3.13) $$\| g - g_n \|_{L^2(\Gamma)} \leqslant n^{-k} \| g \|_{H^k(\Gamma)} \qquad \text{for } g \in H^k(\Gamma),\ k \geqslant 0,$$

as one derives immediately from (2.34b). The eigenvalues of $T := \alpha + 2\beta K$ are $\alpha \pm \beta$, from which one concludes as in (2.33') that

(7.3.14a) $$\| \alpha + 2\beta K \|_{L^2 \leftarrow L^2} \leqslant A := \max\{ |\alpha+\beta|, |\alpha-\beta| \},$$

(7.3.14b) $$\| (\alpha + 2\beta K)^{-1} \|_{L^2 \leftarrow L^2} \leqslant B := \max\{ 1/|\alpha+\beta|, 1/|\alpha-\beta| \}.$$

The solution of the approximating equation

(7.3.15) $$(\alpha + 2\beta K) f_n = g_n$$

is the partial sum of the exact solution f, i.e.,

(7.3.16) $$f_n(\zeta) = \frac{1}{\sqrt{2\pi}} \sum_{\nu=-n}^{n-1} \alpha_\nu \zeta^\nu \qquad \text{with } \alpha_\nu = \tfrac{1}{\sqrt{2\pi}} \langle f, \zeta^\nu \rangle = \frac{\beta_\nu}{\alpha \pm \beta},$$

where the signs «±» holds for $\{{\nu \geqslant 0 \atop \nu < 0}\}$ (cf. (2.25)). From (13) and (14b), we gain the error estimate

(7.3.17) $$\| f - f_n \|_{L^2(\Gamma)} \leqslant B n^{-k} \| g \|_{H^k(\Gamma)} \qquad \text{for } g \in H^k(\Gamma),\ k \geqslant 0.$$

Exercise 7.3.3. Prove that $g_n \to g$ in $L^2(\Gamma)$ for an arbitrary $g \in L^2(\Gamma)$ and $f_n \to f$ for the solution f_n of (15).

While the computation of $\alpha_\nu = \beta_\nu/(\alpha \pm \beta)$ in (16) easily feasible, the Fourier coefficients β_ν from (12) are, in general, not exactly to be determined. A possible approximation is the quadrature by the composite trapezoidal rule:

(7.3.18) $$\widetilde{\beta}_\nu := \frac{1}{\sqrt{2\pi}} \frac{\pi}{n} \sum_{\mu=-n}^{n-1} g(e^{i\mu\pi/n})\, e^{-i\nu\mu\pi/n},$$

which can be evaluated very efficiently for $n = 2^\ell$ by the fast Fourier transform (FFT, cf. Stoer [1, §2.2.2]).

Remark 7.3.4. **(a)** $\widetilde{\beta}_\nu$ is the exact integral of the interpolating trigonometric polynomial $\widetilde{g}_n$ at the support abscissae $\{e^{i\mu\pi/n}: -n\leqslant\mu<n\}$.
(b) For $g\in H^\varkappa(\Gamma)$, $\varkappa>\frac{1}{2}$, convergence $\widetilde{g}_n\to g$ holds in $L^2(\Gamma)$; more precisely:

(7.3.19) $$\|g-\widetilde{g}_n\|_{L^2(\Gamma)} \leqslant C_\varkappa n^{0.5-\varkappa}\|g\|_{H^\varkappa(\Gamma)} \qquad \text{for } g\in H^\varkappa(\Gamma),\ \varkappa>\tfrac{1}{2}.$$

For the Galerkin approximation we choose the orthogonal projection Π_n onto the subspace $X_n:=\operatorname{span}\{\zeta^\nu: -n\leqslant\mu<n\}$. In equation (15), K can be replaced by $K_n:=\Pi_n K$. K_n is characterised by

$$(K_n\zeta^\nu)(z) = \pm\tfrac{1}{2}z^\nu \text{ for } \{{}^{0\leqslant\nu<n}_{-n\leqslant\nu<0}\}, \qquad K_n\zeta^\nu=0 \text{ otherwise.}$$

The Galerkin method (15') below for equation (11) is identical to (15):

(7.3.15') $$(\alpha+2\beta K_n)f_n = \Pi_n g.$$

Remark 7.3.5. Let $\alpha\neq 0$. In $L^2(\Gamma)$ the method $\{K_n\}$ is

- *convergent*, since $f_n\to f$ in $L^2(\Gamma)$,
- *consistent*, since $(K_n-K)\varphi\to 0$ for all $\varphi\in L^2(\Gamma)$,
- *stable*, since $\|(\alpha+2\beta K_n)^{-1}\|\leqslant\max\{B,1/|\alpha|\}$ for all n with

B from (14b). Also for $\alpha=0$, $2\beta K_n$ has all mentioned properties on the subspace $X_n\subset L^2(\Gamma)$.

Note that the previous criteria are not applicable to K, since K is not compact and the approximations $\{K_n\}$ are not collectively compact.

Replacing the orthogonal projection Π_n by the projection $\widetilde{\Pi}_n$ of the trigonometric interpolation, one can represent $\widetilde{g}_n$ from Remark 4 by $\widetilde{\Pi}_n g$. $\widetilde{K}_n:=K\widetilde{\Pi}_n$ describes a collocation method. The solution $\widetilde{f}_n$ of $(\alpha+2\beta\widetilde{K}_n)\widetilde{f}_n=\widetilde{\Pi}_n g$ converges for $g\in H^\varkappa(\Gamma)$, $\varkappa>\frac{1}{2}$, to f (cf. Remark 4b). Nevertheless, $\{\widetilde{K}_n\}$ is neither convergent nor consistent in $L^2(\Gamma)$.

7.3.5 Approximation of the Cauchy Integral on an Arbitrary Curve Γ

Let Γ be an arbitrary closed curve from C_0^2 with of some parametrisation $\Gamma=\{\zeta(\sigma): 0\leqslant\sigma\leqslant 2\pi\}$. The Cauchy integral $K_\Gamma f$ evaluated at $z=\zeta(s)\in\Gamma$ is

(7.3.20) $$(K_\Gamma\hat{f})(s) = \frac{1}{2\pi i}\oint_0^{2\pi}\frac{f(\zeta(\sigma))}{\zeta(\sigma)-\zeta(s)}\zeta'(\sigma)d\sigma \qquad \text{for } \hat{f}:=f\circ\zeta.$$

We split $K_\Gamma\hat{f}$ into

(7.3.21) $$\begin{aligned}(K_\Gamma\hat{f})(s) &= \frac{1}{2\pi i}\int_0^{2\pi}\underbrace{\Big[\frac{\zeta'(\sigma)}{\zeta(\sigma)-\zeta(s)}-\frac{ie^{i\sigma}}{e^{i\sigma}-e^{is}}\Big]}_{k(\sigma,s)}f(\zeta(\sigma))d\sigma+\\ &\quad+\frac{1}{2\pi i}\oint_0^{2\pi}\frac{ie^{i\sigma}}{e^{i\sigma}-e^{is}}f(\zeta(\sigma))d\sigma\\ &= (K_1\hat{f})(s)+(K_0\hat{f})(s) \qquad\qquad \text{with } \hat{f}:=f\circ\zeta.\end{aligned}$$

Exercise 7.3.6. Prove the following: **(a)** The kernel function k defined in (21) is continuous with $k(s,s)=\frac{1}{2}[\zeta''(s)/\zeta'(s)-i]$.
(b) The second integral in (21) is the parametrisation of the Cauchy integral $K_\Omega f_\Omega$ over the unit circle Ω with the function $f_\Omega(\hat{z})=f(z)$ for $z\in\Gamma$ and $\hat{z}=e^{is}=\exp(i\zeta^{-1}(z))$.

Since K_1 has a smooth kernel, $K_1 \in K(X,X)$ is compact in $X = L^2(0,2\pi)$. Choosing the Galerkin method with the same subspace X_n of the trigonometric polynomials as in §7.3.4 for the discretisation of K_1, one obtains the semidiscrete, collectively compact approximation $\{K_{1,n}\}$ (cf. Remark 4.7.9c). Together with the Galerkin approximation $\{K_{0,n}\}$ discussed in Remark 5, we can discretise the integral equation

(7.3.22) $$(\alpha + 2\beta K_\Gamma) f = g \qquad \text{in } L^2(0,2\pi)$$

for $\alpha \neq 0$ by

(7.3.23) $$(\alpha + 2\beta K_{0,n} + 2\beta K_{1,n}) f_n = g_n := \Pi_n g,$$

where Π_n is the orthogonal projection onto X_n. Since, by Remark 5, $\{2\beta K_{0,n}\}$ is convergent and consistent and $\{2\beta K_{1,n}\}$ is collectively compact, one gains from Theorem 4.8.28 stability, convergence, and consistency of $\{2\beta K_{0,n} + 2\beta K_{1,n}\}$ with respect to $X = L^2(0,2\pi)$. These properties are still valid for $\alpha = 0$ with respect to X_n.

The procedure described above uses the same discretisation for K_1 as for the principal part K_0. This is not necessary as described in the following. Let $\Pi_{1,n}$, e.g., be the projection defined by an interpolation (for instance, the piecewise linear) at the support abscissae $\{x_\mu = \pi\mu/n:\ -n \leqslant \mu < n\}$. The semidiscrete collocation operator is $\Pi_{1,n} K_1$. The operator $\Pi_{1,n} K_1 \Pi_{1,n}$ mentioned in Remark 4.3.10a is again collectively compact. This property holds also for

(7.3.24) $$K_{1,n} := \tilde{\Pi}_n K_1 \Pi_{1,n}$$

with the trigonometric interpolation $\tilde{\Pi}_n$ at the support abscissae $\{x_\mu\}$.

As discretisation of equation (22), we obtain an equation the form (23), where the semidiscrete operator $K_{1,n}$ may be based on another discretisation method than that used in $K_{0,n}$.

Criteria for stability of the integral equations (11) or (22) with variable coefficients α, β and further details of these problems can be found, e.g., in the paper of Lamp - Schleicher - Wendland [1]. Concerning the collocation and Galerkin method we refer to Prößdorf-Schmidt [1] and Prößdorf-Rathsfeld [1].

7.3.6 Multi-Grid Methods for Equations of a Special Form

Discretising an equation of the form

(7.3.25) $$(\lambda I + A + B) f = g,$$

yields a discrete version

(7.3.26) $$(\lambda I + \mathbf{A}_n + \mathbf{B}_n) \mathbf{f}_n = \mathbf{g}_n,$$

which can be solved by the multi-grid methods from §5, if $\lambda \neq 0$ and the discrete version $K_n := A_n + B_n$ of $K = A + B$ satisfies the discrete regularity condition (5.2.11c). In practice, this means that $K = A + B$ must be compact. In the case of §7.3.5, only one of the terms in $A + B$ is compact; furthermore, $\lambda = 0$ may happen. Here, the following reformulation of the problem can help.

Let $\lambda I+A$ be bijective. Replacing (25) by the equivalent equation

$$(7.3.27)\qquad (I+C)f = \hat{g} \quad \text{with } C := (\lambda I+A)^{-1}B, \quad \hat{g} := (\lambda I+A)^{-1}g,$$

one is led again to an «equation of the second kind». If $B\in K(X,X)$ is compact and $(\lambda I+A)^{-1}\in L(X,X)$ bounded, C is compact.

For the practical implementation, one applies the reformulations to the discrete equation (26): If the matrix $\lambda I+\mathbf{A}_n$ is stable, it is invertible for $n\geq n_0$ and leads to

$$(7.3.28)\qquad (I+\mathbf{C}_n)\mathbf{f}_n = \hat{\mathbf{g}}_n \quad \text{with } \mathbf{C}_n := (\lambda I+\mathbf{A}_n)^{-1}\mathbf{B}_n, \quad \hat{\mathbf{g}}_n := (\lambda I+\mathbf{A}_n)^{-1}\mathbf{g}_n.$$

For the proof of the discrete regularity condition (5.2.11c), i.e.,

$$\|\mathbf{C}_n\|_{Y_n\leftarrow X_n} \leq C_K,$$

one needs discrete regularity of $\mathbf{B}_n$ and the $\mathbf{Y}_n$-stability $\|(\lambda I+\mathbf{A}_n)^{-1}\|_{Y_n\leftarrow Y_n}\leq$ $\leq$const. In the given case, these spaces may be discrete analogues of $X=L^2(0,2\pi)$ and $Y=H^k(0,2\pi)$ (cf. Theorem 2.27).

An alternative to (28) is the *right-sided transformation* (cf. Hackbusch [3, §8.1.3]): Introduce $\boldsymbol{\varphi}_n := (\lambda I+\mathbf{A}_n)\mathbf{f}_n$ as a new variable. Equation (26) becomes

$$(7.3.29)\qquad (I+\mathbf{C}_n)\boldsymbol{\varphi}_n = \mathbf{g}_n \qquad \text{with } \mathbf{C}_n := \mathbf{B}_n(\lambda I+\mathbf{A}_n)^{-1}.$$

After solving equation (29), one obtains $\mathbf{f}_n = (\lambda I+\mathbf{A}_n)^{-1}\boldsymbol{\varphi}_n$.

The multi-grid method for equation (28) or (29) is practicable only if the multiplication $\mathbf{f}_n\mapsto\mathbf{C}_n\mathbf{f}_n$ is simply feasible, since this operation is involved in the Picard iteration (5.5.3c) as well as in the computation of the defect (5.5.3d). By definition of the operator $\mathbf{C}_n$ appearing here, this means that $\mathbf{f}_n\mapsto(\lambda I+\mathbf{A}_n)^{-1}\mathbf{f}_n$ must be easily computable, i.e., the system of equations $(\lambda I+\mathbf{A}_n)\boldsymbol{a}=\boldsymbol{b}$ is simply solvable.

In the case of the problem from §7.3.5, the values $f_n(x_\mu)$ at the support abscissae $x_\mu:=\pi\mu/n$ are gathered in the vector $\mathbf{f}_n$. As in the second part of §7.3.5, discretise $B=K_1$ by a collocation method, e.g., with piecewise linear interpolation at the support abscissae x_μ. This yields the matrix $\mathbf{B}_n$ with $\beta_{jk}=(K_1\Phi_k)(x_j)$ ($\{\Phi_k\}$: Lagrange basis). Since the semidiscrete operator $K_{0,n}$ is based on the orthogonal projection, the matrix $\mathbf{A}_n$ has the form $\mathbf{F}_n^{-1}\mathbf{D}_n\mathbf{F}_n$, where $\mathbf{F}_n$ is the n-dimensional Fourier transform: $\mathbf{F}_n$ maps the ordinate vector $\{y_\mu: -n\leq\mu\leq n-1\}$ into the coefficient vector $\{c_\nu: -n\leq\mu\leq n-1\}$, i.e., $\sum c_\nu e^{i\nu x\pi}$ is the interpolating trigonometric polynomial.

For binary powers $n=2^\ell$, $\mathbf{F}_n$ as well as $\mathbf{F}_n^{-1}$ can be realised by the fast Fourier transform. The matrix $\mathbf{D}_n$ is the diagonal matrix with the entries $\pm\frac{1}{2}$ (cf. (2.25)). Therefore, the multi-grid method (5.5.3) with $\mathbf{K}_\ell:=\mathbf{C}_n:=$ $(\lambda I+\mathbf{A}_n)^{-1}\mathbf{B}_n=\mathbf{F}_n^{-1}(\lambda I+\mathbf{D}_n)^{-1}\mathbf{F}_n\mathbf{B}_n$ (where $n=n_\ell$) and 1 for λ is easily feasible. In the case of sufficient smoothness of the K_1 kernel function, the convergence speed depends only on the order of the collocation method used for $K_{1,n}$. For piecewise linear interpolation, one obtains the convergence speed $O(h_\ell^2)$ as in Table 5.4.2.

The reformulation (28) into an «equation of the second kind» is exemplified in Hackbusch [1,§16] for different examples from the field of the partial differential equations, but also for an integro-differential equation and an integral equation of the first kind. Concerning the last application we refer, in particular, to §9.3.2.

A particular example of a splitting (25), (26) is induced by the iterated approximation (§4.6.3) discussed by Graham-Atkinson [1].

7.4 Application to the Dirichlet Problem for Laplace's Equation

Here, we derive integral equations for the Laplace equation which we shall discuss again in §8 independently of the Cauchy kernel and for general space dimensions.

7.4.1 The Problem in the Interior Domain

Let $\Omega = \Omega_- \subset \mathbb{R}^2$ be the *interior domain* of the curve $\Gamma \in C^1_{o,stw}$. We are looking for a real-valued solution $u(x_1, x_2) \in C^2(\Omega_-) \cap C(\overline{\Omega}_-)$ satisfying the *Laplace equation*

(7.4.1a) $$\Delta u = 0 \qquad \text{in } \Omega,$$

and the *Dirichlet boundary condition*

(7.4.1b) $$u = \varphi \qquad \text{on } \Gamma \quad (\varphi\text{: real-valued}).$$

Here, $\mathbf{x} = (x_1, x_2)$ is the vector of the spatial variables and

(7.4.1c) $$\Delta = \frac{\partial^2}{\partial x_1^2} + \frac{\partial^2}{\partial x_2^2}$$ the *Laplace operator*.

7.4.2 The Double-Layer Potential

The Laplace equation (1a) is closely connected to the holomorphic function Φ from §7.2.3, if one identifies the complex variable z via $z = x_1 + i x_2$ with $\mathbf{x}$.

Remark 7.4.1. **(a)** The real and imaginary parts of a holomorphic function satisfy the Laplace equation (1a) in its domain of holomorphy.
(b) Vice versa, for any (real) solution u of the Laplace equation (also called *harmonic function*) there exists a further harmonic function v uniquely determined up to a constant, so that $\Phi(z) := u(\mathbf{x}) + i v(\mathbf{x})$ for $z = x_1 + i x_2$ is holomorphic.

We make the following *ansatz* (2) for the solution of problem (1a-c):

(7.4.2) $$u(\mathbf{x}) = -Re\,\Phi(z) \qquad \text{with } z = x_1 + i x_2,\ \Phi(z) = \frac{1}{2\pi i} \oint \frac{f(\zeta)}{\zeta - z}\, d\zeta,$$

where f is real-valued. Φ is the holomorphic function discussed in §7.2.3. Remark 1 guarantees condition (1a). Therefore, it remains to satisfy (1b), i.e., $u = \varphi$. Under the assumption that Φ admits a continuous extension Φ_- onto $\overline{\Omega}_-$, the boundary condition (1b) is identical to

(7.4.3) $$Re\,\Phi_-(z) = -\varphi(\mathbf{x}) \qquad \text{for } z = x_1 + i x_2 \in \Gamma.$$

In the following lemma we represent the real and (for latter use also the) imaginary part of Φ by *real* curvilinear integrals. Note the different definitions (1.10b) and (1.12). Moreover, we recall the *tangential* and *normal directions* $\mathbf{t}(\mathbf{x})$ and $\mathbf{n}(\mathbf{x})$ from (1.11a,b).

Lemma 7.4.2. Let $\mathbf{t}(\mathbf{y})$ and $\mathbf{n}(\mathbf{y})$ be the tangential and normal directions of the curve $\Gamma\in C^1_{0,\mathrm{stw}}$ at the point $\mathbf{y}\in\Gamma$. $f\in C^\lambda(\Gamma)$ is supposed to be real-valued. For $z=x_1+ix_2$ the function $\Phi(z)=\frac{1}{2\pi i}\oint_\Gamma\frac{f(\zeta)}{\zeta-z}d\zeta$ from (2.10) has the real and imaginary parts

$$Re\,\Phi(z) = \frac{1}{2\pi}\oint_\Gamma\frac{\langle\mathbf{n}(\mathbf{y}),\mathbf{y}-\mathbf{x}\rangle}{|\mathbf{y}-\mathbf{x}|^2}f(\mathbf{y})d\Gamma_{\mathbf{y}}, \tag{7.4.4a}$$

$$Im\,\Phi(z) = -\frac{1}{2\pi}\oint_\Gamma\frac{\langle\mathbf{t}(\mathbf{y}),\mathbf{y}-\mathbf{x}\rangle}{|\mathbf{y}-\mathbf{x}|^2}f(\mathbf{y})d\Gamma_{\mathbf{y}}, \tag{7.4.4b}$$

where $|\cdot|:=\|\cdot\|_2$ abbreviates the *Euclidean norm*

$$|\mathbf{x}| = \sqrt{x_1^2+x_2^2}. \tag{7.4.4c}$$

Proof. Let $\zeta(\cdot):[0,L]\to\Gamma$ be an admissible parametrisation of the curve Γ (according to (1.13a), $|\zeta'|$ must be continuous):

$$\Phi(z) = \frac{1}{2\pi i}\oint_0^L\frac{f(\zeta(\sigma))}{\zeta(\sigma)-z}\zeta'(\sigma)d\sigma.$$

The integrand $\frac{f(\zeta(\sigma))}{\zeta(\sigma)-z}\zeta'(\sigma)$ equals $f(\zeta(\sigma))|\zeta(\sigma)-z|^{-2}\zeta'(\sigma)(\overline{\zeta(\sigma)-z})$. We represent ζ and z as follows:

$$\zeta(\sigma)=y_1(\sigma)+iy_2(\sigma),\qquad z=x_1+ix_2,\qquad \mathbf{y}(\sigma)=(y_1(\sigma),y_2(\sigma)),\qquad \mathbf{x}=(x_1,x_2).$$

$f(\zeta(\sigma))|\zeta(\sigma)-z|^{-2}$ becomes $f(\mathbf{y}(\sigma))/|\mathbf{y}(\sigma)-\mathbf{x}|^2$. The complex factor $\zeta'(\sigma)(\overline{\zeta(\sigma)-z})$ is rewritten as

$$[(y_1-x_1)y_1'+(y_2-x_2)y_2']+i[(y_1-x_1)y_2'-(y_2-x_2)y_1'] =$$
$$=\{\langle\mathbf{t}(\mathbf{y}),\mathbf{y}-\mathbf{x}\rangle+i\langle\mathbf{n}(\mathbf{y}),\mathbf{y}-\mathbf{x}\rangle\}|\mathbf{y}'|.$$

Thereby, Φ takes the form

$$\begin{aligned}\Phi(z) &= \frac{1}{2\pi}\oint_0^L\frac{f(\mathbf{y}(\sigma))}{|\mathbf{y}(\sigma)-\mathbf{x}|^2}\{-i\langle\mathbf{t}(\mathbf{y}),\mathbf{y}-\mathbf{x}\rangle+\langle\mathbf{n}(\mathbf{y}),\mathbf{y}-\mathbf{x}\rangle\}|\mathbf{y}'(\sigma)|d\sigma\\ &= \frac{1}{2\pi}\oint_\Gamma\frac{f(\mathbf{y})}{|\mathbf{y}-\mathbf{x}|^2}\{-i\langle\mathbf{t}(\mathbf{y}),\mathbf{y}-\mathbf{x}\rangle+\langle\mathbf{n}(\mathbf{y}),\mathbf{y}-\mathbf{x}\rangle\}d\Gamma_{\mathbf{y}},\end{aligned} \tag{7.4.4d}$$

which contains the real and imaginary parts (4a,b). QED

The assumption $\Gamma\in C^1_{0,\mathrm{stw}}$ ensures that $\mathbf{t}(\mathbf{y})$ and $\mathbf{n}(\mathbf{y})$ are well-defined almost everywhere on Γ. Only at the possibly existing corners, $\mathbf{t}(\mathbf{y})$ and $\mathbf{n}(\mathbf{y})$ are not defined or exist as (different) one-sided derivatives.

For $\mathbf{x}\in\mathbb{R}^2\setminus\Gamma$, the integrals in (4a,b) are obviously without singularity. The case $\mathbf{x}\in\Gamma$ is subject of the next remark.

Remark 7.4.3. **(a)** For $f\in C^\lambda(\Gamma)$, $\lambda>0$, the integrals (4a,b) always exist as Cauchy principal values.
(b) If Γ is locally Hölder continuously differentiable (with exponent $\mu>0$) at $\mathbf{x}\in\Gamma$, then the integral (4a) exists at $\mathbf{x}\in\Gamma$ for all $f\in L^\infty(\Gamma)$ as improper inte-

gral. More precisely, we have

(7.4.4e) $\langle \mathbf{n}(\mathbf{y}), \mathbf{y}-\mathbf{x}\rangle = O(|\mathbf{y}-\mathbf{x}|^{1+\mu})$.

(c) In the case of $\Gamma\in C_0^{1+\mu}$ with $\mu>0$ and $f\in L^\infty(\Gamma)$, the improper integrals (4a) exist uniformly for all $\mathbf{x}\in\mathbb{R}^2$ (cf. §6.1.3).
(d) For open curves $\Gamma\in C^1_{stw}$, the statements (b,c) are also valid, whereas the assertion in (a) requires $f\in C^\lambda_{\partial\Gamma}(\Gamma)$.

Proof. (i) By the definition of the Cauchy principal value, $Re\oint\ldots=\oint Re\ldots$ holds. Since, for $f\in C^\lambda(\Gamma)$, Φ exists as Cauchy principal value (cf. Theorem 1.3), this holds also for the real and imaginary parts (4a,b).
(ii) Let $\mathbf{y}(\sigma)$ be a smooth parametrisation and $\mathbf{x}=\mathbf{y}(\sigma_0)$. By assumption of part (b), $\mathbf{y}(\sigma)$ belongs to $C^{1+\mu}(U)$ for σ in a neighbourhood U of σ_0. Since Γ has no corner at $\mathbf{x}$ (with the angle 0 or 2π), the estimate $|\mathbf{y}(\sigma)-\mathbf{x}|=|\mathbf{y}(\sigma)-\mathbf{y}(\sigma_0)|\geqslant|\sigma-\sigma_0|/C_0$ holds. An expansion around σ_0 yields

$$\mathbf{y}(\sigma) = \mathbf{y}(\sigma_0)+(\sigma-\sigma_0)\mathbf{y}'(\sigma_0)+O(|\sigma-\sigma_0|^{1+\mu}) = \mathbf{x}+c\mathbf{t}(\sigma_0)+O(|\sigma-\sigma_0|^{1+\mu})$$

with $c\in\mathbb{R}$, since $(\sigma-\sigma_0)\mathbf{y}'(\sigma_0)$ coincides with the tangential direction $\mathbf{t}$ up to a factor c. Since this direction is perpendicular to the normal vector $\mathbf{n}(\mathbf{y}(\sigma_0))$, $\langle\mathbf{n}(\mathbf{y}),\mathbf{y}-\mathbf{x}\rangle=O(|\sigma-\sigma_0|^{1+\mu})$ holds for $\mathbf{y}=\mathbf{y}(\sigma)$ and $\sigma\in U$. Together with the previous estimate one obtains (4e) in the form of $\langle\mathbf{n}(\mathbf{y}),\mathbf{y}-\mathbf{x}\rangle/|\mathbf{y}-\mathbf{x}|^2=O(|\sigma-\sigma_0|^{\mu-1})$. This finishes the proof of part (b).
(iii) If $\mathbf{x}$ does not lie on Γ, the integrand is without singularity. If $\mathbf{x}\in\Gamma\in C_0^{1+\mu}$, part (b) applies and proves part (c).
(iv) For the discussion of improper integrals it is irrelevant, whether Γ is closed or not. If $f\in C^\lambda_{\partial\Gamma}(\Gamma)$, Γ can be extended to a closed curve without changing the integral when we set $f=0$ for $\mathbf{x}\notin\Gamma$. QED

To satisfy the *boundary condition* (1b) (written in the form (3)), i.e., $-Re\,\Phi_-=\varphi$, we recall the jump relations (2.12b,c) on Γ: $\Phi_+-\Phi_-=-f$ (f real-valued), $\frac{1}{2}[\Phi_++\Phi_-]=\overline{K}f$, which lead to

$$\Phi_- = \overline{K}f+\tfrac{1}{2}f \qquad \text{on } \Gamma \text{ (cf. (2.12d))}.$$

Together with $-Re\,\Phi_-=\varphi$, we obtain

$$2\,Re\,\overline{K}f + f = -2\varphi \qquad \text{on } \Gamma.$$

$\overline{\Phi}=\overline{K}f$ was defined on Γ as the continuous extension of the Cauchy integral $\Phi=Kf$. At all curve points not being corners, we have $\Phi=Kf=\overline{K}f$. Inserting the real part determined in (4a), we obtain the integral equation

(7.4.5) $$f(\mathbf{x}) = -2\varphi(\mathbf{x})-\frac{1}{\pi}\oint_\Gamma\frac{\langle\mathbf{n}(\mathbf{y}),\mathbf{y}-\mathbf{x}\rangle}{|\mathbf{y}-\mathbf{x}|^2}f(\mathbf{y})\,d\Gamma_y$$
for $\mathbf{x}\in\Gamma$, $\mathbf{x}$ no corner.

In §7.4.4 and §§8.2.6-7, we shall discuss the solvability of the integral equation (5) . Under the assumption that (5) admits a solution $f\in C^\lambda(\Gamma)$, Theorem 8 will show that

$$(7.4.6)\qquad u(\mathbf{x}) := -\frac{1}{2\pi}\int_\Gamma \frac{\langle \mathbf{n}(\mathbf{y}),\mathbf{y}-\mathbf{x}\rangle}{|\mathbf{y}-\mathbf{x}|^2} f(\mathbf{y})\,d\Gamma_y \qquad (\mathbf{x}\in\mathbb{R}^2\setminus\Gamma)$$

(cf. (4a)) represents the desired solution of problem (1a,b).

The function defined in (6) is called the *double-layer potential* generated by the *(double-layer) density* f. Correspondingly,

$$(7.4.7a)\qquad k(\mathbf{x},\mathbf{y}) := -\frac{1}{\pi}\frac{\langle \mathbf{n}(\mathbf{y}),\mathbf{y}-\mathbf{x}\rangle}{|\mathbf{y}-\mathbf{x}|^2} \qquad (\mathbf{x},\mathbf{y}\in\Gamma)$$

is the *double-layer kernel* and

$$(7.4.7b)\qquad (K_1 f)(\mathbf{x}) := -\frac{1}{\pi}\oint_\Gamma \frac{\langle \mathbf{n}(\mathbf{y}),\mathbf{y}-\mathbf{x}\rangle}{|\mathbf{y}-\mathbf{x}|^2} f(\mathbf{y})\,d\Gamma_y \qquad (\mathbf{x}\in\Gamma)$$

defines the operator of the *double-layer operator*.

Another notation for $k(\mathbf{x},\mathbf{y})$ is

$$(7.4.7c)\qquad k(\mathbf{x},\mathbf{y}) = -\frac{1}{\pi}\frac{\cos(\mathbf{n}(\mathbf{y}),\mathbf{y}-\mathbf{x})}{|\mathbf{y}-\mathbf{x}|} \qquad (\mathbf{x},\mathbf{y}\in\Gamma),$$

where $\cos(\mathbf{a},\mathbf{b}) := \langle\mathbf{a},\mathbf{b}\rangle/|\mathbf{a}||\mathbf{b}|$ is the cosine of the angle formed by the vectors $\mathbf{a},\mathbf{b}\in\mathbb{R}^2\setminus\{0\}$. In Remark 3b (cf. (4e)). we have proved under the assumption $\Gamma\in C_0^{1+\mu}$ that $\cos(\mathbf{n}(\mathbf{y}),\mathbf{y}-\mathbf{x}) = O(|\mathbf{x}-\mathbf{y}|^{\mu})$ and, therefore, $k(\mathbf{x},\mathbf{y}) = O(|\mathbf{x}-\mathbf{y}|^{\mu-1})$. If $\Gamma\in C_0^2$, $k(\mathbf{x},\mathbf{y})$ is even continuous at $\mathbf{y}=\mathbf{x}$. The proof of Lemma 4 can be found , e.g., in Walter [1, p. 107].

Lemma 7.4.4. Under the assumption $\Gamma\in C_0^2$, the double-layer kernel is continuous, i.e., $k\in C(\Gamma\times\Gamma)$. As $\mathbf{y}\to\mathbf{x}$, $k(\mathbf{x},\mathbf{y})$ is continuously extendable by the value

$$(7.4.7d)\qquad \lim_{\mathbf{y}\to\mathbf{x}} k(\mathbf{x},\mathbf{y}) = -\varkappa(\mathbf{x})/(2\pi), \qquad \varkappa(\mathbf{x})\text{: curvature (1.11c).}$$

Differentiating the function $\log|\mathbf{x}-\mathbf{y}|$ defined for $\mathbf{x}\neq\mathbf{y}$ with respect to $\mathbf{y}$, one obtains the gradient

$$(7.4.8a)\qquad \nabla_y \log|\mathbf{x}-\mathbf{y}| = \frac{\mathbf{y}-\mathbf{x}}{|\mathbf{y}-\mathbf{x}|^2}.$$

The *directional derivative* of $\varphi:\mathbb{R}^2\to\mathbb{R}$ with respect to the direction $\boldsymbol{\omega}\in\mathbb{R}^2$, $|\boldsymbol{\omega}|=1$, is defined by $\langle\boldsymbol{\omega},\nabla\varphi\rangle$. The particular choice $\boldsymbol{\omega}=\mathbf{n}(\mathbf{y})$ yields the *normal derivative*

$$(7.4.8b)\qquad \frac{\partial\varphi(\mathbf{y})}{\partial n} := \langle\mathbf{n}(\mathbf{y}),\nabla\varphi(\mathbf{y})\rangle.$$

For $\varphi(\mathbf{y}) = \log|\mathbf{y}-\mathbf{x}|$, one obtains a third characterisation of the double-layer kernel:

$$(7.4.8c)\qquad k(\mathbf{x},\mathbf{y}) = -\frac{1}{\pi}\frac{\partial}{\partial n_y}\log|\mathbf{y}-\mathbf{x}|.$$

By definition, $K_1 = -2\,Re\,K$. The continuous extension into the corners leads to

$$\overline{K}_1 = -2\,Re\,\overline{K} \qquad\text{or}\qquad (\overline{K}_1 f)(\mathbf{x}(\sigma)) = (K_1 f)(\mathbf{x}(\sigma\pm 0))$$

(cf. (1.18b)). Therefore, the comment «$\mathbf{x}$ no corner» in (5) can be omitted. The integral equation (5) takes the form (9):

$$(7.4.9)\qquad f = -2\varphi + \overline{K}_1 f \qquad \text{on } \Gamma.$$

Lemma 7.4.5. The image of $f=1$ is $\overline{K}_1 f=-f$, i.e., $\overline{K}_1 1=-1$.

Proof. The assertion is a consequence of $\overline{K}1=1/2$ (cf. (1.17c/18d)). QED

Lemma 5 enables a further formulation of the integral equation: For any fixed $\mathbf{x}\in\Gamma$, add $f(\mathbf{x})$ to the left-hand side of (5) and $\overline{K}_1 c=-c$ with $c(\mathbf{y}):=-f(\mathbf{x})$ (i.e., a constant with respect to the integration over $\mathbf{y}$) to the right-hand side and divide by 2:

$$(7.4.10)\qquad f(\mathbf{x}) = -\varphi(\mathbf{x})-\frac{1}{2\pi}\int_\Gamma \frac{\langle \mathbf{n}(\mathbf{y}),\mathbf{y}-\mathbf{x}\rangle}{|\mathbf{y}-\mathbf{x}|^2}[f(\mathbf{y})-f(\mathbf{x})]\,d\Gamma_{\mathbf{y}} \qquad \text{on } \Gamma.$$

This representation has two advantages:

Remark 7.4.6. **(a)** For functions $f\in C^\lambda(\Gamma)$, the integrand $k(\mathbf{x},\mathbf{y})[f(\mathbf{y})-f(\mathbf{x})]$ in the representation (10) is improperly integrable.
(b) The representation holds for all $\mathbf{x}\in\Gamma$, including the corners.

7.4.3 Uniqueness and Representation Theorem

Exercise 7.4.7. Let Φ be holomorphic on $\Omega\subset\mathbb{C}$. Prove that $u=Re\,\Phi$ is constant if and only if $v=Im\,\Phi$ is also constant. *Hint.* Use the Cauchy-Riemann differential equations $u_x=v_y$, $u_y=-v_x$.

Theorem 7.4.8 (uniqueness). Let $\Gamma\in C^1_{0,stw}$ and $0<\lambda<1$. **(a)** The integral equation (9) admits at most one solution $f\in C^\lambda(\Gamma)$.
(b) If (9) has a (real) solution $f\in C^\lambda(\Gamma)$, the associated double-layer potential (6) is the unique solution u of the boundary value problems (1a,b) in the interior domain Ω. u is globally Hölder continuous, i.e., $u\in C^\lambda(\overline{\Omega})$.

Proof. (i) First, we treat part (b). Here, it suffices to repeat the argument of Subsection 7.4.2. f defines the potential $\Phi(z)=\frac{1}{2\pi i}\oint_\Gamma\frac{f(\zeta)}{\zeta-z}d\zeta$. Since $f\in C^\lambda(\Gamma)$, the continuous extension $\Phi_-\in C^\lambda(\overline{\Omega})$ exists (cf. Theorem 2.8)). Setting $u:=-Re\,\Phi_-\in C^\lambda(\overline{\Omega})$, we obtain a real-valued function satisfying $\Delta u=0$ in Ω (cf. Remark 1a). From $\Phi_-=\overline{K}f+\frac{1}{2}f$ on Γ (cf. (2.12d)) we conclude that $u=-Re\,\Phi_-=-Re(\overline{K}f+\frac{1}{2}f)=\frac{1}{2}(\overline{K}_1 f-f)=\varphi$ (cf. (9)), i.e., the boundary condition (1b) is satisfied on Γ. It is known from the theory of elliptic differential equations that the boundary value problem (1a,b) admits at most one solution (cf. Hackbusch [2, Theorem 2.3.8]).
(ii) If there are two solutions f_1 and f_2 of (9) belonging to $C^\lambda(\Gamma)$, then $f:=f_1-f_2\in C^\lambda(\Gamma)$ satisfies the homogeneous integral equation

$$(7.4.11)\qquad f-\overline{K}_1 f = 0.$$

$2\overline{K}f=-\overline{K}_1 f+ig=-f+ig$ holds on Γ with $g:=2Im(\overline{K}f)$. Let Φ be defined as in (i); $\Phi_\mp$ are the continuous extensions onto the interior and exterior domains $\Omega_\mp$. Formula (2.12d) from Theorem 2.8 states that $\Phi_-=\overline{K}f+\frac{1}{2}f=\frac{1}{2}(-f+ig)+\frac{1}{2}f=\frac{1}{2}ig$ on Γ. This shows $u:=-Re\,\Phi_-=0$ and $v:=Im\,\Phi_-=\frac{1}{2}g$. The real part $u=0$ of a holomorphic function implies constancy of the imaginary part: $v=c$ (c constant, cf. Exercise 7). Hence, $g=2c$ must hold. Formula (2.12d) implies $\Phi_+=\overline{K}f-\frac{1}{2}f=\frac{1}{2}(-f+ig)-\frac{1}{2}f=\frac{1}{2}ig-f=ic-f$ on Γ, so that

$v := Im\Phi_+$ satisfies the Laplace equation $\Delta v = 0$ in Ω_+ and the boundary condition $v = c$. By Lemma 2.7, v is bounded. As will be discussed in the following Exercise 9, $v = c$ is the only solution of this boundary value problem. From $Im\,\Phi_+ = v = c$ in Ω_+ one concludes via Exercise 7 that $Re\,\Phi_+ = \text{const}$. Thereby, also the boundary values $Re\,\Phi_+ = Re\,(ic - f) = f$ are constant on Γ. By Lemma 5, a constant function f yields $\overline{K}_1 f = -f$. With (11) one shows $f = 0$, which proves the uniqueness $f_1 = f_2$. QED

Exercise 7.4.9. **(a)** Let $\Omega \subset \mathbb{R}^d$ be a domain which is a neighbourhood of ∞ with a positive distance from 0. Then $\Omega' := \{\mathbf{x}' \in \mathbb{R}^d : \mathbf{x}' = \mathbf{x}/|\mathbf{x}|^2,\ \mathbf{x} \in \Omega\} \cup \{0\}$ is a bounded domain.
(b) Let u be defined on Ω. The *Kelvin transformation* applied to u yields the function $U(\mathbf{x}') := |\mathbf{x}'|^{2-d} u(\mathbf{x}'/|\mathbf{x}'|^2)$ defined on Ω'. Prove the following: $\Delta U(\mathbf{x}') = |\mathbf{x}'|^{-2-d} \Delta u(\mathbf{x}'/|\mathbf{x}'|^2)$. This states, in particular, that solutions of the Laplace equation become again Laplace solutions (cf. Walter [1]).
(c) Formulate the Kelvin transformation in $\mathbb{C}$ instead of $\mathbb{R}^2$.
(d) Let $\Gamma \in C^1_{0,\mathrm{stw}}$ define an interior domain with $0 \in \Omega_-$ and the exterior domain Ω_+. Prove the following: There is at most one *bounded* solution of the boundary value problem $\Delta u = 0$ in Ω_+ and $u = \varphi$ on Γ. *Hint.* Use (b) and the uniqueness of the boundary value problem in bounded domains.

More generally, the proof of Theorem 8b yields the following statement.

Remark 7.4.10. Let $\Gamma \in C^1_{0,\mathrm{stw}}$ and $f \in C^\lambda(\Gamma)$ be real-valued. Define the functions $\Phi_\pm$ as above. If one of the functions $Re\,\Phi_+$, $Im\,\Phi_+$, $Re\,\Phi_-$, $Im\,\Phi_-$ has constant boundary values on Γ, then all boundary values and $f = -\overline{K}_1 f$ are constant.

Proof. Let, e.g., $Re\,\Phi_-$ be constant on Γ. As in the proof of Theorem 8b, one concludes via Exercise 7 that also $Im\,\Phi_-$ is constant. From (2.12b), i.e., $\Phi_+ - \Phi_- = -f$ on Γ one obtains that $Im\,\Phi_+$ and thereby also $Re\,\Phi_+$ are constant. Hence, $f = \Phi_- - \Phi_+$ is constant. Lemma 5 shows $f = -\overline{K}_1 f$. QED

The following extension of Lemma 5 is also valid:

Corollary 7.4.11. The kernel of $I + \overline{K}_1$ consists only of the constant functions, i.e., $f = \text{const}$ is the only function $f \in C^\lambda(\Gamma)$ with $f = -\overline{K}_1 f$.

Proof. (2.12d) shows $Re\,\Phi_+ = \frac{1}{2} Re\,(2\overline{K} f - f) = -\frac{1}{2}(\overline{K}_1 f + f) = 0$. Remark 10 proves $f = \text{const}$. QED

In the general case, an *existence* guarantee for a solution $f \in C^\lambda(\Gamma)$ of the integral equation (9) is not possible, as shown in

Example 7.4.12. Let Ω be the «L-shaped domain» of Fig. 1, where the re-entrant corner is located at the origin. The function

$$u(r, \vartheta) = r^{2/3} \sin((2\vartheta - \pi)/3) \tag{7.4.12}$$

described in polar coordinates (r, ϑ) satisfies the Laplace equation $\Delta u = 0$ in

Ω. Its boundary values φ on the part Γ_0 in the neighbourhood of the re-entrant corner vanish identically. In the remaining part $\Gamma\backslash\Gamma_0$ they are smooth, so that $\varphi\in C^{\lambda}(\Gamma)$ holds for all $0<\lambda<1$. Choose $\lambda\in(2/3,1)$. If the integral equation (9) for $\varphi\in C^{\lambda}(\Gamma)$ allowed a solution $f\in C^{\lambda}(\Gamma)$, then the Hölder continuity $u\in C^{\lambda}(\overline{\Omega})$ would follow from Theorem 8b. However, u from (12) satisfies $u\in C^{\lambda}(\overline{\Omega})$ *only* for $\lambda\leqslant 2/3$.

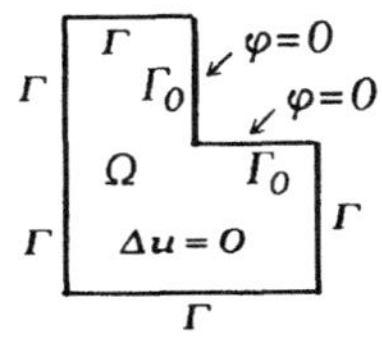

Fig. 7.4.1. L-shaped domain

7.4.4 The Case of a Smooth Boundary Γ

In the following, no corners are admitted. The general assumption is

(7.4.13) $\qquad \Gamma\in C_0^{1+\mu} \qquad$ with $\mu>0$.

As a consequence, the versions $\overline{K}_1$ and K_1 of the double-layer operator coincide. While the singular Cauchy operator $\overline{K}$ could never be compact, we now show *compactness* in $X=C^{\lambda}(\Gamma)$. For $\lambda>1$, we define «$f\in C^{\lambda}(\Gamma)$» by means of an appropriate parametrisation, i.e., $f(\zeta(\cdot))\in C^{\lambda}([0,L])$.

Theorem 7.4.13. Assume (13), i.e., $\Gamma\in C_0^{1+\mu}$ with $\mu\in(0,2)\backslash\{1\}$. Then the double-layer operator K_1 has the following properties:

(7.4.14a) $\qquad K_1\in L(L^{\infty}(\Gamma),C^{\lambda}(\Gamma)) \qquad$ for all $0\leqslant\lambda\leqslant\mu$,

(7.4.14b) $\qquad K_1\in K(C^{\lambda}(\Gamma),C^{\lambda}(\Gamma)) \qquad$ for all $0\leqslant\lambda\leqslant\mu$,

i.e., K_1 is compact on the Banach space $X=C^{\lambda}(\Gamma)$.

Proof. (i) (14a) implies (14b) because of the compact embeddings $C^{\lambda}(\Gamma)\subset L^{\infty}(\Gamma)$ $(\lambda>0)$ or $C^{\mu}(\Gamma)\subset C^{\lambda}(\Gamma)$ $(\lambda<\mu)$, respectively (cf. Remark 3.4.13).
(ii) For the proof of (14a), it suffices to consider the case $\lambda=\mu$. Here, we restrict ourselves to $0<\mu<1$. The case $1<\mu<2$ can be found in Schippers [2]. We write $k(\mathbf{x},\mathbf{y})$ as

$$k(\mathbf{x},\mathbf{y}) = -\frac{1}{\pi}\frac{\ell(\mathbf{x},\mathbf{y})}{|\mathbf{y}-\mathbf{x}|^{1-\mu}} \qquad \text{with } \ell(\mathbf{x},\mathbf{y}) = \frac{\langle\mathbf{n}(\mathbf{y}),\mathbf{y}-\mathbf{x}\rangle}{|\mathbf{y}-\mathbf{x}|^{1+\mu}}.$$

By Remark 3b,c, $\|\ell\|_{\infty}\leqslant C$ holds . Let $\boldsymbol{\gamma}$: $[0,L]\to\Gamma$ be a parametrisation of Γ by the arc length: $\mathbf{y}=\boldsymbol{\gamma}(\sigma)$, $\mathbf{x}=\boldsymbol{\gamma}(\tau)$. Differentiation with respect to τ at $\mathbf{x}=\boldsymbol{\gamma}(\tau)$ yields

$$\frac{d}{d\tau}\,\ell(\mathbf{x},\mathbf{y}) = \frac{-\langle\mathbf{n}(\mathbf{y}),\mathbf{t}(\mathbf{x})\rangle}{|\mathbf{y}-\mathbf{x}|^{1+\mu}} + \frac{\langle\mathbf{n}(\mathbf{y}),\mathbf{y}-\mathbf{x}\rangle}{|\mathbf{y}-\mathbf{x}|^{3+\mu}}\,(1+\mu)\,\langle\mathbf{t}(\mathbf{x}),\mathbf{y}-\mathbf{x}\rangle$$

($\mathbf{t}$: tangential direction). From $\mathbf{t}(\mathbf{x})=\mathbf{t}(\mathbf{y})+O(|\mathbf{y}-\mathbf{x}|^{\mu})$, $\langle\mathbf{n}(\mathbf{y}),\mathbf{t}(\mathbf{y})\rangle=0$ and $\langle\mathbf{n}(\mathbf{y}),\mathbf{y}-\mathbf{x}\rangle=O(|\mathbf{y}-\mathbf{x}|^{1+\mu})$ (cf. (4e)) we conclude that

$$\frac{d}{d\tau}\,\ell(\mathbf{x},\mathbf{y}) = O\Big(\frac{1}{|\mathbf{y}-\mathbf{x}|}\Big) \qquad \text{for } \mathbf{x}\neq\mathbf{y}.$$

Hence, condition (3.4.7f) [more precisely, its counterpart for curvilinear integrals] is satisfied and Theorem 3.4.9 ensures the assertion (14a). QED

Thanks to the compactness (14b), we can guarantee the solvability of the integral equation (9):

Theorem 7.4.14. Assume (13), i.e., $\Gamma \in C_0^{1+\mu}$ with $\mu \in (0,2)\setminus\{1\}$. Then the inverse $(I-K_1)^{-1} \in L(C^\lambda(\Gamma), C^\lambda(\Gamma))$ exists for all $0 \le \lambda \le \mu$. Hence, the integral equation (9) has a unique solution $f = -2(I-K_1)^{-1}\varphi \in C^\lambda(\Gamma)$ for all boundary values $\varphi \in C^\lambda(\Gamma)$ with $\lambda \in [0,\mu]$.

Proof. The uniqueness (injectivity) proved in Theorem 8 together with compactness (14b) implies that $\lambda = 1$ is a regular value, i.e., $I-K_1$ is bijective (cf. Theorem 3.2.1). QED

The question of the existence of a solution of (9) in the case of insufficiently smooth boundaries will be revisited in §8.2.7.

7.4.5 The Double-Layer Potential for Solving the Exterior Problem

So far, we solved the Dirichlet boundary value problem in the interior domain $\Omega_- \subset \mathbb{R}^2$. Now, we consider the corresponding problem in the *exterior domain* $\Omega_+ = \mathbb{R}^2 \setminus \overline{\Omega}_-$. We look for a solution $u \in C^2(\Omega_+) \cap C(\overline{\Omega}_+)$ of *Laplace's equation*

(7.4.15a) $\Delta u = 0$ in Ω_+,

subject to the *Dirichlet boundary condition*

(7.4.15b) $u = \varphi$ on Γ.

Since Ω_+ is unbounded, one needs a further condition,

(7.4.15c) $u(\infty) = 0$, i.e., $u(\mathbf{x}) \to 0$ as $|\mathbf{x}| \to \infty$.

In (15c), the «boundary condition at infinity» is chosen without loss of generality in the form of $u(\infty) = 0$, as can be seen from the following exercise.

Exercise 7.4.15. Instead of $u(\infty) = 0$, consider the following condition (15c'):

(7.4.15c') u behaves asymptotically like $\alpha_0 + \alpha_1 x_1 + \alpha_2 x_2$ as $|\mathbf{x}| \to \infty$

with real coefficients α_0, α_1, α_2. How can the problem (15a,b,c') be reduced to (15a-c)?

The additional condition (15c) inflicts no difficulties, since it is always guaranteed by the ansatz (2).

Remark 7.4.16. The holomorphic function Φ from (2.10) and thereby also its real and imaginary parts satisfy condition (15c), i.e., $\Phi(\infty) = 0$.

Proof. According to (2.11c), Φ tends to zero like $O(1/|\mathbf{x}|)$. QED

Again, we use the ansatz (2) for u. Under the assumption that Φ admits a continuous extension Φ_+ onto $\overline{\Omega}_+$, the boundary condition (15b) is identical to

(7.4.16) $\qquad -Re\ \Phi_+(z) = \varphi \qquad$ on Γ.

Using the jump relations (2.12b,c) on Γ, i.e., $\Phi_+ - \Phi_- = -f$ and $\frac{1}{2}[\Phi_+ + \Phi_-] = \overline{K}f$, we conclude that

$$\Phi_+ = \overline{K}f - \tfrac{1}{2}f \quad \text{on } \Gamma \qquad \text{(cf. (2.12c))}.$$

This result and (16) yield

$$-2\varphi = 2\,Re\,\overline{K}f - f \qquad \text{on } \Gamma.$$

This leads to the following integral equation, which differs from equation (5) only in the sign on the right-hand side:

$$(7.4.17) \qquad f(\mathbf{x}) = 2\varphi(\mathbf{x}) + \frac{1}{\pi}\oint_\Gamma \frac{\langle \mathbf{n}(\mathbf{y}), \mathbf{y}-\mathbf{x}\rangle}{|\mathbf{y}-\mathbf{x}|^2} f(\mathbf{y})\,d\Gamma_y$$

for $\mathbf{x} \in \Gamma$, $\mathbf{x}$ no corner.

Equation (17) is a Fredholm integral equation of the second kind:

(7.4.18) $\qquad \lambda f = g + \overline{K}_1 f \quad$ on $\Gamma \qquad$ with $\lambda = -1$, $g = -2\varphi$.

Next, we want to obtain a representation of the integral equation analogous to (10). For any fixed $\mathbf{x} \in \Gamma$, we subtract $f(\mathbf{x})$ from the left-hand side of (5) and $-\overline{K}_1\varphi = \varphi$ with $\varphi(\mathbf{y}) := f(\mathbf{x})$ (i.e., a constant with respect to the integration over $\mathbf{y}$) from the right-hand side and then divide by 2:

$$(7.4.19) \qquad 0 = -\varphi(\mathbf{x}) - \frac{1}{2\pi}\int_\Gamma \frac{\langle \mathbf{n}(\mathbf{y}), \mathbf{y}-\mathbf{x}\rangle}{|\mathbf{y}-\mathbf{x}|^2}[f(\mathbf{y}) - f(\mathbf{x})]\,d\Gamma_y \qquad \text{on } \Gamma.$$

This representation elucidates the reason why f cannot be determined uniquely: For any solution f, f+const is also a solution. But modulo constants, uniqueness can be stated as follows.

Theorem 7.4.17 (uniqueness). Let $\Gamma \in C^1_{0,\mathrm{stw}}$. Let Ω_+ be the exterior domain.
(a) Two solutions $f \in C^\lambda(\Gamma)$ of the integral equation (17) can differ only by a constant.
(b) If (17) has a solution $f \in C^\lambda(\Gamma)$, the associated double-layer potential (6) represents the unique solution u of the boundary value problem (15a-c). u satisfies global Hölder continuity, i.e., $u \in C^\lambda(\overline{\Omega}_+)$.

Proof. Let f_1 and f_2 be two solutions. $f := f_1 - f_2$ satisfies $f = -\overline{K}_1 f$. By Corollary 11, f is constant. QED

Suppose the case of a smooth curve $\Gamma \in C_0^{1+\mu}$. Because of the compactness (14b), dim kernel$(K_1 + I) = 1$ also implies dim kernel$(K_1' + I) = 1$ for the dual integral operator K_1' with the kernel

$$(7.4.20) \qquad k'(\mathbf{x},\mathbf{y}) := \frac{1}{\pi}\frac{\langle \mathbf{n}(\mathbf{x}), \mathbf{y}-\mathbf{x}\rangle}{|\mathbf{y}-\mathbf{x}|^2} \qquad (\mathbf{x},\mathbf{y} \in \Gamma)$$

(cf. (4.6.5)). Let $\xi \neq 0$ be the eigensolution $K_1'\xi = -\xi$ which is unique up to a scalar factor. Concerning ξ, we shall show in Lemma 19 that

$$(7.4.21\text{a}) \qquad \int_\Gamma \xi(y)\,d\Gamma \neq 0.$$

If the boundary value φ is orthogonal to ξ, i.e.,

$$\int_\Gamma \varphi(y)\xi(y)\,d\Gamma = 0, \tag{7.4.21b}$$

equation (17) has a solution determined up to a constant.

Theorem 7.4.18. Assume (13), i.e., $\Gamma \in C_0^{1+\mu}$ with $\mu \in (0,2)\setminus\{1\}$. Let $\lambda \in [0,\mu]$ and assume that $\varphi \in C^\lambda(\Gamma)$ satisfies (21b). Then the integral equation (17) admits a solution $f \in C^\lambda(\Gamma)$. All further solutions differ from f only by a constant.

Proof. It is a consequence of compactness and the fact that the dimension of kernel(K_1+I) is 1. QED

Lemma 7.4.19. Under the assumptions of Theorem 18, inequality (21a) holds.

Proof. Otherwise, $\varphi=1$ satisfies condition (21b). The unique bounded solution for the boundary value $\varphi=1$ (cf. Exercise 7.4.9d) equals $u=1$ and contradicts condition (15c), which according to Theorem 17b must be satisfied. QED

The restriction of the possible boundary values by (21b) is caused by the additional requirement (15c), i.e, $u(\infty)=0$. As seen below, we may omit (21b), if we replace (15c) by the boundedness at ∞:

$$u(\mathbf{x}) = O(1) \qquad \text{for } \mathbf{x} \to \infty. \tag{7.4.22}$$

Remark 7.4.20. Assume (13) (i.e., $\Gamma \in C_0^{1+\mu}$) with $\mu \in (0,2)\setminus\{1\}$. Let $0 \le \lambda \le \mu$. $\varphi \in C^\lambda(\Gamma)$ can be split into $\varphi = \varphi_0 + c$ with the constant $c := \int_\Gamma \varphi\xi\,d\Gamma / \int_\Gamma \xi\,d\Gamma$. By Theorem 18, there is a solution $f \in C^\lambda(\Gamma)$ of the equation $-f = -2\varphi_0 + K_1 f$. Let u_0 be the solution in the exterior domain generated by the double-layer potential (6) of f. The solution for the boundary value φ satisfying (22) is uniquely given by $u := u_0 + c$.

7.4.6 The Tangential Derivative of the Single-Layer Potential

So far, we used the ansatz $u = -Re\,\Phi$ (cf. (2)). Similarly, the ansatz

$$u(\mathbf{x}) = Im\,\Phi(z) \quad \text{with} \quad z = x_1 + i x_2, \quad \Phi(z) = \frac{1}{2\pi i}\oint \frac{f(\zeta)}{\zeta - z}\,d\zeta \tag{7.4.23}$$

can be made, where f is real-valued. The boundary condition (1b) becomes

$$Im\,\Phi_-(z) = \varphi(\mathbf{x}) \qquad \text{for } z = x_1 + i x_2 \in \Gamma. \tag{7.4.24}$$

From $\Phi_- = \overline{K}f + \frac{1}{2}f$ on Γ one gains the imaginary part $Im\,\Phi_- = (Im\,\overline{K})f$ (cf. (2.12d)). Therefore, the boundary condition (1b) leads to the integral equation

$$Im\,\overline{K}f = \varphi \qquad \text{on } \Gamma. \tag{7.4.25}$$

Using (4b), we rewrite the imaginary part $Im\,\overline{K}f$ as $K_2 f$ with

$$(K_2 f)(\mathbf{x}) := -\frac{1}{2\pi}\oint_\Gamma \frac{\langle \mathbf{t}(\mathbf{y}), \mathbf{y}-\mathbf{x}\rangle}{|\mathbf{y}-\mathbf{x}|^2} f(\mathbf{y})\,d\Gamma_{\mathbf{y}}. \tag{7.4.26}$$

By Remark 2.4c, $K_2 f$ is also continuous at corners of Γ. A generalisation by $\overline{K}_2 := Im\,\overline{K}$ is not necessary. The integral equation (25) reads

(7.4.25') $\quad K_2 f = \varphi \quad$ on Γ

and is an equation of the *first kind*. Written explicitly, (25') becomes

(7.4.25'') $$\varphi(\mathbf{x}) = -\frac{1}{2\pi}\oint_\Gamma \frac{\langle \mathbf{t}(\mathbf{y}), \mathbf{y}-\mathbf{x}\rangle}{|\mathbf{y}-\mathbf{x}|^2} f(\mathbf{y})\,d\Gamma_\mathbf{y}.$$

The kernel of the operator K_2 equals

(7.4.27a) $$k(\mathbf{x},\mathbf{y}) = -\frac{1}{2\pi}\frac{\langle \mathbf{t}(\mathbf{y}), \mathbf{y}-\mathbf{x}\rangle}{|\mathbf{y}-\mathbf{x}|^2}$$

and represents the tangential derivative

(7.4.27b) $$k(\mathbf{x},\mathbf{y}) = -\frac{1}{2\pi}\frac{\partial}{\partial t_\mathbf{y}}\log|\mathbf{y}-\mathbf{x}|$$

of the function $-\frac{1}{2\pi}\log|\mathbf{y}-\mathbf{x}|$. The latter will be discussed again in §8.1 as the kernel of the *single-layer potential*. Here, the *tangential derivative* of a function f is defined as follows:

(7.4.28) $$\partial f(\mathbf{y})/\partial t := \langle \mathbf{t}(\mathbf{y}), \nabla f(\mathbf{y})\rangle.$$

Equation (25') cannot be uniquely solvable, because of

Lemma 7.4.21. $f=1$ has the vanishing image $K_2 f = 0$.

Proof. $\overline{K} f = \frac{1}{2}$ implies $K_2 f = \mathrm{Im}\,\overline{K} f = 0$ (cf. (1.17c/18d)). QED

Different from $\overline{K}_1 = -2\,\mathrm{Re}\,\overline{K}$, K_2 is always strongly singular even in the case of a smooth boundary, so that (25'') has to be understood by means of Cauchy's principal value. Analogously to (10), we can bring the integral equation (25') into the form (29) by using Lemma 21:

(7.4.29) $$\varphi(\mathbf{x}) = -\frac{1}{2\pi}\oint_\Gamma \frac{\langle \mathbf{t}(\mathbf{y}), \mathbf{y}-\mathbf{x}\rangle}{|\mathbf{y}-\mathbf{x}|^2}[f(\mathbf{y})-f(\mathbf{x})]\,d\Gamma_\mathbf{y}.$$

For $f\in C^\lambda(\Gamma)$ with $\lambda>0$, the integral in (29) is an improper one.

Exercise 7.4.22. For a function f defined on Γ, the derivative f' is explained as tangential derivative: $f'(\mathbf{x}) := df(\gamma(s))/ds$, where $\mathbf{x}=\gamma(s)$ and $\gamma: [0,L]\to\Gamma$ is the parametrisation of Γ by the arc length. Assume $f\in C^1(\Gamma)$, i.e., $f'\in C^0(\Gamma)$ {$f\in C^\lambda(\Gamma)$ and $f'\in L^1(\Gamma)$ would already be sufficient}. Show for the parametrisation of Γ by the arc length that

$$\oint_\Gamma \frac{\langle \mathbf{t}(\mathbf{y}), \mathbf{y}-\mathbf{x}\rangle}{|\mathbf{y}-\mathbf{x}|^2} f(\mathbf{y})\,d\Gamma_\mathbf{y} = \oint_0^L f(\gamma(s))\frac{d}{ds}\log|\gamma(s)-\mathbf{x}|\,ds =$$
$$= -\int_0^L \log|\gamma(s)-\mathbf{x}|\frac{d}{ds}f(\gamma(s))\,ds = -\int_\Gamma \log|\mathbf{y}-\mathbf{x}|\,f'(\mathbf{y})\,d\Gamma_\mathbf{y}.$$

The integrals in the second line are improper ones (cf. Exercise 1.7).

Exercise 7.4.23. Let $\Gamma\in C^1_{0,\mathrm{stw}}$. Prove the following identities in $C^\lambda(\Gamma)$, $\lambda>0$:

(7.4.30) $$\overline{K}_1^2 - 4K_2^2 = I, \quad \overline{K}_1 K_2 = -K_2\overline{K}_1.$$

Hint: Use the Poincaré-Bertrand formula (2.22).

7.5 Hypersingular Integrals

The typical singularity of the one-dimensional Cauchy principal value is $1/(x-x_0)$. It may happen that integrands containing $(x-x_0)^{-3}$ or $(x-x_0)^{-2}\operatorname{sign}(x-x_0)$ still admit a Cauchy principal value, but in the case of $f(x)/(x-x_0)^3$ the function f must satisfy very restrictive conditions to ensure the existence of Cauchy's principal value (cf. Exercise 1.7b).

If, in the neighbourhood of the singularity, the integrand has a constant sign, only two cases can occur: Either the improper integral exists or the integral diverges even as Cauchy principal value. The latter situation occurs, e.g., in the integral

(7.5.1) $$\int_{-\infty}^{\infty} \frac{f(x)dx}{(x-\xi)^2} \qquad (\xi\in\mathbb{R}).$$

The convergence problems arise only from the second order singularity at $x=\xi$. The improper integrals over the remaining intervals $(-\infty,\xi-\varepsilon]$, $[\xi+\varepsilon,\infty)$ exist for any bounded $f\in L^\infty(\mathbb{R})$.

In order to assign an «integral value» to the integrand $f(x)/(x-\xi)^2$, split f into the terms

(7.5.2) $$f = f_1 + f_2 \quad \text{with} \quad f_1(x) := f(\xi), \quad f_2(x) := f(x)-f(\xi).$$

The second term $f(x)-f(\xi)$ contains a zero at $x=\xi$, so that the singularity is of *first* order and a Cauchy principal value might exist. Therefore, we define the *regularised integral value* by

(7.5.3) $$\int_{-\infty}^{\infty} \frac{f(x)dx}{(x-\xi)^2} := \oint_{\mathbb{R}} \frac{f(x)-f(\xi)}{(x-\xi)^2}dx,$$

if the Cauchy principal value exists on the right-hand side. The expression (3) is also called an «*integral in the sense of Hadamard*» or «*part finie*», expressing that only that part f_2 of the integrand $f=f_1+f_2$ is taken which leads to a (finite) integral.

Lemma 7.5.1. For $f\in C^{1+\lambda}(\mathbb{R})$ with $\lambda>0$ the value (3) exists for all $\xi\in\mathbb{R}$.

Proof. Since f is bounded, the (improper) integrals over $|x-\xi|\geqslant\varepsilon>0$ exist. In $[\xi-\varepsilon,\xi+\varepsilon]$ we use the Taylor expansion

$$f(x)=f(\xi)+(x-\xi)f'(\xi)+R(x,\xi) \quad \text{with} \quad R(x,\xi)=O(|x-\xi|^{1+\lambda}),$$

where, without loss of generality, we have supposed $\lambda\leqslant 1$. Insertion into the right-hand side of (3) yields

$$\oint_{\xi-\varepsilon}^{\xi+\varepsilon} \frac{f(x)-f(\xi)}{(x-\xi)^2}dx = f'(\xi)\oint_{\xi-\varepsilon}^{\xi+\varepsilon}\frac{dx}{x-\xi} + \int_{\xi-\varepsilon}^{\xi+\varepsilon}\frac{R(x,\xi)}{(x-\xi)^2}dx.$$

The first integral on the right side has a vanishing Cauchy principal value, the second one exists as improper integral, since $C|x-\xi|^{\lambda-1}$ is a majorant. QED

The previous proof leads to the new representation of the hypersingular integral:

$$(7.5.4a)\qquad \int_{-\infty}^{\infty} \frac{f(x)dx}{(x-\xi)^2} := \int_{|x-\xi|\geq\varepsilon} \frac{f(x)-f(\xi)}{(x-\xi)^2}\,dx + \int_{\xi-\varepsilon}^{\xi+\varepsilon} \frac{R(x,\xi)}{(x-\xi)^2}\,dx,$$

$$(7.5.4b)\qquad R(x,\xi) := f(x)-f(\xi)-(x-\xi)f'(\xi).$$

Since $1/(x-\xi)^2 = -\frac{d}{dx}1/(x-\xi)$, formally, the rule of *integration by parts* reads as follows

$$(7.5.5)\qquad \int_{-\infty}^{\infty} \frac{f(x)dx}{(x-\xi)^2} = \oint_{-\infty}^{\infty} \frac{f'(x)}{x-\xi}\,dx\,.$$

Lemma 7.5.2. Let f be continuously differentiable with $f(x)=o(|x|)$ as $|x|\to\infty$. Then, formula (5) holds in the sense that the left-hand side exists as Hadamard integral if and only if the right-hand side exists as Cauchy's principal value.

Proof. Integration of $\int_{\xi+\varepsilon}^{R} \frac{f(x)-f(\xi)}{(x-\xi)^2}dx$ and $\int_{-R'}^{\xi-\varepsilon} ...dx$ by parts yields $\int_{\xi+\varepsilon}^{R} \frac{f'(x)}{x-\xi}dx - \frac{f(R)-f(\xi)}{x-\xi} + \frac{f(\xi+\varepsilon)-f(\xi)}{\varepsilon}$ and $\int_{-R'}^{\xi-\varepsilon} \frac{f'(x)}{x-\xi}dx + \frac{f(-R')-f(\xi)}{-R'-\xi} +$ $+\frac{f(\xi-\varepsilon)-f(\xi)}{\varepsilon}$. The sum of the integrals tends to the Cauchy principal value, if it exist. The sum of the f-differences is $f'(\varepsilon')-f'(\varepsilon'')$ with intermediate values $\xi-\varepsilon<\varepsilon''<\xi<\varepsilon'<\xi+\varepsilon$ and, by assumption, converges to $f'(\xi)-f'(\xi)=0$. The terms involving R, R' tend to zero as $R, R' \to \infty$. QED

Exercise 7.5.3. Prove the following: If $f\in C^{1+\lambda}(\mathbb{R})$ with $\lambda\in(0,1)$, the Hadamard integral (3) defines a function of ξ belonging to $C^{\lambda}(\mathbb{R})$. *Hint:* Use (4a) with $\varepsilon = 2|\xi-\xi'|$.

The next lemma shows that the differentiation of a Cauchy principal value may «almost» be performed under the integral sign.

Lemma 7.5.4. Let $f\in C^{1+\lambda}_{\text{local}}(\mathbb{R})$ with $\lambda\in(0,1)$, $f(x)=o(|x|)$ as $|x|\to\infty$, and suppose that Cauchy's principal value $\oint_{-\infty}^{\infty}\frac{f(x)}{x-\xi}dx$ and the Hadamard integral (3) exist. Then we have

$$(7.5.6)\qquad \frac{d}{d\xi}\oint_{-\infty}^{\infty}\frac{f(x)}{x-\xi}\,dx = \int_{-\infty}^{\infty}\frac{f(x)dx}{(x-\xi)^2}\,.$$

Proof. Let $\varepsilon>0$, $I_\varepsilon := \{x\in\mathbb{R}:\ \varepsilon\leq|x-\xi|\leq 1/\varepsilon\}$, and $F_\varepsilon(\xi) := \oint_{I_\varepsilon}\frac{f(x)}{x-\xi}dx$. For $\varepsilon\to 0$, F_ε tends to Cauchy's principal value. By symmetry reasons, F_ε remains unchanged if f is replaced by $f-f(\xi)$, i.e., $F_\varepsilon = \oint_{I_\varepsilon}\frac{f(x)-f(\xi)}{x-\xi}dx$. The derivative $dF_\varepsilon/d\xi$ of F_ε is

$$\tfrac{1}{\varepsilon}[2f(\xi)-f(\xi-\varepsilon)-f(\xi+\varepsilon)] - \varepsilon[2f(\xi)-f(\xi-\tfrac{1}{\varepsilon})-f(\xi+\tfrac{1}{\varepsilon})] +$$
$$+\int_{I_\varepsilon}\frac{f(x)-f(\xi)}{(x-\xi)^2}dx - \int_{I_\varepsilon}\frac{f'(\xi)}{x-\xi}dx.$$

The latter integral vanishes ($f'(\xi)$ is a constant with respect to integration), the first terms tends to zero as $\varepsilon\to 0$, while the integral in the middle converges to the Hadamard integral of f. By assumption, the derivatives $dF_\varepsilon/d\xi$ tend *uniformly* in ξ to the right-hand side of (6). Therefore, $\lim dF/d\xi = d(\lim F)/d\xi$ proves the assertion. QED

8 The Integral Equation Method

The *(boundary) integral equation method* denotes the transformation of partial differential equations with d spatial variables into an integral equation over a $(d-1)$-dimensional surface. In §7.4, the method has already been introduced for the two-dimensional Laplace equation. There, results concerning the singular Cauchy kernel were used to find integral equation formulations for the Laplace equation (7.4.1a). According to this approach, the Laplace equation with its close connection to holomorphic functions seems to represent a very particular case (cf. Remark 1.1) and the question remains whether other equations can be treated. The *integral equation method* or *boundary integral method* starts from a differential equation $Lu = 0$ with suitable boundary conditions and looks for an equivalent formulation as integral equation. The numerical treatment of the integral equation, which thus arises, is the subject of the next chapter (§9) entitled «boundary element method».

8.1 The Single-Layer Potential

Although, we shall finally restrict ourselves again to the Laplace equation, we will demonstrate in this section how to transform a differential equation into an integral equation. Here, the singularity function plays a fundamental role.

8.1.1 The Fundamental Solution

The partial differential equation

(8.1.1) $\quad Lu = 0 \qquad$ in Ω_+ or Ω_-,

where $\Omega_\mp \subset \mathbb{R}^d$ denote the interior and exterior domains of Γ and L is a differential operator with *constant coefficients*. Examples for L are

(8.1.2a) $\quad L = -\Delta, \qquad$ (Laplace equation)

(8.1.2b) $\quad L = -\Delta - c, \qquad$ (Helmholtz equation)

(8.1.2c) $\quad L = \Delta^2 = \frac{\partial^4}{\partial x^4} + 2\frac{\partial^4}{\partial x^2 \partial y^2} + \frac{\partial^4}{\partial y^4} \qquad$ (biharmonic equation).

The integral equation method is restricted to the case of a homogeneous differential equation as in (1). Moreover, the coefficients (as, e.g., c in (2b))

must be constant. The case of an inhomogeneous equation will be discussed in §9.5.

The *fundamental solution (singularity function)* of L is a function of two variables $\mathbf{x},\mathbf{y}\in\mathbb{R}^d$, which only depends on the Euclidean norm $|\mathbf{x}-\mathbf{y}|$ and, as a function of $\mathbf{x}$, satisfies the equation $Ls=0$ for $\mathbf{x}\neq\mathbf{y}$, while the order of singularity at $\mathbf{y}\in\mathbb{R}^d$ is as weak as possible.

For Laplace's equation $-\Delta u=0$ in $\mathbb{R}^d$ $(d=2,3)$ one finds the following fundamental solution which depends on the dimension d:

$$(8.1.3)\qquad s(\mathbf{x},\mathbf{y}) = \begin{cases} -\frac{1}{2\pi}\log|\mathbf{x}-\mathbf{y}| & \text{for } d=2,\\ \frac{1}{4\pi}\frac{1}{|\mathbf{x}-\mathbf{y}|} & \text{for } d=3\end{cases}$$

(for $d>3$ see Hackbusch [2, §2.2]). $\mathbf{x}=(x_1,\dots,x_d)$ and $\mathbf{y}=(y_1,\dots,y_d)$ are vectors from $\mathbb{R}^d$. $|\cdot|$ is the Euclidean norm (cf. (7.4.4c)). For the proof of (3) one has to verify that for any fixed $\mathbf{y}\in\mathbb{R}^d$, the function $u(\mathbf{x}):=s(\mathbf{x},\mathbf{y})$ solves the Laplace equation $-\Delta u=0$ in $\mathbb{R}^d\setminus\{\mathbf{y}\}$.

The scaling of the fundamental solution is chosen such that the equation

$$(8.1.4)\qquad L_x s(\cdot,\mathbf{y}) = \delta_{\mathbf{y}} \qquad (L_x=L)$$

holds in the sense of distributions. Here, $\delta_{\mathbf{y}}$ is the Dirac functional, i.e., $\delta_{\mathbf{y}}(f)=f(\mathbf{y})$. The notation L_x instead of L indicates that the derivatives in L are to be performed with respect to the variable $\mathbf{x}$.

The fundamental solution is defined above as the function with the weakest singularity at the point $\mathbf{y}$. Functions with stronger singularity can be generated from $s(\mathbf{x},\mathbf{y})$ by differentiation. Let

$$(8.1.5)\qquad D^{\boldsymbol\nu} = D_x^{\boldsymbol\nu} := \partial^{|\boldsymbol\nu|}/\partial x_1^{\nu_1}\,\partial x_2^{\nu_2}\cdots\partial x_d^{\nu_d} \qquad (\nu_j\geq 0)$$

denote the $|\boldsymbol\nu|$-fold partial derivative with respect to the variable $\mathbf{x}$, where $\boldsymbol\nu=(\nu_1,\dots,\nu_d)\in\mathbb{Z}^d$ is the multi-index and $|\boldsymbol\nu|$ equals $\nu_1+\dots+\nu_d$. $D_y^{\boldsymbol\mu}$ stands for the corresponding derivative with respect to $\mathbf{y}$.

Lemma 8.1.1. If s is the fundamental solution corresponding to L, then

$$(8.1.6a)\qquad k(\mathbf{x},\mathbf{y}) := \sum_{\boldsymbol\nu}\sum_{\boldsymbol\mu}\alpha_{\boldsymbol\nu\boldsymbol\mu}(\mathbf{y})\,D_x^{\boldsymbol\nu}D_y^{\boldsymbol\mu}s(\mathbf{x},\mathbf{y})$$

with finite sums over $\boldsymbol\nu$, $\boldsymbol\mu$ and coefficients $\alpha_{\boldsymbol\nu\boldsymbol\mu}(\mathbf{y})$ depending only on $\mathbf{y}$ solves again equation (1) in $\mathbb{R}^d\setminus\{\mathbf{y}\}$.

Proof. Set $D=\sum_{\boldsymbol\nu}\sum_{\boldsymbol\mu}\alpha_{\boldsymbol\nu\boldsymbol\mu}(\mathbf{y})D_x^{\boldsymbol\nu}D_y^{\boldsymbol\mu}$. Interchanging the differentiations, one obtains $L_xk(\mathbf{x},\mathbf{y})=L_xDs(\mathbf{x},\mathbf{y})=DL_xs(\mathbf{x},\mathbf{y})=D0=0$ for $\mathbf{x}\neq\mathbf{y}$. QED

Exercise 8.1.2. Prove the following: **(a)** A particular derivative of the form (6a) is the normal derivative $\partial s(\mathbf{x},\mathbf{y})/\partial n_y$ with respect to $\mathbf{y}$.
(b) The gradient of the fundamental solution s from (3) has the form

$$(8.1.6b)\qquad \nabla_x s(\mathbf{x},\mathbf{y}) = -\frac{1}{\omega_d}\frac{\mathbf{x}-\mathbf{y}}{|\mathbf{x}-\mathbf{y}|^d}$$

(ω_d = surface of the d-dimensional unit sphere, e.g., $\omega_2=2\pi$, $\omega_3=4\pi$).

A fundamental solution need not be discontinuous. It may be its (higher) derivatives which are singular. For example, the *biharmonic* differential equation $\Delta^2 u = 0$ from (2c) in $\mathbb{R}^2$ has the fundamental solution

$$s(\mathbf{x},\mathbf{y}) = \tfrac{1}{8\pi}\,|\mathbf{x}-\mathbf{y}|^2 \log|\mathbf{x}-\mathbf{y}| \qquad (\mathbf{x},\mathbf{y}\in\mathbb{R}^2,\ \mathbf{x}\neq\mathbf{y}) \tag{8.1.7}$$

with continuous zeroth and first derivatives.

In the following subsections, we return to the Laplace equation $-\Delta u = 0$ (also called *potential equation*) and restrict the discussion to the corresponding fundamental solution (3).

8.1.2 Continuity of the Single-Layer Potential

8.1.2.1 Definition

In the case of $d=2$, $\Gamma=\partial\Omega_\pm$ is a closed curve. For $d=3$, Γ is the *surface* of the interior domain $\Omega_-\subset\mathbb{R}^3$. The integration over Γ is always denoted by $\int_\Gamma \dots d\Gamma$. §8.1.2.2 will contain some comments concerning the definition of a surface integral. The curvilinear ($d=2$) or surface ($d\geqslant 3$) integral

$$\Phi(\mathbf{x}) := \int_\Gamma s(\mathbf{x},\mathbf{y})\,f(\mathbf{y})\,d\Gamma_y \tag{8.1.8}$$

over Γ with fundamental solution s (corresponding to L) defines the *single-layer potential* (corresponding to L). According to (3), the single-layer potential corresponding to Laplace's equation has the explicit representation

$$\Phi(\mathbf{x}) = \begin{cases} -\frac{1}{2\pi}\int_\Gamma \log|\mathbf{x}-\mathbf{y}|\,f(\mathbf{y})\,d\Gamma_y & \text{for } d=2,\\ \frac{1}{4\pi}\int_\Gamma \frac{f(\mathbf{y})}{|\mathbf{x}-\mathbf{y}|}\,d\Gamma_y & \text{for } d=3. \end{cases} \tag{8.1.9}$$

The function f in (8) and (9) is called the *density* of the single-layer potential.

The name «potential» is justified since, outside of Γ, Φ satisfies the potential equation (Laplace equation) $\Delta\Phi=0$, as seen in the next lemma.

Lemma 8.1.3. Let $\Gamma\in C^1_{\mathrm{stw}}$. The single-layer potential (9) satisfies the Laplace equation in $\mathbb{R}^d\setminus\Gamma$, i.e.,

$$-\Delta\Phi = 0 \qquad \text{for } \mathbf{x}\notin\Gamma. \tag{8.1.10}$$

Proof. Γ is compact. For fixed $\mathbf{x}\notin\Gamma$, the second derivatives of $s(\mathbf{x},\mathbf{y})$ with respect to $\mathbf{x}$ are uniformly continuous on Γ, so that the Laplacian may be interchanged with the integral sign:

$$-\Delta\Phi = -\Delta\int s(\mathbf{x},\mathbf{y})f(\mathbf{y})\,d\Gamma_y = -\int_\Gamma \Delta_x s(\mathbf{x},\mathbf{y})f(\mathbf{y})\,d\Gamma_y = -\int_\Gamma 0\,d\Gamma_y = 0.$$ QED

The proof shows that a single-layer potential (8) involving the fundamental solution of differential operator L different from Δ satisfies again the corresponding equation $L\Phi=0$.

8.1.2.2 Surface Integrals

In §7, only *contour integrals* appear. Since now we allow the case $d=3$, which of course is very important for practical applications, we recall the definition of a surface integral.

Let a p-dimensional manifold $\Gamma\subset\mathbb{R}^d$ be the image of the parameter coordinates $\mathbf{t}=(t_1,\dots,t_p)\in S$ under a bijective and differentiable mapping $\varphi\colon S\subset\mathbb{R}^p\to\Gamma\subset\mathbb{R}^d$, i.e.,

$$\Gamma = \{\mathbf{x}=\varphi(\mathbf{t})\colon\ \mathbf{t}\in S\}.$$

Let f be a function defined on Γ. The integral of f over Γ is defined by

$$\text{(8.1.11a)}\qquad \int_\Gamma f\,d\Gamma := \int_S f(\varphi(\mathbf{t}))\sqrt{g(\mathbf{t})}\,d\mathbf{t} \qquad (d\mathbf{t}=dt_1\cdot\ldots\cdot dt_p),$$

where g is the *Gram determinant*

$$g = g(\mathbf{t}) = \det\Big(\Big\langle\frac{\partial\varphi}{\partial t_i},\frac{\partial\varphi}{\partial t_j}\Big\rangle_{i,j=1,\dots,p}\Big)$$

(cf. Exercise 4.5.4). Often, it is not possible to construct the whole set Γ as an image of a single parameter set S. If Γ is the disjoint union of different parts Γ_i being images of certain sets S_i under φ_i, then the integral over Γ can be reduced to the integrals over S_i:

$$\int_\Gamma f\,d\Gamma = \sum_i\int_{\Gamma_i} f\,d\Gamma.$$

In the following, $p=d-1$ is the only interesting choice.

In the particular case of a two-dimensional surface Γ in $\mathbb{R}^3$ described by coordinates

$$\left\{\begin{pmatrix}x_1\\x_2\\\eta(x_1,x_2)\end{pmatrix}\colon\ (x_1,x_2)\in S\right\},$$

the definition becomes

$$\text{(8.1.11b)}\qquad \int_\Gamma f\,d\Gamma = \int_S f(x_1,x_2,\eta(x_1,x_2))\sqrt{1+\eta_{x_1}^2+\eta_{x_2}^2}\,dx_1\,dx_2.$$

For curves Γ, we have introduced the classes

$$\Gamma\in C^\alpha,\quad \Gamma\in C^\alpha_{\mathbf{stw}},\quad \Gamma\in C^\alpha_0,\quad \Gamma\in C^\alpha_{0,\mathbf{stw}},$$

as well as C^ν_L, $\hat{C}^\alpha$ etc. (cf. §7.1.2). A general manifold Γ belongs to C^α, if Γ is compact, connected, and double-point free and if for any $\mathbf{x}\in\Gamma$ there is a neighbourhood $U\subset\Gamma$ of $\mathbf{x}$ such that U is the image of a one-to-one mapping $\varphi\colon S\subset\mathbb{R}^p\to U$, where $\varphi\in C^\alpha(S)$. Γ belongs to $\Gamma\in C^\alpha_{\mathbf{stw}}$, if $\Gamma\in C^0$, $\Gamma=\Gamma_1\cup\Gamma_2\cup\ldots\cup\Gamma_n$, and $\Gamma_i\in C^\alpha$ for $1\le i\le n$. In the following, the Γ_i's are called the *smooth components* of Γ. A manifold is called *closed*, if its boundary is the empty set. For a closed manifold $\Gamma\in C^\alpha$ (or $\Gamma\in C^\alpha_{\mathbf{stw}}$) we write $\Gamma\in C^\alpha_0$ (or $\Gamma\in C^\alpha_{0,\mathbf{stw}}$).

8.1.2.3 Improper Integrals on Surfaces

As for curves, one can reduce the definition of improper integrals over $\Gamma\in C^1_{\mathbf{stw}}$ to that of the parametrisation (11a), provided that $\varphi\in C^1_{\mathbf{stw}}$. To justify this, one has to check that the improper integrability does not depend on the choice of the parametrisation on the right-hand side of (11a). Since a neighbourhood of $\mathbf{t}\in S$ with respect to the parameter coordinates maps into a neighbourhood of $\mathbf{x}=\varphi(\mathbf{t})\in\Gamma$ on the manifold, the definition from §6.1.3 is directly applicable:

$$\int_\Gamma f\,d\Gamma = \lim_{k\to\infty}\int_{\Gamma\setminus U_k} f\,d\Gamma,$$

where $U_k\subset U_{k-1}$ are neighbourhoods of the singularities $\sigma\subset\Gamma$ and $\bigcap U_k=\sigma$ holds.

Remark 8.1.4. Let $\Gamma\in C^1_{\mathrm{stw}}$ be a curve or surface in $\mathbb{R}^d$. Assume that a continuous function f defined on $\Gamma\setminus\{\mathbf{x}_0\}$ has a singularity at $\mathbf{x}_0\in\Gamma$. If $|f(\mathbf{x})|\leqslant C|\mathbf{x}-\mathbf{x}_0|^{-\lambda}$ with $\lambda<d-1$, then f is improperly integrable over Γ.

Proof. Apply Remark 6.1.2b to the right-hand side of (11a). QED

Lemma 6.1.5 (continuity with respect to a parameter) remains also valid for improper integrals over $\Gamma\in C^1_{\mathrm{stw}}$. Taking pattern from Exercise 6.1.4c, we mention a criterion for the uniform existence of improper integrals.

Lemma 8.1.5. Let $\Gamma\in C^1_{\mathrm{stw}}$ be a p-dimensional manifold in $\mathbb{R}^d$ and $D\subset\mathbb{R}^d$. $f(\xi,\mathbf{y})$ is supposed to be defined and continuous for variable values $\mathbf{y}\in\Gamma$ and parameter values $\xi\in D\setminus\{\mathbf{y}\}$ and satisfy $|f(\xi,\mathbf{y})|\leqslant C|\mathbf{y}-\xi|^{-\lambda}$ with $\lambda<p$. Then *the improper integrals $\Phi(\xi):=\int_\Gamma f(\xi,\mathbf{y})\,d\Gamma_y$ exist uniformly*, so that Φ is continuous on D.

Proof. (i) Without loss of generality, we may assume $\Gamma\in C^1$, since the following considerations can be applied for any single C^1-piece of $\Gamma\in C^1_{\mathrm{stw}}$. There is an $\varepsilon>0$ such that the following statements (a,b) hold.

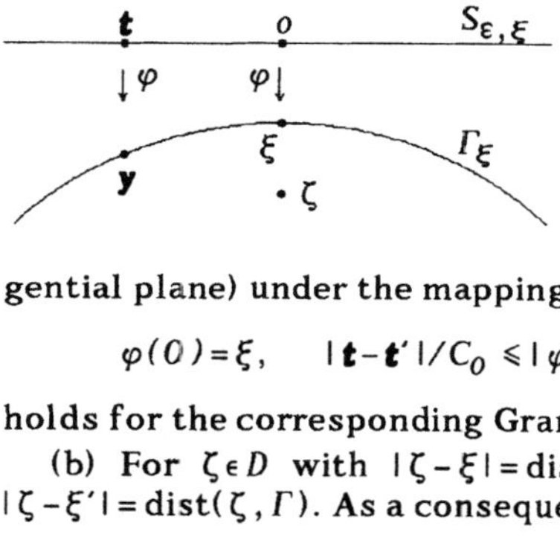

(a) For all $\xi\in\Gamma$, the subset $\Gamma_\xi:=\{\mathbf{y}\in\Gamma: |\mathbf{y}-\xi|\leqslant\varepsilon\}$ of Γ is the image of $S_{\varepsilon,\xi}\subset\{\mathbf{t}\in\mathbb{R}^p: |\mathbf{t}|\leqslant\varepsilon\}$ (e.g., of the tangential plane) under the mapping $\varphi=\varphi_\xi$, where

$$\varphi(0)=\xi,\qquad |\mathbf{t}-\mathbf{t}'|/C_0\leqslant|\varphi(\mathbf{t})-\varphi(\mathbf{t}')|\leqslant C_0|\mathbf{t}-\mathbf{t}'|,\qquad\text{and}\quad C_0^{-2}\leqslant g\leqslant C_0^2$$

holds for the corresponding Gram determinant g and C_0 is independent of x.

(b) For $\zeta\in D$ with $|\zeta-\xi|=\mathrm{dist}(\zeta,\Gamma)\leqslant\varepsilon$, there is no further $\xi'\in\Gamma$ with $|\zeta-\xi'|=\mathrm{dist}(\zeta,\Gamma)$. As a consequence, we have

$$|\mathbf{y}-\xi|\leqslant 2|\mathbf{y}-\zeta|\qquad\text{for all }\mathbf{y}\in\Gamma_\xi,\ \zeta\in D\ \text{ with }\mathrm{dist}(\zeta,\Gamma)=|\zeta-\xi|.$$

(ii) Let $\eta:=\mu(U):=\int_U d\Gamma$ be the measure of $U\subset\Gamma$. In the case of $\mathrm{dist}(\zeta,\Gamma)\geqslant\varepsilon$, we have $\int_U|f(\zeta,\mathbf{y})|\,d\Gamma\leqslant\eta C\varepsilon^{-\lambda}$. For ζ with $\mathrm{dist}(\zeta,\Gamma)<\varepsilon$, there is a unique $\xi\in\Gamma$ with $\mathrm{dist}(\zeta,\Gamma)=|\zeta-\xi|$. The integral over $U\setminus\Gamma_\xi$ is also $\leqslant\eta C\varepsilon^{-\lambda}$. Concerning the integral over $V:=U\cap\Gamma_\xi$, the parametrisation via $\varphi=\varphi_\xi$: $S_{\varepsilon,\xi}\to\Gamma_\xi$ shows that

$$\int_V|f(\zeta,\mathbf{y})|\,d\Gamma\leqslant C\int_V|\mathbf{y}-\zeta|^{-\lambda}d\Gamma_y\leqslant C\,2^\lambda\int_V|\mathbf{y}-\xi|^{-\lambda}d\Gamma_y\leqslant$$

$$\leqslant C\,2^\lambda\int_W\sqrt{g}\,|\varphi(\mathbf{t})-\varphi(0)|^{-\lambda}d\mathbf{t}\leqslant C\,2^\lambda C_0^{\lambda+1}\int_W|\mathbf{t}|^{-\lambda}d\mathbf{t}=:\tau(V),$$

where $W := \varphi^{-1}(V) \subset K_\varepsilon(0) \subset \mathbb{R}^p$ is the pre-image of V. $|\mathbf{t}|^{-\lambda}$ is improperly integrable in $\mathbb{R}^p$: For $\mu(W) \to 0$ also $\tau(V)$ tends to zero. The estimate $1/g \leqslant C_0^2$ of the Gram determinant implies

$$\mu(W) = \int_W d\mathbf{t} = \int_V g^{-1/2} d\Gamma \leqslant \mu(V)/C_0 \leqslant \mu(U)/C_0 = \eta/C_0.$$

For any sequence of neighbourhoods U with $\eta := \mu(U) \to 0$, one concludes from $\mu(W) \to 0$ that $\tau(V) \to 0$, so that $\sup\{\int_U |f(\zeta,\mathbf{y})| d\Gamma: \zeta \in D\} \leqslant \eta C \varepsilon^{-\lambda} + \tau(V) \to 0$ proves the uniform existence of the improper integral. ∎

Exercise 8.1.6. Let the estimate $|f(\xi,\mathbf{y})| \leqslant C|\mathbf{y}-\xi|^{-\lambda}$ $(\lambda<d-1)$ be restricted to $|\xi| \leqslant 2R$ with $R := \sup\{|\mathbf{y}|: \mathbf{y} \in \Gamma\}$; otherwise, the assumptions of Lemma 5 should hold. Prove the following: Boundedness of f as $\xi \to \infty$ implies that also $\Phi(\xi) := \int_\Gamma f(\xi,\mathbf{y}) d\Gamma_y$ is bounded as $\xi \to \infty$. If, in addition, f converges uniformly to $\lim f(\xi,\mathbf{y}) = f(\infty)$ as $\xi \to \infty$, then Φ tends to $f(\infty)\mu(\Gamma)$ as $\xi \to \infty$.

8.1.2.4 Properties of the Single-Layer Potential

Lemma 5 and Exercise 6 imply the first part of the following result.

Theorem 8.1.7. The single-layer potential Φ from (9) over $\Gamma \in C^1_{\mathrm{stw}}$ is continuous in $\mathbf{x} \in \mathbb{R}^d$. For $d \geqslant 3$, Φ is continuous at $\mathbf{x} = \infty$ with $\Phi(\infty) = 0$, i.e.,

(8.1.12) $$\Phi(\mathbf{x}) \to 0 \qquad \text{as } |\mathbf{x}| \to \infty.$$

In the two-dimensional case $d=2$, (12) holds if and only if, in addition, the density f satisfies

(8.1.13) $$\int_\Gamma f\, d\Gamma = 0.$$

Proof of the second part. Let $d=2$. Under the assumption (13), we have $\Phi(\mathbf{x}) = -\int[\log|\mathbf{x}-\mathbf{y}| - \log|\mathbf{x}|] f(\mathbf{y}) d\Gamma_y/(2\pi)$ for $\mathbf{x} \neq 0$. Since the difference $\log|\mathbf{x}-\mathbf{y}| - \log|\mathbf{x}| = \log[|\mathbf{x}-\mathbf{y}|/|\mathbf{x}|]$ tends to zero uniformly in $\mathbf{y}$ as $|\mathbf{x}| \to \infty$, (12) is proved. The reverse direction (12) $\Rightarrow$ (13) is left to the reader. ∎

8.1.3 Derivatives of the Single-Layer Potential

8.1.3.1 The Normal Derivative

For $\mathbf{x} \notin \Gamma$, the derivative of Φ may be taken under the integral sign, since Γ is compact and $\nabla_x s(\mathbf{x},\mathbf{y})$ is uniformly continuous on Γ with respect to $\mathbf{y}$. The gradient equals

(8.1.14a) $$\nabla\Phi(\mathbf{x}) = -\frac{1}{\omega_d}\int_\Gamma \frac{\mathbf{x}-\mathbf{y}}{|\mathbf{x}-\mathbf{y}|^d} f(\mathbf{y})\, d\Gamma_y \qquad (\mathbf{x} \notin \Gamma)$$

(cf. (6b)), where

(8.1.14b) $$\omega_2 = 2\pi, \qquad \omega_3 = 4\pi,$$

(in general, ω_d = surface of the d-dimensional unit sphere). Similarly, one can form higher derivatives at $\mathbf{x} \notin \Gamma$ and finds that Φ is *infinitely often differentiable* outside of Γ, i.e., $\Phi \in C^\infty(\mathbb{R}^d \backslash \Gamma)$. Since the kernel function

$(\mathbf{x}-\mathbf{y})/|\mathbf{x}-\mathbf{y}|^d$ is not improperly integrable over Γ, the gradient $\nabla\Phi(\mathbf{x})$ may be discontinuous across Γ.

Let Ω_- be the interior and Ω_+ the exterior domain of Γ. One is interested, in particular, in the question of whether the single-layer potential Φ admits a *normal derivative* $\partial\Phi/\partial n$ on Γ and how the resulting function (if it exists) is related to the density f. Let $\mathbf{x}_0\in\Gamma$ be a point on the surface, where Γ is Hölder continuously differentiable, i.e., the following inequality is satisfied at $\mathbf{x}_0$:

(8.1.15) $$|\langle\mathbf{n}(\mathbf{x}_0),\mathbf{x}_0-\mathbf{y}\rangle| \leqslant C\,|\mathbf{x}_0-\mathbf{y}|^{1+\lambda} \qquad (\lambda\in(0,1],\text{ for all }\mathbf{y}\in\Gamma).$$

Exercise 8.1.8. Prove the following: **(a)** A sufficient condition for (15) is $\Gamma_0\in C^{1+\lambda}$ for a neighbourhood $\Gamma_0\subset\Gamma$ of $\mathbf{x}_0$.
(b) The weaker condition $\langle\mathbf{n}(\mathbf{x}_0),\mathbf{x}_0-\mathbf{y}\rangle=o(|\mathbf{x}_0-\mathbf{y}|)$ corresponds to the differentiability of Γ at $\mathbf{x}_0$ and ensures the existence of a *tangential plane* $(d=3)$ or a tangent $(d=2)$ at $\mathbf{x}_0$, which is represented by $T:=\{\mathbf{x}\in\mathbb{R}^d\colon\ \mathbf{x}=\mathbf{x}_0+\mathbf{z},\ \mathbf{z}\perp\mathbf{n}(\mathbf{x}_0)\}$.

Theorem 8.1.9. Let (15) be satisfied at a point $\mathbf{x}_0\in\Gamma$. Assume that the density f of the single-layer potential Φ is bounded (i.e., $f\in L^\infty(\Gamma)$) and continuous at $\mathbf{x}_0$. $\mathbf{n}(\mathbf{x}_0)$ denotes the outer normal direction at $\mathbf{x}_0$. Then the one-sided *normal derivatives*

(8.1.16a) $$\frac{\partial\Phi_\pm}{\partial n}(\mathbf{x}_0) := \lim_{\substack{\alpha\to 0\\ \alpha>0}}\langle\mathbf{n}(\mathbf{x}_0),\nabla\Phi(\mathbf{x}_0\pm\alpha\mathbf{n}(\mathbf{x}_0))\rangle$$

exist at $\mathbf{x}_0$ and satisfy the *jump relation*

(8.1.16b) $$\partial\Phi_+(\mathbf{x}_0)/\partial n - \partial\Phi_-(\mathbf{x}_0)/\partial n = -f(\mathbf{x}_0).$$

Their mean value at $\mathbf{x}_0$ equals

(8.1.16c) $$\tfrac{1}{2}[\partial\Phi_+/\partial n+\partial\Phi_-/\partial n] = -\frac{1}{\omega_d}\int_\Gamma\frac{\langle\mathbf{n}(\mathbf{x}_0),\,\mathbf{x}_0-\mathbf{y}\rangle}{|\mathbf{x}_0-\mathbf{y}|^d}f(\mathbf{y})\,d\Gamma_y,$$

where the right-hand side in (16c) represents an improper integral.

Hint. (i) If the inner normal direction $\mathbf{n}(\mathbf{x}_0)$ is used instead of the outer one, the signs of $\partial\Phi_\pm/\partial n$ change. The notation «±» of $\partial\Phi_\pm/\partial n$ is chosen so that $\partial\Phi_-/\partial n$ [$\partial\Phi_+/\partial n$, respectively] is the usual normal derivative of the boundary value problem in the interior [exterior] domain Ω_-, provided that Γ is closed.
(ii) Equation (16c) can also be written in the form (16c'):

(8.1.16c') $$\tfrac{1}{2}[\partial\Phi_+/\partial n+\partial\Phi_-/\partial n] = \int_\Gamma\frac{\partial}{\partial n_x}s(\mathbf{x}_0,\mathbf{y})f(\mathbf{y})\,d\Gamma_y.$$

To prove Theorem 9, we first need the lemma:

Lemma 8.1.10. Let $\varepsilon>0$ and $S_\varepsilon:=\{\mathbf{y}'\in\mathbb{R}^{d-1}\colon\ |\mathbf{y}'|\leqslant\varepsilon\}$. **(a)** The integral

(8.1.17a₁) $$I(\alpha,\varepsilon) := \frac{1}{\omega_d}\int_{S_\varepsilon}\frac{\alpha\,d\mathbf{y}'}{[|\mathbf{y}'|^2+\alpha^2]^{d/2}} \qquad \text{for }\alpha\neq 0$$

depends only on ε/α and has the following one-sided limits at $\alpha=0$:

(8.1.17a$_2$) $\quad I(0\pm 0,\varepsilon) = \pm\frac{1}{2}$ $\qquad$ for all $\varepsilon>0$.

(b) If the function η satisfies the estimate

(8.1.17b) $\quad |\eta(\mathbf{y}')| \leqslant C_\eta |\mathbf{y}'|^{1+\mu}$ $\qquad$ for $\mathbf{y}'\in S_\varepsilon$

with some $\mu>0$, then the integral

(8.1.17c$_1$) $\quad I_\eta(\alpha,\varepsilon) := \frac{1}{\omega_d}\int\limits_{S_\varepsilon} \frac{\alpha-\eta(\mathbf{y}')}{[|\mathbf{y}'|^2+(\alpha-\eta(\mathbf{y}'))^2]^{d/2}}\,d\mathbf{y}'$

has one-sided limits at $\alpha=0$ satisfying

(8.1.17c$_2$) $\quad I_\eta(0+0,\varepsilon)-I_\eta(0-0,\varepsilon) = 1,$

(8.1.17c$_3$) $\quad I_\eta(0+0,\varepsilon)+I_\eta(0-0,\varepsilon) = \frac{2}{\omega_d}\int\limits_{S_\varepsilon} \frac{-\eta(\mathbf{y}')\,d\mathbf{y}'}{[|\mathbf{y}'|^2+\eta(\mathbf{y}')^2]^{d/2}}.$

(c) Assume (17b). Then for all $\varphi\in L^\infty(S_\varepsilon)$ and $\alpha\in\mathbb{R}$, we have

(8.1.17d) $\quad \frac{1}{\omega_d}\int\limits_{S_\varepsilon} \frac{|\varphi(\mathbf{y}')(\alpha-\eta(\mathbf{y}'))|\,d\mathbf{y}'}{[|\mathbf{y}'|^2+(\alpha-\eta(\mathbf{y}'))^2]^{d/2}} \leqslant [\tfrac{1}{2}+O(\varepsilon^\mu)]\,\|\varphi\|_{\infty,S_\varepsilon}.$

(d) Assume (17b) and continuity of the function $\varphi\in L^\infty(S_\varepsilon)$ at $\mathbf{y}'=0$. Then,

(8.1.17e$_1$) $\quad \Phi(\alpha,\varepsilon) := \frac{1}{\omega_d}\int_{S_\varepsilon} \frac{\varphi(\mathbf{y}')(\alpha-\eta(\mathbf{y}'))\,d\mathbf{y}'}{[|\mathbf{y}'|^2+(\alpha-\eta(\mathbf{y}'))^2]^{d/2}}$

has one-sided limits $\Phi(0\pm 0,\varepsilon)$ at $\alpha=0$ satisfying

(8.1.17e$_2$) $\quad \Phi(0+0,\varepsilon)-\Phi(0-0,\varepsilon) = \varphi(0),$

(8.1.17e$_3$) $\quad \Phi(0+0,\varepsilon)+\Phi(0-0,\varepsilon) = \frac{2}{\omega_d}\int_{S_\varepsilon} \frac{-\varphi(\mathbf{y}')\eta(\mathbf{y}')d\mathbf{y}'}{[|\mathbf{y}'|^2+\eta(\mathbf{y}')^2]^{d/2}}.$

Note that the mean values of the one-sided limits of I, I_η, and Φ are obtained by setting $\alpha=0$ in the integrands.

Proof. (i) Making the substitution $\mathbf{y}'=\alpha\mathbf{z}$, we have $d\mathbf{y}'=|\alpha|^{d-1}d\mathbf{z}$ and

$$I(\alpha,\varepsilon) = \frac{1}{\omega_d}\int_{S_\varepsilon/|\alpha|} \frac{\alpha\,|\alpha|^{d-1}d\mathbf{z}}{|\alpha|^d[|\mathbf{z}|^2+1]^{d/2}} =$$

$$= \operatorname{sign}\alpha\,\frac{1}{\omega_d}\int_{S_\varepsilon/|\alpha|} \frac{d\mathbf{z}}{[|\mathbf{z}|^2+1]^{d/2}} = \operatorname{sign}\alpha\; I(1,\varepsilon/|\alpha|).$$

As $\varepsilon/|\alpha|\to\infty$, the function $I(1,\varepsilon/|\alpha|)$ converges to the improper integral

$$I(1,\infty) = \frac{1}{\omega_d}\int_{\mathbb{R}^{d-1}} \frac{d\mathbf{z}}{[|\mathbf{z}|^2+1]^{d/2}}.$$

For the evaluation of the integral one uses polar coordinates in $\mathbb{R}^{d-1}$ ($r=|\mathbf{z}|$, $d\mathbf{z}=r^{d-2}\,dr\,d\Omega$, $\int d\Omega=\omega_{d-1}$, $\omega_1=2$, $\omega_2=2\pi$):

$$I(1,\infty) = \frac{\omega_{d-1}}{\omega_d}\int_0^\infty r^{d-2}[r^2+1]^{-d/2}\,dr.$$

For $d=2$, one has $\omega_{d-1}=2$, $\omega_d=2\pi$, $\int_0^\infty\ldots dr=\int_0^\infty [r^2+1]^{-1}dr=\frac{\pi}{2}$; hence $I(1,\infty)=\frac{1}{2}$. One finds the same result for $d=3$, since $\omega_{d-1}=2\pi$, $\omega_d=4\pi$, and $\int_0^\infty\ldots dr=\int_0^\infty r[r^2+1]^{-3/2}dr=\frac{1}{2}\int_0^\infty [t+1]^{-3/2}dt=1$. This shows part (a) of the lemma.

(ii) For abbreviation, we introduce

$$\alpha(t,\mathbf{y}') := \alpha - t\eta(\mathbf{y}'), \qquad N(t,\mathbf{y}') := |\mathbf{y}'|^2+\alpha(t,\mathbf{y}')^2.$$

The mean value theorem of differential calculus yields

$$\delta(\mathbf{y}') := \frac{\alpha(1,\mathbf{y}')}{N(1,\mathbf{y}')^{d/2}} - \frac{\alpha(0,\mathbf{y}')}{N(0,\mathbf{y}')^{d/2}} = \frac{-\eta(\mathbf{y}')}{N(t,\mathbf{y}')^{d/2}} - d\,\frac{\alpha(t,\mathbf{y}')^2\eta(\mathbf{y}')}{N(t,\mathbf{y}')^{d/2+1}}$$

with an intermediate value $t\in(0,1)$. Since $\alpha(t,\mathbf{y}')\leqslant N(t,\mathbf{y}')^{1/2}$, $|\eta(\mathbf{y}')|\leqslant C_\eta|\mathbf{y}'|^{1+\mu}$, and $N(t,\mathbf{y}')\geqslant|\mathbf{y}'|^2$, the function $\delta(\mathbf{y}')$ has the improperly integrable majorant $C_\eta(d+1)|\mathbf{y}'|^{1+\mu-d}$. $\alpha(0,\mathbf{y}')/N(0,\mathbf{y}')^{d/2}$ and $\alpha(1,\mathbf{y}')/N(1,\mathbf{y}')^{d/2}$ are the integrands in $I(\alpha,\varepsilon)$ and $I_\eta(\alpha,\varepsilon)$, respectively. Hence, $\delta I(\alpha,\varepsilon):=I_\eta(\alpha,\varepsilon)-I(\alpha,\varepsilon)$ exists uniformly as an improper integral and, consequently, it is continuous in α: $\delta I(0,\varepsilon)=\frac{1}{\omega_d}\int_{S_\varepsilon}\frac{-\eta(\mathbf{y}')\,d\mathbf{y}'}{[|\mathbf{y}'|^2+\eta(\mathbf{y}')^2]^{d/2}}$. Together with (17a$_2$), assertion (b) follows.

(iii) Since $|\varphi(\mathbf{y}')|$ is estimated by $\|\varphi\|_{\infty,S_\varepsilon}$, it remains to show

$$\frac{1}{\omega_d}\int_{S_\varepsilon}\frac{|\alpha-\eta(\mathbf{y}')|\,d\mathbf{y}'}{[|\mathbf{y}'|^2+(\alpha-\eta(\mathbf{y}'))^2]^{d/2}} \leqslant \tfrac{1}{2}+O(\varepsilon^\mu)$$

for proving (c). For this inequality, use the considerations from (ii).

(iv) The splitting $\varphi(\mathbf{y}')=[\varphi(\mathbf{y}')-\varphi(0)]+\varphi(0)$ induces the splitting of $\Phi(\alpha,\varepsilon)$ into $\Phi_0(\alpha,\varepsilon)+\varphi(0)\,I_\eta(\alpha,\varepsilon)$, where

$$\Phi_0(\alpha,\varepsilon) := \frac{1}{\omega_d}\int_{S_\varepsilon}\frac{[\varphi(\mathbf{y}')-\varphi(0)]\,(\alpha-\eta(\mathbf{y}'))}{[|\mathbf{y}'|^2+(\alpha-\eta(\mathbf{y}'))^2]^{d/2}}\,d\mathbf{y}'.$$

For the proof of (17e$_{2,3}$), it suffices to show the continuity of Φ_0 with respect to α at $\alpha=0$ and

$$\Phi_0(0,\varepsilon)=\frac{1}{\omega_d}\int_{S_\varepsilon}\frac{-[\varphi(\mathbf{y}')-\varphi(0)]\,\eta(\mathbf{y}')}{[|\mathbf{y}'|^2+\eta(\mathbf{y}')^2]^{d/2}}\,d\mathbf{y}'.$$

Set

$$\Phi(\alpha,\varepsilon',\varepsilon) := \frac{1}{\omega_d}\int_{S_\varepsilon\setminus S_{\varepsilon'}}\frac{[\varphi(\mathbf{y}')-\varphi(0)]\,(\alpha-\eta(\mathbf{y}'))}{[|\mathbf{y}'|^2+(\alpha-\eta(\mathbf{y}'))^2]^{d/2}}\,d\mathbf{y}'$$

for $0\leqslant\varepsilon'\leqslant\varepsilon$. For any $\varepsilon'>0$, $\Phi(\cdot,\varepsilon',\varepsilon)$ belongs to $C(\mathbb{R})$, since the integrand is uniformly continuous on $S_\varepsilon\setminus S_{\varepsilon'}$ with respect to α. On the other hand, (17d) together with

$$|\Phi(\alpha,\varepsilon',\varepsilon)-\Phi(\alpha,\varepsilon'',\varepsilon)|=|\Phi(\alpha,\varepsilon',\varepsilon'')|\leqslant[\tfrac{1}{2}+O(\varepsilon'^{\mu})]\,\|\varphi-\varphi(0)\|_{\infty,S_{\varepsilon'}} \quad (0<\varepsilon''\leqslant\varepsilon'\leqslant\varepsilon)$$

and $\|\varphi-\varphi(0)\|_{\infty,S_{\varepsilon'}}\to 0$ and the continuity of φ at 0 proves the uniform convergence of $\Phi(\cdot,\varepsilon',\varepsilon)$ to $\Phi(\cdot,0,\varepsilon)=\Phi_0(\cdot,\varepsilon)$ for $\varepsilon'\to 0$, so that Φ_0 is continuous in α and Φ_0 has the claimed representation at $\alpha=0$:

$$\Phi_0(0,\varepsilon) = \lim_{\varepsilon'\to 0}\Phi(0,\varepsilon',\varepsilon) = \frac{1}{\omega_d}\int_{S_\varepsilon}\frac{-[\varphi(\mathbf{y}')-\varphi(0)]\,\eta(\mathbf{y}')}{[|\mathbf{y}'|^2+\eta(\mathbf{y}')^2]^{d/2}}\,d\mathbf{y}'. \quad \text{QED}$$

Proof of Theorem 9. (i) Since the quantities to be discussed are invariant with respect to rotations and translations, we may assume without loss of generality that $\mathbf{x}_0=0$ and that the tangential plane T (cf. Exercise 8b) is given by $x_d=0$, i.e., $T=\{\mathbf{x}=\binom{\mathbf{x}'}{0}:\ \mathbf{x}'\in\mathbb{R}^{d-1}\}$. Then the normal direction at $\mathbf{x}_0=0$ is

$$\mathbf{n} := \mathbf{n}(\mathbf{x}_0) \qquad \text{with } n_1=\ldots=n_{d-1}=0,\ n_d=1,$$

if the outer normal direction points upwards (otherwise, $n_d=-1$). For sufficiently small $\varepsilon>0$, $S_\varepsilon:=\{\mathbf{y}'\in\mathbb{R}^{d-1}: |\mathbf{y}'|\leqslant\varepsilon\}$ can be mapped by

$$\varphi:\ \mathbf{y}'\in\mathbb{R}^{d-1}\mapsto\mathbf{y}=\binom{\mathbf{y}'}{\eta(\mathbf{y}')}\in\Gamma$$

onto a neighbourhood $\Gamma_\varepsilon\subset\Gamma$ of $\mathbf{x}_0=0$ (cf. Lemma 10). Because of (15), we have

$$\begin{aligned}&\eta(\mathbf{y}')=\langle\mathbf{n},\mathbf{y}\rangle=-\langle\mathbf{n}(\mathbf{x}_0),\mathbf{x}_0-\mathbf{y}\rangle,\\ \text{(8.1.17f)}\quad&\eta(0)=\eta_{y_i}(0)=0 \qquad (1\leqslant i\leqslant d-1),\\ &|\eta(\mathbf{y}')|\leqslant C|\mathbf{y}|^{1+\lambda}=C[|\mathbf{y}'|^2+\eta(\mathbf{y}')^2]^{(1+\lambda)/2}.\end{aligned}$$

Solving the last inequality with respect to $\eta(\mathbf{y}')$, we are led to (17b) with $\mu:=\lambda$.
(ii) The partition of Γ into Γ_ε and $\Gamma\backslash\Gamma_\varepsilon$ induces a decomposition of the single-layer potential Φ into

$$\Phi=\Phi_0+\Phi_1 \qquad \text{with}\quad \Phi_0=\int_{\Gamma\backslash\Gamma_\varepsilon}s(\cdot,\mathbf{y})f(\mathbf{y})\,d\Gamma_y,\quad \Phi_1=\int_{\Gamma_\varepsilon}s(\cdot,\mathbf{y})\,f(\mathbf{y})\,d\Gamma_y.$$

Since $\mathbf{x}_0=0\notin\Gamma\backslash\Gamma_\varepsilon$, the gradient $\nabla\Phi_0$ and, consequently, also the directional derivative $\langle\mathbf{n},\nabla\Phi_0(\alpha\mathbf{n})\rangle$ are infinitely continuously differentiable in the neighbourhood of $\alpha=0$ and, in particular, continuous. Therefore, a jump of the normal derivative at $\alpha=0$ can be caused by Φ_1 only. For the representation of $\nabla\Phi_1$ on Γ_ε, we can use the parametrisation φ (cf. (11b)):

$$\nabla\Phi_1(\mathbf{x})=-\frac{1}{\omega_d}\int_{S_\varepsilon}f(\mathbf{y})\sqrt{g(\mathbf{y})}\,\frac{\mathbf{x}-\mathbf{y}}{|\mathbf{x}-\mathbf{y}|^d}\,d\mathbf{y}' \qquad (\mathbf{x}\in\Gamma_\varepsilon)$$

with $\quad\mathbf{y}=\varphi(\mathbf{y}')=\binom{\mathbf{y}'}{\eta(\mathbf{y}')},\qquad g(\mathbf{y})=1+\sum_{i=1}^{d-1}\eta_{y_i}(\mathbf{y}')^2.$

For $\mathbf{x}=\alpha\mathbf{n}$, the expressions $\langle\mathbf{n},\mathbf{x}-\mathbf{y}\rangle$ and $|\mathbf{x}-\mathbf{y}|^d$ become $\alpha-\eta(\mathbf{y}')$ and $[|\mathbf{y}'|^2+(\alpha-\eta(\mathbf{y}'))^2]^{d/2}$.

$$\text{(8.1.17g)}\qquad \frac{\partial\Phi_1(\alpha\mathbf{n})}{\partial n}=-\frac{1}{\omega_d}\int_{S_\varepsilon}\frac{f(\mathbf{y})\sqrt{g(\mathbf{y})}\,(\alpha-\eta(\mathbf{y}'))}{[|\mathbf{y}'|^2+(\alpha-\eta(\mathbf{y}'))^2]^{d/2}}\,d\mathbf{y}' \qquad \text{for } \alpha\neq0$$

is the normal derivative at $\mathbf{x}=\mathbf{x}_0+\alpha\mathbf{n}$. Lemma 10d with $\varphi:=-f\sqrt{g}$ proves the jump (16b) by $-f(0)\sqrt{g(0)}=-f(0)$ because of $g(0)=1$ (cf. (17f)). According to ($17e_3$), one obtains the following result for the mean value of the one-sided limits of (17g) at $\alpha=0\pm0$:

$$\frac{1}{\omega_d}\int_{S_\varepsilon}\frac{f(\mathbf{y})\sqrt{g(\mathbf{y})}\,\eta(\mathbf{y}')}{[|\mathbf{y}'|^2+\eta(\mathbf{y}')^2]^{d/2}}\,d\mathbf{y}'=-\frac{1}{\omega_d}\int_{\Gamma_\varepsilon}\frac{f(\mathbf{y})\langle\mathbf{n}(\mathbf{x}_0),\mathbf{x}_0-\mathbf{y}\rangle}{|\mathbf{x}_0-\mathbf{y}|^d}\,d\Gamma_y.$$

Since $\langle\mathbf{n},\nabla\Phi_0\rangle$ is continuous in α, the corresponding integral equals $\int_{\Gamma\backslash\Gamma_\varepsilon}\ldots d\Gamma_y$ with $\alpha=0$ in the integrand. Together, we obtain (16c). QED

8.1.3.2 The Cauchy Principal Value for Surface Integrals

The Cauchy principal value $⨏_\Gamma f\,d\Gamma$ with respect to a singularity at $\mathbf{x}_0\in\Gamma$ is defined by (18), if this limit exists:

$$\text{(8.1.18)}\qquad ⨏_\Gamma f\,d\Gamma:=\lim_{\substack{\varepsilon\to0\\ \varepsilon>0}}\int_{\mathbf{y}\in\Gamma,\,|\mathbf{y}-\mathbf{x}_0|\geqslant\varepsilon}f(\mathbf{y})\,d\Gamma.$$

By analogy with Definition 7.1.15 and (7.1.13d) from Theorem 7.1.17, Cauchy's principal value can be characterised by means of a parametrisation.

Lemma 8.1.11. Assume that at $\mathbf{x}_0 \in \Gamma \subset \mathbb{R}^d$, f is Hölder continuous and Γ is Hölder continuously differentiable, i.e., $|f(\mathbf{x})-f(\mathbf{x}_0)| \le C|\mathbf{x}-\mathbf{x}_0|^\lambda$ $(\lambda>0)$ holds for all $\mathbf{x}\in\Gamma$, and there is a neighbourhood $U\subset\Gamma$ of $\mathbf{x}_0$ with a parametrisation $\varphi\in C^{1+\lambda}(V)$ mapping $V\subset\mathbb{R}^{d-1}$ one-to-one onto $U\subset\Gamma$. Without loss of generality, suppose $\varphi(0)=\mathbf{x}_0$. Outside of U, let $\Gamma\setminus U\in C^1_{\mathrm{stw}}$ and assume that f has at most weak singularities in $\Gamma\setminus U$. Then the Cauchy principal value

$$\text{(8.1.19a)}\qquad \oint_\Gamma f(\mathbf{y})\frac{\langle \mathbf{v},\mathbf{y}-\mathbf{x}_0\rangle}{|\mathbf{y}-\mathbf{x}_0|^d}\,d\Gamma_y \qquad \text{for any } \mathbf{v}\in\mathbb{R}^d$$

exists and coincides with the Cauchy principal value of the parametrisation

$$\text{(8.1.19b)}\qquad \oint_\Gamma f(\mathbf{y})\frac{\langle \mathbf{v},\mathbf{y}-\mathbf{x}_0\rangle}{|\mathbf{y}-\mathbf{x}_0|^d}\,d\Gamma_y =$$

$$\int_{\Gamma\setminus U} f(\mathbf{y})\frac{\langle \mathbf{v},\mathbf{y}-\mathbf{x}_0\rangle}{|\mathbf{y}-\mathbf{x}_0|^d}\,d\Gamma_y + \oint_V f(\varphi(\mathbf{t}))\sqrt{g(\mathbf{t})}\,\frac{\langle \mathbf{v},\varphi(\mathbf{t})-\mathbf{x}_0\rangle}{|\varphi(\mathbf{t})-\mathbf{x}_0|^d}\,d\mathbf{t},$$

where $g(\mathbf{t})$ is the Gram determinant. According to (18), the Cauchy principal value on the right-hand side is defined by

$$\text{(8.1.19c)}\qquad \oint_V \psi(\mathbf{t})\,d\mathbf{t} = \lim_{\substack{\tau\to 0\\ \tau>0}} \int_{V\setminus K_\tau(0)} \psi(\mathbf{t})\,d\mathbf{t}.$$

As usual, $K_\tau(0)=\{\mathbf{t}: |\mathbf{t}|<\tau\}$. Since another substitution $\mathbf{t}=\Phi(\mathbf{t}')$ in (19b) leads to a further parametrisation (19b), Lemma 11 implies that a substitution within the class $C^{1+\lambda}$ does not change the Cauchy principal value (cf. Lemma 7.1.6).

Proof of Lemma 11. (i) In the case of $d=2$, Theorem 7.1.17 and Lemma 7.1.6a can be applied (cf. Definition 7.1.15). Therefore we restrict our considerations to $d=3$. It suffices to consider $\Gamma=U$, so that the first term in (19b) vanishes. In particular, we can choose $U=\Gamma_\varepsilon := \Gamma\cap K_\varepsilon(\mathbf{x}_0)$.
(ii) By definition, the integral

$$I^1_\eta := \int_{\Gamma_\varepsilon\setminus\Gamma_\eta} f(\mathbf{y})\frac{\langle \mathbf{v},\mathbf{y}-\mathbf{x}_0\rangle}{|\mathbf{y}-\mathbf{x}_0|^3}\,d\Gamma_y \qquad (0<\eta\le\varepsilon)$$

tends to the Cauchy principal value as $\eta\to 0$, if this exists. On the other hand, we define

$$I^2_\eta := \int_{V\setminus K_\eta(0)} f(\varphi(\mathbf{t}))\sqrt{g(\mathbf{t})}\,\frac{\langle \mathbf{v},\varphi(\mathbf{t})-\mathbf{x}_0\rangle}{|\varphi(\mathbf{t})-\mathbf{x}_0|^3}\,d\mathbf{t} \qquad (0<\eta,\ K_\eta(0)\subset V).$$

(iii) First, we want to show the existence of $\lim I^2_\eta$ as $\eta\to 0$, i.e., the existence of Cauchy's principal value (19c). Because of the Cauchy criterion, $|I^2_{\eta'}-I^2_{\eta''}|\le\varepsilon$ is to be proved for all sufficiently small $\eta',\eta''\le\eta$. Without loss of generality, let $0<\eta'\le\eta''\le\eta$. Using polar coordinates (r,ϑ), we write the difference as

$$I^2_{\eta'}-I^2_{\eta''} = \int_{K_{\eta''}(0)\setminus K_{\eta'}(0)} \cdots\, d\mathbf{t} =$$

$$= \int_{-\pi}^{\pi}\left\{\int_{\eta'}^{\eta''} r\, f(\hat\varphi(r,\vartheta))\,\sqrt{\hat g(r,\vartheta)}\,\frac{\langle \mathbf{v},\hat\varphi(r,\vartheta)-\mathbf{x}_0\rangle}{|\hat\varphi(r,\vartheta)-\mathbf{x}_0|^3}\,dr\right\}d\vartheta,$$

where $\hat{\varphi}(r,\vartheta)=\varphi(\mathbf{t})$, $\mathbf{t}=r\binom{\cos\vartheta}{\sin\vartheta}$, $\hat{g}(r,\vartheta)=g(\mathbf{t})$. For abbreviation set

$$\hat{k}(r,\vartheta) = f(\hat{\varphi}(r,\vartheta))\sqrt{\hat{g}(r,\vartheta)}, \qquad k(\mathbf{t}) = f(\varphi(\mathbf{t}))\sqrt{g(\mathbf{t})},$$
$$\hat{\ell}(r,\vartheta) = \langle \boldsymbol{\nu}, \hat{\varphi}(r,\vartheta)-\mathbf{x}_0\rangle / |\hat{\varphi}(r,\vartheta)-\mathbf{x}_0|^3 = \ell(\mathbf{t}).$$

Perform the ϑ-integration separately over the intervals $[0,\pi]$ and $[-\pi,0]$:

(8.1.19d) $$I^2_{\eta'}-I^2_{\eta''} = \int_0^\pi \Big\{\int_{\eta'}^{\eta''} r\,[\hat{k}(r,\vartheta)\,\hat{\ell}(r,\vartheta)+\hat{k}(r,\vartheta-\pi)\,\hat{\ell}(r,\vartheta-\pi)]\,dr\Big\}\,d\vartheta.$$

Because of $\mathbf{t}=r\binom{\cos\vartheta}{\sin\vartheta}$ and of the Hölder continuity of f, g, φ, we have

$$|\hat{k}(r,\vartheta)-\hat{k}(r,\vartheta-\pi)| = |k(\mathbf{t})-k(-\mathbf{t})| \leqslant C_0|\mathbf{t}|^\lambda = C_0 r^\lambda,$$
$$|\varphi(\pm\mathbf{t})-\mathbf{x}_0| \geqslant r/C$$

and

$$|\hat{\ell}(r,\vartheta)+\hat{\ell}(r,\vartheta-\pi)| = \left|\frac{\langle\boldsymbol{\nu},\varphi(\mathbf{t})-\mathbf{x}_0\rangle}{|\varphi(\mathbf{t})-\mathbf{x}_0|^3} + \frac{\langle\boldsymbol{\nu},\varphi(-\mathbf{t})-\mathbf{x}_0\rangle}{|\varphi(-\mathbf{t})-\mathbf{x}_0|^3}\right| =$$
$$= \left|-\langle\boldsymbol{\nu},\varphi(-\mathbf{t})-\mathbf{x}_0\rangle\{|\varphi(\mathbf{t})-\mathbf{x}_0|^{-3} - |\varphi(-\mathbf{t})-\mathbf{x}_0|^{-3}\}\right| +$$
$$+\left|\frac{\langle\boldsymbol{\nu},\varphi(\mathbf{t})+\varphi(-\mathbf{t})-2\mathbf{x}_0\rangle}{|\varphi(\mathbf{t})-\mathbf{x}_0|^3}\right| \leqslant C_1\frac{r\,r^{1+\lambda}}{r^4}+C_2\frac{r^{1+\lambda}}{r^3} = C_3 r^{\lambda-2}.$$

Using $\varphi(\pm\mathbf{t})=\varphi(0)\pm\varphi'(0)\mathbf{t}+O(|\mathbf{t}|^{1+\lambda})$, $\xi^{-3}-\zeta^{-3}=-3(\xi-\zeta)/\tilde{\xi}^4$ with $\xi\leqslant\tilde{\xi}\leqslant\zeta$.

$$[\hat{k}(r,\vartheta)\,\hat{\ell}(r,\vartheta) + \hat{k}(r,\vartheta-\pi)\,\hat{\ell}(r,\vartheta-\pi)] =$$
$$[\hat{k}(r,\vartheta)-\hat{k}(r,\vartheta-\pi)]\,\hat{\ell}(r,\vartheta) + \hat{k}(r,\vartheta-\pi)\,[\hat{\ell}(r,\vartheta)+\hat{\ell}(r,\vartheta-\pi)],$$

we find the majorant $O(r^{\lambda-1})$ of the integrand in (19d). Hence,

$$|I^2_{\eta'}-I^2_{\eta''}| \leqslant \text{const}\int_0^\pi\int_{\eta'}^{\eta''} r^{\lambda-1}dr\,d\vartheta = \text{const}\,\tfrac{\pi}{\lambda}(\eta''^{\lambda}-\eta'^{\lambda})\to 0 \qquad \text{for } \eta',\ \eta''\to 0$$

is proved and therewith the existence of the Cauchy principal value (19c).
(iv) For the investigation of (19a), we have to estimate $I^1_\eta - I^2_\eta$. Polar coordinates yield

$$I^1_\eta - I^2_\eta = \int_{-\pi}^{\pi}\Big\{\int_{H(\eta,\vartheta)}^{\eta} r\,\hat{k}(r,\vartheta)\,\hat{\ell}(r,\vartheta)\,dr\Big\}\,d\vartheta,$$

where the lower limit $H(\eta,\vartheta)$ is defined implicitly by the equation $|\varphi(H(\eta,\vartheta),\vartheta)-\mathbf{x}_0|=\eta$, i.e., the inner boundary $|\mathbf{y}-\mathbf{x}_0|=\eta$ of $\Gamma_\varepsilon\backslash\Gamma_\eta$ is described in polar coordinates by $(H(\eta,\vartheta),\vartheta)$. $\varphi\in C^{1+\lambda}(V)$ implies $H\in C^{1+\lambda}([0,\varepsilon]\times[-\pi,\pi])$. It is easily verified that

$$H(0,\vartheta)=0, \qquad H(\eta,\vartheta)=\eta H_\eta(0,\vartheta)+O(\eta^{1+\lambda}), \qquad H_\eta(0,\vartheta)=1/\left|\varphi'(0)\binom{\cos\vartheta}{\sin\vartheta}\right|.$$

The same procedure as in (19d) leads to

$$I^1_\eta - I^2_\eta = \int_0^\pi\Big\{\int_{H(\eta,\vartheta-\pi)}^{\eta} r\,[\hat{k}(r,\vartheta)\,\hat{\ell}(r,\vartheta)+\hat{k}(r,\vartheta-\pi)\,\hat{\ell}(r,\vartheta-\pi)]\,dr\Big\}\,d\vartheta +$$
$$+\int_0^\pi\int_{H(\eta,\vartheta)}^{H(\eta,\vartheta-\pi)} r\,\hat{k}(r,\vartheta)\,\hat{\ell}(r,\vartheta)\,dr\,d\vartheta.$$

Since $H(\eta,\vartheta)=O(\eta)$, the considerations of part (iii) show that the first integral tends to zero. The integrand in the second integral is $\leqslant O(r^{-1})$ and, because of $\lim H(\eta,\vartheta)/\eta=\lim H(\eta,\vartheta-\pi)/\eta=H_\eta(0,\vartheta)$ (see above), leads to

$$\Big|\int_{H(\eta,\vartheta)}^{H(\eta,\vartheta-\pi)} r\,\hat{k}(r,\vartheta)\,\hat{\ell}(r,\vartheta)\,dr\Big| \leqslant \Big|C\int_{H(\eta,\vartheta)}^{H(\eta,\vartheta-\pi)} r^{-1}dr\Big| = C\,\Big|\log\frac{H(\eta,\vartheta-\pi)}{H(\eta,\vartheta)}\Big|\to 0.$$

The just proved fact that $I_\eta^1-I_\eta^2\to 0$ and the convergence of the integral I_η^2 to the Cauchy principal value show that also I_η^1 converges as $\eta\to 0$ and has the same limit. QED

If strongly singular integrands depend on a parameter, one must carefully check whether the Cauchy principal value is a continuous function of this parameter. The following lemma discusses this subject.

Lemma 8.1.12. **(a)** Let $\Omega\subset\mathbb{R}^{d-1}$ be an *open* and bounded subset.

$$F(\mathbf{y}_0,\alpha):=\oint_\Omega\frac{\langle\boldsymbol{v},\mathbf{y}-\mathbf{y}_0\rangle\,d\mathbf{y}}{(|\mathbf{y}-\mathbf{y}_0|^2+\alpha^2)^{d/2}}\qquad\text{with }\boldsymbol{v}\in\mathbb{R}^{d-1}\tag{8.1.20a}$$

is a continuous (even analytical) function of $\mathbf{y}_0\in\Omega$ and $\alpha\in\mathbb{R}$. Here, the integral (20a) is to be interpreted as Cauchy principal value only if $\alpha=0$.

(b) Let $\Gamma\in C^1_{\text{stw}}$ be a (compact) $(d-1)$-dimensional manifold in $\mathbb{R}^d$. Suppose that P is a compact parameter set and that the mapping $p\in P\mapsto\mathbf{y}_p\in\Gamma$ is continuous. Assume that the functions $f(p,\mathbf{y})$ and $g(p,\mathbf{y})$ are continuous in $p\in P$, $\mathbf{y}\in\Gamma\setminus\{\mathbf{y}_p\}$ and that f has a possible singularity at $\mathbf{y}_p$ and admits a Cauchy principal value $F(p)=\oint_\Gamma f(p,\mathbf{y})d\Gamma_y$, which is supposed to be continuous in $p\in P$. Furthermore, the estimate $|f(p,\mathbf{y})-g(p,\mathbf{y})|\leqslant C|\mathbf{y}-\mathbf{y}_p|^{\lambda+1-d}$ must hold with $\lambda>0$. Then, $G(p)=\oint_\Gamma g(p,\mathbf{y})d\Gamma_y$ is also continuous in $p\in P$.

(c) Let $\lambda>0$ and $\varphi\in C(\Omega)$ be a mapping of a compact subset $\Omega\subset\mathbb{R}^{d-1}$ into $\mathbb{R}^{d-1}$. Assume the inequality

$$|\varphi(\mathbf{y})-\varphi(\mathbf{y}_0)|\geqslant|\mathbf{y}-\mathbf{y}_0|/C\qquad\text{for all }\mathbf{y}\in\Omega,\ \mathbf{y}_0\in\Omega_0\subset\Omega,$$

where Ω_0 is open. Furthermore, let $\varphi\in C^{1+\lambda}(\Omega_0)$, $\Phi\in C(\Omega)\cap C^\lambda(\Omega_0)$. The mapping $\boldsymbol{v}$: $\Omega\to\mathbb{R}^{d-1}$ must belong to $C(\Omega)\cap C^\lambda(\Omega_0)$. Then the function (20b) below is continuous in $\mathbf{y}_0\in\Omega_0$ and $\alpha\in\mathbb{R}$:

$$F(\mathbf{y}_0,\alpha)=\oint_\Omega\Phi(\mathbf{y})\frac{\langle\boldsymbol{v}(\mathbf{y}),\varphi(\mathbf{y})-\varphi(\mathbf{y}_0)\rangle}{[|\varphi(\mathbf{y})-\varphi(\mathbf{y}_0)|^2+\alpha^2]^{d/2}}\,d\mathbf{y}\tag{8.1.20b}$$

(d) Let Φ be continuous on Ω and Hölder continuous with exponent $\lambda>0$ at $0\in\Omega$. The function $\varkappa(\mathbf{y},\alpha)$ has to satisfy the inequality

$$|\varkappa(\mathbf{y},\alpha)|\leqslant C[|\mathbf{y}|^2+\alpha^2]^{(1+\lambda)/2}\qquad\text{for }\mathbf{y}\in\Omega,\ \alpha\in\mathbb{R}.\tag{8.1.20c}$$

Then the following function is continuous in $\alpha\in\mathbb{R}$:

$$F(\alpha)=\oint_\Omega\Phi(\mathbf{y})\frac{\langle\boldsymbol{v},\mathbf{y}\rangle}{[|\mathbf{y}|^2+(\alpha-\varkappa(\mathbf{y},\alpha))^2]^{d/2}}\,d\mathbf{y}.\tag{8.1.20d}$$

Proof. (a) Let $\varepsilon>0$ be so small that $K_\varepsilon(\mathbf{y}_0)\subset\Omega$. The integral over $\Omega\setminus K_\varepsilon(\mathbf{y}_0)$ is continuous with respect to α, while the integral (the Cauchy principal value in the case of $\alpha=0$) over $K_\varepsilon(\mathbf{y}_0)$ vanishes by symmetry reasons. Concerning the $\mathbf{y}_0$-dependence, choose the coordinate system such that $\boldsymbol{v}$ is a multiple of the first unit vector and, therefore, $\langle \mathbf{v}, \mathbf{y}-\mathbf{y}_0\rangle = v_1(y_1-y_{0,1})$ holds. Take $\varepsilon>0$ so small that $S_\varepsilon := \{\mathbf{y}\in\mathbb{R}^{d-1}: |y_j-\hat{y}_j|\le\varepsilon\}\subset\Omega$ for some $\hat{\mathbf{y}}\in\Omega$. The integral over $\Omega\backslash S_\varepsilon$ is continuous with respect to $\mathbf{y}_0$, as long as $\mathbf{y}_0\in S_\varepsilon$. The integral over S_ε can be represented explicitly:

$$\oint_{S_\varepsilon}\dots d\mathbf{y} = \tfrac{1}{2}v_1\int_{-\varepsilon}^{\varepsilon}\cdots\int_{-\varepsilon}^{\varepsilon}[J(\hat{y}_1+\varepsilon;\mathbf{y}+\hat{\mathbf{y}}-\mathbf{y}_0)-J(\hat{y}_1-\varepsilon;\mathbf{y}+\hat{\mathbf{y}}-\mathbf{y}_0)]\,dy_2\dots dy_{d-1}$$

for $d>2$, where $J(\eta;\mathbf{z}):=(1-\frac{d}{2})[(\eta-y_{0,1})^2+\alpha^2+z_2^2+\dots+z_{d-1}^2]^{1-\frac{d}{2}}$. The S_ε-integral is obviously continuous (analytical) with respect to $\mathbf{y}_0\in S_\varepsilon$.
(b) $\int[f(p,\mathbf{y})-g(p,\mathbf{y})]d\mathbf{y}$ exists uniformly as an improper integral. This implies continuity in p (cf. Remark 4, Lemma 5).
(c) The proof is divided in four steps. The parameter $p\in P$ is given by $(\mathbf{y}_0,\alpha)\in\Omega_{00}\times\mathbb{R}$, where $\Omega_{00}\subset\Omega_0$ is an arbitrary compact subset.

1st Step: The substitution $\mathbf{z}=\mathbf{y}_0+A(\mathbf{y}_0)(\mathbf{y}-\mathbf{y}_0)$ with continuous and non-singular $A(\mathbf{y}_0)$ shows that $\mathbf{y}-\mathbf{y}_0$ may be replaced in (20a) by $A(\mathbf{y}_0)(\mathbf{y}-\mathbf{y}_0)$. The integral $F(\mathbf{y}_0,\alpha)$ which arises is smooth in all parameters.

2nd Step: Choose $A(\mathbf{y}_0)$ as Jacobian matrix $\varphi'(\mathbf{y}_0)$ and set

$$f(\mathbf{y}_0,\alpha,\mathbf{y}) := \langle\mathbf{v},\varphi'(\mathbf{y}_0)(\mathbf{y}-\mathbf{y}_0)\rangle/(|\varphi'(\mathbf{y}_0)(\mathbf{y}-\mathbf{y}_0)|^2+\alpha^2)^{d/2},$$

$$g(\mathbf{y}_0,\alpha,\mathbf{y}) := \langle\mathbf{v},\varphi(\mathbf{y})-\varphi(\mathbf{y}_0)\rangle\,/\,(|\varphi(\mathbf{y})-\varphi(\mathbf{y}_0)|^2+\alpha^2)^{d/2}.$$

The inequality $|\varphi(\mathbf{y})-\varphi(\mathbf{y}_0)-\varphi'(\mathbf{y}_0)(\mathbf{y}-\mathbf{y}_0)|\le C|\mathbf{y}-\mathbf{y}_0|^{\lambda+1}$ admits the estimate of $f(\mathbf{y}_0,\alpha,\mathbf{y})-g(\mathbf{y}_0,\alpha,\mathbf{y})$ by $C|\mathbf{y}-\mathbf{y}_p|^{\lambda+1-d}$ with $\mathbf{y}_p=\mathbf{y}_0$. Part (b) with $\Gamma:=\Omega$ proves the continuity of the integral over g.

3rd Step: When $\mathbf{v}$ is replaced by a vector function $\mathbf{v}(\mathbf{y}_0)$ and a further factor $\Phi(\mathbf{y}_0)$ is inserted, this does not affect the integration, since these functions can be taken outside of the integral sign. Therefore the integral over $f(\mathbf{y}_0,\alpha,\mathbf{y}):=\Phi(\mathbf{y}_0)\langle\mathbf{v}(\mathbf{y}_0),\varphi(\mathbf{y})-\varphi(\mathbf{y}_0)\rangle/(|\varphi(\mathbf{y})-\varphi(\mathbf{y}_0)|^2+\alpha^2)^{d/2}$ is also continuous in $\mathbf{y}_0$ and α.

4th Step: Let $g(\mathbf{y}_0,\alpha,\mathbf{y})$ be the previous function f with $\Phi(\mathbf{y}_0)$ replaced by $\Phi(\mathbf{y})$. Because of $\Phi(\mathbf{y}_0)-\Phi(\mathbf{y})=O(|\mathbf{y}-\mathbf{y}_0|^\lambda)$, $f-g$ again satisfies the assumption of part (b), so that the integral over g is continuous in $\mathbf{y}_0$ and α. Similarly, $\mathbf{v}(\mathbf{y}_0)$ can be replaced by $\mathbf{v}(\mathbf{y})$.
(d) Let f and g be the integrand of (20d) with and without $\varkappa$-term. By the considerations from above, the integral over g is continuous in α. Inequality (20c) together with the mean value theorem shows $f-g=O(|\mathbf{y}|^{\lambda+1-d})$, so that part (b) yields the assertion. QED

B.1.3.3 Other Directional Derivatives

Let $\mathbf{x}_0\in\Gamma$ be a point of the surface and $\mathbf{t}$ a *tangential direction* at $\mathbf{x}_0$, i.e., $\mathbf{t}\perp\mathbf{n}(\mathbf{x}_0)$ and $|\mathbf{t}|=1$. We shall investigate how the directional derivative (21) behaves when $\mathbf{x}\notin\Gamma$ approaches $\mathbf{x}_0\in\Gamma$:

$$\text{(B.1.21)}\qquad \langle\mathbf{t},\nabla\Phi(\mathbf{x})\rangle = -\frac{1}{\omega_d}\int_\Gamma\frac{\langle\mathbf{t},\mathbf{x}-\mathbf{y}\rangle}{|\mathbf{x}-\mathbf{y}|^d}f(\mathbf{y})\,d\Gamma_y \qquad (\mathbf{x}\notin\Gamma).$$

Lemma 8.1.13. In addition to the assumptions of Theorem 9, let f be Hölder continuous at $\mathbf{x}_0 \in \Gamma$. Then for any sequence $\mathbf{x} \to \mathbf{x}_0$ with

(8.1.22a) $\quad |\mathbf{x}-\mathbf{x}_0| \geqslant \operatorname{dist}(\mathbf{x},\Gamma)/C \qquad$ (C fixed)

the following statement holds:

(8.1.22b) $$\lim_{\mathbf{x}\to\mathbf{x}_0} \langle \mathbf{t}, \nabla\Phi(\mathbf{x})\rangle = -\frac{1}{\omega_d} \oint_\Gamma \frac{\langle \mathbf{t}, \mathbf{x}_0-\mathbf{y}\rangle}{|\mathbf{x}_0-\mathbf{y}|^d} f(\mathbf{y})\,d\Gamma_y .$$

Since $\mathbf{x}$ may be on either side of Γ, any *tangential derivative* $\langle \mathbf{t}, \nabla\Phi(\mathbf{x})\rangle$ is *continuous* when passing $\mathbf{x}_0 \in \Gamma$.

Proof. As in the proof of Theorem 9, we assume, without loss of generality, $\mathbf{x}_0 = 0$ and $n_j = 0\ (1 \leqslant j \leqslant d-1)$, $n_d = 1$. Thereby, the tangential direction has the form $\mathbf{t} = \binom{\mathbf{t}'}{0}$ with $\mathbf{t}' \in \mathbb{R}^{d-1}$. Analogously, we split the vector $\mathbf{x} \in \mathbb{R}^d$ of the sequence $\mathbf{x} \to \mathbf{x}_0$ into the components $\mathbf{x} = \binom{\mathbf{x}'}{x_d}$, $\mathbf{x}' \in \mathbb{R}^{d-1}$, $x_d \in \mathbb{R}$. Using the notation $\mathbf{x}'$, $\mathbf{t}'$ from above, we write (21) in the following form:

$$-\frac{1}{\omega_d}\int_\Gamma \frac{\langle \mathbf{t}', \mathbf{x}'-\mathbf{y}'\rangle}{|\mathbf{x}-\mathbf{y}|^d} f(\mathbf{y})\,d\Gamma_y \qquad \left(\mathbf{x} \notin \Gamma,\ \mathbf{y} = \binom{\mathbf{y}'}{\eta(\mathbf{y}')}\right).$$

$|\mathbf{x}-\mathbf{y}|^d$ can be written as $[\,|\mathbf{x}'-\mathbf{y}'|^2 + |x_d - \eta(\mathbf{y}')|^2]^{d/2}$. Because of inequality (22a), $\mathbf{x}'$ and x_d do not vary independently: We can replace x_d by $\alpha \in \mathbb{R}$ and $\mathbf{x}'$ by $\alpha\mathbf{a}$, where $\mathbf{a} \in \mathbb{R}^{d-1}$ satisfies the condition $|\mathbf{a}| \leqslant C'$. In a neighbourhood of $\mathbf{x}_0 = 0$, the integral from above can be parametrised as follows:

$$-\frac{1}{\omega_d}\int_{S_\varepsilon} \frac{\langle \mathbf{t}', \alpha\mathbf{a}-\mathbf{y}'\rangle}{[\,|\alpha\mathbf{a}-\mathbf{y}'|^2 + (\alpha-\eta(\mathbf{y}'))^2]^{d/2}}\,\Phi(\mathbf{y}')\,d\mathbf{y}',$$

where $\Phi(\mathbf{y}') = f(\mathbf{y})\sqrt{g(\mathbf{y}')}$, $S_\varepsilon = K_\varepsilon(0)$. The substitution $\mathbf{y}' = \mathbf{z} + \alpha\mathbf{a}$ yields the integral over $\Phi(\mathbf{z}+\alpha\mathbf{a})\langle \mathbf{t}', \mathbf{z}\rangle / [\,|\mathbf{z}|^2 + (\alpha - \eta(\mathbf{z}+\alpha\mathbf{a}))^2]^{d/2}$. Lemma 12b is applicable to the difference with $\Phi(0)\langle \mathbf{t}', \mathbf{z}\rangle / [\,|\mathbf{z}|^2 + (\alpha-\eta(\mathbf{z}+\alpha\mathbf{a}))^2]^{d/2}$. In the latter integrand, $\varkappa(\mathbf{z},\alpha) := \eta(\mathbf{z}+\alpha\mathbf{a})$ satisfies condition (20c) because of (17f), so that the continuity in α follows. ∎

Let $\{\mathbf{t}_1, \dots, \mathbf{t}_{d-1}\}$ be an orthonormal basis of the tangential plane at $\mathbf{x}_0 \in \Gamma$. Since $\mathbf{n}$ and $\mathbf{t}_1, \dots, \mathbf{t}_{d-1}$ form an orthonormal basis of $\mathbb{R}^d$, the vector $\mathbf{x}_0 - \mathbf{y}$ can be written as $\mathbf{n}\langle \mathbf{n}, \mathbf{x}_0-\mathbf{y}\rangle + \sum_{j=1}^{d-1} \mathbf{t}_j \langle \mathbf{t}_j, \mathbf{x}_0-\mathbf{y}\rangle$. The combination of Theorem 9 and Lemma 13 yields

Corollary 8.1.14. Let $\mathbf{n}(\mathbf{x}_0)$ be the normal direction at $\mathbf{x}_0 \in \Gamma$ pointing into the exterior domain. Under the assumptions of Lemma 13, the gradient $\nabla\Phi$ of the single-layer potential has the one-sided limits $\nabla\Phi_-$ from inside and $\nabla\Phi_+$ from outside at $\mathbf{x}_0 \in \Gamma$. The *jump relation*

(8.1.23a) $\quad \nabla\Phi_+(\mathbf{x}_0) - \nabla\Phi_-(\mathbf{x}_0) = -f(\mathbf{x}_0)\,\mathbf{n}(\mathbf{x}_0)$

is valid. The mean value equals

(8.1.23b) $$\tfrac{1}{2}[\nabla\Phi_+(\mathbf{x}_0) + \nabla\Phi_-(\mathbf{x}_0)] = -\frac{1}{\omega_d}\oint_\Gamma \frac{\mathbf{x}_0-\mathbf{y}}{|\mathbf{x}_0-\mathbf{y}|^d} f(\mathbf{y})\,d\Gamma_y .$$

The continuity of $\nabla\Phi_\pm$ on Γ is the subject of

Remark 8.1.15. **(a)** Let $\Gamma_0 \subset \Gamma \in C^1_{0,\text{stw}}$ be open with $\Gamma_0 \in C^{1+\lambda}$ $(\lambda > 0)$. Furthermore, suppose $f \in C^\lambda(\Gamma_0) \cap L^\infty(\Gamma)$. Let $U \subset \mathbb{R}^d$ be a simply connected set, which is divided by Γ into two disjoint parts U_+ and U_-, where $\Gamma \cap \overline{U}_\pm \subset \Gamma_0$ must hold. Then, $\nabla\Phi_\pm$ is continuous in $\overline{U}_\pm$, where $\nabla\Phi_\pm$ are defined on $\Gamma_0 \subset \overline{U}_\pm$ as continuous extensions.
(b) $\nabla\Phi(\mathbf{x})$ becomes singular, when $\mathbf{x}$ approaches a point $\mathbf{x}_0 \in \Gamma$ being a corner $(d=2)$ or an edge $(d=3)$ of Γ.
(c) Colton-Kreß [1, Theorem 2.17] prove for a closed smooth surface $\Gamma \in C^2_0$ that the single-layer potential is *globally Hölder continuously differentiable* with the same exponent as the density f:

$$\|\Phi\|_{C^{1+\lambda}(\overline{\Omega}_\pm)} \leqslant C\,\|f\|_{C^\lambda(\Gamma)}.$$

Proof. (a) follows from Lemma 12c. For (b) use the following Example 16. (c) is closely related to Theorem 8.2.10. ∎

Example 8.1.16. Let $\Gamma \subset \mathbb{R}^2$ be the unit square with the corners $(0,0)$, $(1,0)$, $(1,1)$, $(0,1)$. Choose the single-layer density $f=1$. For all $\mathbf{x}=(x_1,x_2)$ different from the corners, $\nabla\Phi(\mathbf{x})$ has the explicit representation

$$\frac{\partial\Phi(\mathbf{x})}{\partial x_1} = \frac{1}{2\pi}\left\{\log\sqrt{\frac{x_2^2+(1-x_1)^2}{x_1^2+x_2^2}\times\frac{(1-x_2)^2+(1-x_1)^2}{x_1^2+(1-x_2)^2}}\; + \right.$$
$$\left. +\operatorname{arctg}\frac{x_2}{x_1} + \operatorname{arctg}\frac{1-x_2}{x_1} + \operatorname{arctg}\frac{1-x_2}{1-x_1} + \operatorname{arctg}\frac{x_2}{1-x_1}\right\},$$
$$\frac{\partial\Phi(x_1,x_2)}{\partial x_2} = \frac{\partial\Phi(x_2,x_1)}{\partial x_1}.$$

Obviously, the gradient $\nabla\Phi$ has a logarithmic singularity at the corners.

8.1.4 Formulation of the Dirichlet Boundary Value Problem as Integral Equation of the First Kind for the Single-Layer Potential

8.1.4.1 The Interior and Exterior Problem of the Laplace Equation

In the following, we need statements about the uniqueness of solutions u of the Laplace equation $\Delta u = 0$ subject to given boundary data. Let $\Gamma \subset \mathbb{R}^d$ be the boundary of the interior domain Ω_-. The *normal derivative* of u is denoted by

$$u_n(\mathbf{x}) = \partial u(\mathbf{x})/\partial n := \langle \mathbf{n}(\mathbf{x}), \nabla u(\mathbf{x})\rangle \qquad (\mathbf{x}\in\Gamma),$$

where, in the case of the interior [exterior] problem, ∇u is the one-sided limit ∇u_- [∇u_+] (cf. (23a)).

We study the following boundary value problems for Laplace's equation:

interior Dirichlet problem:	$\Delta u = 0$	in Ω_-,	$u = \varphi$	on Γ,
interior Neumann problem:	$\Delta u = 0$	in Ω_-,	$u_n = \varphi$	on Γ,
exterior Dirichlet problem:	$\Delta u = 0$	in Ω_+,	$u = \varphi$	on Γ,
exterior Neumann problem:	$\Delta u = 0$	in Ω_+,	$u_n = \varphi$	on Γ.

Green's formula over a bounded domain Ω reads

(8.1.24a) $\int_\Omega \langle \nabla u, \nabla v\rangle d\mathbf{x} = \int_\Gamma \frac{\partial u}{\partial n} v\, d\Gamma - \int_\Omega v\Delta u\, d\mathbf{x}$ $\qquad (\Gamma = \partial\Omega)$.

For a solution of $\Delta u = 0$, the choice $v = u$ or $v = 1$ yields

(8.1.24b) $\int_\Omega \langle \nabla u, \nabla u\rangle d\mathbf{x} = \int_\Gamma \frac{\partial u}{\partial n} u\, d\Gamma,$

(8.1.24c) $0 = \int_\Gamma \frac{\partial u}{\partial n}\, d\Gamma.$

To get valid conclusions from identity (24b), it must implicitly be assumed that only those functions u are considered that lead to a *finite* integral $\int_\Omega \langle \nabla u, \nabla u\rangle d\mathbf{x}$. The latter requirement defines the functions of the Sobolev space $H^1(\Omega)$ (cf. Definition 4.5.23, Hackbusch [2,§6.2.2]). The gradient $\nabla\Phi$ of the single-layer potential of a bounded density is uniformly bounded except the neighbourhood of corners or edges. If r denotes the distance from a corner [edge] point $\mathbf{x}_0 \in \Gamma$, one can show $|\nabla\Phi| = O(|\log r|)$ (cf. Example 16). Consequently, $\int_\Omega \langle \nabla\Phi, \nabla\Phi\rangle d\mathbf{x}$ is finite and Φ belongs to $H^1(\Omega)$.

Corollary 8.1.17. **(a)** *The interior Dirichlet problem* has at most one solution. **(b)** Two solutions of the *interior Neumann problem* differ only by a constant.

Proof. Let u_1, u_2 be two solutions. $u := u_1 - u_2$ is a solution for the respective boundary condition $u = 0$ or $u_n = 0$. In both cases, it follows from (24b) that $\nabla u = 0$ in Ω; hence, u=const. In the case of (a), the boundary condition enforces const$=0$. QED

To apply Green's identity (24b) in the classical sense, one has to remove possible corners $\mathbf{x}_0 \in \Gamma$ in Ω together with an ε-neighbourhood: In $\Omega \backslash K_\varepsilon(\mathbf{x}_0)$, (24b) is applicable in the classical sense. The boundary integral over $\Omega \cap \partial K_\varepsilon(\mathbf{x}_0)$ vanishes as $\varepsilon \to 0$, since $\varepsilon \log \varepsilon \to 0$ (see hint before Corollary 17).

Lemma 8.1.18. **(a)** The *exterior* Dirichlet or Neumann solution is unique if

(8.1.25a) $|u(\mathbf{x})| \leqslant O(1/|\mathbf{x}|)$ $\qquad$ as $|\mathbf{x}| \to \infty$,

(8.1.25b) $|\nabla u(\mathbf{x})| = O(|\mathbf{x}|^{1-d})$ $\qquad$ as $|\mathbf{x}| \to \infty$.

(b) Condition (25b) is always satisfied for the single-layer potential (9).

(c) Condition (25a) holds for the single-layer potential (9), if $d \geqslant 3$ or if for $d = 2$ the additional condition (13) is satisfied.

(d) For all $d \geqslant 2$, (13) implies the strengthened estimate

(8.1.25c) $|\nabla u(\mathbf{x})| = O(|\mathbf{x}|^{-d})$ $\qquad$ as $|\mathbf{x}| \to \infty$.

(e) The *exterior* Dirichlet and Neumann problems have unique solution, if (25c) and $|u(\mathbf{x})| = O(1)$ hold.

Proof. The difference $u := u_1 - u_2$ of two solution satisfies $u = 0$ or $u_n = 0$ on Γ. Let Ω_R be the annular domain $\Omega_R := \{\mathbf{x} \in \Omega_+ : |\mathbf{x}| < R\}$ for $R > \max\{|\mathbf{x}| : \mathbf{x} \in \Gamma\}$. The boundary of Ω_R consists of Γ and $\gamma := \{|\mathbf{x}| = R\}$. Equation (24b) becomes

$$\int_{\Omega_R} \langle \nabla u, \nabla u\rangle d\mathbf{x} = \int_\Gamma u_n u\, d\Gamma + \int_\gamma u_n u\, d\Gamma,$$

where the first term on the right-hand side vanishes because of $u=0$ or $u_n=0$. The estimates (25a,b) together with $\int_\gamma d\Gamma = \mu(\gamma) = R^{d-1}\omega_d$ yield $\int_{\Omega_R} |\nabla u|^2 d\mathbf{x} = O(1/R)$. Hence, $\int_{\Omega_+} |\nabla u|^2 d\mathbf{x} = 0$ implies $\nabla u = 0$, and thereby u=const. $u(\infty)=0$ yields $u=0$ and therefore the uniqueness. Parts (c), (d) are proved similarly as in the second part of Theorem 7. QED

For later use, we formulate a *necessary* condition for the solvability of the Neumann boundary value problem.

Lemma 8.1.19. **(a)** The normal derivative $\varphi = u_n$ of a solution of the interior Laplace equation satisfies the condition

(8.1.26) $$\int_\Gamma \varphi \, d\Gamma = 0.$$

(b) The same holds for the solution of the exterior Laplace equation, if it satisfies (25c).

Proof. (a) follows from (24c). For part (b) replace Ω_+ by Ω_R as before and show $\int_\gamma u_n d\Gamma = O(1/R) \to 0$. QED

8.1.4.2 The Integral Equation of the First Kind

By Lemma 3, the single-layer potential Φ satisfies the Laplace equation. To enforce the Dirichlet boundary condition $\Phi = \varphi$ on the boundary Γ (cf. (7.4.1b)), we have to satisfy the additional condition

(8.1.27) $$\int_\Gamma s(\mathbf{x},\mathbf{y}) f(\mathbf{y}) d\Gamma_y = \varphi(\mathbf{x}) \qquad \text{for } \mathbf{x} \in \Gamma$$

for the interior as well as for the exterior problem. (27) is a *Fredholm integral equation of the first kind* with the weakly singular kernel $s(\mathbf{x},\mathbf{y})$. The uniqueness of a solution f of this integral equation is ensured by

Theorem 8.1.20 (uniqueness). Let $\Gamma \in C^{1+\lambda}_{0,\mathrm{stw}}$ with $\lambda > 0$. Amongst all functions $f \in L^\infty(\Gamma)$ which are continuous almost everywhere (and in the two-dimensional case $d=2$ also satisfy condition (13), i.e., $\int_\Gamma f\, d\Gamma = 0$), there is at most one solution of problem (27).

Proof. If f_1 and f_2 are two solutions of (27), then $f := f_1 - f_2 \in L^\infty(\Gamma)$ solves problem (27) with $\varphi = 0$. Let Φ be the single-layer potential generated by f. It satisfies the Laplace equation $\Delta\Phi = 0$ in the interior and exterior domain $\Omega_\mp$. Because of the continuity of Φ in Γ (cf. Theorem 7), Φ takes the boundary values $\Phi = \varphi = 0$ in Γ and vanishes at ∞. Unique solutions of these interior and exterior problems are $\Phi = 0$ in $\mathbb{R}^d$. In almost all $\mathbf{x}_0 \in \Gamma$, f is continuous and condition (15) is satisfied. Due to Theorem 9, the normal derivatives $\partial\Phi_\pm/\partial n = 0$ imply $f(\mathbf{x}_0) = 0$ (cf. (16b)). QED

Note that $d=2$ plays an exceptional role. Without condition (13), uniqueness does not hold:

Example 8.1.21. Let $d=2$ and $\Gamma\subset\mathbb{R}^2$ be the unit circle. **(a)** The constant density $f=1$ satisfies equation (27) with $\varphi=0$.
(b) The single-layer potential generated by $f=1$ is $\Phi(\mathbf{x}) = \min(0,-\log|\mathbf{x}|)$, i.e., $\Phi=0$ in the interior domain $|\mathbf{x}|\leqslant 1$ and $\Phi=-\log|\mathbf{x}|$ in the exterior domain $|\mathbf{x}|\geqslant 1$.
(c) The only bounded solution in the exterior domain corresponding to the Dirichlet data $\varphi=1$ on Γ is the function $\Phi_+=1$. The unbounded ones are $1+c\log|\mathbf{x}|$ $(c\neq 0)$.
(d) Equation (27) is not solvable for $\varphi=1$.

Proof of (d). Suppose that f is a solution for $\varphi=1$ and set $c:=\int_\Gamma f\,d\Gamma/2\pi$. By part (b), $g:=f-c$ also solves equation (27) with $\varphi=1$. Since g satisfies (13), the single-layer potential generated by g fulfils $\Phi_+(\infty)=0$ (cf. Lemma 18c) in contradiction to $\Phi_+=1$ (cf. part c). QED

Theorem 8.1.22 (existence). Let $\Gamma\in C_0^{1+\lambda}$ with $0<\lambda<1$. In the case of $d\geqslant 3$, the integral equation (27) has a solution $f\in C^\lambda(\Gamma)$ for any $\varphi\in C^{1+\lambda}(\Gamma)$. For $d=2$, φ has to satisfy an additional condition which can be written in the form $\int_\Gamma \xi\varphi\,d\Gamma=0$ with suitable ξ.

Proof. Let $d\geqslant 3$. For the Dirichlet boundary value $\varphi\in C^{1+\lambda}(\Gamma)$, there is a solution $\Phi_-\in C^{1+\lambda}(\Omega_-)$ of the interior problem possessing a normal derivative $\partial\Phi_-/\partial n\in C^\lambda(\Gamma)$ (cf. Hackbusch [2, Theorem 9.1.20]). Similarly, an exterior solution $\Phi_+\in C^{1+\lambda}(\overline{\Omega}_+)$ exists satisfying $\Phi_+(\infty)=0$ and yielding $\partial\Phi_+/\partial n\in C^\lambda(\Gamma)$. Set $f:=-\partial\Phi_+/\partial n+\partial\Phi_-/\partial n\in C^\lambda(\Gamma)$. Let Ψ be the single-layer potential generated by f. By Theorem 9, $f=-\partial\Psi_+/\partial n+\partial\Psi_-/\partial n$ holds. The difference $\Phi-\Psi$ satisfies the Laplace equation in $\Omega_\pm$. In Γ, $\Phi-\Psi$ as well as the normal derivative $\partial(\Phi-\Psi)/\partial n$ are continuous. This implies $\Delta(\Phi-\Psi)=0$ in the whole $\mathbb{R}^d$ (even in the classical sense). Because of $\Phi(\infty)=\Psi(\infty)=0$, the only solution is $\Phi=\Psi$. On Γ, $\Psi=\Phi=\varphi$ holds; hence, f is a solution of (27).

In the case of $d=2$, one finds again a solution $\Phi_+\in C^{1+\lambda}(\Omega_+)$ in the exterior domain, which is continuous in ∞. As soon as $\Phi_+(\infty)=0$ is satisfied, the argumentation from above can be repeated. By the Riesz representation of the functional $\varphi\mapsto\Phi_+(\infty)$ as $\int_\Gamma\xi\varphi\,d\Gamma$ (cf. Hackbusch [2,§6.3.2]), we obtain the assertion. QED

In a certain sense, Theorem 22 is the converse of Remark 15c. The latter implies that $f\in C^\lambda(\Gamma)$ leads to a single-layer potential $\Phi_\pm\in C^{1+\lambda}(\overline{\Omega}_\pm)$. Therefore, the Dirichlet data $\varphi=\Phi_{|\Gamma}$ also belong to $C^{1+\lambda}(\Gamma)$.

8.1.5 Formulation of the Neumann Boundary Value Problem as Integral Equation of the Second Kind for the Single-Layer Potential

In the following, we are looking for the solution u of the Laplace equation in the interior or exterior domain subject to the *Neumann boundary condition*

$$\partial u(\mathbf{x})/\partial n = \varphi(\mathbf{x}) \qquad \text{for } \mathbf{x}\in\Gamma. \tag{8.1.28}$$

Obviously, the normal derivative can be required only where a normal direction exists. If Γ is only piecewise smooth, (28) is not required for the corners and edges. We note that the boundary condition (28) can be formulated without smoothness assumptions on Γ, if the variational formulation is used (cf. Hackbusch [2, §7.4]).

First, we consider the *interior problem* and try to define u as the single-layer potential (9):

(8.1.29) $$u(\mathbf{x}) = \Phi(\mathbf{x}) := \int_\Gamma s(\mathbf{x},\mathbf{y}) f(\mathbf{y})\, d\Gamma_y \quad \text{for } \mathbf{x}\in\Omega_- \text{ (interior domain).}$$

By definition, $\mathbf{n}(\mathbf{x})$ is the outer normal direction at $\mathbf{x}\in\Gamma$. Elimination of $\partial\Phi_+/\partial n$ in the equations (16b,c) yields

$$2\,\frac{\partial\Phi_-(\mathbf{x})}{\partial n} = f(\mathbf{x}) - \frac{2}{\omega_d}\int_\Gamma \frac{\langle \mathbf{n}(\mathbf{x}), \mathbf{x}-\mathbf{y}\rangle}{|\mathbf{x}-\mathbf{y}|^d} f(\mathbf{y})\, d\Gamma_y .$$

The normal derivative $\partial u(\mathbf{x})/\partial n = \varphi(\mathbf{x})$ must coincide with $\partial\Phi_-/\partial n$ from (16a), so that condition (28) leads to

(8.1.30) $$f(\mathbf{x}) = 2\varphi(\mathbf{x}) + \frac{2}{\omega_d}\int_\Gamma \frac{\langle \mathbf{n}(\mathbf{x}),\, \mathbf{x}-\mathbf{y}\rangle}{|\mathbf{x}-\mathbf{y}|^d} f(\mathbf{y})\, d\Gamma_y \qquad (\mathbf{x}\in\Gamma).$$

In (7.4.7a), we have defined the double-layer kernel $k(\mathbf{x},\mathbf{y})$ for $d=2$, which for all $d\geq 2$ takes the form

(8.1.31a) $$k(\mathbf{x},\mathbf{y}) := -\frac{2}{\omega_d}\frac{\langle \mathbf{n}(\mathbf{y}), \mathbf{y}-\mathbf{x}\rangle}{|\mathbf{y}-\mathbf{x}|^d} \qquad (\mathbf{x},\mathbf{y}\in\Gamma).$$

Interchanging the arguments $\mathbf{x}$, $\mathbf{y}$, one obtains the kernel adjoint to (31a), which appears in (30):

(8.1.31b) $$k^*(\mathbf{x},\mathbf{y}) := -\frac{2}{\omega_d}\frac{\langle \mathbf{n}(\mathbf{x}), \mathbf{x}-\mathbf{y}\rangle}{|\mathbf{x}-\mathbf{y}|^d} \qquad (\mathbf{x},\mathbf{y}\in\Gamma),$$

Define K as the integral operator associated with (31a):

(8.1.31c) $$(Kf)(\mathbf{x}) := -\frac{2}{\omega_d}\oint_\Gamma \frac{\langle \mathbf{n}(\mathbf{y}), \mathbf{y}-\mathbf{x}\rangle}{|\mathbf{y}-\mathbf{x}|^d} f(\mathbf{y})\, d\Gamma_y \qquad (\mathbf{x}\in\Gamma).$$

Then the operator defined below is formally adjoint to K:

(8.1.31d) $$(K^*f)(\mathbf{x}) := -\frac{2}{\omega_d}\int_\Gamma \frac{\langle \mathbf{n}(\mathbf{x}),\, \mathbf{x}-\mathbf{y}\rangle}{|\mathbf{x}-\mathbf{y}|^d} f(\mathbf{y})\, d\Gamma_y \qquad (\mathbf{x}\in\Gamma),$$

Equation (30) is a *Fredholm integral equation of the second kind*, which can be rewritten by means of K^*:

(8.1.32a) $$f = 2\varphi - K^* f .$$

Similarly, the Neumann boundary condition $\partial\Phi_+/\partial n = \varphi$ for the *exterior problem* leads to the integral equation

(8.1.32b) $$f = -2\varphi + K^* f .$$

The uniqueness of the problems (32a,b) can be formulated and proved similarly to Theorem 20.

Theorem 8.1.23 (uniqueness). Let $\Gamma\in C^{1+\lambda}_{0,\mathrm{stw}}$ for some $\lambda>0$. There is at most one solution $f\in L^\infty(\Gamma)$ of the respective problems (32a) or (32b), where, in the latter case, the additional condition (13) is required if $d=2$.

Proof. (i) The difference of two solutions $f_1, f_2 \in L^\infty(\Gamma)$ of (32a) leads to a solution $f \in L^\infty(\Gamma)$ of $f = -K^*f$. $(K^*f)(\mathbf{x})$ is continuous on Γ except corners and edges (the proof is left to the reader); hence, also f is almost everywhere continuous. The single-layer potential Φ generated by f has the normal derivative $\partial\Phi_-/\partial n = 0$ almost everywhere. The only solution in the interior domain is $\Phi_- = \text{const}$. Because of the continuity of Φ, the boundary values $\Phi_- = \text{const}$ on Γ also hold for the exterior solution. A possible solution is $u_+ = \text{const}$. The difference $v := u_+ - \Phi$ satisfies $v = 0$ on Γ. Because of condition (13), $\Phi(\infty) = 0$ holds and implies $v = O(1)$. Furthermore, (25c) is valid (cf. Lemma 18d)). The uniqueness of the exterior problem (cf. Lemma 18e)) shows $v = 0$ and therefore, since $0 = v(\infty) = u_+(\infty) - \Phi(\infty) = \text{const}$, also $u_+ = 0$ and $\Phi_+ = 0$ in Ω_+. By $\partial\Phi_+/\partial n = 0$ and Theorem 9, we obtain $f = 0$.
(ii) In the case of equation (32b), one draws the analogous conclusions in the following order: $\partial\Phi_+/\partial n = 0$ on Γ, $\Phi = 0$ on Ω_+, $\Phi_- = 0$ on Γ, $\Phi = 0$ in Ω_-, $\partial\Phi_-/\partial n = 0$ on Γ, $f = 0$ on Γ. QED

Because of Lemma 19, the existence of a solution f of the integral equations (32a,b) cannot be expected for all boundary data φ. To prepare the existence theorem, we show compactness of the operator assuming that Γ is sufficiently smooth.

Theorem 8.1.24 (compactness). Let $\Gamma \in C_0^{1+\lambda}$ for some $\lambda > 0$. The operator K^* from (31d) is compact in $C(\Gamma)$.

Proof. We have to prove the assumptions (3.2.8a,b) of Theorem 3.2.6. Now, (3.2.8a) reads $\int_\Gamma |k^*(\mathbf{x},\mathbf{y})| d\Gamma_y < \infty$ and follows from the improper integrability (cf. Theorem 9). Condition (3.2.8b) requires

$$\int_\Gamma | k^*(\boldsymbol{\xi},\mathbf{y}) - k^*(\mathbf{x},\mathbf{y}) | \, d\Gamma_y \to 0 \qquad \text{for } \boldsymbol{\xi} \to \mathbf{x}$$

and, because of $\varkappa(\mathbf{x},\mathbf{x},\mathbf{y}) = 0$, it is equivalent to the continuity of the following function $\varkappa(\boldsymbol{\xi},\mathbf{x},\mathbf{y})$ with respect to $\boldsymbol{\xi}$:

$$\varkappa(\boldsymbol{\xi},\mathbf{x},\mathbf{y}) := \frac{2}{\omega_d} \int_\Gamma \left| \frac{\langle \mathbf{n}(\boldsymbol{\xi}), \boldsymbol{\xi}-\mathbf{y} \rangle}{|\boldsymbol{\xi}-\mathbf{y}|^d} - \frac{\langle \mathbf{n}(\mathbf{x}), \mathbf{x}-\mathbf{y} \rangle}{|\mathbf{x}-\mathbf{y}|^d} \right| d\Gamma_y .$$

Using the majorant $C[|\boldsymbol{\xi}-\mathbf{y}|^{1+\lambda-d} + |\mathbf{x}-\mathbf{y}|^{1+\lambda-d}]$, one obtains the uniform existence of the integral $\varkappa(\boldsymbol{\xi},\mathbf{x},\mathbf{y})$ as in Lemma 5. Hence, $\varkappa$ is continuous. QED

Theorem 8.1.25 (existence). Let $\Gamma \in C_0^{1+\lambda}$ for some $\lambda > 0$. The integral equations (32a,b) have a solution $f \in C(\Gamma)$ if and only if $\varphi \in C(\Gamma)$ satisfies condition (26).

Proof. Set $X_0 = \{\varphi \in C(\Gamma) : \varphi \text{ satisfies (26)}\}$. Since $2\varphi \in \text{range}(I+K^*)$ is the normal derivative of the single-layer potential in Ω_-, $\text{range}(I+K^*) \subset X_0$ must hold according to Lemma 19a. This proves $k := \dim(C(\Gamma)/\text{range}(I+K^*)) \geq \dim(C(\Gamma)/X_0) = 1$. Since, by Theorem 24, K^* is compact, Theorem 1.3.28b also guarantees $\dim \text{kernel}(I+K^*) = k$. Thanks to Theorem 23, a function $\xi \in \text{kernel}(I+K^*)\backslash\{0\}$ cannot satisfy (13). For two functions $\xi_1, \xi_2 \in \text{kernel}(I+K^*)\backslash\{0\}$ one finds a linear combination $\xi = \alpha\xi_1 + \beta\xi_2$ with $\alpha = \int \xi_2 d\Gamma \neq 0$, $\beta = -\int \xi_1 d\Gamma \neq 0$, which satisfies (13), so that $\xi = 0$ follows. Hence, $\text{kernel}(I+K^*)$ is one-dimensional. $k = 1$ implies $\text{range}(I+K^*) = X_0$. QED

Theorems 24 and 25 both require $\Gamma\in C_0^{1+\lambda}$ for some $\lambda>0$, so that corners are excluded. On the other hand, if corners are present, Corollary 27 shows that a solution, if it exists, has either a zero at the corner or a singularity. The reason is the singular behaviour of K^*f even for smooth f, as exemplified in

Example 8.1.26. Let $f=1$ be the (arbitrarily smooth) density of the single-layer potential for the unit square from Example 16. The normal derivative $\partial\Phi_-/\partial n$ in the interior domain behaves in the neighbourhood of the corner $\mathbf{x}=0$ as

$$\partial\Phi_-(\mathbf{x})/\partial n = \tfrac{1}{2\pi}\log|\mathbf{x}| - \tfrac{5}{8} + O(|\mathbf{x}|).$$

According to (16b,c), the following *logarithmic singularity* results:

$$K^*f(\mathbf{x}) = 2\,\partial\Phi_-(\mathbf{x})/\partial n - 1 = \tfrac{1}{\pi}\log|\mathbf{x}| + O(|\mathbf{x}|) - \tfrac{9}{4}.$$

Corollary 8.1.27. Let $\mathbf{x}_0\in\Gamma\in C^1_{0,\mathrm{stw}}$ be a corner of the curve, while φ is bounded on Γ. If a solution f of (32a) exists, which is Hölder continuous at $\mathbf{x}_0$, it must necessarily satisfy $f(\mathbf{x}_0)=0$.

Proof. Split f into f_0+f_1, where $f_0:=f(\mathbf{x}_0)$ is constant. The remainder $f_1:=f-f_0$ vanishes at $\mathbf{x}_0$: $f_1(\mathbf{x}_0)=0$. Therefore, the second term in $K^*f = K^*f_0+K^*f_1$ is improperly integrable and bounded in the neighbourhood of $\mathbf{x}_0$. If $f_0=f(\mathbf{x}_0)=0$ does not hold, Example 26 shows that the first term and therefore also K^*f are unbounded at $\mathbf{x}_0$. Since f has to satisfy equation (32a) (i.e., $f=2\varphi-K^*f$) and since φ is bounded, f must have a singularity at $\mathbf{x}_0$. QED

Concerning the difficulties caused by corners compare also §8.2.7.

8.2 The Double-Layer Potential

In §7.4.2, the double-layer potential was derived for the two-dimensional Laplace equation from the Cauchy kernel. The following constructions are independent on the dimension.

8.2.1 Definition

In the theory of the single-layer potential, the normal derivative of the fundamental solution $s(\mathbf{x},\mathbf{y})$ with respect to the *first* argument $\mathbf{x}$ appeared. The *double-layer potential* uses the normal derivative with respect to the *second* argument $\mathbf{y}$. The *double-layer kernel* is defined by

$$k(\mathbf{x},\mathbf{y}) = 2\frac{\partial}{\partial n_y}s(\mathbf{x},\mathbf{y}) = 2\langle \mathbf{n}(\mathbf{y}),\nabla_y s(\mathbf{x},\mathbf{y})\rangle \qquad \text{for } \mathbf{y}\in\Gamma. \tag{8.2.1a}$$

Different scalings of this kernel are sometimes used in other references. Since $s(\mathbf{x},\mathbf{y})$ is a function of the difference $\mathbf{x}-\mathbf{y}$, the gradients $\nabla_x s(\mathbf{x},\mathbf{y})$ and $\nabla_y s(\mathbf{x},\mathbf{y})$ differ only in sign. The normal derivative, however, involves in addition the normal direction $\mathbf{n}$, which depends in $\partial/\partial n_x$ on $\mathbf{x}$ and in $\partial/\partial n_y$ on $\mathbf{y}$. The integral operator (identical with (1.31c)) formed by the double-layer kernel (1a) is denoted again by K:

(8.2.1b) $$(Kf)(\mathbf{x}) := 2\oint_\Gamma f(\mathbf{y})\frac{\partial}{\partial n_y}s(\mathbf{x},\mathbf{y})\,d\Gamma_y \quad (\mathbf{x}\in\Gamma),$$

The function f is the *double-layer density*. It generates the *double-layer potential* Φ in $\mathbb{R}^d$:

(8.2.1c) $$\Phi(\mathbf{x}) := \int_\Gamma f(\mathbf{y})\frac{\partial}{\partial n_y}s(\mathbf{x},\mathbf{y})\,d\Gamma_y \qquad (\mathbf{x}\in\mathbb{R}^d).$$

The name «double-layer potential» indicates that Φ can be regarded as a limit of a double single-layer potential, as shown below.

Example 8.2.1. Let Φ_0 be the single-layer potential of the density f on the square $\Gamma=\{\mathbf{x}\in\mathbb{R}^3:\ 0\leq x_1,x_2\leq 1,\ x_3=0\}$. On the square $\Gamma_\varepsilon=\{\mathbf{x}\in\mathbb{R}^3:\ 0\leq x_1,x_2\leq 1,\ x_3=\varepsilon\}$ shifted by $\varepsilon>0$ choose the density equal to the negative copy $f_\varepsilon(x_1,x_2,\varepsilon)=-f(x_1,x_2,0)$ of f. The single-layer potentials Φ_0 and Φ_ε generated on Γ and Γ_ε add to

$$(\Phi_\varepsilon-\Phi_0)(\mathbf{x}) = \int_{\Gamma_\varepsilon} f_\varepsilon(\mathbf{y})s(\mathbf{x},\mathbf{y})\,d\Gamma_y - \int_\Gamma f(\mathbf{y})s(\mathbf{x},\mathbf{y})\,d\Gamma_y =$$
$$= \int_0^1\!\!\int_0^1 f(y_1,y_2,0)\,[s(\mathbf{x},\mathbf{y}+\varepsilon\mathbf{n}(\mathbf{y}))-s(\mathbf{x},\mathbf{y})]\,dy_1\,dy_2$$

with $\mathbf{y}=(y_1,y_2,0)$ and $\mathbf{n}=(0,0,1)$. For $\mathbf{x}\notin\Gamma$, the limit $\varepsilon\to 0$ of the difference quotients exists and yields the double-layer potential Φ on Γ:

$$\Phi(\mathbf{x}) = \lim_{\varepsilon\to 0}\frac{1}{\varepsilon}(\Phi_\varepsilon-\Phi_0)(\mathbf{x}) = \int_\Gamma f(\mathbf{y})\frac{\partial}{\partial n_y}s(\mathbf{x},\mathbf{y})\,d\Gamma_y.$$

As example for $s(\mathbf{x},\mathbf{y})$, we choose again the fundamental solution (1.3) of the *Laplace equation*. The *double-layer kernel* and the corresponding integral operator take the form (2a,b) already mentioned in (1.31a,c):

(8.2.2a) $$k(\mathbf{x},\mathbf{y}) = -\frac{2}{\omega_d}\frac{\langle\mathbf{n}(\mathbf{y}),\mathbf{y}-\mathbf{x}\rangle}{|\mathbf{y}-\mathbf{x}|^d} \qquad (\mathbf{y}\in\Gamma),$$

(8.2.2b) $$(Kf)(\mathbf{x}) = -\frac{2}{\omega_d}\int_\Gamma\frac{\langle\mathbf{n}(\mathbf{y}),\mathbf{y}-\mathbf{x}\rangle}{|\mathbf{y}-\mathbf{x}|^d}f(\mathbf{y})\,d\Gamma_y \qquad (\mathbf{x}\in\Gamma).$$

The *double-layer potential of the Laplace equation* reads

(8.2.2c) $$\Phi(\mathbf{x}) = -\frac{1}{\omega_d}\int_\Gamma\frac{\langle\mathbf{n}(\mathbf{y}),\mathbf{y}-\mathbf{x}\rangle}{|\mathbf{y}-\mathbf{x}|^d}f(\mathbf{y})\,d\Gamma_y \qquad (\mathbf{x}\in\mathbb{R}^d).$$

The integrals in (2b,c) make sense as shown next.

Lemma 8.2.2. Assume $\Gamma\in C^{1+\mu}_{\mathrm{stw}}$ for some $\mu>0$ and let $f\in L^\infty(\Gamma)$ be a bounded density. Then the right-hand side in (2b) exists for all $\mathbf{x}\in\Gamma$ as improper integral.

Proof. Since $\Gamma\in C^{1+\mu}_{\mathrm{stw}}$ consists of finitely many smooth $C^{1+\mu}$ pieces, one may assume without loss of generality $\Gamma\in C^{1+\mu}$. For $\mathbf{x}$ outside of Γ, the integrand is regular. For $\mathbf{x}\in\Gamma$, the nominator has the order $O(|\mathbf{y}-\mathbf{x}|^{1+\mu})$ (cf. Exercise 1.8a). The majorant $O(|\mathbf{y}-\mathbf{x}|^{1+\mu-d})$ is improperly integrable. QED

8.2.2 Regularity Properties of the Double-Layer Integral Operator

By Lemma 2, $(Kf)(\mathbf{x})$ is defined for all $\mathbf{x}\in\Gamma$. A more precise analysis will show that Kf is not only defined but also uniformly bounded on Γ.

Theorem 8.2.3. Let $\mu>0$. **(a)** Under the assumption $\Gamma\in C^{1+\mu}$, K belongs to $L(L^\infty(\Gamma),C(\Gamma))$, i.e., Kf is continuous on Γ for bounded densities f.
(b) Let $\Gamma\in C^{1+\mu}$. The linear mapping, which maps the density $f\in L^\infty(\Gamma)$ into the double-layer potential Φ from (2c), is bounded, i.e., $|\Phi(\mathbf{x})|\leqslant C\|f\|_\infty$ for all $\mathbf{x}\in\mathbb{R}^d$.
(c) Let $\Gamma\in C^{1+\mu}_{\mathbf{stw}}$. Then K is bounded in the Banach space $L^\infty(\Gamma)$, i.e., $K\in L(L^\infty(\Gamma),L^\infty(\Gamma))$. Statement (b) remains valid for such Γ.

For the proof, the following lemma is needed.

Lemma 8.2.4. Let $\Gamma\in C^{1+\mu}_{\mathbf{stw}}$ with $\mu>0$. Then there are global constants C and C_λ such that

(8.2.3a) $$\int_\Gamma |k(\mathbf{x},\mathbf{y})|\,d\Gamma_y \leqslant C \qquad \text{for } \mathbf{x}\in\mathbb{R}^d,$$

(8.2.3b) $$\int_{\Gamma\cap K_\varepsilon(\mathbf{x})}|\mathbf{x}-\mathbf{y}|^{\lambda+1-d}\,d\Gamma_y \leqslant C_\lambda\varepsilon^\lambda \qquad \text{for } \mathbf{x}\in\mathbb{R}^d,\ \lambda>0,\ \varepsilon>0,$$

(8.2.3c) $$\int_{\Gamma\backslash K_\varepsilon(\mathbf{x})}|\mathbf{x}-\mathbf{y}|^{\lambda-d}\,d\Gamma_y \leqslant C_\lambda\varepsilon^{\lambda-1} \qquad \text{for } \mathbf{x}\in\mathbb{R}^d,\ \lambda<1,\ \varepsilon>0.$$

Proof. (i) It is sufficient to assume $\Gamma\in C^{1+\mu}$, since an arbitrary $\Gamma\in C^{1+\mu}_{\mathbf{stw}}$ is a finite sum of such $\Gamma_i\in C^{1+\mu}$, so that the statements (3a-c) carry over to $\Gamma\in C^{1+\mu}_{\mathbf{stw}}$.
(ii) First let $\mathbf{x}\in\Gamma$. There is an $\varepsilon_0>0$ such that $\Gamma\cap K_\varepsilon(\mathbf{x})$ can be parametrised for all $0<\varepsilon\leqslant\varepsilon_0$ by means of a subset $S_\varepsilon(\mathbf{x})\subset\{\mathbf{z}\in\mathbb{R}^{d-1}: |\mathbf{z}|\leqslant\varepsilon\}$, where the $\mathbf{z}$ components are the coordinates in the tangential plane at the point $\mathbf{x}$ and $\mathbf{z}=0$ corresponds to the point $\mathbf{x}$. $\eta(\mathbf{z})$ is the component of $\mathbf{x}\in\Gamma$ in the normal direction $\mathbf{n}(\mathbf{x})$. Hence,

(8.2.3d) $$\int_{\Gamma\cap K_\varepsilon(\mathbf{x})}|k(\mathbf{x},\mathbf{y})|\,d\Gamma_y = \\ = \frac{2}{\omega_d}\int_{S_\varepsilon(\mathbf{x})}\sqrt{g(\mathbf{z})}\,|\langle n_0(\mathbf{z}),\mathbf{z}\rangle + n_d(\mathbf{z})\eta(\mathbf{z})| \,/\, [|\mathbf{z}|^2+\eta(\mathbf{z})^2]^{d/2}\,d\mathbf{z}.$$

Here, $n_0(\mathbf{z})$ is the tangential and $n_d(\mathbf{z})$ the normal component of $\mathbf{n}(\mathbf{z})$. $g(\mathbf{z})=g(\mathbf{z};\mathbf{x})$ denotes the Gram determinant (cf. §8.1.2.2). There is a C independent of the central point $\mathbf{x}\in\Gamma$ such that

(8.2.3e$_1$) $$\|g(\cdot;\mathbf{x})\|_{C^\mu(S_{\varepsilon_0}(\mathbf{x}))}\leqslant C \qquad \text{for all } \mathbf{x}\in\Gamma,$$

(8.2.3e$_2$) $$|n_0(\mathbf{z})|\leqslant C|\mathbf{z}|^\mu,\ |n_d(\mathbf{z})-1|\leqslant C|\mathbf{z}|^\mu \qquad \text{for all } \mathbf{x}\in\Gamma,\ \mathbf{z}\in S_{\varepsilon_0}(\mathbf{x}),$$

(8.2.3e$_3$) $$|\eta(\mathbf{z})|\leqslant C|\mathbf{z}|^{1+\mu} \qquad \text{for all } \mathbf{x}\in\Gamma,\ \mathbf{z}\in S_{\varepsilon_0}(\mathbf{x}).$$

Hence, one obtains the majorant $\frac{4}{\omega_d}C^{3/2}|\mathbf{z}|^{1+\mu-d}$. Enlarging the set of $\mathbf{z}\in S_\varepsilon(\mathbf{x})$ to $|\mathbf{z}|\leqslant\varepsilon$, we can make use of polar coordinates with $r=|\mathbf{z}|$:

(8.2.3f) $$\int_{|\mathbf{z}|\leqslant\varepsilon}\frac{4}{\omega_d}C^{3/2}|\mathbf{z}|^{1+\mu-d}\,d\mathbf{z} = \frac{4}{\omega_d}C^{3/2}\omega_{d-1}\int_0^\varepsilon r^{d-2}r^{1+\mu-d}\,dr.$$

Therefore, the integral of $|k(\mathbf{x},\mathbf{y})|$ over $\Gamma\cap K_{\varepsilon_0}(\mathbf{x})$ is bounded by the constant $C':=4C^{3/2}\omega_{d-1}\varepsilon_0^\mu/[\mu\,\omega_d]$. In $\Gamma\backslash K_{\varepsilon_0}(\mathbf{x})$, the integrand is $\leqslant 2\varepsilon_0^{1-d}/\omega_d$, so that this integral is bounded by $C'' := \mu(\Gamma)\frac{2}{\omega_d}\varepsilon_0^{1-d}$ $\{\mu(\Gamma) = \int_\Gamma d\Gamma = \text{area of } \Gamma\}$. $C'+C''$ is the constant in (3a).

(iii) It remains to show (3a) for $\mathbf{x} \in \mathbb{R}^d \backslash \Gamma$. For sufficiently small $\varepsilon_0 > 0$ one can find for any $\mathbf{x}$ with $\delta := \operatorname{dist}(\mathbf{x}, \Gamma) \leqslant \varepsilon$ an $\mathbf{x}_0 \in \Gamma$ with $\delta = |\mathbf{x} - \mathbf{x}_0|$. We use the parametrisation of $\Gamma \cap K_{\varepsilon_0}(\mathbf{x}_0)$ by means of $S_{\varepsilon_0}(\mathbf{x}_0)$. The integrand is the product of $\sqrt{g(\mathbf{z})}$ and

$$|\langle n_0(\mathbf{z}), \mathbf{z}\rangle + n_d(\mathbf{z})(\eta(\mathbf{z}) - \delta)| \,/\, [|\mathbf{z}|^2 + (\eta(\mathbf{z}) - \delta)^2]^{d/2}.$$

The difference between this and (3d) is $\delta / [|\mathbf{z}|^2 + (\eta(\mathbf{z}) - \delta)^2]^{d/2}$. According to Lemma 1.10c, this additional part can be estimated uniformly with respect to δ and $\mathbf{x}_0$. This concludes the proof of (3a).

(iv) Concerning (3b), we only consider the worst case $\mathbf{x} \in \Gamma$. For the proof of (3b) it suffices to restrict ε by $\varepsilon \leqslant \varepsilon_0$. The parametrisation over $S_\varepsilon(\mathbf{x})$ leads to the integrand $\sqrt{g(\mathbf{z})}\,[|\mathbf{z}|^2 + \eta(\mathbf{z})^2]^{(\lambda+1-d)/2}$. Its integral can be estimated by analogy with (3f) by $C'\varepsilon^\lambda$.

(v) The same considerations in the case of the integrand of (3c) lead to the integral $\int_\varepsilon^{\varepsilon_0} r^{d-2} r^{\lambda-d} dr = [\varepsilon_0^{\lambda-1} - \varepsilon^{\lambda-1}]/(\lambda-1) = O(\varepsilon^{\lambda-1})$ after introducing polar coordinates on $S_{\varepsilon_0}(\mathbf{x}) \backslash S_\varepsilon(\mathbf{x})$. Adding the bounded integrals over $\Gamma \backslash S_{\varepsilon_0}(\mathbf{x})$ does not change $O(\varepsilon^{\lambda-1})$. QED

Proof of Theorem 3. (i) Statement (a) will follow from (4a) in Theorem 5.
(ii) The estimate (3a) proves (b) with the constant $C/2$ instead of C.
(iii) $\Gamma \in C^{1+\mu}_{\mathrm{stw}}$ is the finite sum of pieces $\Gamma_i \in C^{1+\mu}$, each of which leads to double-layer potentials Φ_i. By (b), $\Phi_i \in L^\infty(\mathbb{R}^d)$ holds. Since $\Phi := \sum \Phi_i$ is the double-layer potential corresponding to Γ, $\Phi \in L^\infty(\mathbb{R}^d)$ follows. The restriction of this statement to Γ yields $K \in L(L^\infty(\Gamma), L^\infty(\Gamma))$. QED

The statement $K \in K(C(\Gamma), C(\Gamma))$ becomes false when Γ is *only piecewise* smooth, as Lemma 24 will show. For smooth Γ, however, the statement of Theorem 3a can be improved essentially.

Theorem 8.2.5 (compactness). Assume $\Gamma \in C^{1+\mu}$ with $0<\mu<1$. Then K is compact on $C(\Gamma)$. Moreover, we have

(8.2.4a) $\quad K \in L(L^\infty(\Gamma), C^\lambda(\Gamma)) \quad$ for all $0 \leqslant \lambda \leqslant \mu$,

(8.2.4b) $\quad K \in K(C^\lambda(\Gamma), C^\lambda(\Gamma)) \quad$ for all $0 \leqslant \lambda \leqslant \mu$.

The proof is prepared by Lemma 6, which translates Theorem 3.4.9 onto $(d-1)$-dimensional surfaces. A proof of $K \in L(L^\infty(\Gamma), C^{1+\lambda}(\Gamma))$ for $\Gamma \subset \mathbb{R}^d$ can be found in Schippers [2].

Lemma 8.2.6. Assume $\Gamma \in C^{1+\mu}_{\mathrm{stw}}$ with $\mu>0$. Let the kernel k of the integral operator K satisfy (3a) and assume the representation

(8.2.5a) $\quad k(\mathbf{x},\mathbf{y}) = \ell(\mathbf{x},\mathbf{y}) / |\mathbf{x}-\mathbf{y}|^{d-\mu-1} \quad (\mathbf{x},\mathbf{y} \in \Gamma)$.

The factor ℓ has to satisfy the following conditions:

(8.2.5b) $\quad |\ell(\mathbf{x},\mathbf{y})| \leqslant C_\ell \qquad$ for all $\mathbf{x},\mathbf{y} \in \Gamma$,

(8.2.5c) $\quad |\ell(\mathbf{x},\mathbf{y}) - \ell(\boldsymbol{\xi},\mathbf{y})| \leqslant C_\ell |\mathbf{x}-\boldsymbol{\xi}|/|\mathbf{x}-\mathbf{y}|$
for all $\boldsymbol{\xi},\mathbf{x},\mathbf{y} \in \Gamma$ with $|\mathbf{y}-\mathbf{x}| \geqslant 2|\mathbf{x}-\boldsymbol{\xi}|$.

Then K belongs to $L(L^\infty(\Gamma), C^\mu(\Gamma))$.

Proof. Since (3.4.2a) is identical to (3a), the assertion follows from Theorem 3.4.2 as soon as the second condition (3.4.2b) is shown, i.e.,

(8.2.5d) $\quad \int_\Gamma |k(\mathbf{x},\mathbf{y}) - k(\boldsymbol{\xi},\mathbf{y})|\, d\Gamma_y \leqslant C|\mathbf{x}-\boldsymbol{\xi}|^\mu \qquad$ for $\boldsymbol{\xi},\mathbf{x} \in \Gamma$.

As shown in (3f) (in part (ii) of the proof of Lemma 4), $|k(\mathbf{x},\mathbf{y}) - k(\boldsymbol{\xi},\mathbf{y})| \leqslant$ $\leqslant |k(\mathbf{x},\mathbf{y})| + |k(\boldsymbol{\xi},\mathbf{y})|$ has the bound $O(\varepsilon^\mu)$ for $\mathbf{y} \in \Gamma_0 := \Gamma \cap K_{2\varepsilon}(\mathbf{x})$, where

$$\varepsilon := |\mathbf{x}-\boldsymbol{\xi}|.$$

We now consider the remaining integral over $\Gamma_1 := \Gamma \backslash K_{2\varepsilon}(\mathbf{x})$. Here, the integrand $|k(\mathbf{x},\mathbf{y}) - k(\boldsymbol{\xi},\mathbf{y})|$ is estimated as in (3.4.7g) by

$$|\ell(\mathbf{x},\mathbf{y}) - \ell(\boldsymbol{\xi},\mathbf{y})|/|\mathbf{x}-\mathbf{y}|^{d-\mu-1} + |\ell(\boldsymbol{\xi},\mathbf{y})| \left\{ |\boldsymbol{\xi}-\mathbf{y}|^{\mu+1-d} - |\mathbf{x}-\mathbf{y}|^{\mu+1-d} \right\}.$$

(5c) and (3c) yield the bound $C_\ell \varepsilon C_\mu \varepsilon^{\mu-1} = C\varepsilon^\mu$ for the first term. The application of the mean value theorem of differential calculus shows $\{\ldots\} = O(|\mathbf{x}-\boldsymbol{\xi}|/|\tilde{\mathbf{x}}-\mathbf{y}|^{d-\mu})$ with an intermediate value $\tilde{\mathbf{x}} \in K_\varepsilon(\mathbf{x})$. Since $|\tilde{\mathbf{x}}-\mathbf{y}| \geqslant |\mathbf{x}-\mathbf{y}|/2$ for $|\mathbf{y}-\mathbf{x}| \geqslant 2|\mathbf{x}-\boldsymbol{\xi}| = 2\varepsilon$, one obtains the majorant $O(\varepsilon/|\mathbf{x}-\mathbf{y}|^{d-\mu})$, which by (3c) leads to the integral $O(\varepsilon)C_\lambda \varepsilon^{\mu-1} = O(\varepsilon^\mu)$. For $d-\mu<0$, use $|\tilde{\mathbf{x}}-\mathbf{y}| \geqslant 2|\mathbf{x}-\mathbf{y}|$. Summation of all terms yields $O(\varepsilon^\mu) = O(|\mathbf{x}-\boldsymbol{\xi}|^\mu)$ and proves (5d). QED

Proof of Theorem 5. (i) The factorisation (5a) leads to

(8.2.5e) $\quad \ell(\mathbf{x},\mathbf{y}) := -\frac{2}{\omega_d} \langle \mathbf{n}(\mathbf{y}), \mathbf{y}-\mathbf{x} \rangle / |\mathbf{y}-\mathbf{x}|^{\mu+1}$.

For $\mathbf{x},\mathbf{y} \in \Gamma \in C^{1+\mu}$, the numerator $\langle \mathbf{n}(\mathbf{y}), \mathbf{y}-\mathbf{x} \rangle$ has the order $O(|\mathbf{y}-\mathbf{x}|^{1+\mu})$, as shown by the estimate by $C|\mathbf{z}|^{1+\mu}$ with the $\mathbf{z}$ coordinates (cf. $(3e_{1,2})$). Hence, (5b) is proved. For the proof of (5c), we use the mean value theorem:

(8.2.5f)
$$\begin{aligned} \ell(\mathbf{x},\mathbf{y}) - \ell(\boldsymbol{\xi},\mathbf{y}) &= -\frac{2}{\omega_d} \langle \mathbf{n}(\mathbf{y}), \boldsymbol{\xi}-\mathbf{x} \rangle / |\mathbf{y}-\tilde{\mathbf{x}}|^{\mu+1} - \\ &- \frac{2}{\omega_d}(\mu+1) \langle \mathbf{n}(\mathbf{y}), \mathbf{y}-\tilde{\mathbf{x}} \rangle \langle \mathbf{y}-\tilde{\mathbf{x}}, \boldsymbol{\xi}-\mathbf{x} \rangle / |\mathbf{y}-\tilde{\mathbf{x}}|^{\mu+3}, \end{aligned}$$

where $\tilde{\mathbf{x}}$ lies on the straight line between $\mathbf{x}$ and $\boldsymbol{\xi}$. In the first term, we split $\langle \mathbf{n}(\mathbf{y}), \boldsymbol{\xi}-\mathbf{x} \rangle$ into $\langle \mathbf{n}(\mathbf{x}), \boldsymbol{\xi}-\mathbf{x} \rangle$ and $\langle \mathbf{n}(\mathbf{y})-\mathbf{n}(\mathbf{x}), \boldsymbol{\xi}-\mathbf{x} \rangle$. The first term has the bound $C|\boldsymbol{\xi}-\mathbf{x}|^{\mu+1} \leqslant C|\boldsymbol{\xi}-\mathbf{x}||\mathbf{x}-\mathbf{y}|^\mu$, since the normal component of $\boldsymbol{\xi}$ equals η from $(3e_3)$. The second term is bounded by $|\mathbf{n}(\mathbf{y})-\mathbf{n}(\mathbf{x})||\boldsymbol{\xi}-\mathbf{x}|$. The assumption $\Gamma \in C^{1+\mu}$ yields $|\mathbf{n}(\mathbf{y})-\mathbf{n}(\mathbf{x})| \leqslant C|\mathbf{x}-\mathbf{y}|^\mu$. Together with $|\mathbf{y}-\tilde{\mathbf{x}}| \geqslant |\mathbf{x}-\mathbf{y}|/2$, the bound $O(|\boldsymbol{\xi}-\mathbf{x}|/|\mathbf{x}-\mathbf{y}|)$ follows for the first term in (5f). In the second term, both factors $\langle \mathbf{n}(\mathbf{y}), \mathbf{y}-\tilde{\mathbf{x}} \rangle / |\mathbf{y}-\tilde{\mathbf{x}}|^{\mu+1}$ and $\langle \mathbf{y}-\tilde{\mathbf{x}}, \boldsymbol{\xi}-\mathbf{x} \rangle / |\mathbf{y}-\tilde{\mathbf{x}}|^2$ are estimated by $O(|\boldsymbol{\xi}-\mathbf{x}|/|\mathbf{x}-\mathbf{y}|)$. Since $|\boldsymbol{\xi}-\mathbf{x}| \leqslant$ $\leqslant |\mathbf{x}-\mathbf{y}|$ is supposed in (5c), one factor $O(|\boldsymbol{\xi}-\mathbf{x}|/|\mathbf{x}-\mathbf{y}|)$ can be replaced by $O(1)$. Hence, (5c) is proved. Lemma 6 guarantees statement (4a) of Theorem 5 for $\lambda := \mu$. This implies (4a) for $\lambda < \mu$.
(ii) Remark 3.4.13 with the modification from Remark 3.4.15 allows to conclude from (4a) to (4b). QED

8.2.3 Jump Properties of the Double-Layer Potential

Except for the sign of the jump, the double-layer potential (1c) satisfies the same jump relation on Γ as the normal derivative of the single-layer potential. In the following, the continuous extendability and the jump properties are analysed. The Hölder continuity of the continuously extended potentials Φ_+ and Φ_- will be investigated in §8.2.4.1. In contrast to the normal derivative of the single-layer potential, $\Phi(\mathbf{x})$ has a limit also if $\mathbf{x}$ approaches a point of *discontinuity* $\mathbf{x}_0 \in \Gamma$ of the density f in a certain way. This special case will be discussed in §8.2.4.2.

Let $\mathbf{x}_0 \in \Gamma$ be an *inner* point of $\Gamma \in C^{1+\mu}$ with $\mu > 0$. $\mathbf{n}(\mathbf{x}_0)$ is the normal direction at $\mathbf{x}_0$, which in the case of a closed curve or surface Γ is directed outward. The function

(8.2.6a) $$\varphi(\alpha) := \Phi(\mathbf{x}_0 + \alpha\mathbf{n}(\mathbf{x}_0))$$

describes the «outer» double-layer potential in the neighbourhood of $\mathbf{x}_0$ for $0 < \alpha \leqslant \alpha_0$ and the «inner» one for $-\alpha_0 \leqslant \alpha < 0$ (α_0 sufficiently small). If the straight line $\{\mathbf{x}_0 + \alpha\mathbf{n}(\mathbf{x}_0): \alpha \in \mathbb{R}\}$ has no further intersection point with Γ, one may set $\alpha_0 = \infty$.

Lemma 8.2.7. Under the assumption from above and if the double-layer density $f \in L^\infty(\Gamma)$ is continuous at $\mathbf{x}_0$, the one-sided limits $\varphi(0 \pm 0)$ exist and satisfy

(8.2.6b) $$\varphi(0+0) - \varphi(0-0) = f(\mathbf{x}_0),$$

(8.2.6c) $$\varphi(0+0) + \varphi(0-0) = (Kf)(\mathbf{x}_0).$$

Proof. (i) We split Γ into a suitable neighbourhood Γ_ε of $\mathbf{x}_0$, which can be parametrised by means of $S_\varepsilon := \{\mathbf{y}' \in \mathbb{R}^{d-1}: |\mathbf{y}'| \leqslant \varepsilon\}$, and the rest $\Gamma_0 := \Gamma/\Gamma_\varepsilon$. The double-layer potentials (1c) corresponding to Γ_0 and Γ_ε are denoted by Φ_0 and Φ_1. Since the integrand of Φ_0 has no singularity in the neighbourhood of $\mathbf{x}_0$, $\Phi_0(\mathbf{x})$ is continuous at $\mathbf{x} = \mathbf{x}_0 + \alpha\mathbf{n}(\mathbf{x}_0)$.
(ii) Γ_ε is parametrised via S_ε. For the sake of simplicity, we assume without loss of generality that $\mathbf{x}_0 = 0$ lies in the origin and that the tangent or tangential plane is horizontal, i.e., $\mathbf{n}(\mathbf{x}_0)$ has the components $n_1 = \dots = n_{d-1} = 0$, $n_d = 1$. Γ_1 is represented by $\{(\mathbf{y}', \eta(\mathbf{y}')): \mathbf{y}' \in S_\varepsilon\}$, where $\eta: S_\varepsilon \to \mathbb{R}$ is the function characterised in (1.17b,f). The normal direction $\mathbf{n}(\mathbf{y})$ at $\mathbf{y} = (\mathbf{y}', \eta(\mathbf{y}'))$ is the vector $(\nabla\eta(\mathbf{y}'), 1)/\sqrt{g(\mathbf{y}')}$, where g denotes the Gram determinant. Therefore, the scalar product $\langle\mathbf{n}(\mathbf{y}), \mathbf{y}-\mathbf{x}\rangle$ with $\mathbf{x} = \mathbf{x}_0 + \alpha\mathbf{n}(\mathbf{x}_0)$ takes the form

$$\sqrt{g(\mathbf{y}')}\,\langle\mathbf{n}(\mathbf{y}), \mathbf{y}-\mathbf{x}\rangle = \eta(\mathbf{y}') - \alpha + \langle\nabla\eta(\mathbf{y}'), \mathbf{y}'\rangle,$$

where the scalar product on the left-hand side is d-dimensional, whereas on the right-hand side it is $(d-1)$-dimensional. The denominator of the double-layer potential now equals

$$|\mathbf{y}-\mathbf{x}|^d = [(\alpha - \eta(\mathbf{y}'))^2 + |\mathbf{y}'|^2]^{d/2}.$$

With these replacements, we write the double-layer potential Φ_1 over $\Gamma_1 = \Gamma \cap K_\varepsilon(\mathbf{x}_0)$ as

$$\Phi_1(\mathbf{x}_0 + \alpha\mathbf{n}(\mathbf{x}_0)) = \frac{1}{\omega_d}\int_{S_\varepsilon} f(\mathbf{y}', \eta(\mathbf{y}'))\,\frac{\alpha - \eta(\mathbf{y}') - \langle\nabla\eta(\mathbf{y}'), \mathbf{y}'\rangle}{[(\alpha - \eta(\mathbf{y}'))^2 + |\mathbf{y}'|^2]^{d/2}}\,d\mathbf{y}'.$$

We split Φ_1 into $\Phi_{1,1}+\Phi_{1,2}$:

$$\Phi_{1,1}(\alpha) := \frac{1}{\omega_d}\int_{S_\varepsilon} f(\mathbf{y}',\eta(\mathbf{y}'))\frac{\alpha-\eta(\mathbf{y}')}{[(\alpha-\eta(\mathbf{y}'))^2+|\mathbf{y}'|^2]^{d/2}}\,d\mathbf{y}',$$

$$\Phi_{1,2}(\alpha) := -\frac{1}{\omega_d}\int_{S_\varepsilon} f(\mathbf{y}',\eta(\mathbf{y}'))\frac{\langle\nabla\eta(\mathbf{y}'),\mathbf{y}'\rangle}{[(\alpha-\eta(\mathbf{y}'))^2+|\mathbf{y}'|^2]^{d/2}}\,d\mathbf{y}'.$$

In the numerator of $\Phi_{1,2}$, $|\langle\nabla\eta(\mathbf{y}'),\mathbf{y}'\rangle|\leqslant|\nabla\eta(\mathbf{y}')-\nabla\eta(0)||\mathbf{y}'|$ has the bound $O(|\mathbf{y}'|^{1+\mu})$ because of $\eta\in C^{1+\mu}$ and $\nabla\eta(0)=0$. Therefore, $\Phi_{1,2}$ has the majorant $C|\mathbf{y}'|^{1+\mu-d}$ and the improper integrals $\Phi_{1,2}(\alpha)$, $|\alpha|\leqslant\alpha_0$, exist uniformly, so that $\Phi_{1,2}(\alpha)$ is continuous. One obtains $\Phi_{1,2}(0)$ from the representation from above with $\alpha=0$:

$$\Phi_{1,2}(0) = -\frac{1}{\omega_d}\int_{\Gamma_\varepsilon}\frac{\langle\mathbf{n}(\mathbf{y}),\mathbf{y}-\mathbf{x}_0\rangle}{|\mathbf{y}-\mathbf{x}_0|^d}f(\mathbf{y})\,d\Gamma_{\mathbf{y}}\,.$$

Lemma 1.10d is applicable to the integral $\Phi_{1,1}(\alpha)$: The difference $\varphi(0+0)-\varphi(0-0)$ is given by $\Phi_{1,1}(0+0)-\Phi_{1,1}(0-0)=f(\mathbf{x}_0)$. The mean value results from the evaluation of the integral at $\alpha=0$. The comparison with (2b) shows the coincidence with $(Kf)(\mathbf{x}_0)$. QED

At $\mathbf{x}_0\in\Gamma$, three values of the double-layer potential are defined: Besides $\Phi(\mathbf{x}_0)$, the continuous extensions $\Phi_-(\mathbf{x}_0):=\varphi(0-0)$ from inside and $\Phi_+(\mathbf{x}_0):=\varphi(0+0)$ from outside are defined. The names «inside/outside» make sense only if Γ is closed. Otherwise, the orientation is fixed by the arbitrarily chosen normal direction $\mathbf{n}(\mathbf{x}_0)$: $\mathbf{n}(\mathbf{x}_0)$ is directed to the «outside». By definition, the first mentioned value $\Phi(\mathbf{x}_0)$ coincides with $\frac{1}{2}(Kf)(\mathbf{x}_0)$. Lemma 7 leads to

Theorem 8.2.8. Let $\Gamma\in C^{1+\mu}_{\mathbf{stw}}$ for some $\mu>0$ and assume that $f\in L^\infty(\Gamma)$ is a bounded density f. If $\mathbf{x}_0\in\Gamma$ is an inner point of a smooth component $\Gamma_0\subset\Gamma$, $\Gamma_0\in C^{1+\mu}$, and if f is continuous at $\mathbf{x}_0$, then the double-layer potential Φ has an inner extension $\Phi_-(\mathbf{x}_0)$ at $\mathbf{x}_0$ and an outer one $\Phi_+(\mathbf{x}_0)$, which satisfy the following *jump relations*:

(8.2.7a) $\quad \Phi_+(\mathbf{x}_0)-\Phi_-(\mathbf{x}_0) = f(\mathbf{x}_0),$

(8.2.7b) $\quad \Phi_+(\mathbf{x}_0)+\Phi_-(\mathbf{x}_0) = (Kf)(\mathbf{x}_0).$

Proof. Let Γ be the sum of the smooth curves or surfaces $\Gamma_i\in C^{1+\mu}$. $\mathbf{x}_0$ is an inner point of Γ_{i_0}. The double-layer potentials generated by Γ_i $(i\neq i_0)$ are continuous (even analytic) at $\mathbf{x}_0$, so that they do not influence the jump relations. Because of $\Phi_\pm(\mathbf{x}_0)=\varphi(0\pm 0)$, (7a,b) follow from Lemma 7. QED

Here, the extensions $\Phi_\pm(\mathbf{x}_0)$ are defined as particular limits using the normal direction (cf. (6a)). Furthermore, Theorem 8 makes no statement concerning points $\mathbf{x}_0\in\Gamma$ at corners or edges. The possible generalisations are summarised below.

Remark 8.2.9. (a) Let $\Gamma\in C^{1+\mu}_{\mathbf{stw}}$ for some $\mu>0$ and $f\in C(\Gamma)$ {it is sufficient to assume that a density $f\in L^\infty(\Gamma)$ is continuous in a Γ-neighbourhood of

$\mathbf{x}_0 \in \Gamma \backslash \partial\Gamma\}$. Then there is an $\mathbb{R}^3$-neighbourhood U of $\mathbf{x}_0 \in \Gamma \backslash \partial\Gamma$, which is decomposed by Γ into the connection components $U_\pm$ (cf. Fig. 7.2.1a,b), so that Φ_+ is *uniformly continuous* in $\overline{U}_+$ and Φ_- is so in $\overline{U}_-$. Their values on Γ satisfy the jump relation (7a).
(b) If $\Gamma \in C^{1+\mu}_{0,\mathrm{stw}}$ $(\mu>0)$ is closed and if $f \in C(\Gamma)$, then $\Phi_+ \in C(\overline{\Omega}_+)$ and $\Phi_- \in C(\overline{\Omega}_-)$ are *uniformly continuous* in their domains of definition.
(c) In (a) and (b) points $\mathbf{x}_0 \in \Gamma$ are admitted, which are situated on edges or corners of Γ. Only boundary points $\mathbf{x}_0 \in \partial\Gamma$ are excluded. The uniform continuity in $U_\pm$ or $\Omega_\pm$ implies that the definition of $\Phi_\pm$ does not depend on how $\mathbf{x} \in U_\pm$ approaches the point $\mathbf{x}_0 \in \overline{U}_\pm$.
(d) Hence, at all inner corners or edge points $\mathbf{x}_0 \in \Gamma \backslash \partial\Gamma$ a continuous extension $(\overline{K}f)(\mathbf{x}_0)$ exists to which equation (7b) applies. For $d=2$, $\overline{K}$ is identical to the operator $\overline{K}_1$ from (7.4.9).
(e) If Γ is closed, $\overline{K} \in L(C(\Gamma), C(\Gamma))$ holds and the jump relations (7a,b) are satisfied for all $\mathbf{x}_0 \in \Gamma$.

Proof. Since the statements (b–e) are simple implications of (a), a proof of (a) is sufficient. This is postponed to the end of §8.2.4.3. QED

8.2.4 Further Properties of the Double–Layer Potential

8.2.4.1 Hölder Continuity

Outside of Γ, the double-layer potential Φ is arbitrarily often differentiable and, by Exercise 1.2a, satisfies the Laplace equation, i.e.,

$$\Delta\Phi(\mathbf{x}) = 0 \qquad \text{for } \mathbf{x} \in \mathbb{R}^d \backslash \Gamma. \tag{8.2.8}$$

Due to Remark 9, Φ generated by $f \in C(\Gamma)$ is continuously extendable on Γ, where only $\partial\Gamma$ is to be excluded if Γ open is. The following statement concerning the Hölder continuity of $\Phi_\pm$ is not surprising in view of (4b):

Theorem 8.2.10. **(a)** If $\Gamma \in C^{1+\mu}_0$ $(\mu>0)$ is closed and if $f \in C^\lambda(\Gamma)$ for some $\lambda \in (0,\mu]$, then the respective exterior and interior extensions $\Phi_+ \in C^\lambda(\overline{\Omega}_+)$ and $\Phi_- \in C^\lambda(\overline{\Omega}_-)$ are *globally Hölder continuous* in their domains of definition.
(b) For a general (open) curve or surface $\Gamma \in C^{1+\mu}_{\mathrm{stw}}$, the corresponding statements hold in $\overline{U}_\pm$ (cf. Remark 9a), if the intersection $\gamma := \Gamma \cap \overline{U}_\pm$ is in the interior of a smooth component $\Gamma_0 \in C^{1+\mu}$ of Γ.

Proof. (i) For $d=2$, part (a) is already stated in Theorem 7.2.8 (cf. (7.4.4a)). The case $d \geqslant 3$ is analogous. A complete proof can be found in Colton–Kreß [1, Theorem 2.16].
(ii) The behaviour of Γ in a finite distance from γ does not influence the smoothness of the double-layer potential. QED

We recall that Kf coincides with the definition of Φ on Γ up to the factor 2 and equals the sum of the one-sided limits $\Phi_\pm$ (cf. (7b)). Since Kf for $f \in C(\Gamma)$ is already Hölder continuous, one might also expect Hölder continuity of $\Phi_\pm$ for continuous densities f. However, this cannot be satisfied, since by the jump relation (7a) Hölder continuous extensions $\Phi_\pm$ would imply a Hölder continuous density f.

8.2.4.2 The Potential close to the Jump Discontinuity of the Density

Now we want to investigate how a jump discontinuity of f effects the double-layer potential Φ. To begin with, we restrict ourselves to the case $d=2$. If a piecewise continuous function f jumps at $\mathbf{x}_0 \in \Gamma$ by δ, then f can be regarded as a sum of a continuous function f_1 and the piecewise constant function f_0 with the value 0 for $x \geqslant \mathbf{x}_0$ and δ for $x \geqslant \mathbf{x}_0$. The total potential is the sum $\Phi_0+\Phi_1$. Since the properties of Φ_1 are already described, it suffices to discuss f_0 and Φ_0. Since the shape of the curve at $\mathbf{x}_0$ has no influence as long as it is smooth at $\mathbf{x}_0$, we consider a straight line Γ (cf. Fig. 1):

(8.2.9a) $\qquad \Gamma := \{(y_1,y_2) \in \mathbb{R}^2 : \ y_2=0, \ -1 \leqslant y_1 \leqslant 1\}.$

Let $\mathbf{x}_0=(0,0)$ be the point of the jump discontinuity. Choose the density

(8.2.9b) $\qquad f(y_1,0)=1 \quad \text{for } y_1 \geqslant 0, \qquad f(y_1,0)=0 \quad \text{for } y_1<0.$

The corresponding potential has the representation

(8.2.9c) $\qquad \Phi(\mathbf{x}) = \Phi(x_1,x_2) = -\frac{x_2}{2\pi}\int_0^1 \{(x_1-y_1)^2+x_2^2\}^{-1}\,dy_1,$

if $\mathbf{n}(\mathbf{y})=\binom{0}{-1}$ is chosen as normal direction.

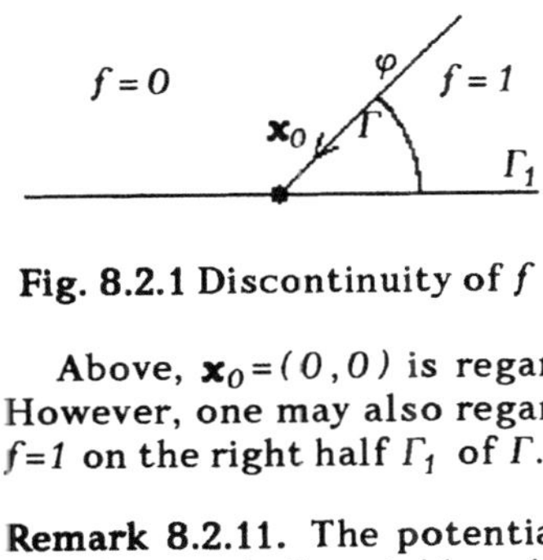

Fig. 8.2.1 Discontinuity of f at $\mathbf{x}_0$

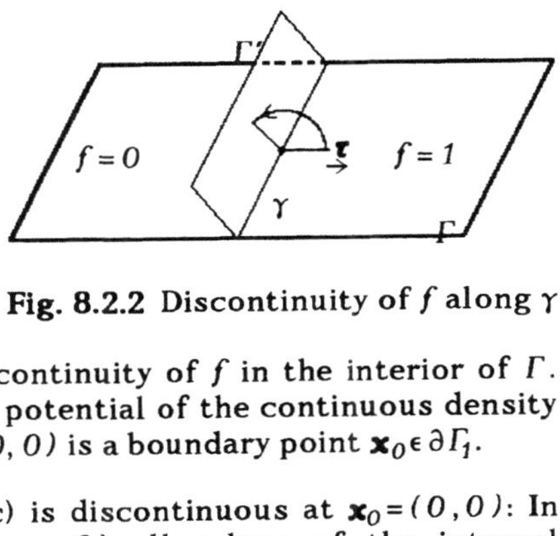

Fig. 8.2.2 Discontinuity of f along γ

Above, $\mathbf{x}_0=(0,0)$ is regarded as a discontinuity of f in the interior of Γ. However, one may also regard (9c) as the potential of the continuous density $f=1$ on the right half Γ_1 of Γ. Then, $\mathbf{x}_0=(0,0)$ is a boundary point $\mathbf{x}_0 \in \partial\Gamma_1$.

Remark 8.2.11. The potential Φ from (9c) is discontinuous at $\mathbf{x}_0=(0,0)$: In any (one-sided) neighbourhood $\{|\mathbf{x}|<\varepsilon : \ x_2>0\}$ all values of the interval $(-\pi/2,0)$ are attained. Nevertheless, Φ is continuous when we approach $\mathbf{x}_0$ from the *radial* direction 8Cf. Fig. 1):

(8.2.9d) $\qquad \Phi(r\cos\varphi, r\sin\varphi) \to -\frac{1}{2\pi}\operatorname{sign}(\varphi)(\pi-|\varphi|) \qquad \text{for } 0<|\varphi|\leqslant\pi, \ r\to 0.$

If one defines $\Phi_\pm$ as in (6a) to be the extension from the normal direction, then the jump relation (7a) holds in the form $\Phi_+(\mathbf{x}_0)-\Phi_-(\mathbf{x}_0)=f_m(\mathbf{x}_0)$, where $f_m(\mathbf{x}_0)$ is the arithmetical mean of the two one-sided limits of f at $\mathbf{x}_0$. In the case of (9b), $f_m(0)=\frac{1}{2}$ holds.

Proof. Evaluation of the integral in (9c) yields $-\frac{1}{2\pi}\{\operatorname{arctg}\frac{1-x_1}{x_2}+\operatorname{arctg}\frac{x_1}{x_2}\}$. For $r\to 0$, $x_2=r\sin\varphi$ tends to zero and hence the first arctg-term approaches $\pm\pi/2$ for the respective cases $\varphi\in(0,\pi)$ or $\varphi\in(-\pi,0)$. The second arctg-term equals $\pm\frac{1}{2}\pi-\varphi$. QED

We remark that continuity of $\Phi(\mathbf{x}(r))$ also holds, if one approaches the point $\mathbf{x}(0)=\mathbf{x}_0$ via a differentiable curve $\mathbf{x}(r)$ with $x_2(r)>0$. Then the limit $\lim \Phi(\mathbf{x}(r))$ is given by the right-hand side of (9d) with the angle φ of the tangent $d\mathbf{x}(0)/dr$: $[dx_2(0)/dr]/[dx_1(0)/dr] = \tan\varphi$.

In the multi-dimensional case $d\geqslant 3$, an analogous statement holds: When approaching a point of a jump discontinuity from a radial direction, Φ has a limit. The radial direction is determined by $d-1$ angles. If, in the case $d=3$, the jump discontinuity is situated along a smooth curve $\gamma\subset\Gamma$, then the limit depends only on *one* angle . Without loss of generality, Γ, γ, and the density f are chosen as follows (cf. Fig. 2):

(8.2.10a) $\quad \Gamma := \{\mathbf{y}\in\mathbb{R}^3:\ -1\leqslant y_1,y_2\leqslant 1,\ y_3=0\}, \qquad \gamma := \{\mathbf{y}\in\Gamma:\ y_1=0\},$

(8.2.10b) $\quad f(y_1,y_2,0) = 1 \quad \text{for } y_1\geqslant 0, \qquad f(y_1,y_2,0) = 0 \quad \text{for } y_1<0.$

Remark 8.2.12. (a) The potential Φ corresponding to (10a,b) is discontinuous in γ. For any unit vector $\boldsymbol{\omega}$ not being perpendicular to $\mathbf{n}(\mathbf{x}_0)$ (i.e., $\omega_3\neq 0$), $\Phi(r\boldsymbol{\omega})$ is continuously extendable at $r=0$. The limit $\Phi_{\boldsymbol{\omega}} := \lim_{r\to 0}\Phi(r\boldsymbol{\omega})$ only depends on $\langle \boldsymbol{\tau},\boldsymbol{\omega}\rangle$, i.e., on the angle between $\boldsymbol{\tau}$ and $\boldsymbol{\omega}$, where the unit vector $\boldsymbol{\tau}$ lies in the tangential plane of Γ at $\mathbf{x}_0$ and is normal to the curve γ (in the example (10a,b), $\boldsymbol{\tau}=(1,0,0)$ holds).
(b) The surface Γ' depicted in Fig. 2 has a constant angle $\langle\boldsymbol{\tau},\mathbf{x}\rangle/|\mathbf{x}|$, $\mathbf{x}\in\Gamma'$. Therefore, the restriction of the potential Φ onto Γ' is continuous and continuously extendable onto γ.

8.2.4.3 The Double-Layer Potential of the Density $f=1$

A corner $\mathbf{x}_0$ of Γ (analogously, an edge in the case of $d=3$) can be viewed as the sum of two curves Γ_1, Γ_2, which meet at the corner. The density f_i on Γ_i coincides with f in the part $\Gamma_i\cap\Gamma$ and is chosen as $f_i=0$ on $\Gamma_i\backslash\Gamma$ (s. Fig. 3). The potential Φ can be represented as sum of the particular potentials $\Phi_1+\Phi_2$ over the respective curves Γ_1, Γ_2. According to Remark 11, Φ_1 and Φ_2 are discontinuous at the corner $\mathbf{x}_0$. The following considerations will show that, nevertheless, the sum $\Phi_1+\Phi_2$ is continuous at $\mathbf{x}_0$, if f is continuous on Γ.

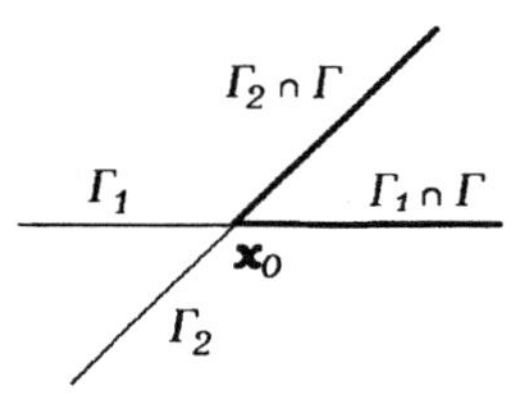

Fig. 8.2.3

Theorem 8.2.13. Let Ω_- and Ω_+ be the respective interior and exterior domains of $\Gamma\in C^1_{0,\mathrm{stw}}$. Then the double-layer density $f=1$ leads to the double-layer potential

(8.2.11a) $\quad \Phi(\mathbf{x}) = \begin{cases} -1 & \text{for } \mathbf{x}\in\Omega_-, \\ 0 & \text{for } \mathbf{x}\in\Omega_+. \end{cases}$

Because of (7b), an immediately consequence of (11a) is

(8.2.11b) $\quad \overline{K}f = -f \quad \text{for } f=1 \qquad$ (concerning $\overline{K}$ compare Remark 9d).

We give two proofs. The first is short and indirect, the second gives further information concerning the double-layer potential.

First Proof. For $\mathbf{x}\in\Omega_+$, $s(\mathbf{x},\cdot)\in C^2(\overline{\Omega}_-)$ is the solution of the Laplace equation in the closed interior domain $\overline{\Omega}_-$. From (1.24c) and Lemma 1.19a we obtain $0=\int_\Gamma \partial s(\mathbf{x},\mathbf{y})/\partial\mathbf{n}_y d\Gamma_y$, i.e., $\Phi_+(\mathbf{x})=0$ for the density $f=1$. Because of the jump relation (7a), $\Phi_-(\mathbf{x})=\Phi_+(\mathbf{x})-f(\mathbf{x})=0-1=-1$ follows on Γ. Unique solution of this Dirichlet boundary value is $\Phi_+=-1$. QED

The second proof is based on

Lemma 8.2.14 (double-layer potential representation in polar coordinates). Let $\mathbf{x}_0\notin\Gamma\in C^1_{stw}$ and $f\in L^\infty(\Gamma)$ be the double-layer density. $\Omega:=\partial K_1(0)$ is the surface of the unit sphere. For any unit vector $\boldsymbol{\omega}\in\Omega$ we define

(8.2.12a) $$F(\boldsymbol{\omega}) := \sum_{\mathbf{y}\in\Gamma,\ \mathbf{y}-\mathbf{x}_0=|\mathbf{y}-\mathbf{x}_0|\boldsymbol{\omega}} f(\mathbf{y})\ \mathrm{sign}(\langle\mathbf{n}(\mathbf{y}),\mathbf{y}-\mathbf{x}_0\rangle)$$

(explanation below). Then the double-layer potential at $\mathbf{x}_0$ has the value

(8.2.12b) $$\Phi(\mathbf{x}_0) = -\frac{1}{\omega_d}\int_\Omega F(\boldsymbol{\omega})d\Omega \qquad (d\Omega\text{: integration over }\Omega).$$

The sum in (12a) involves all $\mathbf{y}\in\Gamma$, which meet the ray $\{\mathbf{x}_0+t\boldsymbol{\omega}:\ t>0\}$. The sign of $(\langle\mathbf{n}(\mathbf{y}),\mathbf{y}-\mathbf{x}_0\rangle)$ is positive if $\boldsymbol{\omega}$ is directed outward, otherwise, negative (cf. case (1) in Fig. 4). It remains to discuss some exceptional cases. In the case (2) in Fig. 4, $\mathbf{y}$ is a corner admitting the two normal directions $\mathbf{n}_\pm(\mathbf{y})$ defined as one-sided limits. In this case the sum (12a) contains $\mathbf{y}$ twice, where the signs of $\langle\mathbf{n}_\pm(\mathbf{y}),\mathbf{y}-\mathbf{x}_0\rangle$ are opposite, so that both terms add to zero and therefore may also be omitted. If, as in case (3), a whole interval of the line $\{\mathbf{x}_0+t\boldsymbol{\omega}:\ t>0\}$ coincides with Γ, the sum (12a) has uncountably many terms which, however, does not matter because of $\mathrm{sign}(\langle\mathbf{n}(\mathbf{y}),\mathbf{y}-\mathbf{x}_0\rangle)=\langle\mathbf{n}(\mathbf{y}),\mathbf{y}-\mathbf{x}_0\rangle=0$. There are still pathological cases, in which the line $\{\mathbf{x}_0+t\boldsymbol{\omega}:\ t>0\}$ intersects Γ at infinitely many points with a non-vanishing angle. However, this case (as well as the cases (2) and (3)) happens only for a subset $\Omega_0\subset\Omega$ of measure zero, so that the definition of $F(\boldsymbol{\omega})$ for such $\boldsymbol{\omega}$ is irrelevant.

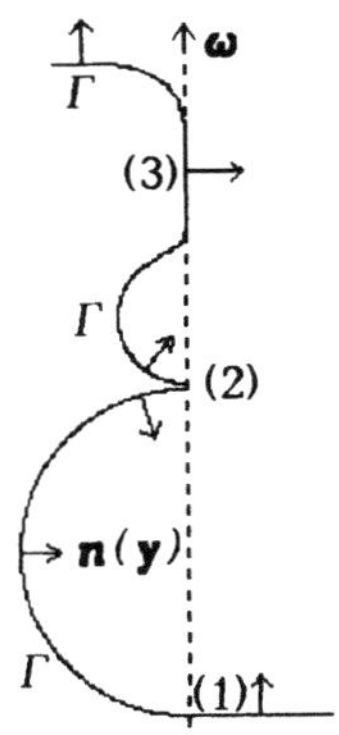

Fig. 8.2.4

Proof of Lemma 14. (i) First we assume that Γ is closed and $\langle\mathbf{n}(\mathbf{y}),\mathbf{y}-\mathbf{x}_0\rangle\geq\varepsilon>0$ is satisfied for all $\mathbf{y}\in\Gamma$. For any «angle» $\boldsymbol{\omega}\in\Omega$ the intersection point $\mathbf{y}=\mathbf{y}(\boldsymbol{\omega})\in\Gamma$ in direction of $\boldsymbol{\omega}$ defines the radius $r=r(\boldsymbol{\omega})$. Hence, for any $\boldsymbol{\omega}\in\Omega$ the sum (12a) contains exactly one term: $F(\boldsymbol{\omega})=f(\mathbf{y}(\boldsymbol{\omega}))$. Γ can be parametrised over Ω by means of $(r(\boldsymbol{\omega}),\boldsymbol{\omega})$. The concrete calculations for the practically interesting cases $d=2,3$ are as follows. For $d=2$, the intersection point $\mathbf{y}\in\Gamma$ has the representation

$$\mathbf{y}(\varphi) = \mathbf{x}_0 + r(\varphi)\binom{\cos\varphi}{\sin\varphi} \qquad \text{for } 0\leq\varphi<2\pi.$$

The derivative $\mathbf{y}_\varphi$ equals $r_\varphi\mathbf{e}_1+r\mathbf{e}_2$ with the orthonormal vectors

$$\boldsymbol{e}_1 := \binom{\cos\varphi}{\sin\varphi}, \qquad \boldsymbol{e}_2 := \binom{-\sin\varphi}{\cos\varphi}.$$

This yields the Gram determinant $g(\varphi)=\langle \mathbf{y}_\varphi, \mathbf{y}_\varphi\rangle = r_\varphi^2 + r^2$. The vector $-r_\varphi \boldsymbol{e}_2 + r\boldsymbol{e}_1$ is perpendicular to the tangential direction $\mathbf{y}_\varphi$ at $\mathbf{y}(\varphi)$. Normalisation yields the normal direction $\boldsymbol{n}(\mathbf{y}) = (-r_\varphi \boldsymbol{e}_2 + r\boldsymbol{e}_1)/\sqrt{g(\varphi)}$. Hence, the scalar product equals

$$\langle \boldsymbol{n}(\mathbf{y}), \mathbf{y}-\mathbf{x}_0\rangle = \langle \boldsymbol{n}(\mathbf{y}), r\boldsymbol{e}_1\rangle = r^2/\sqrt{g(\varphi)}.$$

Together with $|\mathbf{y}-\mathbf{x}_0| = r$ and (1.11a), we obtain the representation

$$\int_\Gamma \frac{\langle \boldsymbol{n}(\mathbf{y}), \mathbf{y}-\mathbf{x}_0\rangle}{|\mathbf{y}-\mathbf{x}_0|^2} f(\mathbf{y})\, d\Gamma_y = \int_0^{2\pi} \sqrt{g(\varphi)}\, f(\mathbf{y}(\varphi))\, \frac{r^2/\sqrt{g(\varphi)}}{r^2}\, d\varphi = \int_0^{2\pi} f(\mathbf{y}(\varphi))\, d\varphi.$$

The factor $-1/\omega_d$ of $\Phi(\mathbf{x}_0)$ proves (12b).

In the three-dimensional case, $\varphi\in[0,2\pi]$ and $\vartheta\in[0,\pi]$ are the angular coordinates. Any $\mathbf{y}=\mathbf{y}(\varphi,\vartheta)$ has the representation $r(\varphi,\vartheta)\boldsymbol{e}_1$, where

$$\boldsymbol{e}_1 = \begin{pmatrix}\cos\varphi\cos\vartheta\\ \cos\varphi\sin\vartheta\\ \sin\varphi\end{pmatrix}, \qquad \boldsymbol{e}_2 = \begin{pmatrix}-\sin\varphi\cos\vartheta\\ -\sin\varphi\sin\vartheta\\ \cos\varphi\end{pmatrix}, \qquad \boldsymbol{e}_3 = \begin{pmatrix}-\sin\vartheta\\ \cos\vartheta\\ 0\end{pmatrix}$$

are orthonormal vectors. The tangential plane is spanned by the derivatives $\mathbf{y}_\varphi = r_\varphi \boldsymbol{e}_1 + r\boldsymbol{e}_2$ and $\mathbf{y}_\vartheta = r_\vartheta \boldsymbol{e}_1 + r\cos(\varphi)\boldsymbol{e}_3$. A vector perpendicular to $\mathbf{y}_\varphi$ and $\mathbf{y}_\vartheta$ is $\boldsymbol{n}(\mathbf{y}) = [\cos(\varphi)(r\boldsymbol{e}_1 - r_\varphi \boldsymbol{e}_2) - r_\vartheta \boldsymbol{e}_3]\, r/\sqrt{g(\varphi,\vartheta)}$. Here, the denominator is the square root of the Gram determinant $g(\varphi,\vartheta) = r^2[\cos^2(\varphi)(r^2 + r_\varphi^2) + r_\vartheta^2]$. With $\langle \boldsymbol{n}(\mathbf{y}), \mathbf{y}-\mathbf{x}_0\rangle = r^3\cos^2(\varphi)/\sqrt{g(\varphi,\vartheta)}$ and $|\mathbf{y}-\mathbf{x}_0| = r$, one obtains (12b) by inserting these quantities.

(ii) In part (i), the condition $\langle \boldsymbol{n}(\mathbf{y}), \mathbf{y}-\mathbf{x}_0\rangle \geqslant \varepsilon > 0$ may be replaced by $\langle \boldsymbol{n}(\mathbf{y}), \mathbf{y}-\mathbf{x}_0\rangle > 0$.

(iii) In the second step, we assume that Γ is open but satisfies $\langle \boldsymbol{n}(\mathbf{y}), \mathbf{y}-\mathbf{x}_0\rangle > 0$ for all $\mathbf{y}\in\Gamma$ (analogous case: $\langle \boldsymbol{n}(\mathbf{y}), \mathbf{y}-\mathbf{x}_0\rangle < 0$ for all $\mathbf{y}\in\Gamma$). One can complete Γ by adding a further piece Γ' to create a closed curve or surface $\Gamma'' := \Gamma + \Gamma'$. Set, in addition, $f(\mathbf{y}) := 0$ for $\mathbf{y}\in\Gamma''$. Then the double-layer potentials over Γ and Γ'' are identical. The function $F(\boldsymbol{\omega})$ remains unchanged by the extension of f on Γ'' so that part (i) shows the assertion.

(iv) Let $\Gamma_0 := \Gamma_+ \cup \Gamma_-$ with $\Gamma_\pm := \{\mathbf{y}\in\Gamma\colon \pm\langle \boldsymbol{n}(\mathbf{y}), \mathbf{y}-\mathbf{x}_0\rangle > 0\}$. For any connection component of Γ_+ and Γ_-, (12b) holds because of (i-iii). Summation over all components yields (12b). QED

Second Proof of Theorem 13. If $\mathbf{x}$ lies in the interior domain Ω_-, the line $\{\mathbf{x}+t\boldsymbol{\omega}\colon t>0\}$ cuts Γ at $k+1$ points with $\langle \boldsymbol{n}(\mathbf{y}), \mathbf{y}-\mathbf{x}\rangle > 0$ and k points with $\langle \boldsymbol{n}(\mathbf{y}), \mathbf{y}-\mathbf{x}\rangle < 0$, where $k\in\mathbb{N}_0$. Since $f=1$, $F(\boldsymbol{\omega})=1$ results. The representation (12b) leads to -1, since ω_d is the surface of the unit sphere. If, however, $\mathbf{x}$ lies in the exterior domain, the case $\langle \boldsymbol{n}(\mathbf{y}), \mathbf{y}-\mathbf{x}\rangle > 0$ happens as often as $\langle \boldsymbol{n}(\mathbf{y}), \mathbf{y}-\mathbf{x}\rangle < 0$ leading to $F(\boldsymbol{\omega})=0$. QED

If $\Gamma\subset\mathbb{R}^3$ is an open surface with a closed curve $\partial\Gamma$ as boundary, which mapped onto the unit sphere Ω yields a double-point free curve γ, then $|F(\boldsymbol{\omega})|$ takes only the values 0 and 1. The subset $\Omega_0 := \{\boldsymbol{\omega}\in\Omega\colon |F(\boldsymbol{\omega})| = 1\}$ or its surface measure $\mu(\Omega_0)$ define the *solid angle* formed by Γ and the centre $\mathbf{x}_0$. The boundary $\partial\Omega_0$ coincides with γ.

For the postponed proof of Remark (9a), one proceeds as follows:

(i) If $\mathbf{x}_0$ is not a corner, repeat the proof of Lemma 7 and use that the convergence of the quantities Φ_0, $\Phi_{1,1}$, $\Phi_{1,2}$ is uniform with respect to $\mathbf{x}_0$ as $\alpha \to 0$, provided that $\mathbf{x}_0$ has a sufficiently large distance from the boundary $\partial\Gamma_i$ of the smooth component $\Gamma_i \subset \Gamma$, $\Gamma_i \in C^{1+\mu}$, since, otherwise, K_ε cannot be used for parametrisation. Since $\Phi_+ = Kf - f$ is continuous in the interior of Γ_i (cf. Theorem 3), the splitting $\Phi(\mathbf{x}) - \Phi_+(\mathbf{x}_0) = [\Phi(\mathbf{x}) - \Phi_+(\boldsymbol{\xi})] + [\Phi_+(\boldsymbol{\xi}) - \Phi_+(\mathbf{x}_0)]$ shows the continuity for $\mathbf{x} \in U_+ \to \mathbf{x}_0$, where $\boldsymbol{\xi} \in \Gamma$ is chosen such that $\mathbf{x} = \boldsymbol{\xi} + \alpha\mathbf{n}(\boldsymbol{\xi})$.

(ii) If $\mathbf{x}_0 \in \partial\Gamma_i$ is a corner, split f into $f_n + f_c$ with $f_c := f(\mathbf{x}_0)$, $f_n := f - f_c$. Without loss of generality, Γ can be supposed to be closed. By Theorem 13, the potential belonging to f_c is constant, hence (one-sided) continuous at $\mathbf{x}_0$. Consider the potentials Φ_i belonging to the different components Γ_i of Γ with the density f_n. Since $f_n(\mathbf{x}_0) = 0$, the integrals over *arbitrary subsets* of K_ε corresponding to $\Phi_{1,1}$ and $\Phi_{1,2}$ from the proof of Lemma 7 have the uniform limit 0. Altogether, Φ_+ is also continuous at $\mathbf{x}_0$. Continuity on a compact set yields the uniform continuity. QED

8.2.5 Derivatives of the Double-Layer Potential

The derivative of the integrand (2c) is defined for all $\mathbf{x} \in \Omega \backslash \Gamma$:

$$\text{(8.2.13a)} \qquad \nabla_x \left\{ -\frac{1}{\omega_d} \frac{\langle \mathbf{n}(\mathbf{y}), \mathbf{y}-\mathbf{x} \rangle}{|\mathbf{y}-\mathbf{x}|^d} \right\} = \frac{1}{\omega_d} \frac{\mathbf{n}(\mathbf{y})}{|\mathbf{y}-\mathbf{x}|^d} - \frac{d}{\omega_d} \frac{\langle \mathbf{n}(\mathbf{y}), \mathbf{y}-\mathbf{x} \rangle (\mathbf{y}-\mathbf{x})}{|\mathbf{y}-\mathbf{x}|^{d+2}}.$$

Therefore, in $\mathbb{R}^d \backslash \Gamma$, the double-layer potential Φ of the density f has the gradient

$$\text{(8.2.13b)} \quad \nabla\Phi(\mathbf{x}) = \frac{1}{\omega_d} \int_\Gamma \left\{ \frac{\mathbf{n}(\mathbf{y})}{|\mathbf{y}-\mathbf{x}|^d} - d \frac{\langle \mathbf{n}(\mathbf{y}), \mathbf{y}-\mathbf{x} \rangle (\mathbf{y}-\mathbf{x})}{|\mathbf{y}-\mathbf{x}|^{d+2}} \right\} f(\mathbf{y}) d\Gamma_y \quad \text{for } \mathbf{x} \notin \Gamma.$$

In the following, we investigate whether $\nabla\Phi$ can be extended to Γ. First, we give a comment concerning the assumption $f \in C^{1+\mu}(\Gamma)$. Since f is only defined on Γ, derivatives exist only in *tangential* directions. Defining formally the derivative in normal direction to be zero, we are able to define all partial derivatives and, therefore, also the gradient ∇f. An equivalent definition reads as follows: Extend f in a sufficiently small neighbourhood of Γ by $f(\mathbf{x}_0 + \alpha\mathbf{n}(\mathbf{x}_0)) := f(\mathbf{x}_0)$ $(\alpha \in [-\varepsilon, \varepsilon],\ \mathbf{x}_0 \in \Gamma)$. On Γ, the gradient in the neighbourhood of Γ exists and coincides with the above defined ∇f. By definition, $\nabla f(\mathbf{x}_0) \perp \mathbf{n}(\mathbf{x}_0)$ holds, i.e., $\nabla f(\mathbf{x}_0)$ lies in the tangential plane.

The following theorem shows that the gradient of the double-layer potential Φ_D behaves in a complementary manner to the gradient $\nabla\Phi_E$ of the single-layer potential: While the tangential derivative $\langle \nabla\Phi_E, \mathbf{t} \rangle$ of the single-layer potential is continuous and the normal derivative $\langle \nabla\Phi_E, \mathbf{n} \rangle$ jumps, the statements are reversed for the gradient of the double-layer potential.

Theorem 8.2.15. Let $\Gamma \in C_0^{1+\mu}$ and $f \in C^{1+\mu}(\Gamma)$ for some $\mu > 0$. Then the gradient $\nabla\Phi$ of the double-layer potential is continuous in Ω_+ and Ω_- and can be extended continuously to $\nabla\Phi_\pm$ on $\overline{\Omega}_\pm$ with the *jump properties*

(8.2.14a) $\quad \nabla\Phi_+(\mathbf{x}) - \nabla\Phi_-(\mathbf{x}) = \nabla f(\mathbf{x}) \qquad$ for $\mathbf{x} \in \Gamma$,

(8.2.14b) $\quad \nabla\Phi_+(\mathbf{x}) + \nabla\Phi_-(\mathbf{x}) = \qquad$ for $\mathbf{x} \in \Gamma$

$$= \frac{2}{\omega_d} \int_\Gamma \left\{ \frac{\mathbf{n}(\mathbf{y})}{|\mathbf{y}-\mathbf{x}|^d} - d\, \frac{\langle \mathbf{n}(\mathbf{y}), \mathbf{y}-\mathbf{x}\rangle\, (\mathbf{y}-\mathbf{x})}{|\mathbf{y}-\mathbf{x}|^{d+2}} \right\} f(\mathbf{y})\, d\Gamma_y,$$

where the integral is to be understood in the sense of Hadamard (cf. §7.5). The *normal derivative* is continuous when crossing Γ, i.e.,

(8.2.14c) $\quad \partial\Phi_+(\mathbf{x})/\partial n = \partial\Phi_-(\mathbf{x})/\partial n \qquad$ for $\mathbf{x} \in \Gamma$,

and has the following representation as Hadamard integral:

(8.2.14d) $\quad \partial\Phi_\pm(\mathbf{x})/\partial n = \int_\Gamma f(\mathbf{y}) \frac{\partial}{\partial n_x} \frac{\partial}{\partial n_y} s(\mathbf{x},\mathbf{y})\, d\Gamma_y = \qquad (\mathbf{x} \in \Gamma)$

$$= \frac{1}{\omega_d} \int_\Gamma \left\{ \frac{\langle \mathbf{n}(\mathbf{y}), \mathbf{n}(\mathbf{x})\rangle}{|\mathbf{y}-\mathbf{x}|^d} - d\, \frac{\langle \mathbf{n}(\mathbf{y}), \mathbf{y}-\mathbf{x}\rangle \langle \mathbf{y}-\mathbf{x}, \mathbf{n}(\mathbf{x})\rangle}{|\mathbf{y}-\mathbf{x}|^{d+2}} \right\} f(\mathbf{y})\, d\Gamma_y .$$

Proof. (i) Since Φ is arbitrarily smooth outside of Γ, it suffices to discuss the limit process $\mathbf{x} \to \mathbf{x}_0 \in \Gamma$ from Ω_+ or Ω_-. Extend f into the neighbourhood U of Γ using the procedure which follows equation (13b). Since the constant density $f(\mathbf{x})$ ($\mathbf{x} \in U$ fixed) leads to a piecewise constant potential Φ_0 possessing a vanishing derivative $\nabla\Phi_0 = 0$ in $\Omega \setminus \Gamma$ (cf. Theorem 13), one may replace the density f by $f(\cdot) - f(\mathbf{x})$ without changing the gradient $\nabla\Phi$:

(8.2.14e) $$\nabla\Phi(\mathbf{x}) = \frac{1}{\omega_d} \int_\Gamma \left\{ \frac{\mathbf{n}(\mathbf{y})}{|\mathbf{y}-\mathbf{x}|^d} - d\, \frac{\langle \mathbf{n}(\mathbf{y}), \mathbf{y}-\mathbf{x}\rangle\, (\mathbf{y}-\mathbf{x})}{|\mathbf{y}-\mathbf{x}|^{d+2}} \right\} [f(\mathbf{y}) - f(\mathbf{x})]\, d\Gamma_y .$$

The Taylor expansion of f yields

(8.2.14f) $$f(\mathbf{y}) - f(\mathbf{x}) = \langle \nabla f(\mathbf{x}), \mathbf{y}-\mathbf{x}\rangle + O(|\mathbf{y}-\mathbf{x}|^{1+\mu}).$$

The $O(|\mathbf{y}-\mathbf{x}|^{1+\mu})$ term leads to a uniformly existing improper integral, which therefore is continuous with respect to $\mathbf{x}$. Inserting the first term into the integrand (14e), we obtain the integrals

(8.2.14g) $$\frac{1}{\omega_d} \int_\Gamma \frac{\mathbf{n}(\mathbf{y})}{|\mathbf{y}-\mathbf{x}|^d} \langle \nabla f(\mathbf{x}), \mathbf{y}-\mathbf{x}\rangle\, d\Gamma_y ,$$

(8.2.14h) $$\frac{d}{\omega_d} \int_\Gamma \frac{\langle \mathbf{n}(\mathbf{y}), \mathbf{y}-\mathbf{x}\rangle\, (\mathbf{y}-\mathbf{x})}{|\mathbf{y}-\mathbf{x}|^{d+2}} \langle \nabla f(\mathbf{x}), \mathbf{y}-\mathbf{x}\rangle\, d\Gamma_y .$$

Since $\mathbf{t} := \nabla f(\mathbf{x})$ is a fixed *tangential* direction (see above), (14g) has the same form as (1.21). Analogously to Lemma 1.13, the limit $\mathbf{x} \to \mathbf{x}_0$ leads to the Cauchy principal value $⨏_\Gamma \mathbf{n}(\mathbf{y}) \langle \nabla f(\mathbf{x}_0), \mathbf{y}-\mathbf{x}_0\rangle / |\mathbf{y}-\mathbf{x}_0|^d\, d\Gamma_y$. Concerning the analysis of (14h), we first consider the tangential component

(8.2.14i) $$-\frac{d}{\omega_d} \int_\Gamma \frac{\langle \mathbf{n}(\mathbf{y}), \mathbf{y}-\mathbf{x}\rangle}{|\mathbf{y}-\mathbf{x}|^d} \frac{\langle \mathbf{y}-\mathbf{x}, \mathbf{t}(\mathbf{x}_0)\rangle}{|\mathbf{y}-\mathbf{x}|} \frac{\langle \nabla f(\mathbf{x}), \mathbf{y}-\mathbf{x}\rangle}{|\mathbf{y}-\mathbf{x}|}\, d\Gamma_y .$$

The first factor is the same as in the double-layer potential, while the second and third factors replace the density $f(x)$, so that analogous one-sided limit exists. The jump turns out to be

$$\mp\frac{d}{\omega_d}\int_0^\infty r^{d-2}[r^2+1]^{-d/2}dr\int_\Omega\langle\nabla f(\mathbf{x}_0),\mathbf{z}\rangle\langle\mathbf{z},\mathbf{t}(\mathbf{x}_0)\rangle d\Omega,$$

where Ω denotes the $(d-1)$-dimensional unit sphere (for $d=2$, we have $\Omega=\{0,2\}$ and we set formally $\int_\Omega\dots d\Omega=2\langle\nabla f(\mathbf{x}_0),\mathbf{t}(\mathbf{x}_0)\rangle$). Since the one-sided continuous extendability of $\langle\nabla\Phi,\mathbf{t}(\mathbf{x}_0)\rangle$ onto Γ is ensured, one need not evaluate the integral from above, but can form the tangential derivative in equation (7a). This yields

(8.2.14j) $$\langle\nabla\Phi_+(\mathbf{x}_0)-\nabla\Phi_-(\mathbf{x}_0),\mathbf{t}(\mathbf{x}_0)\rangle=\langle\nabla f(\mathbf{x}_0),\mathbf{t}(\mathbf{x}_0)\rangle$$

for any tangential direction $\mathbf{t}(\mathbf{x}_0)$ at $\mathbf{x}_0\in\Gamma$.

(ii) Finally, the normal component

(8.2.14k) $$-\frac{d}{\omega_d}\int_\Gamma\frac{\langle\mathbf{n}(\mathbf{y}),\mathbf{y}-\mathbf{x}\rangle\langle\mathbf{y}-\mathbf{x},\mathbf{n}(\mathbf{x}_0)\rangle}{|\mathbf{y}-\mathbf{x}|^{d+1}}\frac{\langle\nabla f(\mathbf{x}),\mathbf{y}-\mathbf{x}\rangle}{|\mathbf{y}-\mathbf{x}|}d\Gamma_y$$

is to be investigated. Parametrising a neighbourhood Γ_ε of $\mathbf{x}_0\in\Gamma$ by means of S_ε as in $(1.17c_1)$, one obtains

$$\frac{d}{\omega_d}\int_{S_\varepsilon}\frac{\{\alpha-\eta(\mathbf{y}')-\langle\nabla\eta(\mathbf{y}'),\mathbf{y}'\rangle\}\ (\alpha-\eta(\mathbf{y}'))}{[(\alpha-\eta(\mathbf{y}'))^2+|\mathbf{y}'|^2]^{(d+1)/2}}\frac{\langle\nabla f(\mathbf{x}_0),\mathbf{y}'\rangle}{[\dots]^{1/2}}\sqrt{g}\,d\Gamma_y,$$

where $\mathbf{x}_0=0$, $\mathbf{x}=\alpha\mathbf{n}(\mathbf{x}_0)$, $\mathbf{y}=\binom{\mathbf{y}'}{\eta(\mathbf{y}')}$ as in (1.17f). Since $\langle\nabla\eta(\mathbf{y}'),\mathbf{y}'\rangle$ is of the order $O(|\mathbf{y}'|^{1+\mu})$, the integrand $\langle\nabla\eta(\mathbf{y}'),\mathbf{y}'\rangle(\alpha-\eta(\mathbf{y}'))\langle\nabla f(\mathbf{x}_0),\mathbf{y}'\rangle/[\dots]^{d/2+1}$ has an improperly integrable majorant. Therefore, the corresponding integral is continuous with respect to $\alpha\to0$. The remaining term equals

(8.2.14l) $$\frac{d}{\omega_d}\int_{S_\varepsilon}\frac{(\alpha-\eta(\mathbf{y}'))^2\langle\nabla f(\mathbf{x}_0),\mathbf{y}'\rangle}{[(\alpha-\eta(\mathbf{y}'))^2+|\mathbf{y}'|^2]^{d/2+1}}\sqrt{g}\,d\Gamma_y$$

and converges as $\alpha\to0$, as one sees after substituting $\mathbf{y}'\mapsto\mathbf{z}=\alpha\mathbf{y}'$. In contrast to the integral $(1.17c_1)$, the limit does not depend on the sign of α, since in (14l) the square $(\alpha-\eta(\mathbf{y}'))^2$ appears instead of $\alpha-\eta(\mathbf{y}')$. Hence, $\langle\nabla\Phi,\mathbf{n}(\mathbf{x}_0)\rangle$ is continuous at $\mathbf{x}_0$. Since $\nabla f(\mathbf{x}_0)\perp\mathbf{n}(\mathbf{x}_0)$, the equations (14j) and $\langle\nabla\Phi_+-\nabla\Phi_-,\mathbf{n}(\mathbf{x}_0)\rangle=0$ imply (14a).

(iii) The Hadamard integral (14b) is the Cauchy principal value of (14e) (cf. §7.5). ∎

Corollary 8.2.16. **(a)** Let $\Gamma_0\subset\Gamma$ be an open subset with $\Gamma_0\in C^{1+\mu}$, assume $f\in C^{1+\mu}_{\mathrm{loc}}(\Gamma_0)$ and weaken the smoothness assumptions outside Γ_0 to $\Gamma\in C^1_{\mathrm{stw}}$ and $f\in L^\infty(\Gamma)$. Then the jump relations (14a,b) also hold locally in Γ_0.

(b) If the Hölder property of Γ and f holds only pointwise at $\mathbf{x}_0$, then (14a,b) still remains true, if one does not approach $\mathbf{x}_0$ tangentially, i.e., if $|\langle\mathbf{x}-\mathbf{x}_0,\mathbf{n}(\mathbf{x}_0)\rangle|\geq\varepsilon|\mathbf{x}-\mathbf{x}_0|$ holds for some $\varepsilon>0$.

(c) Let $\Gamma\in C^1_{0,\mathrm{stw}}$. In $\mathbb{R}^d\setminus\Gamma$, the following estimates hold

$$|\nabla\Phi(\mathbf{x})|\leq C_\Gamma\,\mathrm{dist}(\mathbf{x},\Gamma)^{\mu-1}\qquad\text{for } f\in C^\mu(\Gamma),\ 0\leq\mu<1,$$

$$|\nabla\Phi(\mathbf{x})|\leq C_1\,|\log\mathrm{dist}(\mathbf{x},\Gamma)|\qquad\text{for } f\in C_L(\Gamma).$$

Here, the neighbourhoods of corners or edges are included.
(d) For $\mu>0$, $f\in C^{\mu}(\Gamma)$ on $\Gamma\in C^1_{0,\mathrm{stw}}$ generates a double-layer potential with $|\nabla\Phi|\in L^2(\Omega_\pm)$. For $d\geq 3$, also $\Phi\in H^1(\Omega_\pm)$ holds.
(e) The asymptotical behaviour for $|\mathbf{x}|\to\infty$ is

$$\text{(8.2.15a)}\qquad \Phi(\mathbf{x}) = O(|\mathbf{x}|^{1-d}),$$

$$\text{(8.2.15b)}\qquad |\nabla\Phi(\mathbf{x})| = O(|\mathbf{x}|^{-d}).$$

Proof. The parts (a) and (b) result from the considerations in the proof of Theorem 15. (c) is a consequence of Lemma 4 applied to the representation (13b) of $\nabla\Phi$. This representation (13b) also yields (d). (e) is a consequence of (c) and (d). QED

For $\alpha\neq 0$, the following normal derivative always exists:

$$\text{(8.2.16a)}\qquad \varphi(\alpha) := \langle\nabla\Phi(\mathbf{x}_0+\alpha\mathbf{n}(\mathbf{x}_0)),\mathbf{n}(\mathbf{x}_0)\rangle,$$

but the limit of $\varphi(\alpha)$ for $\alpha\to 0$ may fail to exist if f is only continuous. The following lemma gives sufficient conditions for the equality of the one-sided normal derivatives in an appropriate limiting process.

Lemma 8.2.17. Let $\Gamma_0\subset\Gamma\in C^1_{0,\mathrm{stw}}$, $\Gamma_0\in C^{1+\mu}$ for some $\mu>0$ and $f\in L^\infty(\Gamma)$. Let f be continuous at $\mathbf{x}_0$ (in this case, set $\lambda:=0$) or Hölder continuous, i.e., $|f(\mathbf{x})-f(\mathbf{x}_0)|\leq C|\mathbf{x}-\mathbf{x}_0|^\lambda$ with $\lambda\in(0,1)$. Assume $\mu+\lambda>1$ or $\mu=1$, $\lambda=0$.
(a) Then

$$\text{(8.2.16b)}\qquad \lim_{\substack{\alpha\to 0\\ \alpha>0}} \varphi(\alpha)-\varphi(-\alpha) = 0.$$

(b) If the interior potential Φ_- admits the normal derivative $\partial\Phi_-/\partial n$ at $\mathbf{x}_0$, then also the normal derivative $\partial\Phi_+/\partial n$ of the exterior potential exists and coincides with $\partial\Phi_-/\partial n$.

Proof. Since $\partial\Phi_-/\partial n$ is the limit of $\varphi(\alpha)$ for $0>\alpha\to 0$, part (b) follows from (a). For the proof of part (a) we may assume $f(\mathbf{x}_0)=\mathbf{x}_0=0$ without loss of generality as in the proof of Theorem 15. We set $\mathbf{x}_+:=\mathbf{x}_0+\alpha\mathbf{n}(\mathbf{x}_0)$ and $\mathbf{x}_-:=\mathbf{x}_0-\alpha\mathbf{n}(\mathbf{x}_0)$. After multiplication by $\mathbf{n}(\mathbf{x}_0)$, the first term $\mathbf{n}(\mathbf{y})/|\mathbf{y}-\mathbf{x}|^d$ of the integrand in (14e) leads to the difference

$$\text{(8.2.16c)}\qquad \langle\mathbf{n}(\mathbf{y}),\mathbf{n}(\mathbf{x}_0)\rangle\left[|\mathbf{y}-\mathbf{x}_+|^{-d}-|\mathbf{y}-\mathbf{x}_-|^{-d}\right].$$

We parametrise over $S_\varepsilon=\{\mathbf{y}'\in\mathbb{R}^{d-1}: |\mathbf{y}'|\leq\varepsilon\}$ and suppose that $\mathbf{y}=\binom{\mathbf{y}'}{\eta(\mathbf{y}')}$ with η as in (1.17f). Since $\langle\mathbf{n}(\mathbf{y}),\mathbf{n}(\mathbf{x}_0)\rangle=\langle\mathbf{n}(\mathbf{y})-\mathbf{n}(\mathbf{x}_0),\mathbf{n}(\mathbf{x}_0)\rangle+1=1+O(|\mathbf{y}-\mathbf{x}_0|^\mu)$, it remains to investigate the bracket [...] in (16c). The mean value theorem of differential calculus yields

$$|\mathbf{y}-\mathbf{x}_+|^{-d}-|\mathbf{y}-\mathbf{x}_-|^{-d} = [|\mathbf{y}'|^2+(\alpha-\eta(\mathbf{y}'))^2]^{-d/2}-[|\mathbf{y}'|^2+(\alpha+\eta(\mathbf{y}'))^2]^{-d/2} =$$
$$= d\,\frac{2\eta(\mathbf{y}')[\alpha+\vartheta\eta(\mathbf{y}')]}{[|\mathbf{y}'|^2+(\alpha+\vartheta\eta(\mathbf{y}'))^2]^{(d+2)/2}}\qquad\text{with } |\vartheta|<1.$$

By $|\alpha+\vartheta\eta|\leq[\ldots\text{denominator}\ldots]^{1/2}$ and $|\eta(\mathbf{y}')|\leq C|\mathbf{y}'|^{1+\mu}\leq C[\ldots]^{(1+\mu)/2}$, one obtains the majorant $C[\ldots]^{(-d+\mu)/2}$. Analogously, one proceeds with

the second term of the integral (13b).
(ii) Since without loss of generality $f(\mathbf{x}_0)=\mathbf{x}_0=0$, one finally gains from $|f(\mathbf{y})|=|f(\mathbf{y})-f(\mathbf{x}_0)|\le C|\mathbf{y}-\mathbf{x}_0|^\lambda=C|\mathbf{y}|^\lambda$ the improperly integrable majorant $C|\mathbf{y}|^{-d+\mu+\lambda}$ in the case of $\mu+\lambda>1$. This proves the continuity of $\psi(\alpha):=\varphi(\alpha)-\varphi(-\alpha)$ with respect to α. A continuous and odd function (i.e., $\psi(-\alpha)==-\psi(\alpha)$) must attain the value $\psi(0)=0$ at $\alpha=0$. This implies (16b).
(iii) Using $|\eta(\mathbf{y}')|\le C|\mathbf{y}'|^2$, one estimates $\left|\frac{2\eta(\mathbf{y}')[\alpha+\vartheta\eta(\mathbf{y}')][f(\mathbf{y}_0)-f(\mathbf{x}_0)]}{[|\mathbf{y}'|^2+(\alpha+\vartheta\eta(\mathbf{y}'))^2]^{(d+2)/2}}\right|$ in the case of $\mu=1$, $\lambda=0$ by

$$C\,\frac{|\alpha+\vartheta\eta(\mathbf{y}')||f(\mathbf{y}_0)-f(\mathbf{x}_0)|}{[|\mathbf{y}'|^2+(\alpha+\vartheta\eta(\mathbf{y}'))^2]^{d/2}}\le C\|f(\cdot)-f(\mathbf{x}_0)\|_{\infty,K_\rho}\frac{|\alpha+\vartheta\eta(\mathbf{y}')|}{[\dots]^{d/2}}$$

for all $\mathbf{y}'\in K_\rho(\mathbf{x}_0)$. Since f is continuous, the maximum norm $\|f(\cdot)-f(\mathbf{x}_0)\|_{\infty,K_\rho}$ converges to zero as $\rho\to0$. The integral over $|\alpha+\vartheta\eta(\mathbf{y}')|/[\dots]^d$ is uniformly bounded with respect to α (cf. Lemma 1.10c). Splitting $\psi(\alpha):=\varphi(\alpha)-\varphi(-\alpha)$ into $\psi_1+\psi_2$, where $\psi_i=\varphi_i(\alpha)-\varphi_i(-\alpha)$ arises from integration over the respective parts $\Gamma\cap K_\rho(\mathbf{x}_0)$ and $\Gamma\setminus K_\rho(\mathbf{x}_0)$, we can conclude from the continuity of f at $\mathbf{x}_0$ that $|\psi_1(\alpha)|\le\varepsilon/2$ for sufficiently small ρ and all $\alpha\in\mathbb{R}$. Since ψ_2 is arbitrarily smooth in α, $|\psi_2(\alpha)|\le\varepsilon/2$ holds for sufficiently small α. These results then imply the continuity of ψ at $\alpha=0$, i.e., $\lim_{\alpha\to0}\psi(\alpha)=0$. QED

In Theorem 15 we have discussed the one-sided continuity of the derivative of Φ. Moreover, *global Hölder continuously differentiability* holds under slightly stronger assumptions. Colton-Kreß [1, Theorem 2.23] prove the following result. Assume $\Gamma\in C_0^2$ and $f\in C^{1+\lambda}(\Gamma)$. Then the double-layer potential is globally Hölder continuously differentiable in the respective domains $\overline{\Omega}_\pm$ and satisfies the estimate

$$\|\Phi\|_{C^{1+\lambda}(\overline{\Omega}_\pm)}\le C\,\|f\|_{C^{1+\lambda}(\Gamma)}.$$

8.2.6 Integral Equations with the Double-Layer Operator

8.2.6.1 Formulation of the Dirichlet Boundary Value Problem as Integral Equation of the Second Kind with the Double-Layer Operator

We use the double-layer potential ansatz (2c) for the solution u of the boundary value problem

$$\Delta u=0\ \text{ in }\Omega,\qquad u=\varphi\ \text{ on }\Gamma, \tag{8.2.17}$$

since Φ satisfies the Laplace equation (cf. (8)). In the case of the *interior problem*,

$$\Phi_-=\varphi$$

must hold. Then equations (7a,b) yield $2\Phi_-=Kf-f$; hence,

$$f=-2\varphi+\overline{K}f \qquad \text{(interior problem)}. \tag{8.2.18}$$

Concerning $\overline{K}$, recall Remark 9d,e. This second kind Fredholm integral equation for the double-layer density f is adjoint to (1.32).

In the case of the *exterior problem*, one has to require

$$\Phi_+=\varphi.$$

The sum of the equations (7a,b) leads to

(8.2.19) $f = 2\varphi - \overline{K}f$ (exterior problem).

Theorem 8.2.18 (uniqueness). Let $\varphi \in L^\infty(\Gamma)$ and $\Gamma \in C^{1+\mu}_{0,\mathrm{stw}}$ for some $\mu > 1/2$.
(a) The integral equation (18) has at most one solution $f \in L^\infty(\Gamma)$.
(b) Any two solutions of equation (19) can differ only by a constant.

Proof. (i) If equation (18) has two solutions, then the difference leads to $f \in L^\infty(\Gamma)$ with $f = \overline{K}f$. Theorem 10b shows that $\overline{K}f$ (and therefore also f) is Hölder continuous with exponent $\lambda := \mu > 1/2$ except corner and edge points. According to Theorem 8, the corresponding double-layer potential Φ belonging to f satisfies the jump relations (7a,b), implying $\Phi_- = 0$ on Γ. The unique solution of the Laplace equation corresponding to boundary values $\Phi_- = 0$ in Ω_- is $\Phi = 0$. It satisfies $\partial\Phi_-/\partial n = 0$ on Γ. By Lemma 17b and because of $\mu + \lambda = 2\mu > 1$, $\partial\Phi_+/\partial n = 0$ must also hold for Φ in the *exterior domain.* The only solution of the exterior problem subject to the Neumann condition $\partial\Phi_-/\partial n = 0$ is $\Phi = 0$ in Ω_-, as can be seen from the following Lemma 19. From $\Phi = 0$ in $\mathbb{R}^d \setminus \Gamma$, one concludes $f = 0$ because of (7a).
(ii) The difference of two solutions of (19) leads to a solution $f \in L^\infty(\Gamma)$ of $f = -\overline{K}f$, which must be Hölder continuous almost everywhere. As in (i), one derives that $\Phi_+ = 0$ on Γ. Together with (20), one concludes that $\Phi = 0$ in Ω_- and $\partial\Phi_\pm/\partial n = 0$ on Γ. Since $\Phi = \text{const}$ in Ω_- describes all solutions of the boundary condition $\partial\Phi_-/\partial n = 0$, (7a) proves that $f = \text{const}$. QED

Lemma 8.2.19. **(a)** At infinity, any double-layer potential behaves as follows:

(8.2.20) $|\Phi(\mathbf{x})| \leqslant C|\mathbf{x}|^{1-d}, \qquad |\nabla\Phi(\mathbf{x})| \leqslant C|\mathbf{x}|^{-d}.$

(b) The Neumann boundary value problem for the exterior Laplace problem $\Delta\Phi = 0$ in Ω_- and $\partial\Phi_-/\partial n = \varphi$ on Γ, which, in addition, is assumed to satisfy condition (20), has a unique solution (cf. Lemma 1.18).

Corollary 8.2.20. To enforce uniqueness of problem (19), the augmented equation (21a,b) can be posed according to (4.8.15a,b):

(8.2.21a) $f = 2\varphi - \overline{K}f + \alpha$ ($\alpha \in \mathbb{R}$: unknown constant),

(8.2.21b) $\int_\Gamma f\,d\Gamma = \beta$ ($\beta \in \mathbb{R}$: given constant).

Another modification ensuring uniqueness will be mentioned in §8.5.1, where one regains the Laplace equation for $\varkappa = 0$.

Theorem 8.2.21 (solvability). Assume $\Gamma \in C^{1+\mu}_0$ for some $\mu \in (0,1)$ and $\varphi \in C^\lambda(\Gamma)$ for some $\lambda \in [0,\mu]$. Then equation (18) for the interior problem and equation (21a,b) for the exterior problem both admit a unique solution $f \in C(\Gamma)$, which moreover belongs to $C^\lambda(\Gamma)$.

Proof. By Theorem 5, K is compact in $C(\Gamma)$. Because of the Riesz-Schauder theory (cf. Theorem 1.3.28), uniqueness (injectivity) implies solvability (surjectivity). For $\mu>\frac{1}{2}$, the uniqueness follows from Theorem 18. Otherwise, one uses the bijectivity of the dual integral operator proved in Theorem 1.23. Finally, (4a) implies $f\in C^{\lambda}(\Gamma)$. QED

8.2.6.2 Formulation of the Neumann Boundary Value Problem as Integral Equation of the Second Kind with the Double-Layer Operator

If u is a (sufficiently smooth) solution of the Laplace equation in the interior domain Ω_-, it satisfies *Green's representation formula*:

$$u(\mathbf{x}) = \int_{\Gamma}\{s(\mathbf{x},\mathbf{y})\frac{\partial}{\partial n}u(\mathbf{y}) - u(\mathbf{y})\frac{\partial}{\partial n_y}s(\mathbf{x},\mathbf{y})\}d\Gamma_y \qquad (\mathbf{x}\in\Omega_-) \tag{8.2.22}$$

(cf. Hackbusch [2, Theorem 2.2.21]). Inserting the particular fundamental solution (1.3) for $d\geq 3$ into the right-hand side of (22), we obtain the potential

$$\Phi(\mathbf{x}) = \frac{1}{\omega_d}\int_{\Gamma}\frac{\varphi(\mathbf{y})}{|\mathbf{y}-\mathbf{x}|^{d-2}}d\Gamma_y + \frac{1}{\omega_d}\int_{\Gamma}\frac{\langle\mathbf{n}(\mathbf{y}),\mathbf{y}-\mathbf{x}\rangle}{|\mathbf{y}-\mathbf{x}|^{d}}u(\mathbf{y})\,d\Gamma_y, \tag{8.2.23}$$

where the first integral makes use of the *Neumann boundary condition*

$$\frac{\partial}{\partial n}u(\mathbf{x}) = \varphi(\mathbf{x}) \qquad \text{on } \Gamma. \tag{8.2.24}$$

The right-hand side in (23) is a linear combination of the single-layer potential Φ_E for the density φ and of the double-layer potential Φ_D with the Dirichlet boundary values u on Γ as density, i.e., $\Phi=\Phi_E-\Phi_D$. For the posed Neumann boundary value problem, the Dirichlet data are unknown, so that formula (22) is not an *explicit* but only an *implicit* representation.

Following (22), Φ must coincide with u in the interior domain Ω_-. Outside, for $\mathbf{x}\in\Omega_+$, the right-hand side of (22) vanishes, since integration by parts leads to $\int_\Omega[s\Delta u-u\Delta_y s]d\mathbf{y}=0$. Hence, the mean value of $\Phi_\pm$ equals $u/2$ on Γ. Φ_E is continuous in $\mathbb{R}^d$, so that this part of $\Phi=\Phi_E-\Phi_D$ can directly be evaluated on Γ. By Theorem 8, the mean value of the two continuous extensions $\Phi_{D,\pm}$ of Φ_D yield the value $\frac{1}{2}Ku$, where K is the double-layer operator (2b). Together, one obtains the following condition on Γ for the unknown Dirichlet values of u:

$$u(\mathbf{x}) = g(\mathbf{x}) - (Ku)(\mathbf{x}) \qquad \text{for } x\in\Gamma, \tag{8.2.25a}$$

where, according to (1.3),

$$g(\mathbf{x}) := \frac{2}{\omega_d}\int_{\Gamma}\frac{\varphi(\mathbf{y})}{|\mathbf{y}-\mathbf{x}|^{d-2}}\,d\Gamma_y \qquad \text{for } d\geq 3, \tag{8.2.25b}$$

$$(\varphi \text{ from (24)})$$

$$g(\mathbf{x}) := -\frac{1}{\pi}\int_{\Gamma}\varphi(\mathbf{y})\log|\mathbf{y}-\mathbf{x}|\,d\Gamma_y \qquad \text{for } d=2. \tag{8.2.25c}$$

The integral equation (25a) differs from equation (19) only in its inhomogeneous parts. Formally, both are identical. The question of uniqueness (up to multiples of the constant) and existence of a solution is already answered by the Theorems 18 and 21. To obtain uniqueness including the constant, one can again make use of the formulation corresponding to (21a,b).

The solution u of the partial differential equation $\Delta u=0$ is the *direct* result of the integral equation (25a) and one need not to insert the solution f of the integral equation into the double-layer ansatz (2c) as in the case of equation (18) or (19). Therefore, equation (25a) is also called the «*direct method*» or the «*direct integral equation method*». Other approaches are *indirect methods*.

The direct method (25a) was derived from Green's representation (22). However, one can also prove directly that the solution u of equation (25a) is the solution of the Neumann boundary value problem.

Exercise 8.2.22. Let $\Gamma\in C^{1+\mu}_{0,\mathrm{stw}}$ for some $\mu>1/2$. Let equation (25a) have a solution $u\in C(\Gamma)$ for the Neumann boundary values $\varphi\in L^{\infty}(\Gamma)$. Prove the following: **(a)** φ and u are locally Hölder continuous with exponent μ up to exceptional points (corners, edges).
(b) The potential (23) defined by the solution u satisfies the following properties: (i) $\Delta\Phi=0$ in $\mathbb{R}^d\setminus\Gamma$, (ii) $\Phi_-=u$ on Γ, (iii) $\Phi_+=0$ in the exterior domain Ω_+, (iv) $\partial\Phi_-/\partial n=\varphi$, so that $\Phi=u$ is the solution of the Neumann boundary value problem. *Hint for (ii):* (7a,b), *for (iv):* (1.16b) and Lemma 17b for $\partial\Phi_+/\partial n-\partial\Phi_-/\partial n=-\varphi$, $\partial\Phi_+/\partial n=0$ according to (iii).

The previous discussion was concerned with the interior problem. For the *exterior problem*, Green's representation formula becomes

$$u(\mathbf{x}) = -\int_{\Gamma}\Big\{s(\mathbf{x},\mathbf{y})\frac{\partial}{\partial n}u(\mathbf{y})-u(\mathbf{y})\frac{\partial}{\partial n_y}s(\mathbf{x},\mathbf{y})\Big\}d\Gamma_y \qquad (\mathbf{x}\in\Omega_+), \tag{8.2.26}$$

where $\partial/\partial n$ describes again the derivative in the *outer* normal direction. Corresponding to the change of the sign in (26), we now obtain the integral equation

$$u(\mathbf{x}) = -g(\mathbf{x})+(Ku)(\mathbf{x}) \qquad \text{for } x\in\Gamma \text{ with } g \text{ from (25b,c).} \tag{8.2.27}$$

Theorem 18 guarantees the uniqueness of the solution. The existence of a solution is investigated in Theorem 21.

Green's representations (22) and (26) enable also a *direct* method *of the first kind* for determining the Neumann values from given Dirichlet values, in order to determine afterwards u from (22/26).

Exercise 8.2.23. We pose the *Dirichlet* problem $u=\psi$ of the Laplace equation $\Delta u=0$. Insert $u=\psi$ into the respective equation (22) or (26) and derive an integral equation of the first kind for $\varphi:=\partial u/\partial n$. Rework the Theorems 1.20 and 1.22 for this equation.

8.2.7 Non-Smooth Curves or Surfaces

While uniqueness of the integral equation (18) could be shown for piecewise smooth curves or surfaces, the existence statement in Theorem 21 requires *global* smoothness. Similarly for the exterior problem (21a,b). The proof of Theorem 21 is based on the compactness of K. First, we want to demonstrate that the global smoothness $\Gamma\in C^{1+\mu}_0$ is not only sufficient (as shown in Theorem 5), but also necessary for the compactness of K in $C(\Gamma)$. To simplify the notation, we restrict ourselves to the two-dimensional case $d=2$. Nonetheless, the behaviour of K at a corner of $\Gamma\subset\mathbb{R}^2$ carries over to an edge of a surface $\Gamma\subset\mathbb{R}^3$.

Lemma 8.2.24. If $\Gamma\in C^{1+\mu}_{0,stw}$ $(\mu>0)$ contains at least one corner, then the double-layer operator K is not compact in $L^\infty(\Gamma)$ or $C(\Gamma)$.

Proof. Without loss of generality, consider the situation from Fig. 5a: The corner $\mathbf{x}_0=0$ has an angle $\alpha\in(0,\pi)$ with the both sides Γ_0 and Γ_1. The functions φ_n defined in the following have their support in the part $\Gamma_0=\{\mathbf{x}=\binom{x}{0}: 0\leqslant x\leqslant 1\}\subset\Gamma$:

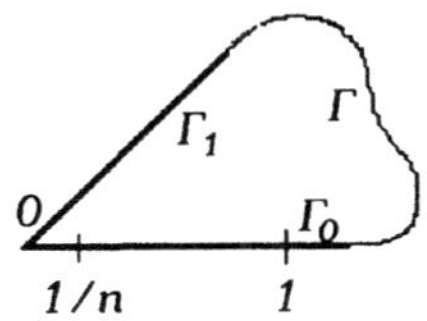

Fig. 8.2.5a

$$\varphi_n(\mathbf{x}) = \begin{cases} 1 & \text{for } \mathbf{x}=\binom{x}{0}\in\Gamma_0,\ 1/n\leqslant x\leqslant 1, \\ 0 & \text{otherwise.}\end{cases}$$

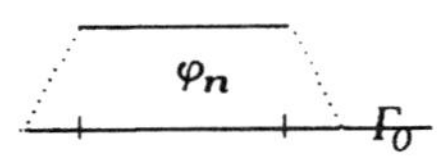

Fig. 8.2.5b

φ_n belongs to $L^\infty(\Gamma)$ and has the norm $\|\varphi_n\|_\infty=1$. If K were compact, $\psi_n:=K\varphi_n$ would contain a uniformly convergent subsequence $\{\psi_{n_j}=K\varphi_{n_j}: j\in\mathbb{N}\}$. Since $\psi_n=K\varphi_n$ coincides in the upper part $\Gamma_1=\{\mathbf{x}=r\binom{\cos\alpha}{\sin\alpha}: 0\leqslant r\leqslant 1\}$ with the double-layer potential generated on Γ_0 by φ_n up to the factor 2, one can evaluate ψ_n as in (9c):

$$\psi_n\binom{r\cos\alpha}{r\sin\alpha} = -\frac{r\sin\alpha}{\pi}\int_{1/n}^{1}\{(r\cos\alpha-y_1)^2+r^2\sin^2\alpha\}^{-1}dy_1.$$

ψ_n and thus also the subsequence ψ_{n_j} converge *pointwise* to

$$\psi\binom{r\cos\alpha}{r\sin\alpha} = -\frac{r\sin\alpha}{\pi}\int_0^1\{(r\cos\alpha-y_1)^2+r^2\sin^2\alpha\}^{-1}dy_1.$$

Since ψ_n is continuous on Γ_1, also the limit function $\psi=\lim\psi_{n_j}$ must be continuous because of the uniform convergence. Following Remark 11, $\psi(0)=-\frac{\pi-\alpha}{\pi}\neq 0$ holds and contradicts $\psi_n(0)=0$. Hence, K is not compact in $L^\infty(\Gamma)$.
(ii) To show the non-compactness in $C(\Gamma)$ instead of $L^\infty(\Gamma)$, one can smooth the functions φ_n as indicated in Fig. 5b. One obtains continuous functions and performs the proof in the same way. QED

Example 7.4.12 already contains a counterexample for the statement that equation (18) would admit a solution $f\in C^\lambda(\Gamma)$ for $\varphi\in C^\lambda(\Gamma)$. The contradiction, however, could be shown only for $\lambda>2/3$. Even without compactness, equation (18) can still be solvable in $C(\Gamma)$ or even $C^\lambda(\Gamma)$ with suitable $\lambda>0$, i.e., $I-K$ is bijective.

Remark 8.2.25. If a non-compact operator admits a decomposition

(8.2.28) $\qquad K = K_0+K_1, \qquad \|K_0\|_{C(\Gamma)\leftarrow C(\Gamma)}\leqslant q<1, \qquad K_1\in K(C(\Gamma),C(\Gamma)),$

the Fredholm alternative from Theorem 1.3.28a still holds for $I\pm K$. If injectivity is shown as, e.g., by Theorem 18, (28) is a sufficient condition for bijectivity of $I\pm K$.

Proof. By Theorem 1.3.10, $I \pm K_0$ is bijective and $(I \pm K_0)^{-1}K_1$ is compact because of Theorem 1.3.23a. Theorem 1.3.28 can be applied to

$$I+T = I \pm (I \pm K_0)^{-1}K_1.$$

The bijectivity of $I \pm (I \pm K_0)^{-1}K_1$ is equivalent to that of $I \pm K$. QED

The inverse $1/q_0$ of the infimum q_0 of all possible values q from (28) is called the *Fredholm radius* of K. Hence, the condition of Remark 25 can be formulated as follows: The Fredholm radius of K must exceed one.

In the case of a corner at $\mathbf{x}_0 \in \Gamma$, one constructs a suitable decomposition (28) as follows. Let $\mathbf{x}_0$ correspond to the curve parameter $t=0$. For $\varepsilon > 0$, there exists a function

$$\chi \in C^\infty(\mathbb{R}) \text{ with } \chi = 1 \text{ in } [-\varepsilon, \varepsilon], \qquad \chi \geqslant 0 \text{ in } \mathbb{R}, \qquad \chi = 0 \text{ for } t \geqslant 2\varepsilon.$$

We set

$$K_0 f := K(\chi f), \quad K_1 f := K((1-\chi)f).$$

Since $(1-\chi)f$ vanishes in the neighbourhood of the corner, K_1 is compact. For K_0 one finds by means of Lemma 14 that for

$$\|f\|_\infty \leqslant 1 \text{ and } |\mathbf{x} - \mathbf{x}_0| < \varepsilon$$

the value $|(K_0 f)(\mathbf{x})|$ can be estimated by

$$q_\varepsilon := |\pi - |\alpha||/\pi + o(1) \qquad (\text{for } \varepsilon \to 0).$$

Hence, for angles $\alpha \neq 0$, $\alpha \neq \pi$ one finds an $\varepsilon > 0$ such that $q_\varepsilon < 1$ satisfies the condition (28). For $\alpha = \pi$, no corner appears so that K is compact anyway. Even in the case $\alpha = 0$ (s. Fig. 6) one can derive the compactness in $C(\Gamma)$ from $K \in L(C(\Gamma), C^\lambda(\Gamma))$. In the latter case, one must use that the density is continuous when approaching the corner $\mathbf{x}_0$ from either side of Γ, whereas the normal directions have opposite signs.

Fig. 8.2.6 $\alpha = 0$

The three-dimensional case, where corners, edges and even conical points of all kind may appear, cannot be treated analogously. For certain corners the Fredholm radius can be smaller than 1. In many practically important cases, the introduction of another norm equivalent to the supremum norm may help (cf. Král -Wendland [11]). The just mentioned paper contains also many further interesting comments and references to literature concerning the subject of this subsection.

8.3 The Hypersingular Integral Equation

We return to the Neumann problem for the Laplace equation:

(8.3.1a) $\Delta u = 0$ in Ω_+ and Ω_-, $\partial u/\partial n = \varphi$ on Γ,

and want to solve the interior and exterior problem simultaneously. φ must satisfy the necessary condition (1.26) (cf. Lemma 1.19):

(8.3.1b) $$\int_\Gamma \varphi \, d\Gamma = 0.$$

Again, we use the double-layer potential ansatz (2.2c) with the unknown density f. Since, by Theorem 2.15, it has identical normal derivatives from inside and outside, it is indeed possible to solve the interior and exterior problem (1a) simultaneously. Combining the representation (2.14d) of the normal derivative with the boundary condition (1a), we obtain the equation

(8.3.2) $$\int_\Gamma f(\mathbf{y}) \frac{\partial}{\partial n_x} \frac{\partial}{\partial n_y} s(\mathbf{x},\mathbf{y}) d\Gamma_y = \varphi(\mathbf{x}) \qquad \text{for } \mathbf{x} \in \Gamma.$$

Because its simpler form, we prefer the representation (2) instead of the following explicit description (2'):

(8.3.2') $$\frac{1}{\omega_d} \int_\Gamma \left\{ \frac{\langle \mathbf{n}(\mathbf{y}), \mathbf{n}(\mathbf{x}) \rangle}{|\mathbf{y}-\mathbf{x}|^d} - d \frac{\langle \mathbf{n}(\mathbf{y}), \mathbf{y}-\mathbf{x} \rangle \langle \mathbf{y}-\mathbf{x}, \mathbf{n}(\mathbf{x}) \rangle}{|\mathbf{y}-\mathbf{x}|^{d+2}} \right\} f(\mathbf{y}) d\Gamma_y = \varphi(\mathbf{x}).$$

Equation (2) is an integral equation of the first kind for determining the unknown density f. Since the integral in (2) exists only in the sense of Hadamard, we have a *hypersingular integral equation.*

By definition of the Hadamard integral, a constant f leads to the integral value zero, so that equation (2) is not uniquely solvable. Indeed, $f = \text{const}$ yields the double-layer potential $\Phi = 0$ in Ω_+ and $\Phi = \text{const}$ in Ω_- (cf. Theorem 2.13) and, therefore, solves the problem (1a,b) for any value of $f = \text{const}$. To avoid this non-uniqueness, we require a normalisation of the solution f by

(8.3.3) $$\int_\Gamma f \, d\Gamma = 0 \qquad \text{(cf. (1.13), (2.21b))}.$$

The question concerning the existence and uniqueness of a solution f of (2) with the side condition (3) is apparently more difficult than the previous problems because of the Hadamard integral. However, the answer is relatively easy if one uses a suitable function space for f.

First, let Γ and all appearing densities be sufficiently smooth. The densities f and g, which should satisfy (3), generate the double-layer potentials Φ^f and Φ^g. We multiply the normal derivative of Φ^f by g and integrate over Γ:

(8.3.4a) $$\int_\Gamma g \, \partial\Phi^f/\partial n \, d\Gamma = \int_\Gamma \{\Phi^g_+ - \Phi^g_-\} \, \partial\Phi^f/\partial n \, d\Gamma.$$

The right-hand side in (4a) is obtained from the jump relation (2.7a). Since $\nabla\Phi$ is square-integrable in the exterior domain and satisfies the estimates (2.15a,b) (cf. Corollary 2.16d,e), Green's formula (1.24a) is applicable with $u = \Phi^f$ and $v = \Phi^g_\pm$ in the interior and exterior domains $\Omega = \Omega_\pm$ and yields (4b,c) because of $\Delta u = 0$:

(8.3.4b) $$\int_\Gamma \Phi_-^g \,\partial\Phi^f/\partial n \,d\Gamma = \int_{\Omega_-} \langle \nabla\Phi^g, \nabla\Phi^f\rangle \,d\mathbf{x},$$

(8.3.4c) $$\int_\Gamma \Phi_+^g \,\partial\Phi^f/\partial n \,d\Gamma = -\int_{\Omega_+} \langle \nabla\Phi^g, \nabla\Phi^f\rangle \,d\mathbf{x}.$$

Insertion into (4a) yields

(8.3.4d) $$-\int_\Gamma g \,\partial\Phi^f/\partial n \,d\Gamma = \int_{\Omega_-} \langle \nabla\Phi^g, \nabla\Phi^f\rangle \,d\mathbf{x} + \int_{\Omega_+} \langle \nabla\Phi^g, \nabla\Phi^f\rangle \,d\mathbf{x}.$$

We define a *bilinear form* $a(\cdot,\cdot)$ by means of

(8.3.5) $$a(f,g) := -\int_\Gamma g(\mathbf{x}) \int_\Gamma f(\mathbf{y}) \frac{\partial}{\partial n_x}\frac{\partial}{\partial n_y} s(\mathbf{x},\mathbf{y})\,d\Gamma_y\,d\Gamma_x .$$

Obviously, $a(\cdot,\cdot)$ is linear in both arguments and therefore justifies the name «bilinear form». Since the inner integral coincides with $\partial\Phi^f/\partial n$, (4d) shows that

(8.3.6) $$a(f,g) = \int_{\Omega_-} \langle \nabla\Phi^g, \nabla\Phi^f\rangle \,d\mathbf{x} + \int_{\Omega_+} \langle \nabla\Phi^g, \nabla\Phi^f\rangle \,d\mathbf{x}.$$

Lemma 8.3.1. The bilinear form $a(\cdot,\cdot)$ is symmetric and positive for functions satisfying condition (3), i.e., $a(f,g)=a(g,f)$ and $a(f,f)>0$ for all $f \neq 0$ with (3).

Proof. The symmetry can be seen from the right-hand side in (6). (6) also implies $a(f,f)\geq 0$ for all $g=f$. Suppose $a(f,f)=0$. (6) shows $\nabla\Phi^f=0$ in Ω_- and Ω_+. As a consequence, $\Phi^f=c_+$ holds in Ω_+ and $\Phi^f=c_-$ in Ω_- with possibly different constants $c_\pm$. The jump relation $f=\Phi_+-\Phi_-=c_+-c_-$ proves $f=\text{const}$. From (3) one concludes that const$=0$; hence, $f=0$, so that $a(f,f)>0$ for all $f\neq 0$ with (3). QED

For deriving further properties of $a(\cdot,\cdot)$, we reformulate the double integral in (5). By definition of the Hadamard integral, $f(\mathbf{y})$ may be replaced in the inner integral by $f(\mathbf{y})-f(\mathbf{x})$:

(8.3.7a) $$a(f,g) = \int_\Gamma g(\mathbf{x}) \int_\Gamma [f(\mathbf{x})-f(\mathbf{y})] \frac{\partial}{\partial n_x}\frac{\partial}{\partial n_y} s(\mathbf{x},\mathbf{y})\,d\Gamma_y\,d\Gamma_x .$$

Since $\partial^2 s(\mathbf{x},\mathbf{y})/\partial n_x \partial n_y$ is symmetric with respect to the variables $\mathbf{x}$ and $\mathbf{y}$, interchanging of the names $\mathbf{x}\leftrightarrow\mathbf{y}$ yields the representation

(8.3.7b) $$a(f,g) = \int_\Gamma g(\mathbf{y}) \int_\Gamma [f(\mathbf{y})-f(\mathbf{x})] \frac{\partial}{\partial n_x}\frac{\partial}{\partial n_y} s(\mathbf{x},\mathbf{y})\,d\Gamma_x\,d\Gamma_y .$$

For $f\in C^{1+\mu}(\Gamma)$, $\mu>0$, the inner integral in (7a,b) exists as Cauchy principal value. Hence, $2a(f,g)$ is the limit $\varepsilon\to 0$ of the double integrals

$$\int_\Gamma g(\mathbf{x}) \int_{|\mathbf{x}-\mathbf{y}|\geq\varepsilon} [f(\mathbf{x})-f(\mathbf{y})] \frac{\partial}{\partial n_x}\frac{\partial}{\partial n_y} s(\mathbf{x},\mathbf{y})\,d\Gamma_y\,d\Gamma_x +$$

$$+\int_\Gamma g(\mathbf{y}) \int_{|\mathbf{x}-\mathbf{y}|\geq\varepsilon} [f(\mathbf{y})-f(\mathbf{x})] \frac{\partial}{\partial n_x}\frac{\partial}{\partial n_y} s(\mathbf{x},\mathbf{y})\,d\Gamma_x\,d\Gamma_y =$$

$$= \iint_{|\mathbf{x}-\mathbf{y}|\geq\varepsilon} [g(\mathbf{x})-g(\mathbf{y})][f(\mathbf{x})-f(\mathbf{y})] \frac{\partial}{\partial n_x}\frac{\partial}{\partial n_y} s(\mathbf{x},\mathbf{y})\,d\Gamma_y\,d\Gamma_x .$$

The integrand has the improperly integrable majorant $O(|\mathbf{x}-\mathbf{y}|^{2\lambda-d})$, if $g, f \in C^{\lambda}(\Gamma)$ with $\lambda > \frac{1}{2}$. Therefore, we obtain the following representation of $a(f,g)$ as an improper integral:

$$(8.3.8) \qquad a(f,g) = \frac{1}{2} \iint_{\Gamma\times\Gamma} [g(\mathbf{x})-g(\mathbf{y})][f(\mathbf{x})-f(\mathbf{y})] \frac{\partial^2 s(\mathbf{x},\mathbf{y})}{\partial n_x \partial n_y} d\Gamma_y \, d\Gamma_x .$$

The bilinear form (8) is well-defined on $C^{\lambda}(\Gamma) \times C^{\lambda}(\Gamma)$ if $\lambda > \frac{1}{2}$. In §7.2.5, the Sobolev space $H^{1/2}(\Gamma)$ was introduced for the unit circle Γ. On a general curve or surface Γ, we define the *Sobolev-Slobodeckij norm*

$$(8.3.9) \qquad \|f\|_{1/2} := \sqrt{\iint_{\Gamma\times\Gamma} [f(\mathbf{x})-f(\mathbf{y})]^2 \, |\mathbf{x}-\mathbf{y}|^{-d} \, d\Gamma_y \, d\Gamma_x}$$

(cf. Hackbusch [2,§6.2.4]). In general, the term $\int |f|^2 d\Gamma$ has to be added to the radicand in (9), since, otherwise, the function f=const leads to $\|f\|_{1/2}=0$ violating the norm axiom (1.3.1a). In this section, however, constant functions are excluded because of (3). Hence, the norm (9) is well-defined for $f \in C^{\lambda}(\Gamma)$ with $\lambda > \frac{1}{2}$. The space

$$\widetilde{X} := \{ f \in C^{\lambda}(\Gamma) \colon \lambda > \tfrac{1}{2}, f \text{ satisfies (3)} \} \qquad \text{endowed with the norm (9)}$$

is a normed space, but not complete. Completion with respect to the norm (9) defines the Sobolev subspace

$$(8.3.10) \qquad X := \{ f \in H^{1/2}(\Gamma) \colon f \text{ satisfies (3)} \} \qquad \text{with } \|\cdot\|_X = \|\cdot\|_{1/2}$$

of $H^{1/2}(\Gamma)$. The side condition (3) can also be interpreted as forming the quotient space $H^{1/2}(\Gamma)/\mathbb{R}$, where $\mathbb{R}$ denotes the (isomorphic) subspace of the constant functions.

The kernel $\partial^2 s(\mathbf{x},\mathbf{y})/\partial n_x \partial n_y$ can be estimated by $C|\mathbf{x}-\mathbf{y}|^{-d}$, so that the Schwarz inequality leads to

$$|a(f,g)| = a(f,g) \leqslant \frac{C}{2} \iint_{\Gamma\times\Gamma} |g(\mathbf{x})-g(\mathbf{y})| \, |f(\mathbf{x})-f(\mathbf{y})| \, |\mathbf{x}-\mathbf{y}|^{-d} d\Gamma_y \, d\Gamma_x \leqslant$$
$$\leqslant \frac{C}{2} \|f\|_{1/2} \|g\|_{1/2} .$$

This inequality describes that the *bilinear form* $a(\cdot,\cdot)$ is *bounded* on $X \times X$ (or equivalently: *continuous*). Moreover, the following theorem holds.

Theorem 8.3.2. The bilinear form $a(\cdot,\cdot)$ defined in (8) is X-elliptic, i.e.,

$$(8.3.11) \qquad a(f,f) \geqslant c\|f\|_{1/2}^2 \qquad \text{for some } c>0 \text{ and all } f \in X \qquad (X \text{ from (10)}).$$

Proof. The jump relation (2.7a) (i.e., $f = \Phi_+ - \Phi_-$) admits the estimate $\|f\|_{1/2} \leqslant \|\Phi_+\|_{1/2} + \|\Phi_-\|_{1/2}$, where $\Phi_\pm = \Phi_\pm|_\Gamma$ are the traces of $\Phi_\pm$ on Γ. The boundary values ψ on Γ and any extension $\Psi \in H^1(\Omega_\pm)$ of these values into the domain $\Omega_\pm$ are connected by $\|\psi\|_{1/2} \leqslant C\|\Psi\|_{H^1(\Omega_\pm)}$ (cf. Hackbusch [2, Theorems 6.2.28 and 6.2.40a]). Choosing $\Psi = \Phi_\pm$ and $\psi = \Phi_\pm|_\Gamma$, we obtain $\|\Phi_\pm\|_{1/2} \leqslant C\|\Phi_\pm\|_{H^1(\Omega_\pm)}$. The first inequality and the representation (6) of

$a(f,f)$ imply $\|f\|^2_{1/2} \leq 2C(\|\Phi_+\|^2_{H^1(\Omega_+)} + \|\Phi_-\|^2_{H^1(\Omega_-)}) = 2Ca(f,f)$. For a detailed analysis, we refer to Giroire - Nédélec [1]. QED

Multiplying both sides in equation (2) by g and integrating over Γ, we obtain

$$\int_\Gamma g(\mathbf{x}) \int_\Gamma f(\mathbf{y}) \frac{\partial}{\partial n_\mathbf{x}} \frac{\partial}{\partial n_\mathbf{y}} s(\mathbf{x},\mathbf{y})\, d\Gamma_\mathbf{y}\, d\Gamma_\mathbf{x} = \int_\Gamma g\,\varphi\, d\Gamma.$$

The left-hand side equals $-a(f,g)$ (cf. (5)). Hence, we are led to the problem

$$(8.3.12) \qquad a(f,g) = -\int_\Gamma g\,\varphi\, d\Gamma \qquad \text{for all } g \in X.$$

The right-hand side in (12) can be written as scalar product $-\langle g,\varphi\rangle$. For $g\in X$, $\langle g,\varphi\rangle$ is defined even if φ is in the dual space

$$X' = \{\varphi\in H^{-1/2}(\Gamma):\ \varphi \text{ satisfies (1b)}\} \subset \{\varphi\in L^2(\Gamma):\ \varphi \text{ satisfies (1b)}\}$$

equipped with the dual norm

$$\|\varphi\|_{-1/2} := \sup\{|\langle g,\varphi\rangle| :\ g\in X,\ \|g\|_{1/2} = 1\}.$$

Equations of the form (12) belong to the standard problems for elliptic boundary value problems. The solvability of problem (12) is guaranteed by the property (11) (cf. Hackbusch [2,§6.5]):

Theorem 8.3.3. For any Neumann boundary value $\varphi\in X'$, problem (12) has a unique solution $f\in X$. It satisfies the estimate $\|f\|_{1/2} \leq \|\varphi\|_{-1/2}/c$ (c given in (11)). The double-layer potential belonging to f solves the Neumann boundary value problem (1a,b).

As explained in Hackbusch [2,§6.5], problem (12) is equivalent to

$$(8.3.12') \qquad Kf = \varphi,$$

where $a(f,g) = -\langle Kf,g\rangle$. By Theorem 3, the operator $K\in L(X,X')$ is invertible, i.e., $K^{-1}\in L(X',X)$. Equation (12') is the short notation for the first kind integral equation (2).

If φ belongs to a better (smoother) space than X', the following regularity result can be obtained: $K^{-1}\in L(H^{\varkappa-1}/\mathbb{R}, H^{\varkappa}/\mathbb{R})$ holds for all $\varkappa\in\mathbb{R}$, if Γ is arbitrarily smooth. The formation of the quotient «$/\mathbb{R}$» is necessary because of (1b) and (3).

Remark 8.3.4. Since K^{-1} increases the differentiability order by one, while the application of $K\in L(H^{\varkappa}/\mathbb{R}, H^{\varkappa-1}/\mathbb{R})$ decreases the order by one, K is said to have the *order 1*. According to Theorem 7.2.27, the Cauchy kernel has order 0. An integral operator, e.g., with the kernel $k(\mathbf{x},\mathbf{y}) = |\mathbf{x}-\mathbf{y}|^{-1/2}$ on $\Gamma\subset\mathbb{R}^2$ has the order $-1/2$ because of $K\in L(H^{\varkappa}, H^{\varkappa+1/2})$, $K^{-1}\in L(H^{\varkappa+1/2}, H^{\varkappa})$.

8.4 Synopsis: Integral Equations for the Laplace Equation

In the previous chapters 8.1-3 we have see that there are several possibilities to translate the Laplace equation into an integral equation. These are summarised in the following table. Four different kernels are involved:

$$k_0(\mathbf{x},\mathbf{y}) = s(\mathbf{x},\mathbf{y}) \qquad \text{(cf. (8.1.3))},$$

$$k_1(\mathbf{x},\mathbf{y}) = 2\,\partial s(\mathbf{x},\mathbf{y})/\partial n_y \qquad \text{(cf. (8.1.31a))},$$

$$k_1^*(\mathbf{x},\mathbf{y}) = 2\,\partial s(\mathbf{x},\mathbf{y})/\partial n_x \qquad \text{(cf. (8.1.31b))},$$

$$k_2(\mathbf{x},\mathbf{y}) = \partial^2 s(\mathbf{x},\mathbf{y})/\partial n_x \partial n_y \qquad \text{(cf. (8.3.2))}.$$

Here, $s(\mathbf{x},\mathbf{y})$ is the fundamental solution (1.3) of the Laplace equation. k_0 and k_1 are improperly integrable. The same holds for the adjoint kernel k_1^* up to corners and edges. The kernel k_2 is hypersingular.

In the following, we use the abbreviation $\int k_i f d\Gamma$ for $\int_\Gamma k_i(\mathbf{x},\mathbf{y}) f(\mathbf{y}) d\Gamma_y$ $(i=0,1,2)$. The symbols mean

- D: Dirichlet boundary value problem
- N: Neumann boundary value problem
- I: interior problem
- E: exterior problem
- SL: the solution of the Laplace equation is represented by the single-layer potential (1.8/9).
- DL: the solution of the Laplace equation is represented by the double-layer potential (2.1c).
- direct: direct method, i.e., the boundary values of the Laplace equation solution are determined directly. The values in the domain are given by (2.22) or (2.26).

case	where?	integral equation	problem	representation
(A)	(8.1.27)	$\int k_0 f d\Gamma = \varphi$	D, I+E	SL
(B)	Exerc. 8.2.23	$\int k_0 u_n d\Gamma = \varphi$	D, I+E	direct
(C)	(8.1.30/32a)	$f = 2\varphi + \int k_1^* f d\Gamma$	N, I	SL
(D)	(8.1.32b)	$f = -2\varphi - \int k_1^* f d\Gamma$	N, E	SL
(E)	(8.2.18)	$f = -2\varphi + \int k_1 f d\Gamma$	D, I	DL
(F)	(8.2.19)	$f = 2\varphi - \int k_1 f d\Gamma$	D, E	DL
(G)	(8.2.25a)	$u = g - \int k_1 u d\Gamma$	N, I	direct
(H)	(8.2.27)	$u = -g + \int k_1 u d\Gamma$	N, E	direct
(I)	(8.3.2)	$\int k_2 f d\Gamma = \varphi$	N, I+E	DL

The equations are ordered with respect to the kernels. (A), (B), and (I) are integral equations of the first kind, the others of the second kind.

8.5 The Integral Equation Method for Other Differential Equations

8.5.1 Differential Equations of Second Order

Starting point of the integral equation method is the fundamental solution of the differential equation. As alternative to the Laplace equation, we discuss the Helmholtz equation

(8.5.1) $$\Delta u + \varkappa^2 u = 0 \qquad \text{in } \Omega.$$

Here, $\varkappa$ is real or complex with $Im\,\varkappa \geqslant 0$. In the three-dimensional case, the corresponding fundamental solution is given by

(8.5.2) $$s(\mathbf{x},\mathbf{y}) = \frac{1}{4\pi}\frac{e^{i\varkappa|\mathbf{x}-\mathbf{y}|}}{|\mathbf{x}-\mathbf{y}|} \qquad (d=3).$$

The order of the differential equation (1) equals 2 as for the Laplace equation. Since this number (together with the dimension d) determines the order of the singularity, (2) has the same singularity behaviour as s from (1.3). In principle, all previous statements concerning the existence and representation of the single- and double-layer potentials and their derivatives remain valid.

As first example we discuss the solution of the interior problem by means of the double-layer potential. We make the ansatz $u=\Phi$ with

(8.5.3a) $$\Phi(\mathbf{x}) = \int_\Gamma \frac{\partial s(\mathbf{x},\mathbf{y})}{\partial n_y} f(\mathbf{y})\,d\Gamma_y$$

and define the double-layer operator by

(8.5.3b) $$(Kf)(\mathbf{x}) = 2\int_\Gamma \frac{\partial s(\mathbf{x},\mathbf{y})}{\partial n_y} f(\mathbf{y})\,d\Gamma_y.$$

Exercise 8.5.1. Show that $\dfrac{\partial s(\mathbf{x},\mathbf{y})}{\partial n_y} = \dfrac{1}{4\pi}\dfrac{e^{i\varkappa|\mathbf{x}-\mathbf{y}|}}{|\mathbf{x}-\mathbf{y}|^3}(i\varkappa|\mathbf{x}-\mathbf{y}|^2-1)\langle n(\mathbf{y}),\mathbf{y}-\mathbf{x}\rangle.$

Lemma 8.5.2. The jump relations read again (2.7a,b), i.e. $\Phi_+-\Phi_-=f$ and $\Phi_++\Phi_-=Kf$ on Γ with Φ and K from (3a,b).

Proof. The expansion $e^{i\varkappa|\mathbf{x}-\mathbf{y}|} = 1 + O(|\mathbf{x}-\mathbf{y}|)$ induces a decomposition of s into the fundamental solution of the Laplace equation plus a smoother remainder causing no additional jump. QED

The interior problem of the Helmholtz equation with Dirichlet boundary values $u=\varphi$ leads to $\Phi_-=\varphi$ and therefore to

(8.5.4) $$f = -2\varphi + Kf$$

as in (2.18) with the only difference that K is now given by (3b). The existence and uniqueness statements are analogous to those in §8.2.6.1, provided that $-\varkappa^2$ is no eigenvalue of the Laplace equation.

Of more interest is the *exterior problem*, i.e., equation (1) in $\Omega=\Omega_+$. The solution has to obey the following «*radiation condition* or Sommerfeld condition» at ∞:

(8.5.5) $u(\mathbf{x}) = O(1/r), \quad (\frac{\partial}{\partial r} - i\varkappa)u(\mathbf{x}) = o(1/r) \quad$ for $r = |\mathbf{x}| \to \infty$.

Following Brakhage-Werner [1], we use a linear combination of the single- and double-layer potential:

(8.5.6a) $\Phi(\mathbf{x}) = \int_\Gamma f(\mathbf{y}) [\frac{\partial}{\partial n_y} - i\eta] \, s(\mathbf{x},\mathbf{y}) \, d\Gamma_y \qquad$ with

(8.5.6b) $\eta = 1$ for $Re\,\varkappa \geq 0, \quad \eta = -1$ for $Re\,\varkappa < 0$.

Since Φ from (6a) satisfies the same jump relations as in Lemma 2, the Dirichlet problem $u = \Phi_+ = \varphi$ lead to the integral equation

(8.5.7) $f = 2\varphi - Kf.$

Lemma 8.5.3 (Brakhage - Werner [1]). The ansatz (6a,b) satisfies the condition (5). The integral equation (7) has at most one solution.

Using the compactness of K, one then obtains the unique solvability of (7).

A direct integral equation method is described, e.g., by Kleinman - Wendland [1]. Compare also Dautray - Lions [1, chapter XI, §3].

8.5.2 Equations of Higher Order

With increasing order of the differential equation, the smoothness of the fundamental solution increases, too. In (1.7), we already mentioned the fundamental solution

(8.5.8) $s(\mathbf{x},\mathbf{y}) = \frac{1}{8\pi} |\mathbf{x}-\mathbf{y}|^2 \log|\mathbf{x}-\mathbf{y}| \qquad (\mathbf{x}, \mathbf{y} \in \mathbb{R}^2)$

of the two-dimensional *biharmonic differential equation*

(8.5.9) $\Delta^2 u \equiv \left\{ \frac{\partial^4}{\partial x^4} + 2\frac{\partial^4}{\partial x^2 \partial y^2} + \frac{\partial^4}{\partial y^4} \right\} u = 0.$

The increased smoothness of s implies that the «potentials» formed by analogy with (1.8) and (2.1c) and their derivatives contain no jumps.

Exercise 8.5.4. Prove the following: The potential (1.8) formed with s from (8) belongs to $C^2(\mathbb{R}^2)$, whereas the potential (2.1c) belongs to $C^1(\mathbb{R}^2)$.

The name «potential» is used only in the generalised sense, since the generated function Φ does not solve the potential equation (=Laplace equation).

A further distinction is due to the nature of the elliptic boundary value problems of higher order: Equations of the order $2m$ $(m \in \mathbb{N})$ require m boundary conditions. In the case of the biharmonic equation $(m=2)$, two boundary conditions are needed, e.g.,

(8.5.10a) $u = \varphi_1, \quad \frac{\partial}{\partial n} u = \varphi_2,$

or

(8.5.10b) $\Delta u = \varphi_3, \quad \frac{\partial}{\partial n} \Delta u = \varphi_4 \qquad$ (cf. Hackbusch [2,§5.3]).

Since two boundary conditions are to be satisfied, also two densities are to be determined. Therefore, the ansatz must be of the form

$$\Phi(\mathbf{x}) = \int_\Gamma [f(\mathbf{y}) D_{1,y} s(\mathbf{x},\mathbf{y}) + g(\mathbf{y}) D_{2,y} s(\mathbf{x},\mathbf{y})] \, d\Gamma_y$$

with unknown densities f and g. D_1 and D_2 mean differentiation operators with respect to y (e.g., $D=I$ {zeroth derivative}, $D=\partial/\partial n$, $D=\Delta$, $D=\frac{\partial}{\partial n}\Delta$).

Example (10b) shows that the boundary conditions may involve higher derivatives than in the case of the Laplace equation. The potential $\Phi(\mathbf{x}) = \int f(\mathbf{y}) s(\mathbf{x},\mathbf{y}) \, d\Gamma_y$ has a jump in its third derivative $\frac{\partial}{\partial n}\Delta\Phi$, which can be used for deriving suitable integral equations of the second kind.

An integral equation of the first kind for the biharmonic equation is described, e.g., by Costabel - Stephan - Wendland [1] . There, the boundary condition (10a) appears in a somewhat modified form. Differentiating $u=\varphi_1$ in tangential direction, one obtains $\partial u/\partial t = \varphi_1'$. From this directional derivative and the second boundary condition $\frac{\partial}{\partial n}u = \varphi_2$ one obtains a boundary condition $\nabla u = \boldsymbol{\varphi}$ for the gradient of the desired solution.

8.5.3 Systems of Differential Equations

The differential equations discussed so far were scalar equations. On the other hand, *systems of differential equations* are very important. An example is the *Lamé equations*

(8.5.11) $$\mu\,\Delta\mathbf{u} + (\lambda+\mu)\,\nabla\operatorname{div}\mathbf{u} = 0 \qquad \text{in } \Omega\subset\mathbb{R}^d,$$

where $\mathbf{u}$ is a vector function with d components. $\Delta\mathbf{u}$ denotes the componentwise application of Δ. «div» abbreviates «divergence»:

$$\mathbf{u} = \begin{bmatrix} u_1 \\ \vdots \\ u_d \end{bmatrix}, \qquad \Delta\mathbf{u} = \begin{bmatrix} \Delta u_1 \\ \vdots \\ \Delta u_d \end{bmatrix}, \qquad \operatorname{div}\mathbf{u} = \sum_{i=1}^{d} \frac{\partial u_i}{\partial x_i}.$$

The Neumann boundary condition $\frac{\partial}{\partial n}u = \varphi$ of Laplace's equation corresponds to

(8.5.12a) $$T\mathbf{u} = \boldsymbol{\varphi}$$

in the case of the Lamé system, where the operator T is defined componentwise by

(8.5.12b) $$(Tu)_i := \lambda n_i \operatorname{div}\mathbf{u} + \mu\frac{\partial}{\partial n}u_i + \mu\left\langle\frac{\partial\mathbf{u}}{\partial x_i}, \mathbf{n}\right\rangle \qquad (1\le i\le d).$$

Here, n_i are the components of the outer normal direction vector $\mathbf{n}$.

In the case of a system, the fundamental solution becomes a matrix, here a $d\times d$ matrix. Let s be the fundamental solution (1.3) of the Laplace equation and ω_d as in (1.14b). The «singularity matrix» of the Lamé system equals

(8.5.13) $$\mathbf{S}(\mathbf{x},\mathbf{y}) = \frac{\lambda+3\mu}{2(\lambda+2\mu)}\left\{ s(\mathbf{x},\mathbf{y})I + \frac{\lambda+\mu}{\omega_d(\lambda+3\mu)} \frac{(\mathbf{x}-\mathbf{y})(\mathbf{x}-\mathbf{y})^T}{|\mathbf{x}-\mathbf{y}|^d}\right\}.$$

I is the $d\times d$ identity matrix. The matrix $(\mathbf{x}-\mathbf{y})(\mathbf{x}-\mathbf{y})^T$ has the entries $(x_i-y_i)(x_j-y_j)$. In (12b). T is applied to a vector function. Applying T to all columns of the matrix $\mathbf{S}$ and performing the differentiation with respect to $\mathbf{y}$, one obtains a new matrix $T_y \mathbf{S}(\mathbf{x},\mathbf{y})$. Denote the transposed matrix by

$$\mathbf{T}(\mathbf{x},\mathbf{y}) := \left(T_y \mathbf{S}(\mathbf{x},\mathbf{y})\right)^T .$$

In the Lamé case, Green's representation formula (2.22) corresponds to the *Betti formula*

(8.5.14) $$\pm \mathbf{u}(\mathbf{x}) = \int_\Gamma \{\mathbf{S}(\mathbf{x},\mathbf{y})T(\mathbf{u})(\mathbf{y}) - \mathbf{T}(\mathbf{x},\mathbf{y})\mathbf{u}(\mathbf{y})\}\, d\Gamma_y ,$$

where the signs «±» hold for the interior and exterior domains $\Omega_\mp$. Because of the boundary condition (12a), one can replace $T(\mathbf{u})(\mathbf{y})$ by the boundary values $\boldsymbol{\varphi}(\mathbf{y})$. Together with the corresponding jump relations, which read as in §8.2.6.2, one gains integral equations of the form (2.25a) and (2.27) for the interior or exterior problem. In contrast to the Laplace equation, the integral operator K with the kernel $\mathbf{T}(\mathbf{x},\mathbf{y})$ which now appears is strongly singular.

In the same way, as the solution of Neumann's boundary value problem for $\Delta u = 0$ is determined only up to a constant, the equation (12a) fixes the solution of the Lamé equation only up to the so-called rigid body motions. Augmenting the system according to (4.8.15a,b), one reaches a uniquely solvable system of integral equations.

Details for the treatment of the Lamé equation can be found, e.g., in Wendland [2] or Schatz-Thomée-Wendland [1,III,§1,4]

A further example of a system is the *Stokes equations*

(8.5.15a) $$-\Delta \mathbf{u} + \nabla p = 0 \qquad \text{in } \Omega,$$

(8.5.15b) $$\operatorname{div} \mathbf{u} = 0 \qquad \text{in } \Omega$$

(cf. Hackbusch [2,§12]). The vector function $\mathbf{u}$ consists of d components. In the two-dimensional case $d = 2$, the Stokes equation (15a,b) can be rewritten as the biharmonic equation (9) just mentioned in §8.5.2 (cf. Hackbusch [2, Remark 12.2.5]).

In the three-dimensional case, one has to determine four unknown functions u_1, u_2, u_3, and p. However, only three boundary values are given:

(8.5.15c) $$\mathbf{u} = \boldsymbol{\varphi} \qquad \text{on } \Gamma .$$

The singularity matrix of the Stokes system is

(8.5.16) $$\mathbf{S}(\mathbf{x},\mathbf{y}) = -\frac{1}{8\pi}\begin{bmatrix} S_3 & \dfrac{2(\mathbf{x}-\mathbf{y})}{|\mathbf{x}-\mathbf{y}|^3} \\ \dfrac{2(\mathbf{x}-\mathbf{y})^T}{|\mathbf{x}-\mathbf{y}|^3} & 8\pi\,\delta(\mathbf{x}-\mathbf{y}) \end{bmatrix}$$

with the 3×3 submatrix $S_3 = \frac{1}{|\mathbf{x}-\mathbf{y}|} I + \frac{(\mathbf{x}-\mathbf{y})(\mathbf{x}-\mathbf{y})^T}{|\mathbf{x}-\mathbf{y}|^3}$ and the Dirac function $\delta(\mathbf{x}-\mathbf{y})$. We refer to Dautray-Lions [1, chapter XI, §5], Hsiao-Kreß [1], Hsiao-Kopp-Wendland [2] and Hebeker [1] for the definition of the single- and double-layer potentials and the formulation of second kind integral equation corresponding to (5.7).

Concerning the system of Maxwell's equations we refer to MacCamy - Stephan [1].

9 The Boundary Element Method

9.1 Construction of the Boundary Element Method

9.1.1 Definition of the Boundary Element Method

Boundary value problems like the Laplace, Helmholtz, biharmonic, Lamé, or Stokes equations discussed in §8 can be approximated in their original domain of definition by various discretisation methods. Besides the (finite) difference method there is, in particular, the finite element method («FEM»; cf. Hackbusch [2]).

Translating the boundary value problem by means of the boundary integral method into an integral equation over the boundary Γ, we can discretise this integral equation by the methods discussed in §4. The Galerkin and collocation method use ansatz spaces, for which basis functions with small support have turned out to be advantageous. Such basis functions are called «finite elements». Discretisations of integral equations, which use these finite elements, could be named again «finite element methods», but since this name is associated with the direct approximation of partial differential equations, the combination of the boundary integral method with the discretisation by finite elements is called the «boundary element method» (abbreviation: «BEM»).

The specific *advantage of the boundary element method* is the transition from a d-dimensional domain onto the $(d-1)$-dimensional boundary, thus reducing the dimension by one. If a comparable step size h is used for the discretisation of the partial differential equation on the one hand and for the integral equation on the other hand, then the discretisations lead to systems of $O(h^{-d})$ or $O(h^{1-d})$ equations, respectively. Further comments concerning the computational work can be found in §9.7.

A particular advantage of the boundary element method is the easy treatment of *exterior problems*, since the discretisation of boundary value problems in unbounded domains lead to considerable additional problems.

Since the development of BEM, the number of papers about applications in the engineering sciences has grown rapidly. Concerning this aspect, we refer to the collection in the reference book of Mackerle - Brebbia [1].

9.1.2 Galerkin Method

In §9.2 we shall discuss the ansatz functions $\{\Phi_{1,n},\dots,\Phi_{n,n}\}$, the «finite elements» which now can also be called «boundary elements». According to the notation in §8, we write the integral equation as

(9.1.1) $\lambda f = g + Kf.$

The application of the Galerkin discretisation yields the system (4.5.8), i.e.,

(9.1.2a) $(\lambda A_n - B_n)a_n = b_n$

for the coefficients α_k of the semidiscrete solution $f_n = \sum \alpha_k \Phi_{k,n}$. The entries of the matrices are given in (4.5.7b,c):

(9.1.2b) $\alpha_{jk} = \langle \Phi_{k,n}, \Phi_{j,n} \rangle, \qquad \beta_{jk} = \langle K\Phi_{k,n}, \Phi_{j,n} \rangle.$

According to the name «finite elements», $T_k = \text{support}(\Phi_{k,n})$ has the length or surface area $O(h^{d-1})$, where h is the underlying step size. The coefficients α_{jk} vanish, if T_k and T_j are disjoint or if T_k and T_j intersect only in their boundaries.

Remark 9.1.1. In general, B_n is a full matrix. The coefficients β_{jk} result from a double integration over T_k and T_j:

(9.1.2c) $\beta_{jk} = \int_{T_j} [\int_{T_k} k(\mathbf{x},\mathbf{y})\Phi_{k,n}(\mathbf{y})d\Gamma_y] \Phi_{j,n}(\mathbf{x})\, d\Gamma_x.$

Concerning the Galerkin method for first kind integral equations compare §9.1.5. The numerical treatment of the integral (2c) is considered in §9.4.5.

9.1.3 Collocation Method

The ansatz functions used in the following are Lagrange functions, so that the collocation method leads to a system (2a) with the identity matrix $A_n = I$ (cf. Remark 4.4.3). The matrix coefficients of B_n contain only one integration:

(9.1.3) $\beta_{jk} = \int_{T_k} k(\boldsymbol{\xi}_{j,n}, \mathbf{y})\Phi_{k,n}(\mathbf{y})d\Gamma_y \qquad (1 \leqslant j,k \leqslant n).$

Here, $\{\boldsymbol{\xi}_{j,n} : 1 \leqslant j \leqslant n\}$ are suitable collocation points.

In the case of the double-layer operators (8.1.31a) with piecewise constant ansatz functions $\Phi_{k,n}$ on T_k, the value (3) can be interpreted geometrically.

Remark 9.1.2. Let the ansatz function $\Phi_{k,n}$ have the value 1 on T_k and 0 on $\Gamma \backslash T_k$. For the kernel $k(\mathbf{x},\mathbf{y}) = -\frac{2}{\omega_d}\langle \mathbf{n}(\mathbf{y}), \mathbf{y}-\mathbf{x}\rangle / |\mathbf{y}-\mathbf{x}|^d$ of the double-layer operators (8.2.2b), the coefficient β_{jk} in the two-dimensional case $d=2$ equals the $(-\frac{1}{\pi})$-fold of the angle φ_{jk}, which the arc T_k forms at the collocation point $\boldsymbol{\xi}_{j,n}$ (cf. Fig. 1). In the three-dimensional case $d=3$, β_{jk} is the $(-\frac{1}{2\pi})$-fold of the spatial angle of T_k at $\boldsymbol{\xi}_{j,n}$ (cf. §8.2.4.3).

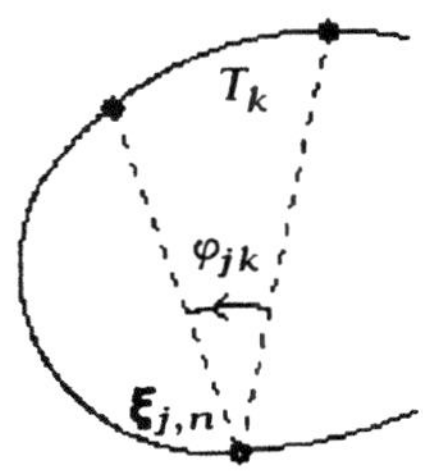

Fig. 9.1.1

The property $-K1 = 1$ of the double-layer operator (cf. (8.2.11b)) allows us to represent the equation $\lambda f = g + Kf$ at $\mathbf{x} = \boldsymbol{\xi}_{j,n}$ in the form

(9.1.4) $$(\lambda+1)f(\boldsymbol{\xi}_{j,n}) = g(\boldsymbol{\xi}_{j,n}) + \int_\Gamma k(\boldsymbol{\xi}_{j,n},\mathbf{y})[f(\mathbf{y})-f(\boldsymbol{\xi}_{j,n})]\,d\Gamma_y .$$

Exercise 9.1.3. Let $\{\Phi_{1,n},\dots,\Phi_{n,n}\}$ be a Lagrange basis. Prove the following: If we replace f in equation (4) by the ansatz $f_n=\sum\alpha_k\Phi_{k,n}$, we obtain a system (2a) with coefficients β_{jk} as in (3) for $j\neq k$, while for $j=k$ we have

(9.1.5) $$\beta_{j,j} := -\sum_{j\neq k}\beta_{j,k}.$$

9.1.4 Convergence in the Compact Case

If K is compact, the convergence results of Sections 4.4 and 4.5 apply. Concerning the order of convergence, one needs statements about the smoothness of the solution f of $\lambda f=g+Kf$. In general, one has to prove smoothness of f via regularity properties of the operator K using that g is smooth. In the two-dimensional case, e.g., Theorem 7.4.13 allows to gain the property $f\in C^\mu(\Gamma)$ from $\Gamma\in C^{1+\mu}$, $\mu\in(0,2)\backslash\{1\}$, and $g\in C^\mu(\Gamma)$. This ensures convergence of the order $O(h^\mu)$ with respect to the maximum norm (cf. Theorem 4.3.15). Better estimates might be possible with respect to weaker norms or at discrete points according to the results in §4.6.2 and §4.6.4.

The convergence of the boundary element method is discussed, e.g., by Wendland [1,2,3] and Kleinman-Wendland [1].

9.1.5 Convergence in the Case of Elliptic Bilinear Forms

One of the examples of integral equations for solving the Laplace's equation is the equation $Kf=\varphi$ of the first kind with the hypersingular kernel $k=\partial^2 s/\partial n_x\partial n_y$ (cf. (8.3.12') in §8.3). As shown there, this equation can be cast into the form

(9.1.6) $$a(f,g) = -\langle\varphi,g\rangle \qquad \text{for all } g\in H^{1/2}(\Gamma)$$

with an $H^{1/2}(\Gamma)$-elliptic bilinear form a. $\langle\cdot,\cdot\rangle$ is the $L^2(\Gamma)$ scalar product.

Let $X_n=\operatorname{span}\{\Phi_{1,n},\dots,\Phi_{n,n}\}$ be a subspace of $X=H^{1/2}(\Gamma)$. An example for X_n are the piecewise linear functions. Note that the piecewise constant functions do not belong to $H^{1/2}(\Gamma)$. For an equation of the first kind the Galerkin approximation (4.5.3) must be written as

$$K_n f_n = \varphi_n \qquad \text{with } K_n=\Pi_n K\Pi_n,\quad \varphi_n=\Pi_n\varphi,$$

where Π_n denotes the orthogonal projection onto the subspace X_n. Translate this equation by means of $\langle K_nf_n,g\rangle=\langle\Pi_nK\Pi_nf_n,g\rangle=\langle K\Pi_nf_n,\Pi_ng\rangle=\langle Kf_n,g_n\rangle=-a(f_n,g_n)$ into a formulation by the bilinear form: Determine $f_n\in X_n$ with

(9.1.7) $$a(f_n,g_n) = -\langle\varphi,g_n\rangle \qquad \text{for all } g\in X_n.$$

The system $-B_na_n=b_n$ for the coefficients α_k in $f_n=\sum\alpha_k\Phi_{k,n}$ is that from (2a) with $\lambda=0$ and

$$\beta_{jk} = a(\Phi_{k,n},\Phi_{j,n}) \qquad (1\le j,k\le n).$$

The coefficients of b_n are $\beta_j=\langle\varphi,\Phi_{j,n}\rangle$ (cf. Hackbusch [2, Theorem 8.1.3]).

The error analysis of the Galerkin discretisation (7) is not repeated here, since it is completely identical to the error estimates in the case of elliptic boundary value problems, which in their weak formulation are also described by means of a bilinear form. The following theorem and its proof is contained, e.g., in Hackbusch [2, Theorem 8.2.1, Corollary 8.2.3].

Theorem 9.1.4. Suppose that the bilinear form a is bounded by the constant C_S: $|a(f,g)| \leqslant C_S \|f\|_X \|g\|_X$, and X-elliptic with the constant $c>0$ in (8.3.11). Then the solution f of (6) and f_n of (7) fulfil the error estimate (8a) with respect to the X-norm:

$$\|f - f_n\|_X \leqslant (1 + C_S/c) \inf\{\|f-\psi\|_X : \psi \in X_n\}. \tag{9.1.8a}$$

In the paragraph preceding Theorem 8.3.2, the constant C_S is denoted by $C/2$. The expression $\inf\{\|f-\psi\|_X : \psi\in X_n\}$ denotes the distance $\mathrm{dist}(f, X_n)$ and is the *optimal* approximation of f in X_n. The estimate of this expression depends on the one hand on X_n and on the other hand on the regularity of the solution f, i.e., of the differentiability order $s \geqslant \frac{1}{2}$ such that $f \in H^s(\Gamma)$. For example, for $f \in H^s(\Gamma)$ and X_n consisting of *piecewise linear functions*, the following inequality (8b) holds with $t=2$:

$$\inf\{\|f-\psi\|_{H^{1/2}(\Gamma)} : \psi \in X_n\} \leqslant C h^{s-1/2} \|f\|_{H^s(\Gamma)} \qquad \text{for } \tfrac{1}{2} \leqslant s \leqslant t. \tag{9.1.8b}$$

The mapping $f \mapsto \varphi = Kf \mapsto f_n$ according to (7) defines the operator S_n: $f \mapsto f_n \in X_n$, which is called the «*Ritz projection*». The projection property is the subject of the next exercises.

Exercise 9.1.5. Prove the following: **(a)** $a(S_n f, g) = a(f,g)$ for all $g \in X_n$.
(b) S_n is a projection onto X_n.

Exercise 9.1.6. Under the assumption that the bilinear form a on X is bounded, symmetric, and X-elliptic, show the following:
(a) $|||f||| := \sqrt{a(f,f)}$ defines a Hilbert norm equivalent to $\|\cdot\|_X$.
(b) S_n is the orthogonal projection on X_n with respect to the norm $|||\cdot|||$.
(c) The operator adjoint to S_n is $S_n^* = K S_n K^{-1}$.

The following lemma can also be formulated for unsymmetric K. However, since the bilinear forms appearing here (and therefore also K) are symmetric, we restrict ourselves to this case.

Lemma 9.1.7. Let $a(f,g) = -\langle Kf, g\rangle$ be a symmetric $H^{1/2}(\Gamma)$-elliptic bilinear form. Assume the following error estimate (which follows from (8a,b)):

$$\|I - S_n\|_{H^{1/2}(\Gamma) \leftarrow H^s(\Gamma)} \leqslant C h^{s-1/2} \qquad \text{for some } s \geqslant \tfrac{1}{2}. \tag{9.1.8c}$$

Furthermore, let K be s-regular, i.e., $K^{-1} \in L(H^{s-1}(\Gamma), H^s(\Gamma))$. Then the estimate (8d) holds with doubled order of convergence:

$$\|I - S_n\|_{H^{1-s}(\Gamma) \leftarrow H^s(\Gamma)} \leqslant C' h^{2s-1}. \tag{9.1.8d}$$

The inequality (8d) implies the following error estimate for $\sigma = s$:

(9.1.8e) $\|f-f_n\|_{H^{1-s}(\Gamma)} \leq C' h^{\sigma+s-1} \|f\|_{H^\sigma(\Gamma)}$ for $\frac{1}{2} \leq \sigma \leq s$, $f \in H^\sigma(\Gamma)$.

Concerning the s-regularity, we add that the $\frac{1}{2}$-regularity is a direct consequence of the $H^{1/2}(\Gamma)$-ellipticity. The s-regularity for higher s requires a corresponding smoothness of Γ. However, 1-regularity holds already for $\Gamma \in C_{0,L}$ (cf. Costabel [1,$\sigma=\frac{1}{2}$ in (2.14)]). For a negative exponent $-\tau$, $H^{-\tau}(\Gamma)$ is by definition the dual space to $H^\tau(\Gamma)$.

Proof. Because of the boundedness of a, $K \in L(H^{1/2}(\Gamma), H^{-1/2}(\Gamma))$ holds:

(9.1.8f) $\|K\|_{H^{-1/2}(\Gamma) \leftarrow H^{1/2}(\Gamma)} \leq C_1$.

(8c) implies the estimate (8g) for the dual operator with the same constant C:

(9.1.8g) $\|I-S_n^*\|_{H^{-s}(\Gamma) \leftarrow H^{-1/2}(\Gamma)} \leq C h^{s-1/2}$.

Because of the symmetry of K, the s-regularity $K^{-1} \in L(H^{s-1}(\Gamma), H^s(\Gamma))$ is equivalent to $K^{-1} \in L(H^{-s}(\Gamma), H^{1-s}(\Gamma))$, i.e.,

(9.1.8h) $\|K^{-1}\|_{H^{1-s}(\Gamma) \leftarrow H^{-s}(\Gamma)} \leq C_2$.

Together, the inequalities (8f-h) yield

(9.1.8i) $\|K^{-1}(I-S_n^*)K\|_{H^{1-s}(\Gamma) \leftarrow H^{1/2}(\Gamma)} \leq C_1 C_2 C h^{s-1/2}$.

According to Exercise 6c, $K^{-1}(I-S_n^*)K$ coincides with $I-S_n$:

(9.1.8i') $\|I-S_n\|_{H^{1-s}(\Gamma) \leftarrow H^{1/2}(\Gamma)} \leq C_1 C_2 C h^{s-1/2}$.

As S_n and also $I-S_n$ is a projection, estimating the right factor in $I-S_n=(I-S_n)^2$ by (8c) and the left one by (8i'), one obtains inequality (8d) with $C' := C_1 C_2 C^2$. The somewhat more general estimate (8e) follows from the previously proved estimates by interpolation techniques (cf. Hackbusch [1, (1.4.10b)]). QED

Choosing the optimal values $s=\sigma=t=2$ in (8e) for the case of piecewise linear functions, we obtain from Lemma 7 convergence of the order 3 with respect to the $H^{-1}(\Gamma)$ norm. When approximating f_n, we still generate further errors which will be discussed in §9.2.2 and §9.4. For a complete error analysis of the examples from §8.3 we refer to Giroire - Nédélec [1].

The hypersingular integral operator from above has the order 1. An equation of the first kind with a (single-layer) integral operator of the order -1 is the equation $Kf=\varphi$ from (8.1.27) with the kernel $k(\mathbf{x},\mathbf{y})=s(\mathbf{x},\mathbf{y})$. For the case $d=3$, the integration of (8.1.27) with a test function g leads to the problem

$$a(f,g) = \langle \varphi, g \rangle \qquad \text{with } a(f,g) := \frac{1}{4\pi} \int_\Gamma \int_\Gamma \frac{f(\mathbf{x}) g(\mathbf{y})}{|\mathbf{x}-\mathbf{y}|} d\Gamma_x d\Gamma_y .$$

One can prove that the bilinear form a is bounded, symmetric, and X-elliptic with respect to the dual space $H^{-1/2}(\Gamma)$ (cf. Nédélec [1]). Considerations analogous to Lemma 7 lead to the error

$$\|f-f_n\|_{H^{-1-s}(\Gamma)} \leq C' h^{\sigma+s+1} \|f\|_{H^\sigma(\Gamma)} \quad \text{for } \tfrac{1}{2} \leq \sigma \leq s \leq t,\ f \in H^\sigma(\Gamma).$$

For the piecewise linear functions, $t=2$ holds as above. But since the piecewise constant functions also belong to $L^2(\Gamma)$ and therefore, in particular, to $H^{-1/2}(\Gamma)$, one may also build the subspace X_n by these functions. Then, t takes the value $t=1$. In this case, the optimal order of convergence is $\sigma+s+1=3$ for $\sigma=s=t=1$.

9.2 The Boundary Elements

9.2.1 Elements in the Two-Dimensional Case

In the preceding subchapter we mentioned piecewise linear or constant functions. Since the choice of the subspace X_n is essential for the practical implementation of the collocation or Galerkin method, we shall discuss the details of the implementation.

To start, we first discuss the simpler two-dimensional case. In contrast to §9.2.2, we suppose that the curve Γ is described completely by means of a practically realisable *parametrisation* φ: $[0,L]\to\Gamma$. For a piecewise interpolation, decompose Γ into disjoint arcs $\Gamma_{k,n}$ $(1\leqslant k\leqslant n)$. Denote the endpoints of $\Gamma_{k,n}$ by $\{\mathbf{x}_{k,n}=\varphi(t_{k,n}): 0\leqslant k\leqslant n\}$, where $t_{0,n}=0$ and $t_{n,n}=L$ correspond to the same curve point $\mathbf{x}_{n,n}$. On $[0,L]$ with the interval partition $0=t_{0,n}<t_{1,n}<\dots<t_{n,n}=L$, the piecewise interpolation can be defined by constant, linear, or even higher order splines. Of course, the interpolating functions must be L-periodic.

The integrations for determining the matrix coefficients are performed over the parameter interval according to (7.1.10b):

$$(9.2.1)\qquad \int_\Gamma k(\mathbf{x},\mathbf{y})\psi(\mathbf{y})d\Gamma_y=\int_0^L k(\mathbf{x},\varphi(t))\psi(\varphi(t))|d\varphi(t)/dt|\,dt.$$

For piecewise interpolation, $\psi(\varphi(t))$ must be piecewise constant, linear, etc. with respect to t. For example, in the linear case, we have

$$\begin{aligned}\Psi(t):=\psi(\varphi(t))= &\qquad\qquad \text{for } t_{k-1,n}\leqslant t\leqslant t_{k,n}\\ &=[(t-t_{k-1,n})\psi(\varphi(t_{k,n}))+(t_{k,n}-t)\psi(\varphi(t_{k-1,n}))]/[t_{k,n}-t_{k-1,n}]\\ &=[(t-t_{k-1,n})\psi(\mathbf{x}_{k,n})+(t_{k,n}-t)\psi(\mathbf{x}_{k-1,n})]/[t_{k,n}-t_{k-1,n}].\end{aligned}$$

Retransforming this ansatz into functions defined on Γ, one obtains a function $\psi(\mathbf{y})=\Psi(\varphi^{-1}(\mathbf{y}))$ which is not piecewise linear in $\mathbf{y}$ but piecewise linear with respect to $t=\varphi^{-1}(\mathbf{y})$. This gives rise to the following remark.

Remark 9.2.1. Even if the support abscissae $\{\mathbf{x}_{k,n}\in\Gamma: 1\leqslant k\leqslant n\}$ are fixed, the definition of the subspace of the piecewise linear functions depends on the chosen parametrisation.

In the case of the collocation method, one has to choose collocation points $\xi_{k,n}\in\Gamma$. For linear functions, they coincide with the interpolation points $\mathbf{x}_{k,n}$. For constant ones, however, the midpoints $\tau_{k,n}=\frac{1}{2}(t_{k,n}+t_{k-1,n})$ are advantageous (cf. §4.4.3). They correspond to the collocation points $\xi_{k,n}=\varphi(\tau_{k,n})$ on Γ. Also here, the definition of the $\xi_{k,n}$ depends on the chosen parametrisation, although the class of piecewise constant functions only depends on the support abscissae $\mathbf{x}_{k,n}$. For the construction of collocation methods with higher order splines, we refer to Wendland [4].

9.2.2 Geometric Discretisation

The discretisation described in §9.2.1 requires the integration of the product $k(\mathbf{x},\varphi(t))\psi(\varphi(t))|\nabla\varphi(t)|$, which by no means is a simple task, since only the factor $\psi(\varphi(t))$ represents a convenient (piecewise constant/linear) function. Note that, in general, the kernel k also contains the normal direction $\mathbf{n}(\varphi(t))$.

A further difficulty of the implementation according to §9.2.1 can be the fact that *the curve Γ is given only by discrete points* $\{\mathbf{x}_k\colon 1\leqslant k\leqslant n\}$. This statement holds, in particular, for many practical three-dimensional problems, where the surface is not defined by an analytical description, but only by discretely measured surface points. Using some reasonable interpolation, one can reconstruct from these discrete points $\mathbf{x}_k\in\Gamma$ only an approximative continuous curve or surface $\tilde{\Gamma}$.

In the two-dimensional case, the simplest interpolation, which still leads to a closed curve $\tilde{\Gamma}$, is the *linear* one. It replaces the arcs Γ_k from $\mathbf{x}_{k-1}$ to $\mathbf{x}_k$ by the corresponding chords $\tilde{\Gamma}_k$. The interior domain enclosed by $\tilde{\Gamma}$ is a polygon (cf. Fig. 1).

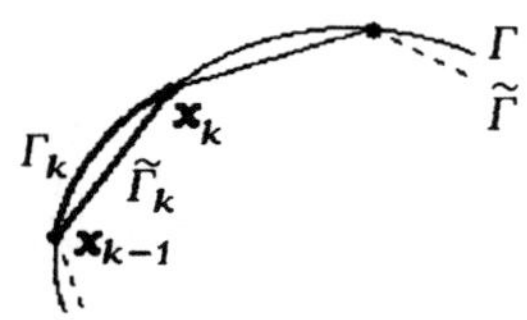

Fig. 9.2.1 Polygon $\tilde{\Gamma}$

The replacement of the curve by the polygon $\tilde{\Gamma}$ does not contradict to what we did in §9.2.1. One can use the sides $\tilde{\Gamma}_k := \{\mathbf{x}_{k-1}+\tau(\mathbf{x}_k-\mathbf{x}_{k-1})\colon 0\leqslant\tau\leqslant 1\}$ of the polygon (more precisely, the parameter $\tau\in[0,1]$) for the parametrisation of the corresponding arc Γ_k. However, by the ***geometric discretisation*** we mean the replacement of Γ by $\tilde{\Gamma}$, so that the integrals over Γ_k are to be replaced by those over $\tilde{\Gamma}_k$, e.g.,

$$\begin{aligned}(9.2.2)\qquad &\int_{\tilde{\Gamma}_k} k(\mathbf{x},\mathbf{y})\psi(\mathbf{y})d\Gamma_y = \\ &= |\mathbf{x}_k-\mathbf{x}_{k-1}|\int_0^1 k(\mathbf{x},\mathbf{x}_{k-1}+\tau(\mathbf{x}_k-\mathbf{x}_{k-1}))\psi(\mathbf{x}_{k-1}+\tau(\mathbf{x}_k-\mathbf{x}_{k-1}))d\tau.\end{aligned}$$

Formally, the integrand $k(\mathbf{x},\mathbf{y})$ in the integral from above is not defined, since, in general, the pair $(\mathbf{x},\mathbf{y})$ does not lie on $\Gamma\times\Gamma$ (the domain of definition of the kernel function k). On the other hand, the kernel functions constructed by means of the integral equation method are the derivatives of the fundamental solution. These, however, are defined everywhere.

Remark 9.2.2. Define the polygon $\tilde{\Gamma}$ by (geometrically) piecewise linear interpolation of Γ. The advantages of this approximation are: **(a)** The integrals reduce to those over the unit interval $[0,1]$ (cf. (2)).

(b) In contrast to Remark 1, the piecewise linear functions are also linear over the side $\tilde{\Gamma}_k$ of the polygon.

(c) A secondary result concerns the definition of an approximative normal direction $\tilde{\mathbf{n}}(\mathbf{y})$ on $\tilde{\Gamma}$: $\tilde{\mathbf{n}}(\mathbf{y})$ is constant on each chord $\tilde{\Gamma}_k$.

9.2.3 Elements in the Three-Dimensional Case

The partition of the surface into «finite elements» can be done in different ways. Starting from a (at least pointwise given) transformation φ of a parameter set $E \subset \mathbb{R}^2$ onto a piece $\Gamma_0 \subset \Gamma$ of the surface, one can decompose E into rectangles or triangles and map this rectangular or triangular grid by means of φ onto Γ_0 (cf. Figs. 2 and 3). Note that only the images of the corners

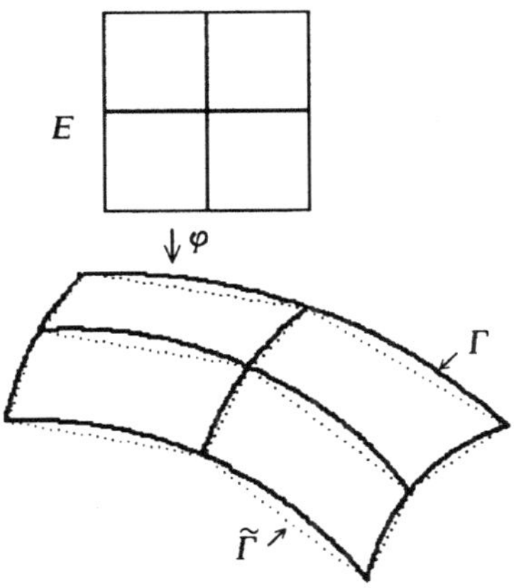

Fig. 9.2.2 rectangular elements

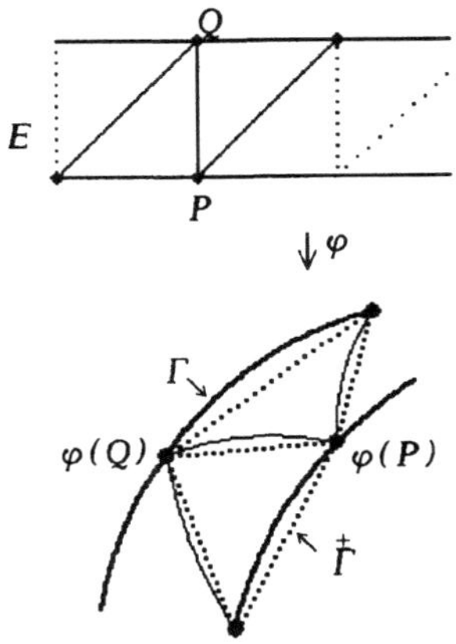

Fig. 9.2.3 triangular elements

are used. The original triangle side PQ of the parameter plane is mapped onto an arc $\varphi(P)\varphi(Q)$ lying on Γ. The polyhedron $\tilde{\Gamma}$ consists of *plane* triangles, which have as triangle side among others the straight line $\varphi(P)\varphi(Q)$ (see. Fig. 3). This means that, in general, Γ and $\tilde{\Gamma}$ may coincide only at the corners of the triangles. We add that the construction of the polyhedron $\tilde{\Gamma}$ can be performed by means of a triangulation directly on Γ without involving a mapping φ from a parameter plane.

Remark 9.2.3. **(a)** If P, Q, R are the corners of a triangle Δ_k of the polyhedron $\tilde{\Gamma}$, then a piecewise linear ansatz function u on Δ_k is uniquely determined by its values $u(P), u(Q), u(R)$.
(b) The normal direction $\tilde{\mathbf{n}}(\mathbf{y})$ on a triangle $\Delta_k \subset \tilde{\Gamma}$ is piecewise constant.
(c) In the case of collocation with piecewise linear interpolation, the interpolation and collocation points lie on Γ (since the corners of $\tilde{\Gamma}$ lie on Γ).

Also Remark 2a can be carried over. Let $\Delta_k \subset \tilde{\Gamma}$ be a face of the polyhedral surface $\tilde{\Gamma}$. Then one can map the unit triangle $D = \{\mathbf{x} \in \mathbb{R}^2 :\ 0 \leq x_1,\ 0 \leq x_2,\ x_1 + x_2 \leq 1\}$ by means of an affine mapping φ_k onto Δ_k. The surface integral over Δ_k equals

$$(9.2.3) \qquad \int_{\Delta_k} k(\mathbf{x},\mathbf{y})\psi(\mathbf{y})\,d\Gamma_y = \sqrt{g}\int_D k(\mathbf{x},\varphi_k(\boldsymbol{\xi}))\,\Psi(\boldsymbol{\xi})\,d\boldsymbol{\xi},$$

where the Gram determinant g is constant on D. $\Psi = \psi \circ \varphi_k$ is linear, if ψ is so.

In this case,

$$\Psi(\xi) = \Psi(0,0)+\xi_1[\Psi(1,0)-\Psi(0,0)] + \xi_2[\Psi(0,1)-\Psi(0,0)] =$$
$$= \Psi(P)+\xi_1[\Psi(Q)-\Psi(P)]+\xi_2[\Psi(R)-\Psi(P)],$$

holds, if $P=\varphi_k(0,0)$, $Q=\varphi_k(1,0)$, $R=\varphi_k(0,1)$ are the corners of Δ_k.

Exercise 9.2.4. The scalar product $\langle \mathbf{n}(\mathbf{y}), \mathbf{y}-\mathbf{x}\rangle$ appearing in the double-layer kernel is piecewise linear on a triangle $\Delta_k \subset \tilde{\Gamma}$. Moreover, it vanishes on Δ_k, if $\mathbf{x} \in \Delta_k$.

The chords introduced by the geometric discretisation in the two-dimensional case and the triangles or quadrangles in the three-dimensional case are often called *panels* in engineering applications. The hereupon based collocation method is called the *panel method.*

9.2.4 Error Considerations

The geometric discretisation is a further reason for errors, which will be discussed in the following. We concentrate on the case of the double-layer kernels and investigate first the two-dimensional case. Concerning Γ we suppose $\Gamma \in C^1_{0,L}$. The value $(\tilde{K}f)(\mathbf{x})$ of the double-layer operator over $\tilde{\Gamma}$ is the sum of the integrals over the polygon sides $\tilde{\Gamma}_k$, which replace the arcs Γ_k.

Lemma 9.2.5. In the case of a constant density f on $\tilde{\Gamma}_k$, the integral value $\int_{\tilde{\Gamma}_k} k(\mathbf{x},\tilde{\mathbf{y}}) f \, d\Gamma_{\tilde{y}}$ is identical to the integral $\int_{\Gamma_k} k(\mathbf{x},\mathbf{y}) f \, d\Gamma_y$ over the original curve with the constant f, as long as $\mathbf{x}$ lies outside of the segment between Γ_k and $\tilde{\Gamma}_k$ (cf. Figs. 1 and 4).

Proof. Γ_k and $\tilde{\Gamma}_k$ form the same angle φ at $\mathbf{x}$ (cf. Remark 1.2). QED

In the piecewise constant case, the geometric discretisation can only introduce additional errors if the collocation points (midpoints of $\tilde{\Gamma}_k$) do not lie on Γ, but differ from the collocation points used in §9.1. The deviation is at most $O(h^2)$, $h=\text{length}(\tilde{\Gamma}_k)$.

The latter error does not appear for piecewise linear elements because of Remark 3c. Instead, the integrals mentioned in Lemma 5 differ if f is linear. Extend a linear density $\tilde{f}$ on $\tilde{\Gamma}_k$, e.g., by means of $f(\mathbf{y}) := \tilde{f}(\tilde{\mathbf{y}})$ for $\langle \mathbf{n}(\mathbf{y}), \mathbf{y}-\tilde{\mathbf{y}}\rangle = 0$ to $\mathbf{y} \in \Gamma_k$. By Lemma 8.2.14, the double-layer integrals over $\tilde{\Gamma}_k$ and Γ_k evaluated at $\mathbf{x}$ can both be represented as integrals over the angle ψ from Fig. 4, where the projected values $\tilde{F}(\omega)$ and the true values $F(\omega)$ from (8.2.12a) differ by $O(h/[h+\sin(\varphi)])$ (φ according to Fig. 4, h: length of $\tilde{\Gamma}_k$, $\Gamma \in C^1_{0,L}$). From $\psi = O(\frac{h}{r}[h+\sin(\varphi)])$ one concludes that

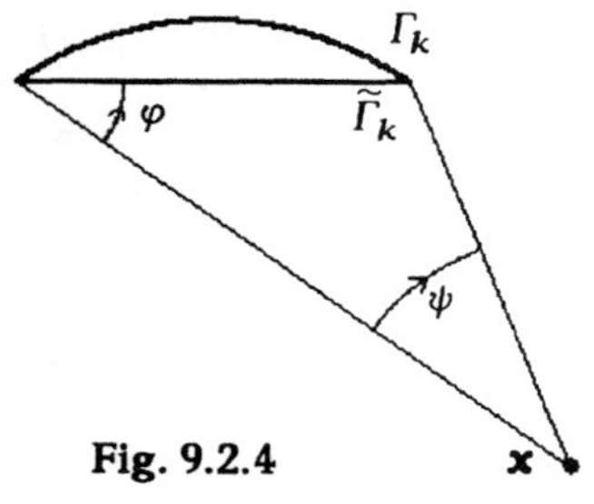

Fig. 9.2.4

$$\int_{\tilde{\Gamma}_k} k(\mathbf{x},\tilde{\mathbf{y}}) \tilde{f}(\tilde{\mathbf{y}}) d\Gamma_{\tilde{y}} - \int_{\Gamma_k} k(\mathbf{x},\mathbf{y}) f(\mathbf{y}) d\Gamma_y = O(h^2/r),$$

where $r = r_k := \mathrm{dist}(\mathbf{x}, \tilde{\Gamma}_k)$. After summation over k in the quasi-uniform case, the sum of all r_k^{-1} behaves like $O(\sum_{0\leq\nu\leq 1/h} \frac{1}{\nu h}) = O(h^{-1}|\log h|)$, so that the matrices $\tilde{B}_n$ and B_n (cf. (4.4.9c)) for the collocation method with and without geometric discretisation differ in the row-sum norm by $\|\tilde{B}_n - B_n\|_\infty \leq O(h|\log h|)$. Assuming stability with respect to $\|\cdot\|_\infty$, this error estimate carries over to the corresponding discrete solutions.

The discussed estimate can be improved, if we assume a sufficiently smooth solution. Let $\mathbf{x}_{k-1}$ and $\mathbf{x}_k$ be the end-points of $\tilde{\Gamma}_k$ as well as Γ_k (cf. Fig. 1). The above mentioned deviation $|\tilde{F}(\omega) - F(\omega)|$ is also proportional to $|f(\mathbf{x}_k) - f(\mathbf{x}_{k-1})|$, since, by Lemma 5, the constant part introduces no error. For $f \in \hat{C}^\mu(\Gamma)$, $0 \leq \mu \leq 1$, one thus obtains the error estimate: $|\tilde{f}_n(\mathbf{x}) - f_n(\mathbf{x})| \leq$ $\leq O(h^{1+\mu}|\log h|)$ at the common collocation points $\mathbf{x} \in \Gamma \cap \tilde{\Gamma}$.

The three-dimensional case is more complicated, since here also the solid angles $\tilde{\omega}$ and ω differ for the triangle $\Delta_k \subset \tilde{\Gamma}$ and the corresponding arc triangle on Γ. However, because of $|\tilde{\omega} - \omega| = O(h^3/r)$ with $r = r_k$ and $\sum_k r_k^{-1} = O(h^{-2})$, one reaches again the estimate $\|\tilde{B} - B_n\|_\infty \leq O(h)$, which can be improved for Hölder continuous solutions to $|\tilde{f}_n(\mathbf{x}) - f_n(\mathbf{x})| \leq O(h^{1+\mu})$.

The geometric discretisation can also regarded as quadrature error. The exact integration over $\tilde{\Gamma}$ one can be interpreted as approximate quadrature over Γ (cf. §9.4).

Concerning error considerations of the geometric discretisation in the case of the Galerkin method for integral equations of the first kind mentioned in §9.1.5, we refer to Nédélec [1] and Giroire -Nédélec [1].

9.3 Multi-Grid Methods

9.3.1 Equations of the Second Kind

In principle, the multi-grid methods described in Chapter 5 are applicable to integral equations of the second kind. A characteristic assumption for the multi-grid convergence is the discrete regularity condition (5.2.11c) (cf. Theorem 5.5.7). The involved discrete spaces $\mathbf{X}_n$, $\mathbf{Y}_n$ correspond, e.g., to the continuous spaces $X = C(\Gamma)$, $Y = C^\mu(\Gamma)$, for which $K \in L(X, Y)$ must hold. The size of the exponent $\mu > 0$ decides upon the convergence speed. For $\mu > 0$, the compact embedding $Y \subset X$ implies the compactness of K. Since in the presence of corners or edges the double-layer operator K is not compact (cf. Lemma 8.2.24), the assumption for the multi-grid convergence is not ensured in this case. Indeed, in numerical experiments one observes a drastic loss of convergence speed as soon as corners appear.

One remedy is a better «smoothing procedure» than the Picard iteration (5.4.7c) (cf. Hackbusch [1,§16.5], Schippers [1]). A second possibility is a splitting of the discrete operator $\mathbf{K}_n$ into $\mathbf{K}_{n,0} + \mathbf{K}_{n,1}$ as already mentioned in §7.3.6. Let, e.g., χ be a smooth function on Γ, which takes the value $\chi(\mathbf{x}) = 1$ $(\mathbf{x} \in U)$ in a neighbourhood U of the corners and the value $\chi(\mathbf{x}) = 0$ $(\mathbf{x} \in \Gamma \backslash U')$ outside a further neighbourhood $U' \supset U$. Then,

$$K_0 f := K(\chi f), \qquad K_1 f := K((1-\chi)f)$$

can be discretised by $\mathbf{K}_{n,0}$, $\mathbf{K}_{n,1}$. Writing $f_n + \mathbf{K}_n f_n = g_n$ as $f_n + \mathbf{K}_{n,0} f_n = g_n - \mathbf{K}_{n,1} f_n$, one gains the equivalent formulation by

(9.3.1a) $\quad f_n = \hat{K}_n f_n + \hat{g}_n \quad$ with $\hat{K}_n := (I+K_{n,0})^{-1} K_{n,1}, \quad \hat{g}_n := (I+K_{n,0})^{-1} g_n$

or

(9.3.1b) $\quad \hat{f}_n = \hat{\hat{K}}_n \hat{f}_n + g_n \quad$ with $\hat{\hat{K}}_n := K_{n,1}(I+K_{n,0})^{-1}, \quad \hat{f}_n := (I+K_{n,0})\, f_n.$

Both reformulations lead to an equation of the second kind, provided that $(I+K_{n,0})^{-1}$ exists. If $K_{n,1}$ satisfies the discrete regularity condition (5.2.11c) (i.e., $\|K_{n,1}\|_{Y_n \leftarrow X_n} \leq C$) and the inverse $(I+K_{n,0})^{-1}$ is uniformly bounded in Y_n (i.e., $\|(I+K_{n,0})^{-1}\|_{Y_n \leftarrow Y_n} \leq C$), then also $\hat{K}_n$ satisfies the discrete regularity condition. In the case from above, $\|(I+K_{n,0})^{-1}\|_{Y_n \leftarrow Y_n} \leq C$ is hard to fulfil, but in the case of $\hat{\hat{K}}_n$ it already suffices to require the estimate

$$\|(I+K_{n,0})^{-1}\|_{X_n \leftarrow X_n} \leq C$$

in X_n, in order to ensure $\|\hat{\hat{K}}_n\|_{Y_n \leftarrow X_n} \leq C$.

When applying the multi-grid method (5.5.3), one has to perform the multiplication $f_n \mapsto \hat{K}_n f_n$ (or $\hat{\hat{K}}_n f_n$, respectively). This consists of the part $f_n \mapsto \varphi_n := K_{n,1} f_n$ and the solution of $(I+K_{n,0})\psi_n = \varphi_n$ leading to $\hat{K}_n f_n := \psi_n$.

9.3.2 Equations of the First Kind

The equations $Kf = g$ of the first kind are to be distinguished with respect to whether the operator K has a positive or non-positive order. For example, the hypersingular equation treated in §9.1.5, has the positive order 1. In this case, multi-grid methods as they are already applied for a long time to discrete elliptic boundary value problems can be carried over without any change. The analysis from Hackbusch [1,§§6-7] and the convergence considerations of the standard multi-grid method is applicable without any modification, since only the general properties of the bilinear form are involved (cf. Lemma 8.3.1). If, as in the case of Lemma 8.3.1, the bilinear form is X-elliptic, a very simple situation arises. Here, e.g., the Richardson iteration $f_n \mapsto f_n - \omega h^\alpha (K_n f_n - \varphi_n)$ can be used as «smoothing method» of the equation $K_n f_n = \varphi_n$. The exponent $\alpha > 0$ is the order of the integral operator ($\alpha = 1$ in §9.1.5) and $\omega := 1/\|h^\alpha K_n\|$ ($\|\cdot\|$: spectral norm).

A completely different situation arises if the order of K is negative. For example, the single-layer operator in (8.1.27) has the order $\alpha = -1$. The opposite sign of α causes that now the large (small) eigenvalues of K correspond to smooth (non-smooth) eigenfunctions. The iterations denoted before as «smoothing methods» no longer work. To remedy the situation, one may try to transform the equation $K_n f_n = g_n$ into a «preconditioned» equation

(9.3.2) $\quad C_n K_n f_n = C_n g_n$

such that $C_n K_n$ has convenient properties. For example, one can split K_n as in §9.3.1 into a sum $K_{n,0} + K_{n,1}$ and choose $C_n := K_{n,0}^{-1}$. Then we obtain

(9.3.3) $\quad f_n = \hat{K}_n f_n + \hat{g}_n \quad$ with $\hat{K}_n := -K_{n,0}^{-1} K_{n,1}, \quad \hat{g}_n := K_{n,0}^{-1} g_n.$

One proposal for choosing $\mathbf{K}_n = \mathbf{K}_{n,0} + \mathbf{K}_{n,1}$ is due to Hsiao-Kopp-Wendland [2], although there the splitting was used for another purpose. First, we explain the decomposition on the continuous level. Let $\Gamma \subset \mathbb{R}^2$ be a smooth curve parametrised over $[0, 2\pi]$, i.e., φ: $[0, 2\pi] \to \Gamma$. The integral $\int_\Gamma k(\mathbf{x}, \mathbf{y}) f(\mathbf{y}) d\Gamma$ with the logarithmic kernel $\log|\mathbf{x}-\mathbf{y}|$ evaluated at $\mathbf{x} = \varphi(t)$ becomes

$$(Kf)(\varphi(t)) = \int_0^{2\pi} \log|\varphi(t)-\varphi(s)|\, f(\varphi(s))\, |\varphi'(s)|\, ds .$$

We set

$$(K_0 f)(\varphi(t)) = \int_{t-\pi}^{t+\pi} \log|t-s|\, f(\varphi(s))\, |\varphi'(s)|\, ds .$$

The difference $K_1 := K - K_0$ equals

$$(K_1 f)(\varphi(t)) = \int_{t-\pi}^{t+\pi} \log\left|\frac{\varphi(t)-\varphi(s)}{t-s}\right| f(\varphi(s))\, |\varphi'(s)|\, ds .$$

Instead of solving with respect to f, take $F(s) := f(\varphi(s))|\varphi'(s)|$ as the function to be determined. Then the kernels of K_0 and K_1 are

$$k_0(t,s) = \log|t-s|, \qquad k_1(t,s) = \log\left|\frac{\varphi(t)-\varphi(s)}{t-s}\right| .$$

For a sufficiently smooth curve Γ, also the kernel k_1 is smooth. At $t=s$, it has the value $\log|\varphi'(s)|$. The logarithmic singularity is contained in k_0, not in k_1. Note that k_0 is a convolution kernel.

For the discretisation by a Galerkin or collocation method, one can analogously split the matrix $\mathbf{K}_n$ into the sum $\mathbf{K}_{n,0} + \mathbf{K}_{n,1}$. Then $\mathbf{K}_{n,1}$ satisfies a certain discrete regularity condition depending on the smoothness of k_1 (and, hence, on the smoothness of Γ). The following property of the matrix $\mathbf{K}_{n,0}$ is essential.

Remark 9.3.1. Use an *equidistant* partition of the interval $[0, 2\pi]$ with step size $h = 2\pi/n$. The basis functions Φ_k of the Galerkin or collocation method should differ only by a periodic shift, i.e., $\Phi_k(s-kh) = \Phi_j(s-jh)$, where the arguments are understood modulo 2π. The kernel k is supposed to be of convolution type. Then the matrix A_n from (4.5.7b) or (4.4.9b) is a periodic *Toeplitz matrix*, i.e., $a_{jk} = a_{j+\ell, k+\ell}$ (indices modulo n).

Since $\mathbf{K}_{n,0}$ is a periodic Toeplitz matrix, it is easy to invert by means of the fast Fourier transform (cf. §7.3.6). Although one loses one differentiability order by the application of $\mathbf{K}_{n,0}^{-1}$, the product $\hat{\mathbf{K}}_n := -\mathbf{K}_{n,0}^{-1}\mathbf{K}_{n,1}$ still satisfies the discrete regularity condition. This shows that the multi-grid method (5.5.3) is applicable to the equivalent equation (3).

Graham-Atkinson [1] use the kernel $-\frac{1}{\pi}\log|2\sin((t-s)/2)/e|$ instead of $\log|t-s|$. The interesting fact is that the eigenvalues of the corresponding operator K_0 are $1/(\sqrt{2\pi}\,|\nu|)$, $\nu \in \mathbb{Z}$.

The construction of $\mathbf{K}_{n,0}$ indicates that the generation of a suitable matrix C_n just described is restricted to the two-dimensional case $d = 2$.

9.4 Integration and Numerical Quadrature

For the Galerkin method as well as for the collocation method many integrals must be computed or approximated. In the case $d=3$ which is of particular practical importance, a considerable amount of computational work arises. If the geometric discretisation from §9.2.2 has replaced the true surface by a polyhedron, it is still possible to compute the single- and double-layer integrals exactly. While in the first German edition of this book, we described the exact quadrature in more detail, we now prefer the numerical quadrature.

9.4.1 General Considerations

In the following we assume that the surface $\tilde{\Gamma}$ consists of *plane triangular pieces* $\Delta_k \subset \tilde{\Gamma}$. If there are faces of $\tilde{\Gamma}$ being quadrangles or more general polygons, these are to be treated as sums of triangles. The single- and double-layer integrals over a triangle $\Delta \subset \tilde{\Gamma}$ evaluated at $\mathbf{x} \in \mathbb{R}^3$ are

(9.4.1a) $$\frac{1}{4\pi}\int_\Delta f(\mathbf{y})/|\mathbf{x}-\mathbf{y}|\,d\Gamma_y \quad\text{and}\quad -\frac{1}{4\pi}\int_\Delta f(\mathbf{y})\langle \mathbf{n}(\mathbf{y}),\mathbf{y}-\mathbf{x}\rangle/|\mathbf{y}-\mathbf{x}|^3\,d\Gamma_y .$$

Because of the choice of the ansatz functions, f restricted to Δ is a polynomial in $\mathbf{y}$, e.g., a piecewise linear function on Δ.

Remark 9.4.1. The integrals (1a) are invariant under rotation and translation. Therefore, the Cartesian coordinate system (x_1, x_2, x_3) in $\mathbb{R}^3$ can be chosen such that the triangle Δ lies in the x_1–x_2 plane. The x_3-axis has the same direction as the outer normal direction, i.e., $\mathbf{n}$ has the constant value $(0,0,1)$ on Δ. The orthogonal projection of the evaluation point $\mathbf{x} \in \mathbb{R}^3$ onto the x_1–x_2 plane determines the origin, i.e., we can assume that $\mathbf{x}$ has the coordinates $(0,0,x_3) = x_3\mathbf{n}$.

Standard methods for the numerical integration of $\int_\Delta k(\mathbf{x},\mathbf{y})\varphi(\mathbf{x})d\Gamma_y$ should be used only if the triangle Δ has a fixed distance from the singularity $\mathbf{x}$. If the quadrature method has the order $\alpha \in \mathbb{N}$ and $O(|\mathbf{x}-\mathbf{y}|^{-1})$ describes the singularity of k, then the relative error equals $O(d^{-\alpha-1}h^{\alpha})$, where

(9.4.1b) $$h = \text{diameter of } \Delta, \qquad d = \operatorname{dist}(\mathbf{x},\Delta),$$

since the α-fold derivatives of $|\mathbf{x}-\mathbf{y}|^{-1}$ behave like $O(|\mathbf{x}-\mathbf{y}|^{-\alpha-1})$ (cf. (1.4.15')). This shows that even in the case when $\mathbf{x}$ does not lie in Δ but in a neighbouring element, the relative error $O(h^{-1})$ is not satisfactory. In a distance of $\sqrt{h}$ around $\mathbf{x}$, the error is not better than $O(h^{(\alpha-1)/2})$.

In principle, a small quadrature error can be achieved by composite formulae, i.e., by a partition of the surface elements into sufficiently small parts. This, however, is a costly procedure.

9.4.2 Weakly Singular Integrals

Consider the case that the singularity $\mathbf{x}$ is in the interior or on the boundary of a (plane) face of the polyhedron. The partition in Fig. 1a shows how to reduce the integration to that over triangles Δ such that one of their corners coincides with the singularity $\mathbf{x}$. Without loss of generality, we may assume that Δ is the unit triangle with the corners $\binom{0}{0}$, $\binom{1}{0}$, $\binom{1}{1}$ and $x = \binom{0}{0}$ is the singularity (cf. Remark 1). The rectangle $R=\{(\xi,\eta): 0<\xi,\eta<1\}$ is mapped one-to-one by

$$\Phi: \binom{\xi}{\eta} \mapsto \binom{\xi}{\eta\xi}$$

onto the interior of Δ. The inverse $\Phi^{-1}: \binom{x}{y} \mapsto \binom{x}{y/x}$ is singular at $x=\binom{0}{0}$. The determinant $\det\Phi' = \det\begin{pmatrix} 1 & 0 \\ \eta & \xi \end{pmatrix} = \xi$ arising from the substitution

$$\int_\Delta \frac{\varphi(\mathbf{y})}{|\mathbf{x}-\mathbf{y}|}\, d\Gamma_y = \int_R \frac{|\xi|}{|\Phi(\xi,\eta)|}\, \varphi(\Phi(\xi,\eta))\, d\xi\, d\eta = \int_R \frac{\varphi(\Phi(\xi,\eta))}{\sqrt{1+\eta^2}}\, d\xi\, d\eta$$

removes the singularity of the integrand, so that the new integral over R is easy to approximate by standard methods. The transformation Φ is named after Duffy [1] and also recommended by Atkinson [4] (cf. Kieser [1]).

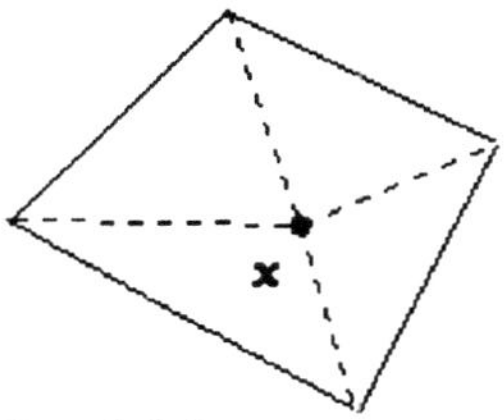

Fig. 9.4.1a

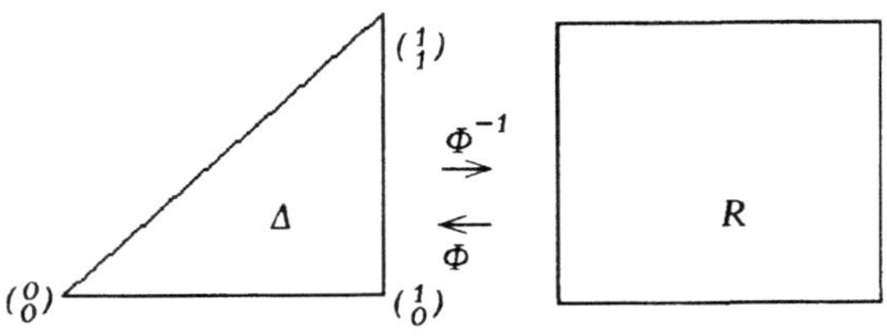

Fig. 9.4.1b

Remark 9.4.2. The use of polar coordinates with the singularity as centre has the same effect as the Duffy coordinates.

Proof. The transformation between both coordinate systems is smooth. ∎

9.4.3 Nearly Singular Integrals

Consider the case that the singularity $\mathbf{x}$ is close to the integration domain. We restrict our discussion to the practically important case of a surface (i.e., dimension $d=3$). Since we can reduce the integration domain to (a union of) triangles, we consider the general integral

$$\text{(9.4.2)} \qquad \int_\Delta \frac{\varphi(x,y-x)}{|y-x|^{s+t}}\, d\Gamma_y \qquad (\Delta\text{: triangle}),$$

where φ is a smooth function with a zero of order t at $y=x$, i.e.,

$$\text{(9.4.3a)} \qquad \varphi(x,z) = O(|z|^t),$$

so that s describes the order of the singularity of the integrand. The integral is regular since y lies outside of Δ. However, the distance $d = \operatorname{dist}(\mathbf{x},\Delta)$ (cf. (1b)) is assumed to be small, e.g., between 0 and the triangle size:

$$\text{(9.4.3b)} \qquad 0 < d \leqslant Ch \qquad (d, h \text{ from (1b)}).$$

Let x_0 be the orthogonal projection of x onto the plane of the triangle. If x_0 lies inside of Δ, split Δ into smaller triangles having x_0 as a corner (cf. Fig. 1a). Hence, we assume without loss of generality that x_0 is not in the interior of the triangle. By suitable rotations and translations (cf. Remark 1), we achieve the situation depicted in Fig. 2:

(9.4.4a) Δ lies in the y_1-y_2-plane (i.e., $y_3=0$),

(9.4.4b) the origin $P_1=(0,0,0)$ is one of the corners of Δ,

(9.4.4c) among all corners of Δ, P_1 has a minimal distance from x_0,

(9.4.4d) the source point equals $x=(-\delta_\Delta, 0, \delta_\perp)$ with $\delta_\Delta=|x_0|$.

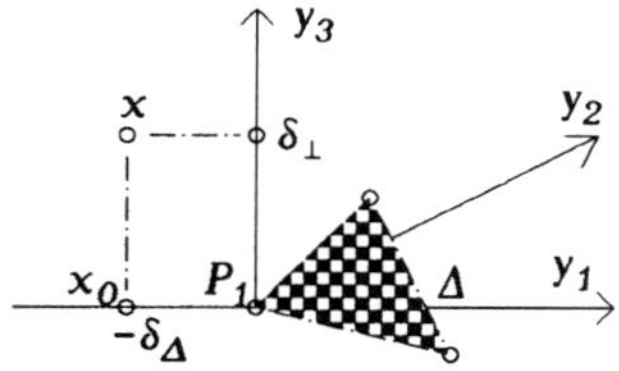

Fig. 9.4.2

The corresponding unitary transformation $y-x = T(u+\delta_\Delta, v, -\delta_\perp)$ replaces the integral (2) by

(9.4.5a) $$\int_\Delta \frac{\Phi(x,u,v)}{[(u+\delta_\Delta)^2+v^2+\delta_\perp^2]^{(s+t)/2}}\,du\,dv$$

with $\Phi(x,u,v)=\varphi(x,T(u+\delta_\Delta,v,-\delta_\perp))$. Usually, φ is a polynomial in z (since the ansatz functions and the geometric factors like $\langle \mathbf{n}(\mathbf{y}),\mathbf{y}-\mathbf{x}\rangle$ are piecewise polynomials). Together with (3a), a finite expansion

(9.4.5b) $$\varphi(x,z) = \sum_{|\nu|\geq t} c_\nu(x)\, z^\nu$$

results (ν: three-dimensional multi-index). Accordingly, the transformed function $\Phi(x,u,v)$ is a polynomial in $u+\delta_\Delta$, v, and $\delta_\perp$:

(9.4.5c) $$\Phi(x,u,v) = \sum_{|\nu|\geq t} \tilde{c}_\nu(x)\,(u+\delta_\Delta)^{\nu_1} v^{\nu_2} \delta_\perp^{\nu_3}.$$

Combining (5a) and (5c), we are left with the problem how to evaluate the particular integrals

(9.4.6) $$I_\nu = \int_\Delta \frac{(u+\delta_\Delta)^{\nu_1} v^{\nu_2} \delta_\perp^{\nu_3}}{[(u+\delta_\Delta)^2+v^2+\delta_\perp^2]^{(s+t)/2}}\,du\,dv \qquad (|\nu|\geq t).$$

Because of $\delta_\Delta^2+\delta_\perp^2>0$, it is guaranteed that the denominator is always positive. Next, we introduce polar coordinates together with $\rho = r+\delta_\Delta\cos\alpha$:

$$u = r\cos\alpha = (\rho-\delta_\Delta\cos\alpha)\cos\alpha, \qquad v = r\cos\alpha = (\rho-\delta_\Delta\cos\alpha)\sin\alpha.$$

They transform the integral I_ν from (6) into

$$I_\nu = \delta_\perp^{\nu_3} \int_{\alpha_{min}}^{\alpha_{max}} \int_{\delta_\Delta\cos\alpha}^{R(\alpha)+\delta_\Delta\cos\alpha} \frac{(\rho\cos\alpha+\delta_\Delta\sin^2\alpha)^{\nu_1}\,(\rho-\delta_\Delta\cos\alpha)^{1+\nu_2}\sin^{\nu_2}\alpha}{[(\rho^2+\delta_\Delta^2\cos^2\alpha+\delta_\perp^2]^{(s+t)/2}}\,d\rho\,d\alpha,$$

where $[\alpha_{min},\alpha_{max}]$ is the angular interval of Δ at the origin and $[0,R(\alpha)]$ is the range of r in $\{r\binom{\cos\alpha}{\sin\alpha}\in\Delta\}$. Expanding $(\rho\cos\alpha+\delta_\Delta\sin^2\alpha)^{\nu_1}$ and $(\rho-\delta_\Delta\cos\alpha)^{1+\nu_2}$, we obtain

$$I_\nu = \delta_\perp^{\nu_3} \sum_{i=0}^{\nu_1+\nu_2+1} \delta_\Delta^{\nu_1+\nu_2+1-i} \int_{\alpha_{min}}^{\alpha_{max}} c(\nu,i,\alpha) \int_{\delta_\Delta\cos\alpha}^{R(\alpha)+\delta_\Delta\cos\alpha} \frac{\rho^i\,d\rho}{[(\rho^2+\delta_\Delta^2\cos^2\alpha+\delta_\perp^2]^{(s+t)/2}}\,d\alpha,$$

with suitable bivariate polynomials $c(\nu,i,\alpha)=c_{\nu_1,\nu_2,i}(\sin\alpha,\cos\alpha)$ in $\sin\alpha$ and $\cos\alpha$. Introducing $\delta=\delta(\alpha)$ with $\delta^2=\delta_\Delta^2\cos^2\alpha+\delta_\perp^2$, we can simplify the denominator to $[\rho^2+\delta^2]^{(s+t)/2}$. The inner (radial) integral is of the form

$$(9.4.7) \qquad \eta_{n,s}(\alpha) := \int_{\delta_\Delta \cos\alpha}^{R(\alpha)+\delta_\Delta\cos\alpha} \frac{\rho^n \, d\rho}{[(\rho^2+\delta^2]^{s/2}} .$$

Its antiderivative is of the explicit form

$$p_{n,s}(\delta, r)\,[\rho^2+\delta^2]^{1-s/2} + \varkappa_{n,s}\log(\pm\rho + [\rho^2+\delta^2]^{1/2}) \qquad (\varkappa_{n,s}=0 \text{ for odd } n).$$

Therefore, the inner integral $\eta_{n,s}(\alpha)$ from (7) can be evaluated. The coefficients $p_{n,s}$ and $\varkappa_{n,s}$ are described in Hackbusch-Sauter [2]. The remaining angular integral in I_ν has the (non-singular) holomorphic integrand

$$(9.4.8) \qquad \delta_\perp^{\nu_3} \sum_{i=0}^{\nu_1+\nu_2+1} \delta_\Delta^{\nu_1+\nu_2+1-i} \, c(\nu,i,\alpha)\,\eta_{n,s+t}(\alpha).$$

This allows the effective use of the standard Gauß quadrature on the interval $[\alpha_{min}, \alpha_{max}]$. Since (8) is also analytic in an ellipse around this interval, one can obtain derivative-free estimates of the quadrature error. It is proved in Hackbusch-Sauter [2] that a Gauß formula with N quadrature points applied to (8) yields an error of the size $ch^\varkappa$ (e.g., $\varkappa$ = consistency order of the method), if

$$N > c_1 + c_2[(\varkappa-1-|\nu|)|\log h| + (s+t-1)|\log d| + \log(1+|\log d|)]$$

with h and d from (1b).

So far, we have considered plane elements. For the case of curved elements see §6 in Hackbusch-Sauter [2].

9.4.4 Strongly Singular Integrals

Using certain regularisations, we can reformulate a Cauchy singular integral as an improper one (cf. §7.1.1). Similarly, hypersingular integrals can be reduced to Cauchy principal values. Another approach for a Hadamard integral is its direct computation. Note that in this case the integral over the surface $\tilde{\Gamma}$ cannot be split into the sum of integral over each element (triangle). If , e.g., the singularity $\mathbf{x}$ is a vertex of the polyhedron $\tilde{\Gamma}$, one has to study the integral over all triangles which meet at $\mathbf{x}$. The required approximation methods are described by Guiggiani [1] and Kieser-Schwab-Wendland [1].

Another possibility circumventing the evaluation of Hadamard (or Cauchy) integrals is the use of the Galerkin discretisation, which involves a double integral, i.e., a $2(d-1)$-dimensional integration. The advantage, however, is that corners and edges do not generate problems. This approach leads to the question how to evaluate efficiently the double integrals of the Galerkin discretisation.

9.4.5 Treatment of Double Integrals Arising from the Galerkin Method

The usual numerical technique for approximating a double integral like (1.2c) over the triangle pair $T_j \times T_k$ is a tensorial quadrature: Apply a formula Q_j with respect to the x integration over T_j and a formula Q_k with respect to the y integration over T_k. This approach is not the appropriate one for singular integrands! If, e.g., the Galerkin formulation is used for hypersingular integrals, an application of Q_j with respect to x requires the evaluation (approximation) of the remaining y integral. This, however, is again hypersingular.

A similar failure happens with two-dimensional integrals with the weak singularity $O(|\mathbf{x}-\mathbf{y}|^{-1})$, e.g., over a rectangle. A one-dimensional quadrature with respect to x leaves one-dimensional integrals with the strong(!) singularity $O(|\mathbf{x}-\mathbf{y}|^{-1})$.

A quadrature adapted to the nature of the singularities arising in BEM must take case of the fact that the singularity mainly depends on the difference x − y. In the following we show how this can be achieved in the case of curves (each of the integrations in (1.2c) is one-dimensional). In this case, the elements are intervals. Furthermore, we assume that the integrand is a sum of the form

$$\varphi(x,y) = \sum \varphi_\nu(x,y)\,\varkappa_\nu(y-x) \qquad (\varphi_\nu \text{ polynomial}). \tag{9.4.9}$$

Let the elements be the intervals $[a,b]$ (with respect to x) and $[c,d]$ (with respect to y). The substitution $\eta = y-x$ and a subsequent interchanging of the integrations yield

$$\int_a^b \Big[\int_c^d \varphi(x,y)\,dy\Big]\,dx = \int_a^b \Big[\int_{c-x}^{d-x} \varphi(x,\eta+x)\,d\eta\Big]\,dx = \tag{9.4.10a}$$

$$= \int_{c-b}^{c-a} \Big[\int_{c-\eta}^{b} \ldots\Big]\,d\eta + \int_{c-a}^{d-b} \Big[\int_a^b \ldots\Big]\,d\eta + \int_{d-b}^{d-a} \Big[\int_a^{d-\eta} \ldots\Big]\,d\eta. \tag{9.4.10b}$$

Each of the inner integrals is of the form

$$\int_{\alpha_1(\eta)}^{\alpha_2(\eta)} \varphi(x,\eta+x)\,dx \qquad \text{with affine bounds } \alpha_i = \alpha_{0,i}+\alpha_{1,i}\eta. \tag{9.4.10c}$$

Inserting the representation (9), we obtain integrands $\varphi_\nu(x,\eta+x)\varkappa_\nu(\eta)$. The singular part $\varkappa_\nu(\eta)$ can be taken outside the integral. The antiderivative of the polynomial $\varphi_\nu(x,\eta+x)$ can be determined explicitly. The evaluation at the affine bounds yields again a polynomial $\pi_\nu(\eta)$. Therefore, the integral (10c) with $\varphi_\nu(x,\eta+x)\,\varkappa_\nu(\eta)$ instead of φ equals $\pi_\nu(\eta)\varkappa_\nu(\eta)$. It remains to perform the outer integration with respect to η. Either this can be done explicitly because of the special nature of the singularity $\varkappa_\nu(\eta)$ or one can apply a Gauß formula with the weight function $\varkappa_\nu(\eta)$. From the practical point, it is very important that the involved intervals are fixed. In the first y-integral in (10a) the interval is fixed but the singularity is moving, while in the second η-integral the singularity is fixed but the interval is moving.

The same procedure can be performed in the case of double integrals over surface elements, i.e., for (1.2c) with triangles T_j, T_k. The analogue of the substitution and integration interchange yields a sum of six integrals instead of the three in (10b). Again, the inner integration can be performed explicitly, while the outer integrals are of the form discussed in §9.4.2. For details we refer to Sauter [1] and Hackbusch-Sauter [1].

Another proposal for approximating the integrals in the Galerkin method with complete error estimates can be found in Johnson-Scott [1].

9.5 Solution of Inhomogeneous Equations

The integral equation method has been derived for the homogeneous equation $Lu=0$, since the potential can satisfy only this equation. Nevertheless, also the inhomogeneous differential equation

(9.5.1) $$Lu = \varphi$$

can be treated. The *volume potential*

(9.5.2) $$V(\mathbf{x}) := \int_{\mathbb{R}^d} s(\mathbf{x},\mathbf{y})\varphi(\mathbf{y})d\Gamma_y \qquad (\mathbf{x}\in\mathbb{R}^d)$$

with the fundamental solution s corresponding to L satisfies the equation (1). Hence, the desired solution is the sum $u = U + V$ consisting of the inhomogeneous solution V and one solution of $LU=0$ satisfying a certain boundary condition. If, e.g., $u=\psi$ on Γ is required, one obtains $U=\psi_0 := \psi - V$ on Γ. For details about volume potentials we refer to Vainikko [1].

In this form, the described method is less recommendable since the computation of V requires volume quadratures (cubatures) which are by one order more expensive than the usual integrations over Γ. Nevertheless, it has to be stated that the grid used for the cubature may be much coarser than a grid in the same volume discretising the original partial differential equation.

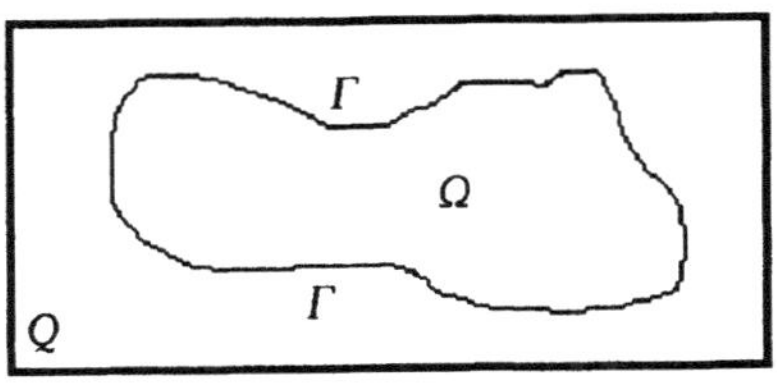

Instead, we recommend the following approach. Let Q be a "brick" $(d=3)$ or rectangle $(d=2)$, containing Γ in its interior: $\Gamma\subset Q$. In Q we approximate the inhomogeneous differential equation (1) with an *arbitrary boundary condition* on ∂Q by means of a usual (spatial) discretisation method. For example, one may use a difference method defining a solution v_h for the step size h on a set Q_h of grid points. By interpolation, one can obtain values $v_{h,\Gamma}$ on Γ (more precisely, at the collocation or nodal points on Γ) from v_h. To satisfy the boundary condition $u=\psi$, subsequently one generates a solution U (the homogeneous equation) corresponding to the boundary data $\psi - v_{h,\Gamma}$ by means of the boundary element method. $U+v_h$ is defined at all grid points Q_h in the interior domain Ω and represents the approximation of (1). The discretisation error of $U+v_h$ consists of the error of the difference solution v_h, of the interpolation error of $v_{h,\Gamma}$, and of the error of U.

The additional solution of the inhomogeneous difference equations can be performed with an operation count of order $O(h^{-d})$ (cf. Hackbusch [1]). Note that the difference solution v_h is essentially simpler to compute than equation (1) discretised in the d-dimensional space. While (1) is to be solved in the interior domain Ω, v_h only requires the solution in the geometrically much simpler brick.

9.6 Computation of the Potential

9.6.1 Evaluation of the Potential

When solving the integral equation, we first approximate the density f. Knowing f, one might be interested in determining the potential Φ, which is the desired solution of the basic differential equation. In principle, the evaluation of the single- or double-layer potential Φ requires the same quadrature as it already appeared in the single- or double-layer integral operators. If $\mathbf{x}$ is far from Γ, the problem of the weakly singular integrand has disappeared. For $\mathbf{x}$ close to Γ, this difficulty is still present. The numerical integration for the double-layer potential outside but close to Γ is even more difficult than the evaluation on Γ. We have discussed this case in §9.4.3.

9.6.2 Evaluation of the Derivatives

The potential Φ is not necessarily the quantity of primary interest in physics. The relevant quantities may be the derivatives $\nabla\Phi$. Therefore, often the problem arises to compute $\nabla\Phi$ instead of Φ. Two approaches are possible.

The derivative $\partial\Phi(\mathbf{x})/\partial x_i$ can be approximated by the difference quotient $[\tilde{\Phi}(\mathbf{x}+\varepsilon\mathbf{e}_i)-\tilde{\Phi}(\mathbf{x})]/\varepsilon$ ($\mathbf{e}_i$: ith unit vector) or similar approximations. This requires only the evaluation of $\tilde{\Phi}$ according to §9.6.1. Note that non-systematic errors of $\tilde{\Phi}$ are amplified by a factor $1/\varepsilon$. By «non-systematic» errors we means those which may depend discontinuously on $\mathbf{x}$. If, however, an asymptotical error expansion $\tilde{\Phi}(\mathbf{x})=\Phi(\mathbf{x})+h^{\alpha}\Phi_{\alpha}(\mathbf{x})+\ldots$ with smooth coefficients Φ_{α} holds (cf. §9.6.3), the term $h^{\alpha}\Phi_{\alpha}(\mathbf{x})$ causes a contribution $O(h^{\alpha})$ in the difference quotient independent of ε.

Second, one can apply the derivation formulae (8.1.1a-c) for (8.2.13b) for the single- or double-layer potential. This avoids additional discretisation errors but, on the other hand, one has to evaluate integrals with a stronger singularity.

9.6.3 Error Considerations

Assume in the following that the potential Φ is to be evaluated in the interior and/or exterior domain in a distance of at least $\mathrm{dist}(\mathbf{x},\Gamma)\geq d$ from the boundary. An error δf of the density f leads to an error $\delta\Phi$ of the potential. Note that $\delta\Phi$ is arbitrarily smooth and its derivatives in $\{x\in\mathbb{R}^d\colon \mathrm{dist}(\mathbf{x},\Gamma)\geq d\}$ are uniformly bounded by

$$|D^{\boldsymbol{\nu}}\delta\Phi(\mathbf{x})| \leq C_{|\boldsymbol{\nu}|}(d)\,\|\delta f\|_{\infty} \qquad \text{for all } \boldsymbol{\nu} \text{ and } \mathrm{dist}(\mathbf{x},\Gamma)\geq d>0.$$

The maximum norm $\|\delta f\|_{\infty}$ used above one can replaced by an arbitrarily *weak norm* - e.g., by the dual norms $\|\cdot\|_{-k}$ corresponding to $\|\cdot\|_k=\|\cdot\|_{H^k(\Gamma)}$ (cf. §4.6.2) - since $s(\mathbf{x},\cdot)$ and all its derivatives are well-defined on Γ for $\mathrm{dist}(\mathbf{x},\Gamma)\geq d>0$. Hence, one obtains error estimates $O(h^{\varkappa})$ for Φ and its derivatives, if δf satisfies the estimate $\|\delta f\|_{-k}\leq Ch^{\varkappa}$ with respect to *any (weak) norm*. This is of interest since in §4.6.2 we proved that often weaker norms lead to better error estimates than with respect to the maximum norm.

9.6.4 Extrapolation

The potential $\Phi = \Phi_h$ computed for the step size h by a collocation or Galerkin method may allow an asymptotical expansion of the form

$$\Phi_h(\mathbf{x}) = \sum_{\nu=0}^{\ell-1} h^{\gamma_\nu} \Phi_{(\nu)}(\mathbf{x}) + h^{\gamma_\ell} \Phi_{(\ell)}(\mathbf{x};h)$$

for $\mathrm{dist}(\mathbf{x},\Gamma) \geqslant d > 0$ with bounded remainder term $|\Phi_{(\ell)}(\mathbf{x};h)| \leqslant C_\ell(d)$, although a corresponding expansion (4.8.4a-c) for the density f_n may not hold. Saranen [2] proves the expansion of $\Phi_h(\mathbf{x})$ and describes the exponents γ_ν.

9.7 The Panel Clustering Algorithm

9.7.1 Introduction

The boundary element method (BEM) leads to a system of linear equations with a *full matrix*, while the finite element method for boundary value problems yields *sparse* matrices. This property of the BEM seems to lead to much computational work for the definition of the matrix, for the solution of the system, and, in particular, for the matrix-vector multiplication, which always occurs as an elementary operation. The panel clustering algorithm is a method for approximating the matrix-vector multiplication with much less arithmetical work. In addition, the storage requirements can substantially be reduced.

The method has been developed in 1986 and is described in Hackbusch-Nowak [2], Hackbusch [7], and Sauter [1]. The method of Rokhlin [1] uses similar ideas. Another approach to the «condensation» of the full matrix by means of wavelets is presently under consideration (cf. Dahmen-Prößdorf-Schneider [1], Schneider-v.Petersdorff-Schwab [1]).

We consider the semidiscrete operator

$$(K_n u_n)(\mathbf{x}) = \int_\Gamma k(\mathbf{x},\mathbf{y})\, u_n(\mathbf{y})\, d\Gamma_y, \tag{9.7.1}$$

where n denotes the number of elements (panels) used in the approximation of Γ. u_n belongs to a subspace, whose dimension is denoted by q. Usually, $O(n) = O(q)$ holds (exception: p-version of BEM).

The collocation matrix B is a *full* matrix of the size $q \times q$ and therefore the usual matrix-vector multiplication $a \mapsto Ba$ requires $O(q^2)$ arithmetical operations. The matrix multiplication arises, e.g., in the iterative solution (cf. §5) or in the evaluation of the integral.

Assume a «finite element basis» (i.e., the support consists of only few elements). Then the evaluation of one entry B_{ij} requires the integration over $O(1)$ elements. The total work in the evaluation phase amounts to $O(n^2) = O(q^2)$ elementary integrals.

Often these facts are put forward as a drawback of BEM compared with the field methods which lead to sparse matrices. We shall show below that a vector sufficiently close to Ba can be computed by only $O(n \log^{d+1} n)$ operations (cf. Theorem 3). It may be even more important that computational work and storage is saved also in the definition phase, since it is not necessary to compute and store all q^2 entries of B. It suffices to evaluate and store $O(n\log^{d+1} n)$ quantities (cf. Theorem 4).

Similar considerations are valid for the Galerkin discretisation instead of collocation (cf. Sauter [1], Hackbusch-Sauter [1]).

9.7.2 Panels

The polyhedron $\tilde{\Gamma}$ is the union of so-called *panels* or *boundary elements*. In the following, the n panels are denoted by $\pi_1, \ldots, \pi_n$. The set of panels is denoted by

$$P := \{\pi_1, \ldots, \pi_n\}, \qquad n := \# P.$$

The integral in (1) is taken over

$$\tilde{\Gamma} := \bigcup_{j=1}^{n} \pi_j = \bigcup_{\pi \in P} \pi .$$

Since the collocation subspace V_n has dimension q, there are also q different collocation points $\Xi = \{\xi^1, \ldots, \xi^q\}$. We recall that the coefficients of the collocation solution $u_n = \sum_{j=1}^{q} \alpha_j b_j$ (b_j: basis functions in V_n) satisfy the linear system $\lambda a - Ba = b$. The matrix entries of B are

$$\beta_{ij} = (K_n b_j)(\xi^i) = \int_{\tilde{\Gamma}} k(\xi^i, \mathbf{y})\, b_j(\mathbf{y})\, d\Gamma_y \qquad \text{(cf. (4.4.9c))}. \tag{9.7.2}$$

9.7.3 The Panel Clustering Method

9.7.3.1 The Far Field Expansion

The algorithm has two requirements: (i) The kernel function must be approximated by an expansion and (ii) a so-called cluster set has to be constructed from the panels. We start with the first requirement. Let the kernel $k(\mathbf{x}, \mathbf{y})$ be approximated by a finite expansion of the form

$$\tilde{k}(\mathbf{x},\mathbf{y}) = \sum_{\iota \in I_m} \varkappa_\iota(\mathbf{x}; \mathbf{y}_0)\, \Phi_\iota(\mathbf{y}), \tag{9.7.3}$$

where $\Phi_\iota(\mathbf{y})$ for $\iota \in I_m$ are functions independent of the other quantities $\mathbf{x}$ and $\mathbf{y}_0$. The point $\mathbf{x}$ will be replaced, e.g., by the collocation points ξ^j. The quantity $\mathbf{y}_0$ denotes the centre of the expansion and may vary in the applications, since $k(\mathbf{x},\mathbf{y})$ is considered over different subsets of the surface $\tilde{\Gamma}$. In the simplest case, $\tilde{k}(\mathbf{x},\mathbf{y})$ is the Taylor expansion around $\mathbf{y}_0$ of degree $m-1$. In this case, the index set I_m is the set $\{\iota \in \mathbb{Z}^d : \iota_1 + \ldots + \iota_d \leq m-1\}$ of multi-indices, where d is the dimension (e.g., $d=3$). In order to make the functions $\Phi_\iota(\mathbf{y})$ independent of the centre $\mathbf{y}_0$, we may choose the monomials $\Phi_\iota(\mathbf{y}) = \mathbf{y}^\iota$. In this case, the number of indices in the set I_m is bounded by (4) with $C_I = 1$:

$$\# I_m \leq C_I\, m^d. \tag{9.7.4}$$

9.7.3.2 Panel Clustering

Assembling several neighbouring boundary elements (panels) $\pi \in P$, we can form first *clusters*. The union of neighbouring clusters yield even larger clusters. This process can be continued until the complete surface $\tilde{\Gamma}$ is obtained as the largest cluster (cf. Fig. 1).

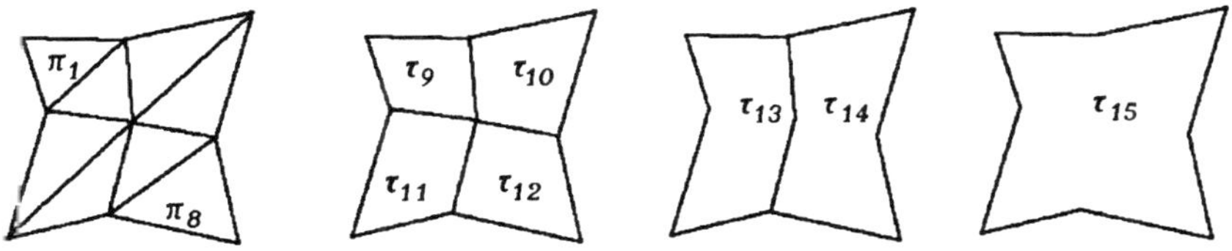

Fig. 9.7.1 Construction of clusters $\tau_9, \ldots, \tau_{15}$ from the panels $\pi_j = \tau_j$ $(1 \leqslant j \leqslant 8)$

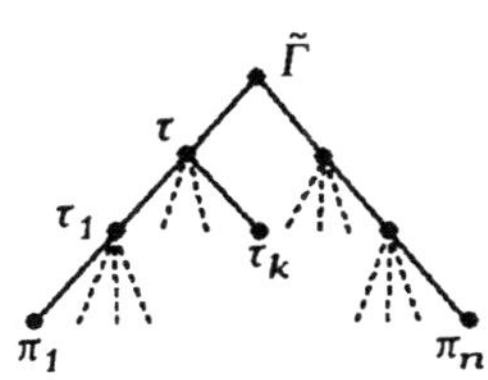

Fig. 9.7.2 The tree T

One may consider this process also from the opposite point of view (domain decomposition). The surface $\tilde{\Gamma}$ is divided into smaller parts (clusters), which are divided further until only the panels remain as the trivial clusters.

The process of clustering (or repeated domain decomposition) is represented by the tree T of clusters with the following properties:

- The cluster set T has a connected tree structure: $\tilde{\Gamma} \in T$ is the *root* of T.
- The panels of P belong to T; they are the *leaves* of T, i.e. there is no $\tau' \in T$ with $\tau \supset \tau'$, $\tau \neq \tau'$.
- If $\tau \in T$ is an inner vertex, there are sons $\tau_1, \ldots, \tau_k$ (k depending on τ) with
  i) $\tau = \tau_1 \cup \ldots \cup \tau_k$,
  ii) for any $\tau' \in T$ with $\tau' \subset \tau$, $\tau' \neq \tau$ there is an index $j \in \{1, \ldots, k\}$ such that $\tau' \subset \tau_j$.

A simple consequence of the tree structure is the following remark.

Remark 9.7.1. Let T as above. Then the number of clusters $\tau \in T$ is at most $2n-1$, where n is the number of panels: $\#T \leqslant 2\#P - 1 = 2n - 1$. The number of non-trivial clusters (i.e., clusters not being panels) is bounded by $n-1$.

9.7.3.3 Admissible Clusters and Admissible Coverings

For the integration of $k(\mathbf{x}, \mathbf{y})u(\mathbf{y})$ over a cluster τ (cf. (2)) we shall use the expansion $\tilde{k}(\mathbf{x}, \mathbf{y})$ from (3) instead of k. This requires that the function k is smooth enough on τ, i.e. $|\mathbf{x} - \mathbf{y}|$ must be sufficiently large.

We name $\tau \in T$ an *admissible cluster* with respect to some source point $\mathbf{x} \in \mathbb{R}^d$ if

$$(9.7.5) \qquad \rho_\tau \leqslant \eta\,|\mathbf{x} - \mathbf{z}_\tau|.$$

Here ρ_τ is the size of the cluster and $\mathbf{z}_\tau$ its centre point. The parameter η has to be chosen appropriately. It will turn out that η should be a constant independent of the panel size h (cf. (13)).

From inequality (5) we conclude the following statement: The larger the distance between $\mathbf{x}$ and the cluster, the larger may be the cluster.

A set of clusters $\mathscr{C} = \{\tau_1, \ldots, \tau_\sigma\} \subset T$ is called a *covering* (of $\tilde{\Gamma}$) if the clusters overlap at most by their edges and satisfy

(9.7.6) $$\tilde{\Gamma} = \bigcup_{j=1}^{\sigma} \tau_j .$$

There are two trivial coverings: We may choose $\mathscr{C}=\{\tau_1\}$ with $\tau_1 := \tilde{\Gamma}$. The second choice is the standard one: $\mathscr{C}=P$. By definition, the panels of the set P cover $\tilde{\Gamma}$. These two possibilities are the extreme positions: In the first case the clusters are as large as possible (but their number is minimal), in the second case the clusters are as small as possible (but their number is maximal).

In the following, we are looking for a covering, which (from the computational point of view) should consist of a small number of clusters and which on the other hand has clusters of reasonable size. The size ρ_τ will be reasonable, if the cluster is admissible. This leads us to the following definition.

Definition 9.7.2. We call $\mathscr{C}=\{\tau_1,\dots,\tau_\sigma\}\subset T$ an *admissible covering* (of $\tilde{\Gamma}$) with respect to $\boldsymbol{x}$ if it satisfies (6) and if each cluster satisfies the following: Either $\tau_j \in P$ or τ_j is admissible with respect to $\boldsymbol{x}$.

If $\boldsymbol{x}\in\tilde{\Gamma}$, there is no covering (6) of $\tilde{\Gamma}$ consisting only of admissible clusters. The second condition in Definition 2 states that the only non-admissible clusters can be the panels. The number of clusters in a covering should be as small as possible. The optimal choice is discussed below.

Proposition 9.7.3. For each $\boldsymbol{x}\in\mathbb{R}^d$ there is a *unique* admissible covering $\mathscr{C}(\boldsymbol{x})$ with respect to $\boldsymbol{x}$ with minimal number $\sigma=\sigma(\boldsymbol{x}) := \#\mathscr{C}(\boldsymbol{x})$ of clusters.

In order to prove this proposition, the size ρ_τ of a cluster τ and its centre $\boldsymbol{z}_\tau$ have to be defined carefully (cf. (5)). To be precise, one has to change inequality (5) into $\rho_\tau \leqslant \bar{\eta}\,\mathrm{dist}(\boldsymbol{x},\tau)$ with $\bar{\eta} := \eta/(1+\eta^2)^{1/2}$ in order to ensure that subsets of admissible clusters are again admissible. Details can be found in Hackbusch-Nowak [2].

The minimal admissible covering $\mathscr{C}(\boldsymbol{x})$ with respect to $\boldsymbol{x}$ can easily be computed by

(9.7.7a) $\mathscr{C} := \emptyset$; $\mathit{Divide}(\tilde{\Gamma},\mathscr{C})$; **comment** the result is $\mathscr{C}=\mathscr{C}(\boldsymbol{x})$;

where *Divide* is the following recursive procedure

(9.7.7b)
```
procedure Divide(τ,𝒞);
comment τ is a cluster, 𝒞 is a subset of T;
begin integer i, s;
        if (τ is admissible w.r.t. x) then 𝒞 := 𝒞 ∪ {τ} else
        if τ ∈ P then 𝒞 := 𝒞 ∪ {τ} else
        begin determine the sons τ_1,...,τ_s of τ;
              for i := 1 step 1 until s do Divide(τ_i,𝒞)
end     end;
```

9.7.3.4 The Algorithm for Matrix-Vector Multiplication

Partition into Near and Far Field. Instead of computing the matrix entries, we compute the *far field coefficients* in *Phase I*. In *Phase II* we evaluate an approximation of $K_n u_n$ at $\boldsymbol{x}\in\mathbb{R}^d$ (e.g., at the collocation points $\xi\in\Xi$). We recall that the desired result of the matrix-vector multiplication $b = Ba$ is the sum

(9.7.8) $$\beta_i = \sum_{j=1}^{q} \beta_{ij}\alpha_j = \sum_{j=1}^{q} \alpha_j \int_{\tilde\Gamma} k(\xi^i,\mathbf{y})b_j(\mathbf{y})d\Gamma_y$$

(cf. (2)). Let $\mathscr{C}(\xi^i)$ be the minimal admissible covering determined in (7b). We split $\mathscr{C}(\xi^i)$ into a near and far field part according to

$$\mathscr{C}_{near}(\xi^i) := \{\tau\in\mathscr{C}(\xi^i): \tau \text{ is a panel}\},$$
$$\mathscr{C}_{far}(\xi^i) := \{\tau\in\mathscr{C}(\xi^i): \tau \text{ is not a panel}\}.$$

The integral in (8) can be split into $\int_{\tilde\Gamma}\ldots = \sum_{\tau\in\mathscr{C}_{near}(\xi^i)}\int_\tau\ldots + \sum_{\tau\in\mathscr{C}_{far}(\xi^i)}\int_\tau\ldots$.

This induces an analogous splitting of β_i from (8) into $\beta_i = \beta_i^{near} + \beta_i^{far}$. We shall approximate β^{far} by $\tilde\beta^{far}$.

The Near Field Part. The near field part of β_i is computed by

(9.7.9) $$\beta_i^{near} = \sum_{\pi\in\mathscr{C}_{near}(\xi^i)} \sum_j \alpha_j \int_\pi k(\xi^i,\mathbf{y})b_j(\mathbf{y})d\Gamma_y.$$

Since the support of the basis functions b_j is small, there is only a constant number of indices j such that the panel π intersects with $\mathrm{supp}(b_j)$. Hence, the sum $\sum_j$ has only $O(1)$ terms. The number of panels in $\mathscr{C}_{near}(\xi^i)$ will turn out to be bounded by $O(\log n)$.

The Far Field Part. Replacing the exact kernel k in the definition of β_i^{far} by $\tilde k(\mathbf{x},\mathbf{y})$ given in (3), we obtain the approximation $\tilde\beta_i^{far}$:

$$\tilde\beta_i^{far} := \sum_{\tau\in\mathscr{C}_{far}(\xi^i)} \sum \alpha_j \int_\tau \tilde k(\xi^i,\mathbf{y})\, b_j(\mathbf{y})\, d\Gamma_y =$$
$$= \sum_{\tau\in\mathscr{C}_{far}(\xi^i)} \sum_j \alpha_j \int_\tau \sum_{\iota\in I_m} \varkappa_\iota(\xi^i;z_\tau)\,\Phi_\iota(\mathbf{y})\, b_j(\mathbf{y})\, d\Gamma_y.$$

Summation and integration can be interchanged:

$$\tilde\beta_i^{far} = \sum_{\tau\in\mathscr{C}_{far}(\xi^i)} \sum_{\iota\in I_m} \varkappa_\iota(\xi^i;z_\tau) \sum_j \alpha_j \int_\tau \Phi_\iota(\mathbf{y})\, b_j(\mathbf{y})\, d\Gamma_y.$$

The integrals

(9.7.10) $$J_\tau^\iota(b_j) := \int_\tau \Phi_\iota(\mathbf{y})\, b_j(\mathbf{y})\, d\Gamma_y \qquad (\iota\in I_m,\ \tau\in T,\ 1\le j\le q)$$

will be called *far field coefficients*. Of particular interest are the far field coefficients corresponding to panels:

(9.7.10') $$J_\pi^\iota(b_j) := \int_\pi \Phi_\iota(\mathbf{y})\, b_j(\mathbf{y})\, d\Gamma_y \qquad (\iota\in I_m,\ \pi\in P,\ 1\le j\le q).$$

They are evaluated in a first phase, since they are independent of the special vector a in the matrix-vector multiplication $b = Ba$. Note that there is only a fixed number of panels π intersecting with the support of b_j. Otherwise, the coefficients $J_\pi^\iota(b_j) = 0$ need not be computed.

As soon as the far field coefficients (10') are computed, the sums

$$J_\pi^\iota := \sum_j \alpha_j\, J_\pi^\iota(b_j) \qquad (\iota\in I_m,\ \pi\in P)$$

can be computed in $O(n*I_m)$ additions. The summation of the panel coefficients should make use of the tree structure:

$$J_\tau^\iota := \sum_{\tau' \text{ son of } \tau} J_{\tau'}^\iota \qquad \text{for } \tau \in T \setminus P.$$

The coefficients J_τ^ι represent the sum $J_\tau^\iota = \sum \alpha_j J_\tau^\iota(b_j)$. Hence the quantities $\tilde{\beta}_i^{\text{far}}$ can easily be computed from the sum

$$(9.7.11) \qquad \tilde{\beta}_i^{\text{far}} = \sum_{\tau \in \mathscr{C}_{\text{far}}(\xi^i)} \sum_{\iota \in I_m} \varkappa_\iota(\xi^i; z_\tau)\, J_\tau^\iota \qquad (1 \leq i \leq q).$$

Since the number of clusters $\tau \in \mathscr{C}_{\text{far}}(\xi^i)$ is expected to be much smaller than the number n of all panels, the representation (11) should be advantageous.

9.7.4 The Additional Quadrature Error

The replacement of β_i^{far} by $\tilde{\beta}_i^{\text{far}}$ can be regarded as an additional quadrature error. The error of the expansion (3) depends on the order m and on the distance $|\mathbf{y}-\mathbf{y}_0|$ from the centre. The exact requirements on $\tilde{k}$ are as follows. Let $\eta_0 \in (0,1)$ be given.

(9.7.12a) There are constants C_1, C_2 such that for all $\mathbf{x}, \mathbf{y}_0 \in \mathbb{R}^d$, all $0<\eta\leq\eta_0<1$, and all $m \in \mathbb{N}$ there are expansions $\tilde{k}_n$ of the form (3) satisfying

(9.7.12b) $|k(\mathbf{x},\mathbf{y}) - \tilde{k}(\mathbf{x},\mathbf{y})| \leq C_1 (C_2 \eta)^m |k(\mathbf{x},\mathbf{y})|$ for all $|\mathbf{y}-\mathbf{y}_0| \leq \eta |\mathbf{x}-\mathbf{y}_0|$.

Inequality (12b) provides an estimation of the *relative error* of $\tilde{k}$. In Hackbusch-Nowak [2], inequality (9.7.12b) is proved for standard examples and the values of C_1, C_2, and η_0 are determined. By (12a,b), the error is of order $O(\eta^m)$. To make the error comparable with the consistency error $O(h^\varkappa)$ (h: panel size), we choose η in (5) by

$$(9.7.13) \qquad \eta = \eta(n) := O((n^{-\varkappa/(d-1)})^{1/m}) \leq \eta_0, \qquad \varkappa \leq m,$$

Then $O(\eta^m)$ equals $O(h^\varkappa)$. Since m will be chosen as $O(\log n)$, the quantity η becomes independent of h.

9.7.5 Complexity of the Algorithm

The complexity of the algorithm depends mainly on the number $\sigma(\mathbf{x})$ of clusters in the minimal admissible covering $\mathscr{C}(\mathbf{x})$. Under natural conditions (cf. Hackbusch-Nowak [2]) there is a constant C_σ with

$$(9.7.14) \qquad \sigma(\mathbf{x}) \leq \sigma(\eta,n) := C_\sigma (\tfrac{1}{\eta})^{d-1} \log(2+\eta^{d-1} n) \qquad \text{for all } \mathbf{x} \in \mathbb{R}^d$$

with d the dimension of $\mathbb{R}^d$, η from (5), and $n = \#P$. The logarithmic factor can even be omitted if $\mathbf{x} \notin S_n$.

Inserting (13) into (14), we obtain the following estimate for the number $\sigma(\mathbf{x})$ of clusters in $\mathscr{C}(\mathbf{x})$:

(9.7.15) $\sigma(\boldsymbol{x}) \leqslant \sigma(\eta, n) = C\, n^{\varkappa/m} \log(2 + C\, n^{1-\varkappa/m})$ for all $\boldsymbol{x} \in \mathbb{R}^d$.

The optimal choice of the expansion order m turns out to be

$$m := \left\lfloor \tfrac{\varkappa}{d+1} \log n \right\rfloor \qquad (\lfloor x \rfloor := \text{largest integer } i \text{ with } i \leqslant x).$$

Then $n^{\varkappa/m}$ (and η) is a constant and we obtain the estimate

(9.7.16) $\sigma(\boldsymbol{x}) \leqslant C \log n$.

Therefore, we have to deal with only $O(\log n)$ clusters instead of n panels.

In detail, the operation count is as follows:

Algorithm (7a,b) (determination of the minimal admissible covering $\mathscr{C}(\boldsymbol{x})$) requires $O(\sigma(\boldsymbol{x}))$ operations per point $\boldsymbol{x}$. Because of (16) and since there are $q = O(n)$ different collocation points $\boldsymbol{x} = \xi^j$, the total amount of work in this part is $O(n \log n)$.

The computation of β_i^{near} in (9) needs $O(\log n)$ operations per index i; therefore, in total $O(n \log n)$ operations, provided that integrals of the form $\int_\pi k(\boldsymbol{x},\boldsymbol{y}) b_j(\boldsymbol{y})\, d\Gamma_y$ can be evaluated by a constant number of arithmetical operations (cf. §9.4).

The far field coefficients $J_\pi^\iota(b_j)$ for the panels $\pi \in P$ and all indices $\iota \in I_m$ can be computed by $O(n\, m^{d+1})$ operations under the assumption (4) and the assumption that an integral of the form $\int_\pi \Phi_\iota(\boldsymbol{y}) b_j(\boldsymbol{y})\, d\Gamma_y$ can be evaluated by $O(m)$ operations. This is possible for the standard case of polynomials of degree $O(m)$ as integrands. Hence, we obtain an operation count of $O(n \log^{d+1} n)$.

The far field coefficients J_τ^ι for the nontrivial clusters $\tau \in T \setminus P$ and for all indices $\iota \in I_m$ can be summed up by $O(n\, \# I_m) = O(n \log^d n)$ additions.

For standard cases, it is possible to compute the coefficients $\varkappa_\iota(\xi^i; z_\tau)$ by $O(m)$ operations per ι, i, and $\tau \in \mathscr{C}_{\text{far}}(\xi^i)$. In total, $O(n \log^{d+2} n)$ operations are necessary. If these coefficients are stored, this part of the computation has to be performed only once.

Phase II of the far field computation with the determination of $\tilde{\beta}_i^{\text{far}}$ from (11) requires the summation over $\tau \in \mathscr{C}_{\text{far}}(\xi^i)$ and $\iota \in I_m$. This takes only $O(n \log^{d+1} n)$ additions. Gathering all results, we obtain the following result.

Theorem 9.7.3. Each matrix-vector multiplication $a \mapsto Ba$ can be approximated up to an error of size $O(h^\varkappa)$ by $O(n \log^{d+1} n)$ operations. The initialisation phase requires $O(n \log^{d+2} n)$ operations.

A similar estimate can be made for the storage requirements. Only the coverings $\mathscr{C}(\xi^j)$, the non-zero integrals $\int_\pi k(\xi^i, \boldsymbol{y}) b_j(\boldsymbol{y})\, d\Gamma_y$ from (9), the far field coefficients $J_\pi^\iota(b_j)$, and the coefficients $\varkappa_\iota(\xi^i; z_\tau)$ are to be stored. The latter fact needs the largest storage.

Theorem 9.7.4. The required storage is of size $O(n \log^{d+1} n)$.

Bibliography

AMINI, S. and I. H. SLOAN: [1] Collocation methods for second kind integral equations with non-compact operators. *J. Int. Eq. Appl. 2* (1989) 1–30

ANSELONE, Ph. M.: [1] *Collectively compact operator approximation theory and applications to integral equations.* Prentice-Hall, Englewood Cliffs, 1971

ANSELONE, Ph. M. and R. H. MOORE: [1] Approximate solution of integral and operator equations. *J. Math. Anal. Appl. 9* (1964) 268–277

ANTES, H. and P. D. PANAGIOTOPOULOS: [1] *The boundary integral approach to static and dynamic contact problems.* ISMN 108, Birkhäuser, Basel, 1992

ARNOLD, D. N.: [1] A spline-trigonometric Galerkin method and an exponentially convergent boundary integral method. *Math. Comp. 41* (1983) 383–397

ARNOLD, D. N. and W. L. WENDLAND: [1] On the asymptotic convergence of collocation methods. *Math. Comp. 41* (1983) 349–381

ARNOLD, D. N. and W. L. WENDLAND: [2] The convergence of spline collocation for strongly elliptic equations on curves. *Numer. Math. 47* (1985) 317–341

ATKINSON, K.: [1] Iterative variants of the Nyström method for the numerical solution of integral equations. *Numer. Math. 22* (1973) 17–31

ATKINSON, K.: [2] An automatic program for linear Fredholm integral equations of the second kind. *ACM Trans. Math. Software 2* (1976) 154–171

ATKINSON, K.: [3] *A survey of numerical methods for the solution of Fredholm integral equations of the 2nd kind.* SIAM, Philadelphia, 1976

ATKINSON, K.: [4] A survey of boundary integral equation methods for the numerical solution of Laplace's equation in three dimensions. In: M. Golberg (ed.): Numerical solution of integral equations. Plenum Press, New York, 1990

ATKINSON, K. and A. BOGOMOLNY: [1] The discrete Galerkin method for integral equations. *Math. Comp. 48* (1987) 596–616

ATKINSON, K. and I. G. GRAHAM: [1] Iterative solution of linear systems arising from the boundary integral method. *SIAM J. Sci. Stat. Comput. 13* (1992) 694–722

ATKINSON, K.: see Graham et al.

BAKER, C. T. H.: [1] *The numerical treatment of integral equations.* Oxford University Press, London, 1977

BAKER, C. T. H. and MILLER, G. F. (eds.): [1] *Treatment of integral equations by numerical methods.* Academic Press London, 1982

BALLMANN, J., R. EPPLER, and W. HACKBUSCH (eds.) [1] *Panel method in fluid mechanics with emphasis on aerodynamics.* Notes on Numerical Fluid Mechanics, vol 21. Vieweg, Braunschweig, 1988

BAMBERGER, L. and G. HÄMMERLIN: [1] Spline-blended substitution kernels of optimal convergence. In: Baker - Miller [1] 47-57

BERTHOLD, D. and B. SILBERMANN: [1] Corrected collocation methods for periodic pseudodifferential equations. Submitted

BOGOMOLNY, A.: see Atkinson et al.

BRAKHAGE, H.: [1] Über die numerische Behandlung von Integralgleichungen nach der Quadraturformelmethode. *Numer. Math. 2* (1960) 183-196

BRAKHAGE, H. and P. WERNER: [1] Über das Dirichletsche Außenraumproblem für die Helmholtzsche Schwingungsgleichung. *Arch. Math. 16* (1965) 325-329

BRANCA, H.W.: [1] The nonlinear Volterra equation of Abel's kind and its numerical treatment. *Computing 20* (1978) 307-324

BRUNNER, H. and P. J. VAN DER HOUWEN: [1] *The numerical solution of Volterra equations.* North-Holland, Amsterdam, 1986

BULIRSCH, R.: [1] Bemerkungen zur Romberg-Integration. *Numer. Math. 6* (1964) 6-16

CHANDLER, G. A. and I. H. SLOAN: [1] Spline qualocation methods for boundary integral equations. *Numer. Math. 58* (1990) 537-567 - Erratum: *Numer. Math. 62* (1992) 295

CHEN, G. and J. ZHOU: [1] *Boundary element methods.* Academic Press. New York 1992

COLTON, D. and R. KREß: [1] *Integral equation methods in scattering theory.* John Wiley, New York 1983

COSTABEL, M.: [1] Boundary integral operators on Lipschitz domains: elementary results. *SIAM J. Math. Anal. 19* (1988) 613-626

COSTABEL, M., E. STEPHAN, and W. L. WENDLAND: [1] On boundary integral equations of the first kind for the bi-Laplacian in a polygonal plane domain. Annali Scuola Normale Superiore Pisa Ser. IV *10* (1983) 197-241

COSTABEL, M. and W. L.WENDLAND: [1] Strong ellipticity of boundary integral operators. *J. Reine u. Angew. Math. 372* (1986) 34-63

DAHMEN, W., S. PRÖßDORF, and R. SCHNEIDER: [1] Wavelet approximation methods for pseudodifferential equations. II: Matrix compression and fast solution. *Advances in Computational Mathematics 1* (1993) 259-335

DAUTRAY, R. and J.-L. LIONS (eds.): *Mathematical analysis and numerical methods for science and technology. Vol. 4: Integral equations and numerical methods.* Springer-Verlag, Berlin 1990

DUFFY, M. G.: [1] Quadrature over a pyramid or cube of integrands with a singularity at a vertex. *SIAM J. Numer. Anal. 19* (1982) 1260–1262

ESSER, R.: [1] Numerische Behandlung einer Volterraschen Integralgleichung. *Computing 19* (1978) 269–284

FABER, G.: [1] Über die interpolatorische Darstellung stetiger Funktionen. *Jahresber. der Deutschen Math. Ver. 23* (1914) 190–210

FAERMANN, B.: [1] *Lokale a-posteriori-Fehlerschätzer bei der Diskretisierung von Randintegralgleichungen.* Dissertation, Universität Kiel 1993

GIROIRE, J. and J. C. NÉDÉLEC: [1] Numerical solution of an exterior Neumann problem using a double layer potential. *Math. Comp. 32* (1978) 973–990

GUIGGIANI, M.: [1] Direct evaluation of hypersingular integrals in 2D BEM. In: Hackbusch [9] 23–34

GÜNTER, N. M.: [1] *Die Potentialtheorie und ihre Anwendung auf Grundaufgaben der Mathematischen Physik.* Teubner, Leipzig, 1957

GRAHAM, I. G.: [1] Singularity expansion for the solutions of second kind Fredholm integral equations with weakly singular convolution kernels. *J. Integral Eqs 4* (1982) 1–30

GRAHAM, I. G.: [2] Galerkin methods for second kind integral equations with singularities. *Math. Comp. 39* (1982) 519–533

GRAHAM, I. G. and K. ATKINSON: [1] On the Sloan iteration applied to integral equations of the first kind. *IMA J. Numer. Analysis 13* (1993) 29–41

GRAHAM, I. G., LIN QUN, and XIE RUI-FENG: [1] Extrapolation of Nyström solutions of boundary integral equations on non-smooth domains. *J. Computational Mathematics 10* (1992) 231–244

HACKBUSCH, W.: [1] *Multi-Grid Methods and Applications.* Springer-Verlag, Berlin, 1985

HACKBUSCH, W.: [2] *Elliptic differential equations.* Springer-Verlag, Berlin, 1992 – *Theorie und Numerik elliptischer Differentialgleichungen.* Teubner, Stuttgart 1986

HACKBUSCH, W.: [3] *Iterative solution of large sparse systems.* Springer-Verlag, Berlin, 1994 – *Iterative Lösung großer schwachbesetzter Gleichungssysteme.* Teubner, Stuttgart 1991 – second edition 1993

HACKBUSCH, W.: [4] On the fast solving of parabolic boundary control problems. *SIAM J. Control Optim. 17* (1979) 231–244

HACKBUSCH, W.: [5] Die schnelle Auflösung der Fredholmschen Integralgleichung zweiter Art. *Beiträge Numer. Math. 9* (1981) 47–62

HACKBUSCH, W.: [6] Multi-grid solutions to linear and nonlinear eigenvalue problems for integral and differential equations. *Rostock Math. Colloq. 25* (1984) 79–98

HACKBUSCH, W.: [7] The solution of large systems of BEM equations by the multi-grid and panel clustering technique. Proceedings, Juni 1990, Turin. *Rend. Sem. Mat. Univ. Pol. Torino, Fasc. Speciale 1991, Numerical Methods.* pages 163–187.

HACKBUSCH, W. (ed.): [8] *Robust Multi-Grid Methods.* Notes on Numerical Fluid Mechanics, vol 23. Vieweg, Braunschweig, 1988

HACKBUSCH, W. (ed.): [9] *Numerical techniques for boundary element methods.* Proceedings, Kiel, Jan. 1991. Notes on numerical fluid mechanics 33. Vieweg, Braunschweig 1992

HACKBUSCH, W. and Z. P. NOWAK: [1] A multilevel discretisation and solution method for potential flow problems in three dimensions. In: Hirschel [1] 71–89

HACKBUSCH, W. and Z. P. NOWAK: [2] On the fast matrix multiplication in the boundary element method by panel clustering. *Numer. Math.* **54** (1989) 463–491

HACKBUSCH, W. and S. SAUTER: [1] On the efficient use of the Galerkin method to solve Fredholm integral equations. *Applications of Mathematics* **38** (1993) 301–322

HACKBUSCH, W. and S. SAUTER: [2] On numerical cubatures of nearly singular surface integrals arising in BEM collocation. *Computing* **52** (1994) 139–159

HACKBUSCH, W.: see Ballmann et al.

HADAMARD, J.: *Lectures on Cauchy's problem.* Yale University Press, 1923

HÄMMERLIN, G.: see Bamberger et al.

HÄMMERLIN, G. and L. L. SCHUMAKER: [1] Procedures for kernel approximation and solution of Fredholm integral equations of the second kind. *Numer. Math.* **34** (1980) 125–141

HEBEKER, F. K.: [1] On the numerical treatment of viscous flows against bodies with corners and edges by boundary element and multigrid methods. *Numer. Math.* **52** (1988) 81–99

HEBEKER, F. K.: [2] On multigrid methods of the first kind for symmetric boundary integral equations of nonnegative order. In: Hackbusch [6] 128–138

HEMKER, P. W. and H. SCHIPPERS: [1] Multiple grid methods for the solution of Fredholm integral equations of the second kind. *Math. Comp.* **36** (1981) 215–232

HEUSER, H.: [1] *Funktionalanalysis.* 2nd edition, Teubner Stuttgart, 1986

HIRSCHEL, E. H. (ed.) [1] *Finite Approximations in Fluid Mechanics.* Notes on Numerical Fluid Mechanics, vol 14. Vieweg, Braunschweig, 1986

HSIAO, G. C. and R. KREß: [1] On an integral equation for the two-dimensional exterior Stokes problem. *Appl. Numer. Math.* **1** (1985) 77–93

HSIAO, G. C., P. KOPP, and W. L. WENDLAND: [1] A Galerkin collocation method for some integral equations of the first kind. *Computing* **25** (1980) 89–130

HSIAO, G. C., P. KOPP, and W. L. WENDLAND: [2] Some applications of a Galerkin-collocation method for boundary integral equations of the first kind. *Math. Meth. in the Appl. Sci. 6* (1984) 280-325

HSIAO, G. C. and R. KREß [1]: On an integral equation for the two-dimensional exterior Stokes problem. *Appl. Numer. Math. 1* (1985) 77-93

HSIAO, G. C. and S. PRÖßDORF: [1] A generalization of the Arnold-Wendland lemma to collocation methods for boundary integral equations in $\mathbb{R}^3$. *Math. Nachr. 163* (1993) 133-144

HSIAO, G. C. and W. L. WENDLAND: [1] A finite element method for some integral equations of the first kind. *J. Math. Anal. Appl. 58* (1977) 449-481

HSIAO, G. C. and W. L. WENDLAND: [2] The Aubin-Nitsche lemma for integral equations. *Journal of Integral Equations 3* (1981) 299-315

JASWON, M. A. and G. T. SYMM: [1] *Integral equation methods in potential theory and elastostatics.* Academic Press, London, 1977

JOHNSON, C. G. L. and L. R. SCOTT: [1] An analysis of quadrature errors in second-kind boundary integral methods. *SIAM J. Numer. Anal. 26* (1989) 1356-1382

KIESER, R.: [1] The triangle-to-square transformation for finite-part integrals. In: Hackbusch [9] 72-78

KIESER, R., C. SCHWAB, and W. L. WENDLAND: [1] Numerical evaluation of singular and finite-part integrals on curved surfaces using symbolic manipulation. *Computing 49* (1992) 279-301

KLEINMAN, R. and W. L. WENDLAND: [1] On Neumann's method for the exterior Neumann problem for the Helmholtz equation. *J. Math. Anal. Appl. 57* (1977) 107-202

KOPP, P.: see Hsiao et al.

KRÁL, J.: [1] *Integral operators in potential theory.* Lecture Notes in Mathematics *823*, Springer-Verlag, Berlin, 1980

KRÁL, J. and W. L. WENDLAND: [1] On the applicability of the Fredholm-Radon method in potential theory and the panel method. In: Ballmann - Eppler - Hackbusch [1] 120-136

KREß, R.: [1] *Linear Integral Equations.* Springer-Verlag, Berlin, 1989

KREß, R.: [2] A Nyström method for boundary integral equations in domains with corners. *Numer. Math. 58* (1990) 145-161

KREß, R. and I.H. SLOAN: [1] On the numerical solution of a logarithmic integral equation of the first kind for the Helmholtz equation. *Numer. Math. 66* (1993) 199-214

KREß, R.: see Colton et al. and Hsiao et al.

LAMP, U., T. SCHLEICHER, E. STEPHAN, and W. L. WENDLAND: [1] Galerkin collocation for an improved boundary element method for a plane mixed boundary value problem. *Computing 33* (1984) 269-296

LAMP, U., T. SCHLEICHER, and W. L. WENDLAND: [1] The fast Fourier transform and the numerical solution of one-dimension boundary integral equations. *Numer. Math.* *47* (1985) 15-38

LEIS, R.: [1] *Vorlesungen über partielle Differentialgleichungen zweiter Ordnung*. B.I., Mannheim 1967

LIN, Q., I. H. SLOAN, and R. XIE: [1] Extrapolation of the iterated collocation method for integral equations of the second kind. *SIAM J. Numer. Anal.* *27* (1990) 1535-1541

LIONS, J.-L.: see Dautray et al.

LUBICH, Ch.: [1] Discretized fractional calculus. *SIAM J. Math. Anal.* *17* (1986) 704-719

LUBICH, Ch.: [2] Fractional linear multistep methods for Abel-Volterra integral equations of the first kind. *IMA J. Numer. Anal.* *7* (1987) 97-106

MACCAMY, R. C. and E. STEPHAN: [1] A boundary element method for an exterior problem for three-dimensional Maxwell's equations. *Applicable Analysis* *16* (1983) 141-163

MACKERLE, J. and C. A. BREBBIA (eds.): [1] *The boundary element reference book*. Springer-Verlag, Berlin, 1988

MANDEL, J: [2] On multilevel iterative methods for integral equations of the second kind and related problems. *Numer. Math.* *46* (1985) 147-157

MARTENSEN, E.: [1] *Potentialtheorie*. Teubner, Stuttgart 1968

MEISTER, E.: [1] *Integraltransformationen mit Anwendungen auf Probleme der mathematischen Physik*. Verlag P. Lang, Frankfurt a.M., 1983

MEISTER, E.: [2] *Randwertaufgaben der Funktionentheorie*. Teubner, Stuttgart, 1983

MIKHLIN, S. G.: [1] *Integral equations*. Pergamon Press, London, 1957

MIKHLIN, S. G. and S. PRÖßDORF: [1] *Singular integral operators*. Springer, Berlin, 1986

MILLER, G.F.: see Baker et al.

MOORE, R.H.: see Anselone et al.

MUSKHELISHVILI, N. I.: [1] *Singular integral equations*. Noordhoff, Groningen, 1958 - *Singuläre Integralgleichungen*. Akademie-Verlag, Berlin, 1965

NÉDÉLEC, J. C.: [1] Curved finite element methods for the solution of singular integral equations on surfaces in $\mathbb{R}^3$. *Comput. Methods Appl. Mech. Engrg.* *8* (1976) 61-80

NÉDÉLEC, J. C. and J. PLANCHARD: [1] Une méthode variationelle d'élément finis pour la résolution numérique d'un problème extérieur dans $\mathbb{R}^3$. *RAIRO* *7* (1973) R3, 105-129

NÉDÉLEC, J. C.: see Giroire et al.

NOBLE, B.: [1] Error analysis of collocation methods for solving Fredholm integral equations. In: *Topics in Numerical Analysis* (ed.: J.J.H. Miller), Academic Press, London, 1973

NOWAK, Z.P.: [1] A new type of higher-order boundary integral approximation for potential flow problems in three dimensions. In: Hirschel [1] 218-231

NOWAK, Z.P.: [2] Panel clustering technique for lifting potential flows in the three space dimensions. In: Ballmann - Eppler - Hackbusch [1] 166-178

NOWAK, Z.P.: see Hackbusch et al.

NYSTRÖM, E. J.: [1] Über die praktische Auflösung von linearen Integralgleichungen mit Anwendungen auf Randwertaufgaben der Potentialtheorie. *Soc. Sci. Fenn. Comment. Phys.-Math.* *4,15* (1928) 1-52

NYSTRÖM, E. J.: [2] Über die praktische Auflösung von linearen Integralgleichungen mit Anwendungen auf Randwertaufgaben. *Acta Mathematica* *54* (1930) 185-204

PIPKIN, A. C.: [1] *A course on integral equations.* Springer-Verlag, New York, 1991

PRÖßDORF, S.: [1] *Einige Klassen singulärer Gleichungen.* Akademie-Verlag, Berlin, 1974

PRÖßDORF, S. and A. RATHSFELD: [1] On spline Galerkin methods for singular integral equations with piecewise continuous coefficients. *Numer. Math.* *48* (1986) 99-118

PRÖßDORF, S. and G. SCHMIDT: [1] A finite element collocation method for singular integral equations. *Math. Nachr.* *100* (1981) 33-60

PRÖßDORF, S. and B. SILBERMANN: [1] *Projektionsverfahren und die näherungsweise Lösung singulärer Gleichungen.* Teubner, Leipzig, 1977

PRÖßDORF, S. and B. SILBERMANN: [2] *Numerical analysis for integral and related operator equations.* Birkhäuser, Basel 1991

PRÖßDORF, S.: see Dahmen et al., Hsiao et al., and Mikhlin et al.

RANNACHER, R. and W. L. WENDLAND: [1] On the order of pointwise convergence of some boundary element methods. Part I: Operators of negative and zero order. *Math. Modelling and Numerical Analysis* *19* (1985) 65-88; Part II: Operators of positive order. *Math. Modelling and Numerical Analysis* *22* (1988) 343-362

RATHSFELD, A.: see Prößdorf et al.

ROKHLIN, V.: [1] Rapid solutions of integral equations of classical potential theory, *J. Comput. Phys.* *60* (1985) 187-207.

ROUTSALAINEN, K. and J. SARANEN: [1] Some boundary element methods using Dirac's distributions as trial functions. *SIAM J. Numer. Anal.* *24* (1987) 816-827

ROUTSALAINEN, K. and J. SARANEN: [2] A dual method to the collocation method. *Math. Meth. in the Appl. Sc.* (1988) 439-445

ROUTSALAINEN, K. and W. L. WENDLAND: [1] On the boundary element method for some nonlinear boundary value problems. *Numer. Math.* *53* (1988) 299-314

SARANEN, J.: [1] On the effect of numerical quadrature in solving boundary integral equations. In: Ballmann-Eppler-Hackbusch [1] 196-209

SARANEN, J.: [2] Extrapolation methods for spline collocation solutions of pseudodifferential equations on curves. *Numer. Math.* ***56*** (1989) 385-407

SARANEN, J.: [3] Projection methods for a class of Hammerstein equations. *SIAM J. Numer. Anal.* *27* (1990) 1445-1449

SARANEN, J.: see Routsalainen et al.

SARANEN, J. and W. L. WENDLAND: [1] On the asymptotic convergence of collocation methods with spline functions of even degree. *Math. Comp.* ***45*** (1985) 91-108

SAUTER, S: [1] *Über die effiziente Verwendung des Galerkinverfahrens zur Lösung Fredholmscher Integralgleichungen.* Dissertation, Universität Kiel, 1992

SAUTER, S: see Hackbusch et al.

SCHATZ, A. H., V. THOMÉE, and W. WENDLAND: [1] *Mathematical theory of finite and boundary element methods.* DMV 15, Birkhäuser, Basel, 1990

SCHIPPERS, H.: [1] Application of multigrid methods for integral equations to two problems from fluid dynamics. *J. Comput. Phys.* ***48*** (1982) 441-461

SCHIPPERS, H.: [2] On the regularity of the principal value of the double-layer potential. *J. Engineering Math.* ***16*** (1982) 59-76

SCHIPPERS, H.: [3] Multigrid methods for boundary integral equations. *Numer. Math.* ***46*** (1985) 351-363

SCHIPPERS, H.: see Hemker et al.

SCHLEICHER, T.: see Lamp et al.

SCHMIDT, G.: see Prößdorf et al.

SCHNEIDER, C.: [1] Product integration for weakly singular integral equations. *Math. Comp.* ***36*** (1981) 207-213

SCHNEIDER, R., T. VON PETERSDORFF, and C. SCHWAB: [1] Multiwavelet for second kind integral equations. To appear in *SIAM J. Numer. Anal.*

SCHNEIDER, R.: see Dahmen et al.

SCHWAB, C. and W. L. WENDLAND: [1] On the numerical integration of singular surface integrals in the BEM. In: Hackbusch [9] 177-190

SCHWAB, C. and W. L. WENDLAND: [2] On the numerical cubature of singular surface integrals in boundary element methods. *Numer. Math.* ***62*** (1992) 343-369

SCHWAB, C.: see Kieser et al.

SCOTT, R. L.: see Johnson et al.

SILBERMANN, B.: see Berthold et al., Prößdorf et al.

SLOAN, I. H.: [1] Quadrature methods for integral equations of the second kind over infinite intervals. *Math. Comp.* ***36*** (1981) 551-523

SLOAN, I. H.: [2] Analysis of general quadrature methods for integral equations of the second kind. *Numer. Math.* ***38*** (1981) 263-278

SLOAN, I. H.: [3] A quadrature-based approach to improving the collocation method. *Numer. Math.* ***54*** (1988) 41-56

SLOAN, I. H.: [4] Error analysis for a class of degenerate kernel methods. *Numer. Math.* **25** (1976) 231-238

SLOAN, I. H. and A. SPENCE: [1] Projection methods for integral equations on the halfline. *IMA J. Numer. Anal.* **6** (1986) 153-172

SLOAN, I. H.: see Amini et al., Chandleret al., and Lin et al.

STEPHAN, E.: see Costabel et al., Lamp et al., and MacCamy et al.

STOER, J. [1]: *Einführung in die Numerische Mathematik I.*, Springer-Verlag, Berlin, 1989

STOER, J. and R. BULIRSCH: [1] *Introduction to numerical analysis.* Springer-Verlag, Berlin 1980

SYMM, G. T.: see Jaswon et al.

SZEGÖ, G.: [1] *Orthogonal polynomials.* AMS, New York, 1959

THOMÉE, H., V.: see Schatz et al.

VAINIKKO, G.: [1] *Multidimensional weakly singular integral equations.* Lect. Notes in Math. **1549**. Springer-Verlag, Berlin, 1993

WALTER, W.: [1] *Einführung in die Potentialtheorie.* Bibliographisches Institut, Mannheim, 1971

WENDLAND, W.: [1] Die Behandlung der Randwertaufgaben im $\mathbb{R}_3$ mit Hilfe von Einfach- und Doppelschichtpotentialen. *Numer. Math.* **11** (1968) 380-404

WENDLAND, W.: [2] Asymptotic accuracy and convergence for point collocation methods. Chapter 9 in: C.A. Brebbia (ed.): *Topics in Boundary Element Research*, vol 2, 230-257. Springer-Verlag, Berlin 1985

WENDLAND, W.: [3] On some mathematical aspects of boundary element methods for elliptic problems. In: *The Mathematics of Finite Elements and Applications V* (ed.: J. R. Whiteman), pages 193-227. Academic Press, London 1985

WENDLAND, W.: [4] On applications and the convergence of boundary integral methods. In: Baker - Miller [1] 463-476

WENDLAND, W.: see Arnold, Costabel, Kieser, Kleinman, Král, Lamp, Hsiao, Rannacher, Routsalainen, Saranen, Schatz et al.

WERNER, P.: see Brakhage et al.

WILKINSON, J. H.: [1] *Rounding errors in algebraic processes.* Her Majesty's Stationery Office, London, 1963 - *Rundungsfehler.* Springer-Verlag, Berlin, 1969

WLOKA, J. [1] *Partial differential equations.* Cambridge University Press, Cambridge 1987 - *Partielle Differentialgleichungen.* Teubner, Stuttgart, 1982

XIE, R.: see Lin et al.

YOSIDA, K.: [1] *Functional Analysis.* Springer-Verlag, Berlin, 1974

Index

Numbers in bold face indicate the first important appearance of the entry (definition, etc.). An arrow (→) indicates other headings under which this entry appears.

A

B

C

D

I

J

K

L

M

Q

R

S

ISNM

INTERNATIONAL SERIES OF NUMERICAL MATHEMATICS
BIRKHÄUSER VERLAG

Titles previously published in the series

ISNM 81 **L. Collatz, G. Mainardus, G. Nürnberger (Eds.):** Numerical Methods of Approximation Theory, Vol. 8: Workshop on Numerical Methods of Approximation Theory Oberwolfach, September 28-October 4, 1986 (ISBN 3-7643-1855-4)

ISNM 82 **B. Heinrich:** Finite Difference Methods on Irregular Networks 1987 (ISBN 3-7643-1880-5)

ISNM 83 **J. Albrecht, L. Collatz, W. Velte, W. Wunderlich (Eds):** Numerical Treatment of Eigenvalue Problems, Vol. 4 Workshop in Oberwolfach, November 30-December 6, 1986, 1987 (ISBN 3-7643-1856-2)

ISNM 84 **K.-H. Hoffmann, J.-B. Hiriart–Urruty, J. Zowe, C. Lemarechal (Eds):** Trends in Mathematical Optimization: 4th French-German Conference on Optimization, 1988 (ISBN 3-7643-1919-4)

ISNM 85 **H. Brass, G. Hämmerlin (Eds.):** Numerical Integration III. Proceedings of the Conference held at the Mathematisches Forschungsinstitut, Oberwolfach, November 8-14, 1987, 1988 (ISBN 3-7643-2205-5)

ISNM 86 **R.P. Agarwal, Y.M. Chow, S.J. Wilson (Eds.):** Numerical Mathematics Singapore 1988. Proceedings of the International Conference on Numerical Mathematics held at the National University of Singapore, May 31-June 4, 1987, 1988 (ISBN 3-7643-2255-1)

ISNM 87 **J.P. Penot (Ed.):** New Methods in Optimization and their Industrial Uses. State of the art, recent advances, perspectives. Proceedings of the Symposia held in Pau, October 19-29, 1987, and Paris, November 1987, 1989 (ISBN 3-7643-2286-1)

ISNM 88 **J.F. Rodriguez (Ed.):** Mathematical Models for Phase Change Problems. Proceedings of the European Workshop held at the Obidos, Portugal, October 1-3, 1988, 1989 (ISBN 3-7643-2309-4)

ISNM 89 **K. Gürlebeck, W. Sprössig:** Quaternionic Analysis and Elliptic Boundary Value Problems, 1990 (ISBN 3-7643-2382-5)

ISNM 90 **C.K. Chui, W. Schempp, K. Zeller (Eds.):** Multivariate Approximation Theory IV. Proceedings of the Conference at the Mathematical Research Institute at Oberwolfach, Black Forest, February 12-18, 1989, 1989 (ISBN 3-7643-2384-1)

ISNM 91 **F. Kappel, K. Kunisch, W. Schappacher (Eds.):** Control and Estimation of Distributed Parameter Systems. 4th International Conference on Control of Distributed Parameter Systems, Vorau, July 10-16, 1988, 1989 (ISBN 3-7643-2345-0)

ISNM 92 **H.D. Mittelmann, D. Roose, (Eds.):** Continuation Techniques and Bifurcation Problems, 1990 (ISBN 3-7643-2397-3)

ISNM 93 **R.E. Bank, R. Bulirsch, K. Merten (Eds.):** Mathematical Modelling and Simulation of Electric Circuits and Semiconductor Devices, 1990 (3-7643-2439-2)

ISNM 94 **W. Haussmann, K. Jetter (Eds.):** Multivariate Approximation and Interpolation, 1990 (3-7643-2450-3)

ISNM 95 **K.-H. Hoffmann, J. Sprekels (Eds.):** Free Boundary Value Problems, 1990 (3-7643-2474-0)

ISNM 96 **J. Albrecht, L. Collatz, P. Hagedorn, W. Velte (Eds.):** Numerical Treatment of Eigenvalue Problems, Vol. 5, 1991 (3-7643-2575-5)

ISNM 97 **R.U. Seydel, F.W. Schneider, T.G. Küpper, H. Troger (Eds.):** Bifurcation and Chaos: Analysis, Algorithms, Applications, 1991 (3-7643-2593-3)

ISNM 98 **W. Hackbusch, U. Trottenberg (Eds.):** Multigrid Methods III, 1991 (3-7643-2632-8)

ISNM 99 **P. Neittaanmäki (Ed.):** Numerical Methods for Free Boundary Problems, 1991 (3-7643-2641-7)

ISNM 100 **W. Desch, F. Kappel, K. Kunisch (Eds.):** Estimation and Control of Distributed Parameter Systems, 1991 (3-7643-2676-X)

ISNM 101 **G. Del Piero, F. Maceri (Eds.):** Unilateral Problems in Structural Analysis IV, 1991 (3-7643-2487-2)

ISNM 102 **U. Hornung, P. Kotelenez, G. Papanicolaou (Eds.):** Random Partial Differential Equations, 1991 (3-7643-2688-3)

ISNM 103 **W. Walter (Ed.):** General Inequalities 6, 1992 (3-7643-2737-5)

ISNM 104 **E. Allgower, K. Böhmer, M. Golubitsky (Eds.):** Bifurcation and Symmetry, 1992 (3-7643-2739-1)

ISNM 105 **D. Braess, L.L. Schumaker (Eds.):** Numerical Methods in Approximation Theory, Vol. 9, 1992 (3-7643-2746-4)

ISNM 106 **S.N. Antontsev, K.-H. Hoffmann, A.M. Khludnev (Eds.):** Free Boundary Problems in Continuum Mechanics, 1992 (3-7643-2784-7)

ISNM 107 **V. Barbu, F.J. Bonnans, D. Tiba (Eds.):** Optimization, Optimal Control and Partial Differential Equations, 1992 (3-7643-2788-X)

ISNM 108 **H. Antes, P.D. Panagiotopoulos:** The Boundary Integral Approach to Static and Dynamic Contact Problems. Equality and Inequality Methods, 1992 (3-7643-2592-5)

ISNM 109 **A.G. Kuz'min:** Non-Classical Equations of Mixed Type and their Applications in Gas Dynamics, 1992 (3-7643-2573-9)

ISNM 110 **H.R.E.M. Hörnlein, K. Schittkowski (Eds.):** Software Systems for Structural Optimization, 1992 (3-7643-2836-3)

ISNM 111 **R. Burlisch, A. Miele, J. Stoer, K.H. Well:** Optimal Control, 1993 (3-7643-2887-8)

ISNM 112 **H. Braess, G. Hämmerlin (Eds.):** Numerical Integration IV. Proceedings of the Conference at the Mathematical Research Institute at Oberwolfach, November 8-14, 1992, 1993 (3-7643-2922-X)

ISNM 113 **L. Quartapelle:** Numerical Solution of the Incompressible Navier-Stokes Equations, 1993 (3-7643-2935-1)

ISNM 114 **J. Douglas, U. Hornung (Eds.):** Flow in Porous Media. Proceedings of the Oberwolfach Conference, June 21-27, 1992 (ISBN 3-7643-2949-1)

ISNM 115 **R. Bulirsch, D. Kraft (Eds.):** Computational Optimal Control, 1994 (ISBN 3-7643-5015-6)

ISNM 116 **P.W. Hemker, P. Wesseling (Eds.):** Multigrid Methods IV. Proceedings of the Fourth European Multigrid Conference, Amsterdam, July 6-9, 1993 (ISBN 3-7643-5030-X)

ISNM 117 **R.E. Bank, R. Bulirsch, H. Gajewski, K. Merten (Eds.):** Mathematical Modelling and Simulation of Electrical Circuits and Semiconductor Devices, 1994 (ISBN 3-7643-5053-9)

ISNM 118 **W. Desch, F. Kappel, K. Kunisch (Eds.):** Control and Estimation of Distributed Parameter Systems: Nonlinear Phenomena. International Conference on Control and Estimation of Distributed Parameter Systems, Vorau, July 18-24, 1993, 1994 (ISBN 3-7643-5098-9)

ISNM 119 **R. Zahar (Ed.):** Approximation and Computation: A Festschrift in Honor of Walter Gautschi, 1995 (ISBN 0-8176-3753-2)

ISNM 120 **W. Hackbusch:** Integral Equations: Theory and Numerical Treatment, 1995 (ISBN 3-7643-2871-1)

GPSR Compliance

The European Union's (EU) General Product Safety Regulation (GPSR) is a set of rules that requires consumer products to be safe and our obligations to ensure this.

If you have any concerns about our products, you can contact us on ProductSafety@springernature.com

In case Publisher is established outside the EU, the EU authorized representative is:

Springer Nature Customer Service Center GmbH
Europaplatz 3
69115 Heidelberg, Germany

Batch number: 10577878

Printed by Printforce, the Netherlands